# Mike Holt's Illustrated Guide

# Understanding The National Electrical Code®

Based on the **2002**

National Electrical Code®

Volume 1
Articles 80 - 460

# Mike Holt's Illustrated Guide

# *Understanding the National Electrical Code*

# *2002 Edition*

# *Volume 1*
# *Articles 80 – 460*

*Mike Holt*

## NOTICE TO THE READER

**Cover Design: Paul Wright - wright1156@myacc.net**

Design, Layout, and Typesetting: Paul Wright - wright1156@myacc.net
Graphics: Mike Culbreath

Last revised July 29, 2002

Printed August, 2002.

For more information, contact:

Mike Holt Enterprises, Inc.
7310 West McNab Road
Suite 201
Tamarac, Florida 33321
1.888.NEC.CODE
www.NECcode.com

*Printed in Canada*

# Table of Contents

## Chapter 3 – Wiring Methods and Materials

# Chapter 4 – Equipment for General Use

## Introduction

*Understanding the National Electrical Code*, 2002 Edition, is intended to provide insight into the technical rules of the 2002 *National Electrical Code, (NEC)®*, 49th edition. (The *NEC* was first issued in 1897.) The writing style is intended to be informal and relaxed, and this book contains many graphics and examples that apply to today's electrical systems.

This book cannot clear up confusing, conflicting, or controversial *Code* rules, but it does try to put these rules into sharper focus to help you understand their intended purpose. The *NEC* is updated every three years because of new products, technologies, materials, and installation techniques. Fortunately, the *Code* allows the Authority Having Jurisdiction (AHJ) the flexibility to waive specific requirements of the NEC or permit alternative methods when assured that equivalent objectives for establishing and maintaining effective safety can be achieved [80.13 and 90.4].

This illustrated book also contains tips on proper electrical installations, advice or cautions about possible conflicts or confusing *Code* rules, and warnings of dangers related to improper electrical installations.

Keeping up with the *NEC* is the goal of all those who are involved in electrical safety, such as the installer, contractor, owner, inspector, architect, engineer, and instructor. To get the most out of this book, you should answer the 1200+ practice questions contained in the *Understanding the National Electrical Code Workbook*, which is available from our office (1.888.NEC.CODE) or the Internet (www.NECcode.com).

## About This Book

This book covers the rules in Articles 80 though 460, specifically for those installations that operate at a voltage of 120/240, 208Y/120, or 480Y/277.

## How to Use This Book

This book is intended as a study guide to be used with the NEC, not as a replacement for the *NEC*. Keep a copy of the 2002 *National Electrical Code* handy. As you read through this book, review the graphics and examples in relation to your *Code* book and discuss with others. This book contains many cross-references to other related *Code* rules to help you develop a better understanding of the *Code* and includes author's comments on the rules covered. These cross-references are identified by a *Code* Section number in brackets, such as "90.4," which would look like "[90.4]."

> *Note. This book follows the NEC format but it does not cover all of the NEC rules. In addition, you may notice that not all Articles, Sections, Subparts, Exceptions, and FPNs, have been included in this book. Do not be concerned if you see that Exception No. 2 and Exception No. 4 are explained, but not Exception No.1 and No. 3.*

As you progress through this book, you might find some formulas, rules or comments that you don't understand. Don't get frustrated. Highlight the section in this book and in your *Code* book that you are having a problem with. Discuss it with your boss, inspector, co-worker, instructor, etc. Maybe they'll have some additional feedback. Once you have completed the book, review those highlighted sections again and see if you can understand them.

> *CAUTION: Text intended to represent the 2002 NEC in this book has been paraphrased to comply with copyright requirements. In addition, for ease of reading, some titles to Articles and Sections are different than they appear in the Code. Please keep this in mind when comparing the paraphrased text of the 2002 NEC in this book against the actual NEC.*

## Errors

I have taken great care in researching the *Code* rules in this book, but I'm not perfect. If you feel that I have made an error, please let me know by contacting me directly at Mike@NECcode.com, or 1.888.NEC.CODE.

## Attitude and Gray Areas

A certain attitude is needed to actually understand the *National Electrical Code*. I hope this book will inspire you to learn more about the *Code*. I have to say that the more I learn, the more I realize how much there still is to learn.

As you progress through this book you should develop a much greater insight into these rules. Yes, there are gray areas in the NEC, but generally the meaning is quite clear.

## Different Interpretations

Electricians, contractors, some inspectors, and others love arguing *Code* interpretations and discussing *Code* requirements. As a matter of fact, discussing the *NEC* and its application is a great way to increase your knowledge of the *Code* and its intended use. The best way to discuss *Code* requirements with others is by referring to a specific section in the *Code*, rather than by talking in vague generalities.

## About the Author

Mike Holt worked his way up through the electrical trade from an apprentice electrician to a master electrician and electrical inspector. He did not complete high school due to unusual circumstances and dropped out after completing the 11th grade. Realizing that success depends on one's education, Mike immediately attained his GED, and 10 years later attended the University of Miami's Graduate School to obtain his Master's in Business Administration (MBA).

Mike is nationally recognized as one of America's most knowledgeable electrical trainers and has touched the lives of thousands of electricians, inspectors, contractors and engineers.

Mike Holt resides in central Florida, is the father of seven children, and has many outside interests and activities. He is a former National Barefoot Waterskiing Champion (1988 and 1999), having set five barefoot waterski records. Mike enjoys motocross racing, white-water rafting, racquetball, playing guitar, and especially spending time with his family. His commitment to God has helped him develop a lifestyle that balances family, career, and self.

## Acknowledgements

To my beautiful wife, Linda, and my seven children: Belynda, Melissa, Autumn, Steven, Michael, Meghan, and Brittney – thank you for loving me.

I would like to thank all the people in my life who believed in me and those who spurred me on. Thanks to the Electrical Construction and Maintenance magazine for my first "big break" in 1980, and especially to Joe McPartland, my mentor, who was there to help and encourage me from 1980 to 1992. Joe, I'll never forget to help others as you've helped me. In addition, I would like to thank Joe Salimando, the former publisher of Electrical Contractor magazine, and Dan Walters of the National Electrical Contractors Association (NECA) for my second "big break" in 1995.

I would like to extend a special thank you to the staff at the National Fire Protection Association, in particular Jeff Sargent for his assistance in answering my many *NEC* questions. Jeff, let me say publicly that you are one of the finest individuals I have had the opportunity to work with in this trade, and I admire your dedication and commitment to helping others understand the *Code*. I would like to also thank John Caloggero, Joe Ross, and Dick Murray for all your help in the past.

I would like to thank Phil Simmons, former Executive Director of the International Association of Electrical Inspectors, for his help in the small details and especially in grounding. Other people who have been important in my development include James Stallcup, Dick Loyd, Mark Ode, DJ Clements, Morris Trimmer, Tony Selvestri, and Marvin Weiss.

A personal thank you goes to Sarina, my long-time friend and Office Manager. Thank you for covering the office for me while I spend so much time writing books, doing seminars, and producing videos. Your love and concern for me and my family has helped me through many difficult times.

There are many people who played a role in the development and production of this book. I would like to start with Mike Culbreath, Master Electrician, who has been with me for over 15 years helping me transform my words and thoughts into lifelike graphics. Also, a thank-you goes to Paul Wright of Digital Design Group for the fine electronic layout of this book.

Thanks are also in order for the following individuals who reviewed the manuscript and offered invaluable suggestions and feedback.

**Victor M. Ammons**
*P.E., Director of Engineering*
*The Prisco Group, Hopewell, NJ*

**Martin Anspach**
*President*
*M.E.T.S. Inc. St. Joe, IN*

**Tom Baker**
*Electrical Instructor*
*Puget Sound Electrical Training, Bremerton, WA*

**Scott Baltic**
*Freelance Writer*
*Chicago, IL*

**Gary Beckstrand**
*Instructor*
*Salt Lake City, UT*

**L. W. Brittian**
*Mechanical & Electrical Instructor*
*Lott, TX*

**Mike Culbreath**
*Graphic Designer*
*Mike Holt Enterprises, Inc.*

**Ted Kessner**
*Electrical Contractor/Instructor*
*Autryville, NC*

**John P. McComb**
*P.E., Power Quality Engineer*
*Florida Power and Light, Miami, FL*

**Sarina Snow**
*Office Manager*
*Mike Holt Enterprises, Inc.*

**Erik Spielvogel**
*Project Manager*
*GSL Electric, Inc. Sandy, UT*

**Steven D. Stack**
*National Electrical Code Training*
*Waxhaw, NC*

**Brooke Stauffer**
*Executive Director of Standards and Safety*
*National Electrical Contractors Association, Bethesda, MD*

**Eric Stromberg**
*Electrical Engineering Specialist*
*The Dow Chemical Company, Freeport, TX*

**James Thomas**
*Electrical/Electronics Instructor*
*James Sprunt Community College, Kenansville, NC*

**Joseph Wages Jr.**
*Instructor*
*Siloam Springs, AR*

Finally, I would like to thank the following individuals who worked tirelessly to proofread and edit the final stages of this publication: Toni Culbreath, Amanda Mullins, Barbara Parks and Sarina Snow. Their attention to detail and dedication to this project is greatly appreciated.

## The National Electrical Code

The *NEC* is to be used by experienced persons having an understanding of electrical terms, theory, safety procedures, and trade practices. These individuals include electrical contractors, electrical inspectors (Authorities Having Jurisdiction/AHJ's) electrical engineers, designers, and qualified persons. The *Code* was not written to serve as an instructive or teaching manual [90.1(C)] for untrained individuals.

Learning to use the *NEC* is somewhat like learning to play chess. You must first learn the names for the game pieces, how each piece moves and how the pieces are placed on the board. Once you understand these fundamentals, you're ready to start playing the game; but, all you can do is make crude moves, because you really don't understand what you're doing. To play the game well, you will need to study the rules, learn subtle and complicated strategies, and then practice, practice, practice.

Electrical work is similar in some ways, though it's certainly not a game; it must be taken very seriously. Learning the terms, concepts, and basic layout of the *NEC* gives you just enough knowledge to be dangerous.

There are thousands of specific and sometimes unique applications of electrical installations and the *Code* does not have a rule for every one of them. To properly apply the *NEC*, you must understand how each rule affects not only the electrical, but also the safety aspects of the installation. I suggest that you have a copy of the 2002 *NEC* nearby, as I will make specific references to it throughout this text.

## NEC Terms and Concepts

The *NEC* contains many technical terms, so it is crucial that the user understands their meanings. If you do not understand a term used in a *Code* rule, it will be next to impossible to properly apply the rule.

It's not only the technical words that require close attention, because even the simplest of words can make a big difference in the intent of a rule. The word "or" can imply alternate choices for equipment wiring methods, while "and" can mean an additional requirement.

**Note:** Electricians, engineers, and other trade-related professionals use terms and phrases (slang or jargon) that are not used in the *NEC*. The problem with slang is that not everybody understands the meaning and it can lead to additional confusion.

Understanding the safety-related concepts behind many of the *NEC* rules requires one to have a good foundation of electrical fundamentals. For example:

- How is it possible that a bird can sit on an energized power line safely?
- Why can't a single-circuit conductor be installed within a metal raceway?
- Why must we reduce the ampacity of a current-carrying conductor if we install four or more in a raceway?
- Why does one section of the *NEC* require a 40A circuit breaker to protect a 12 AWG conductor, yet in another it allows a 20A circuit breaker?
- Why are bonding jumpers sometimes required for metal raceways containing 480Y/277V circuits, but not for 120/240V circuits?

If you have a good understanding of alternating current and impedance fundamentals, you'll find that understanding and properly applying the requirements of the *NEC* much easier.

## The NEC Style

Contrary to popular belief, the *NEC* is fairly well organized. Understanding the *NEC* structure and writing style is extremely important in learning how to use the *Code* effectively. The *National Electrical Code* is organized into 11 components.

1. Table of contents
2. Chapters (major categories)
3. Articles (individual subjects)
4. Parts (divisions of an Article)
5. Sections, Lists, and Tables (Code rules)
6. Exceptions (Code rules)
7. Fine Print Notes (explanatory material, not mandatory *Code* language)
8. Definitions (Code rules)
9. Marginal Notations, *Code* Changes (|)
10. Index
11. Annexes

**1. Table of Contents.** The Table of Contents displays the layout of the Chapters, Articles, and Parts as well as their page numbers.

**2. Chapters.** There are nine chapters, each of which contain chapter-specific articles. The nine chapters fall into four groupings; general requirements, specific requirements, communications systems, and tables.

**General Requirements: Chapters 1 through 4**

Chapter 1. General

Chapter 2. Wiring and Protection

Chapter 3. Wiring Methods and Materials

Chapter 4. Equipment For General Use

**Specific Requirements and Conditions: Chapters 5 through 7**

Chapter 5. Special Occupancies

Chapter 6. Special Equipment

Chapter 7. Special Conditions

Chapter 8. Communications Systems (Telephone, Radio/Television, and Cable TV Systems.)

Chapter 9. Tables (Conductor and raceway specifications and properties.)

**3. Articles.** The *NEC* contains approximately 140 Articles, each of which covers a specific subject. For example:

Article 110. General Requirements

Article 250. Grounding

Article 300. Wiring Methods

Article 430. Motors

Article 500. Hazardous (Classified) Locations

Article 680. Swimming Pools

Article 725. Class 1, Class 2, and Class 3 Remote-Control, Signaling, and Power-Limited Circuits

Article 800. Communications Wiring

**4. Parts.** Many articles are subdivided into Parts. For example, Article 250 is so large that it has been divided into nine parts, such as:

Part I. General

Part II. Circuit and System Grounding

Part III. Grounding Electrode System

*CAUTION: The "Parts" of a Code article are not included in the Section numbers. Because of this, we have a tendency to forget what "Part" the Code rule is relating to. For example, Table 110.34 gives the working space clearances in front of electrical equipment. If we aren't careful, we might think this table applies to all electrical installations, but Table 110.34 is contained in Part III, Over 600V, Nominal of Article 110. The rule for working clearance in front of electrical equipment for systems under 600V is in Table 110.26, which is in Part II, 600V, Nominal, or less.*

**5. Sections, Lists, and Tables.**

**Sections.** Each *Code* rule is called a Section, such as 225.26. A *Code* Section may be broken down into subsections by letters in parentheses, such as (A), numbers in parentheses may further break down each subsection, and lower-case letters to the third level. For example, the rule requiring all receptacles in a dwelling unit bathroom to be GFCI-protected is contained in 210.8(A)(1).

*Note: Many in the electrical industry incorrectly use the term "Article" when referring to a Code Section. For example, they will say "Article 210.8," when they should be saying "Section 210.8."*

From this point on in the book, I will use section references without using the word "Section" to follow the 2002 format. For example, "Section 250.118" will be referred to as just "250.118." Where I use a section number in text as a reference, it may be inside brackets. For example, Section 90.2(A)(4) would be [90.2(A)(4)].

**Lists.** If a rule contains a list of items, then the items are identified as (1), (2), (3), (4), etc. See 250.118. If a list is part of a numeric subsection, then the items are listed as a., b., c., etc. See 250.114(4).

**Tables.** Many *Code* requirements are contained within Tables, which are lists of *Code* rules placed in a systematic arrangement.

Many times notes are provided with each table; be sure to read them, as they are also part of the *Code*. For example, Note 1 for Table 300.5 explains how to measure the cover when burying cables and raceways, and Note 5 explains what to do if solid rock is encountered.

**6. Exceptions.** Exceptions are *Code* rules that provide an alternative method to a specific requirement. There are two types of exceptions: mandatory and permissive. When a rule has several exceptions, those exceptions with mandatory requirements are listed before the permissive exceptions.

**Mandatory Exception.** A mandatory exception uses the words "shall" or "shall not." The word "shall" in an exception means that if you are using the exception, you are required to do it in a particular way. The term "shall not" means it is not permitted; for an example, see 230.24(A) Ex. 1.

**Permissive Exception.** A permissive exception uses such words as "shall be permitted," which means that it is acceptable to do it in this way. See 230.24(A) Ex. 2.

**7. Fine Print Note (FPN).** A Fine Print Note contains explanatory material intended to clarify a rule or give assistance, but it is not a *Code* requirement. FPNs often use the word "may," but never "must." See 90.1(B) FPN.

**8. Definitions.** Definitions of specific words and terms are listed in Article 100, within some Articles, and in Parts of some articles.

**Article 100.** Definitions are listed in Article 100 and throughout the *NEC*. In general, the definitions listed in Article 100 apply to more than one *Code* Article, such as "Branch Circuit," which is used in many Articles.

**Definitions contained in Articles.** Definitions at the beginning of an Article apply only to that specific Article. For example, the definition of "Swimming Pool" is contained in 680.2 because this term applies only to the requirements contained in Article 680 – Swimming Pools.

**Part.** Definitions located in a Part of an article apply only to that Part of the Article. For example, the definition of "motor control circuit" in 430.71 applies only to the rules in Part VI, Motor Control Circuits of Article 430. When a definition is listed in the general definitions of an article, it can be correctly applied to the entire article, but when it is found within a part of an Article, it is to be correctly applied only to that part of the Article in which it is found.

**Section.** Definitions located in a Section apply only to that section. For example, the definition of a "hallway" in 210.52(H)

only applies to that Section.

**9. Changes.** Changes to the *NEC* are identified in the margins of the 2002 *NEC* by a vertical line (|). For example, see 90.1(D) Relation to International Standards.

Many rules in the 2002 *NEC* were relocated. The place from which the *Code* rule was removed has no identifier and the place where the rule was inserted has no identifier. For example, the rules for receptacles that were located in Section 210-7 of the 1999 *NEC* were relocated to 406.3, without any identifying marks.

The *NEC* does not contain a cross-index for rules that have been relocated, but you can access mine free at www.NECcode.com.

**10. Index.** The 30-page index in the 2002 *NEC* is excellent and should be very helpful to locate the rule in question.

**11. Annexes.** Annexes are not a part of the *NEC*, and are included for informational purposes only. They include the following:

| | |
|---|---|
| Annex A. | Product Safety Standards |
| Annex B. | Conductor Ampacity Under Engineering Supervision |
| Annex C. | Raceway Fill Tables |
| Annex D. | Examples |
| Annex E. | Types of Construction |
| Annex F. | Cross-Reference Tables |

### How to Locate Specific Material In the Code

How you go about finding what you are looking for in the *NEC* depends, to some degree, on your experience with it. Many experienced *Code* users will consult the Table of Contents instead of the Index to locate specific information.

For example, what *Code* rule specifies the maximum number of disconnects permitted for a service? If you're an experienced *Code* user, you'll know that Article 230 applies to "Services" and it contains a Part VI, "Disconnection Means." With this knowledge, you can quickly go to page 2 of the Table of Contents and see that the rule is on page 70-81.

You might wonder why the number 70 precedes all of the page numbers. This is because the NFPA has numbered all of its standards, and the *NEC* is standard number 70. Since I will be referencing only the *NEC*, I will drop the 70 in front of the page numbers.

If you used the Index (subjects listed in alphabetical order) and looked up the term "service disconnect," you would see that there is no such listing. If you tried "disconnecting means," then "services," you would find that the Index specifies that the rule is at 230VI! Because the *NEC* does not give a page number in the Index, you'll need to use the Table of Contents to get the page number, or flip through the page of Article 230 until you find Part VI, which is on page 81 (this does not apply to the NEC handbook).

Many people complain that the *Code* only confuses them by taking them in circles. As you gain experience in using the *Code* and deepen your understanding of *Code* words, terms, principles and practices, you will find the *Code* much easier to understand. You may even come to appreciate the value of the multiple section references contained in the articles.

## Customizing Your Code Book

One way for you to increase your comfort level with the *Code* book is to customize it to meet your needs. You can do this by highlighting and underlining important *Code* rules, and by attaching convenient page tabs.

**Highlighting.** As you read this book, be sure you highlight those rules in the *Code* that are important to you. Use yellow for general interest and orange for important rules you want to find quickly. Be sure to highlight terms in the Index and Table of Contents as you use them.

Because of the size of the 2002 *NEC*, I strongly recommend you highlight in green the Parts of larger Articles that are important for your application, particularly:

Article 230. Services

Article 250. Grounding

Article 410. Luminaires

Article 430. Motors

**Underlining.** Underline or circle key words and phrases in the *NEC* with a red pen (not a lead pencil) and use a 6 in. ruler to keep lines straight and neat. This is a very handy way to have important words or terms stand out. A small 6-in. ruler also comes in handy when trying to locate specific information in the many tables in the *Code*.

**Tabbing the NEC.** Placing tabs on important *Code* Articles, Sections, and Tables will make it very easy to access important *Code* rules. However, too many tabs will defeat the purpose. You can easily order a set of *Code* tabs online at www.mikeholt.com or by phone at 1.888.NEC.Code (1.888.632.2633).

## Global Changes to the 2002 NEC

Some changes were made in the *NEC* to facilitate its international adoption.

**Format** – The format system for the 2002 *NEC* was changed from the "hyphen" system to the scientific method of notion typically called the "dot" system. For example, Section 110-26 will now be 110.26. The first letter will be capitalized and the following characters will be lower case. For example: 110.26(A), 110.26(A)(1), 110.26(A)(1)(a).

**New Articles** – Many new Articles were added, for example:

Article 80. Administration and Enforcement

Article 285. Transient Voltage Surge Suppressors (TVSSs)

Article 406. Receptacles

Article 647. Sensitive Electronic Equipment

Article 692. Fuel Cells

**Metric Units** – Units of measurement are now listed with the metric unit first, followed by the foot/pound measurement in parentheses. See 90.9.

**Parts** – Article parts in the 1999 *NEC* were identified by an upper case letter, such as Part A, Part B, and Part C. The 2002 *NEC* identifies article parts by a Roman numeral, such as Part I, Part II, and Part III.

**Sections Relocated** – Many *Code* sections were relocated. For example, the requirements for receptacles contained in Articles 210 and 410 were moved to the new Article 406 covering the requirements for receptacles only.

To achieve a uniform numbering system, all wiring method Articles had their sections renumbered. For example, the Section numbers for all raceways will be as follows:

Part I. General.
    3XX.1 Scope
    3XX.2 Definitions
    3XX.3 Other Articles
    3XX.6 Listing Requirements
Part II. Installation
    3XX.10 Uses Permitted
    3XX.12 Uses Not Permitted
    3XX.14 Dissimilar Metals
    3XX.16 Temperature Limits
    3XX.20 Size
    3XX.22 Number of Conductors
    3XX.24 Bends–How Made
    3XX.26 Bends–Number in One Run
    3XX.28 Reaming and Threading (Metal)
    3XX.28 Trimming (Nonmetallic)
    3XX.30 Securing and Supporting
    3XX.40 Boxes and Fittings
    3XX.42 Coupling and Connectors
    3XX.44 Expansion Fittings
    3XX.46 Bushings
    3XX.48 Joints
    3XX.50 Conductor Terms
    3XX.56 Splices and Taps
    3XX.60 Grounding

Author's Comment: 3XX represents the actual article number, which could be Articles: 320 – Armored Cable, 334 – Nonmetallic-Sheathed Cable, 348 – Flexible Metal Conduit, or 358 – Electrical Metallic Tubing.

**Exceptions** – Many exceptions were rewritten into complete sentences and converted into positive text.

**New Definitions** – New definitions were added to Article 100 as well as other Articles. For example, the term "luminaire" replaces "fixture," and "lighting fixture."

While the *NEC* does have a specific article titled "Definitions," it is not intended to be a dictionary. The official dictionary of the NFPA is the IEEE dictionary. When a technical definition is needed, only an approved technical dictionary such as the IEEE dictionary should be consulted.

**Wire Identifiers** – The common "No." is no longer used to describe conductor size throughout the *NEC*. For example, a No. 8 wire is now identified as 8 AWG.

**Article Numbers** – The article numbers for all wiring methods were renumbered.

| New | Old | |
|---|---|---|
| 312 | 373 | Cabinets and Cutout Boxes |
| 314 | 370 | Outlet, Device, Pull, and Junction Boxes; Conduit Bodies; Fittings; and Manholes |
| 320 | 333 | Armored Cable – Type AC |
| 324 | 328 | Flat Conductor Cable |
| 330 | 334 | Metal Clad Cable – Type MC |
| 332 | 330 | Mineral Insulated Cable – Type MI |
| 334 | 336 | Nonmetallic-Sheathed Cable – Type NM |
| 336 | 340 | Tray Cable – Type TC |
| 338 | 338 | Service Entrance Cable – Type SE |
| 340 | 339 | Underground Feeder Cable – Type UF |
| 342 | 345 | Intermediate Metal Conduit - IMC |
| 344 | 346 | Rigid Metal Conduit - RMC |
| 348 | 350 | Flexible Metal Conduit – FMC |
| 350 | 351A | Liquidtight Flexible Metal – LFMC |
| 352 | 347 | Rigid Nonmetallic Conduit – RNC |
| 354 | 343 | Nonmetallic Underground Conduit with Conductors – NUCC |
| 356 | 351B | Liquidtight Nonmetallic Conduit – LFNC |
| 358 | 348 | Electrical Metallic Tubing – EMT |
| 362 | 331 | Electrical Nonmetallic Tubing – ENT |
| 366 | 374 | Auxiliary Gutters |
| 368 | 364 | Busways |
| 370 | 365 | Cablebus |
| 372 | 358 | Cellular Concrete Raceway |
| 374 | 356 | Cellular Metal Floor Raceway |
| 376 | 362A | Metal Wireways |
| 378 | 362B | Nonmetallic Wireways |
| 380 | 353 | Multioutlet Assembly |
| 382 | 342 | Nonmetallic Extensions |
| 384 | 352C | Strut-Type Channel Raceways |
| 386 | 352A | Surface Metal Raceways |
| 388 | 352B | Surface Nonmetallic Raceways |
| 390 | 354 | Underfloor Raceways |
| 392 | 318 | Cable Trays |
| 394 | 324 | Concealed Knob-and-Tube |
| 396 | 321 | Messenger Supported Wiring |
| 398 | 320 | Open Wiring on Insulators |
| 527 | 305 | Temporary Installations |

*I dedicate this book to Jesus Christ,*
*my mentor and teacher*

# Article 80
# Administration and Enforcement

Article 80 was added to the 2002 *NEC* to provide a model administrative and enforcement ordinance. The requirements contained in this article are informative and cannot be enforced by the Authority Having Jurisdiction (AHJ), unless it is specifically adopted by the local jurisdiction [80.5].

Article 80 is intended to promote uniform administrative provisions throughout the United States and other countries that have adopted, or will be adopting, the *NEC*. In addition, this article provides model administrative and enforcement requirements for jurisdictions that are adopting the *Code* for the first time.

# Article 90
# Introduction

## 90.1 Purpose

**(A) Practical Safeguarding.** The purpose of the *NEC* is to ensure that electrical systems are installed in a manner that protects people and property by minimizing the risks associated by the use of electricity.

**(B) Adequacy.** The *Code* contains rules that are considered necessary for a safe electrical installation. It is expected that when an electrical installation is installed in compliance with the *NEC*, it will be essentially free from electrical hazards.

The *NEC* is a safety standard, not a design guide. It is not intended to ensure that the electrical system will be efficient, convenient, or adequate for good service or future expansion. Specific items of concern such as electrical energy management, maintenance and power quality are typically not within the scope of the *Code*. Figure 90-1

> FPN: Hazards in electrical systems often occur because circuits are overloaded or were never properly installed in accordance with the NEC. Often the initial wiring did not provide reasonable provisions for system changes or for the expected increase in the use of electricity.

**AUTHOR'S COMMENT:** The *NEC* does not require electrical systems to be installed to accommodate future expansion of electrical use. However, the electrical designer, typically an electrical engineer, should be concerned with electrical safety and proper system design. This dictates that electrical systems be designed above the NEC requirements so they are efficient and convenient, and provide sufficient capacity for future use.

**(C) Intention.** The *Code* is to be used by individuals who understand electricity, electrical systems and electrical installation practices. It is not a "how-to" book and is not intended as a design specification or instruction manual for untrained individuals.

Occasionally the separation between a design standard and the *NEC* can become gray. When that happens, it may help to go back and review the stated purpose of the *NEC*. That helps to see the Code's safety requirements as safety and not design.

**(D) Relation to International Standards.** The requirements of the *NEC* address the fundamental safety principles contained in the International Electrotechnical Commission Standard. Figure 90–2

> FPN: IEC Standard 60364-1, Section 131, contains safety protection principles to protect against electric shock, thermal effects, overcurrent, fault currents, and overvoltage. The NEC addresses all of these potential hazards.

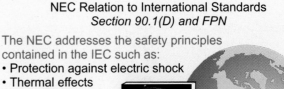

**Figure 90-1**

**Figure 90–2**

## 90.2 Scope

**(A)    What is Covered.** The *NEC* contains rules necessary for the proper electrical installation for power and lighting conductors and equipment, signaling and communications conductors and equipment, as well as fiber-optic cables and raceways for the following: Figure 90–3

(1)    Public and private premises including buildings or structures, mobile homes, recreational vehicles, and floating buildings.

(2)    Yards, lots, parking lots, carnivals, and industrial substations.

(3)    Conductors and equipment that connect to the utility supply.

(4)    Installations in electric utility office buildings, warehouses, garages, machine shops, recreational buildings, or other electric utility buildings that are not an integral part of the utility's generating plant, substation, or control center. Figure 90–4

**(B)    What is not Covered.**

(1)    Installations in cars, trucks, boats, ships and watercraft, planes, electric trains, or underground mines.

(2)    Self-propelled mobile surface mining machinery and its attendant electrical trailing cables.

(3)    Railway power, signaling and communications wiring.

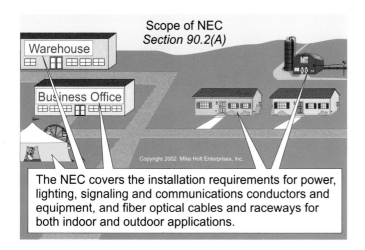

*The NEC covers the installation requirements for power, lighting, signaling and communications conductors and equipment, and fiber optical cables and raceways for both indoor and outdoor applications.*

**Figure 90-3**

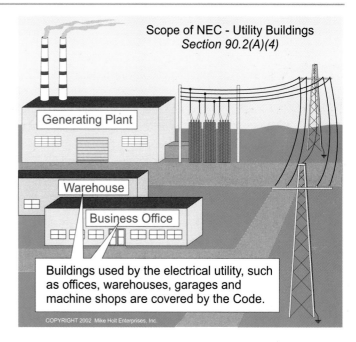

*Buildings used by the electrical utility, such as offices, warehouses, garages and machine shops are covered by the Code.*

**Figure 90-4**

(4)    **Communications Utilities.** The *NEC* does not apply to installations under the exclusive control of a communications utility located outdoors or in building spaces used exclusively for such use. However, the interior and exterior wiring of phone, communications, and CATV not under the exclusive control of communications utilities shall comply with the requirements contained in Chapter 8. Figure 90-5

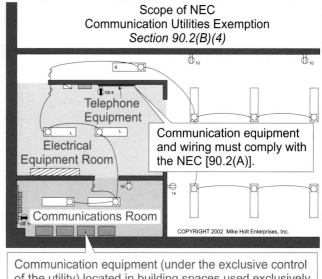

*Communication equipment and wiring must comply with the NEC [90.2(A)].*

*Communication equipment (under the exclusive control of the utility) located in building spaces used exclusively for such installations is not governed by the NEC.*

**Figure 90-5**

**(5) Electric Utilities.** The *NEC* does not apply to installations under the exclusive control of an electric utility, including service drops or service laterals, or wiring for the purpose of communications, metering, generation, control, transformation, transmission, or distribution of electric energy on legally established easements, right-of-ways, or by other agreements recognized by public/utility regulatory agencies, or property owned or leased by the electric utility. Figures 90–6 and 90–7

**AUTHOR'S COMMENT:** Service laterals not under the exclusive control of an electric utility installed by the electrical contractor shall be installed in accordance with the requirements of the NEC and not the National Electrical Safety Code (NESC), which applies to electrical systems typically under the exclusive control of a utility type organization.

Lighting equipment located in legally established easements, or right-of-ways, such as at poles supporting transmission or distribution lines, are exempt from the requirements of the *NEC*. However, if the electric utility provides site and public lighting on private property, then the installation shall comply with *NEC* requirements [90.2(A)(4)]. Figure 90–8

## 90.3 Code Arrangement

The *Code* is divided into an Introduction and nine chapters.

**General Requirements.** Chapters 1, 2, 3 and 4 apply to all installations.

**Special Requirements.** Chapters 5, 6 and 7 apply to special occupancies, special equipment or other special conditions, and they can supplement or modify the requirements in Chapters 1 through 4.

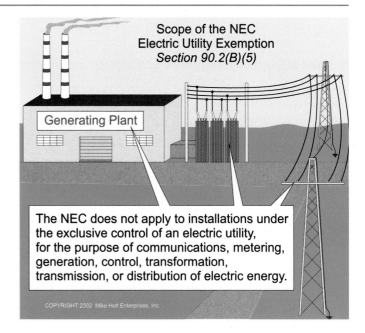

**Figure 90-7**

**Communications Systems.** Chapter 8 contains the requirements for communications systems, such as telephone, antenna wiring, CATV, and network-powered broadband systems. Communications systems are not subject to the general requirements contained in Chapters 1 – 4 or the special requirements of Chapters 5 – 7, except where there is a specific reference in Chapter 8 to a rule in one of those chapters.

**Tables.** Chapter 9 consists of tables used for raceway sizing, conductor fill and voltage drop.

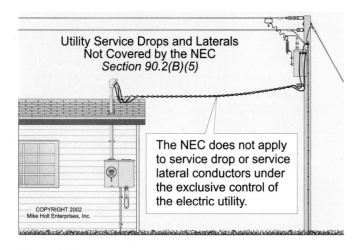

**Figure 90-6**

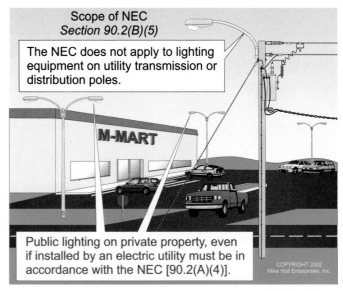

**Figure 90–8**

**Annexes.** Annexes are not part of the *Code*, but they are included in the *NEC* for informational purposes. They are:

**Annex A.** Product Safety Standards

**Annex B.** Conductor Ampacity Under Engineering Supervision

**Annex C.** Raceway Size Tables

**Annex D.** Examples

**Annex E.** Types of Construction

**Annex F.** Cross-Reference Tables

## 90.4 Enforcement

This *Code* is intended to be suitable for enforcement by governmental bodies that exercise legal jurisdiction over electrical installations for power, lighting, signaling and communications systems within the scope of this *Code*, such as: Figure 90–9

**Signal Circuits**

**Article 725.** Class 1, Class 2, and Class 3 Remote-Control, Signaling, and Power-Limited Circuits

**Article 760.** Fire Alarm

**Article 770.** Optical Fiber Cables and Raceways

**Communications Circuits**

**Article 800.** Communications Circuits (Telecommunications)

**Article 810.** Radio and Television Equipment (Satellite and Antenna)

**Article 820.** Community Antenna Television and Radio Distribution Systems (CATV and CCTV)

**Article 830.** Network-Powered Broadband Communications Systems

The enforcement of the NEC falls under the Authority Having Jurisdiction (AHJ), who has the responsibility of interpreting the rules, approving equipment and materials, waiving Code requirements, and ensuring that equipment is installed in accordance with manufacturers' instructions. Figure 90-10

**Interpretation of the NEC Rules.** The AHJ is responsible for interpreting the *NEC*, but his or her decisions shall be based on a specific requirement. If an installation is rejected, the AHJ has a responsibility to inform the installer as to the specific *NEC* rule violated.

---

**AUTHOR'S COMMENT:** The art of getting along with the AHJ consists of doing good work and knowing what the Code says, as opposed to what you only think it says. It's also useful to know how to choose your battles when the inevitable disagreement does occur.

---

**Approval of Equipment and Materials.** The AHJ decides what equipment and materials are considered suitable for use. Typically, equipment listed by an approved testing laboratory such as Underwriters Laboratories, Inc. (UL) is accepted. See 90.7, 110.2, 110.3 and the definitions in Article 100 for Approved, Identified, Labeled, and Listed.

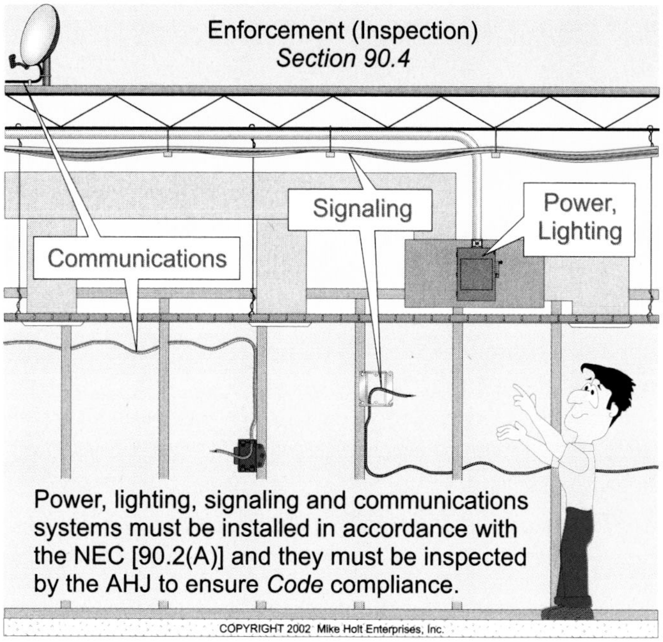

Enforcement (Inspection)
Section 90.4

Communications

Signaling

Power, Lighting

Power, lighting, signaling and communications systems must be installed in accordance with the NEC [90.2(A)] and they must be inspected by the AHJ to ensure *Code* compliance.

COPYRIGHT 2002 Mike Holt Enterprises, Inc.

**Figure 90-9**

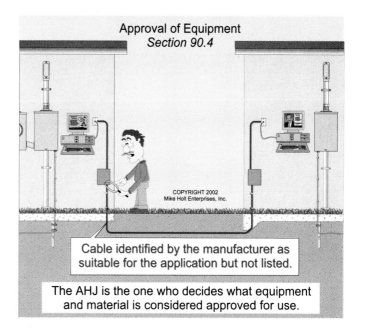

Approval of Equipment
Section 90.4

COPYRIGHT 2002
Mike Holt Enterprises, Inc.

Cable identified by the manufacturer as suitable for the application but not listed.

The AHJ is the one who decides what equipment and material is considered approved for use.

**Figure 90-10**

**AUTHOR'S COMMENT:** The AHJ can prohibit the use of listed equipment or approve the use of non-listed equipment. Given our highly litigious society, approval for the installation of nonlisted equipment is becoming increasingly difficult to obtain.

**Waiver of Rules.** By special permission, the AHJ can waive specific requirements in this *Code* or permit alternative methods where it is assured that equivalent objectives can be achieved by establishing and maintaining effective safety.

**AUTHOR'S COMMENT:** Special permission is defined as the written consent of the AHJ [Article 100].

**Waiver of New Product Requirements.** If the 2002 *Code* requires products that are not yet available at the time the *Code* is adopted, the AHJ can allow products that were acceptable in the 1999 *Code* to continue to be used.

**AUTHOR'S COMMENT:** Sometimes it takes years before testing laboratories establish product standards for new NEC requirements, then it takes time before manufacturers can redesign, manufacture, and distribute these products to meet the new Code requirements.

**Compliance with Manufacturers' Instructions.** It is the AHJ's responsibility to ensure that electrical equipment is installed in accordance with equipment listing and/or labeling instructions. (The *NEC* does not address the maintenance of electrical equipment, as it is an installation standard not a maintenance standard.) In addition, the AHJ can reject the installation of equipment modified in the field. See 90.7, 110.2, and 110.3(B) for more information.

## 90.5 Mandatory Rules and Explanatory Material

**(A) Mandatory Rules.** Rules that are specifically required or prohibited are characterized by the use of the terms "shall" or "shall not." For example:

**110.3(B) Installation and Use.** Listed or labeled equipment shall be installed and used in accordance with any instructions included in the listing or labeling.

**(B) Permissive Rules.** Rules that identify actions that are allowed but not required are characterized by the use of the terms "shall be permitted" or "shall not be required." A permissive rule is often an exception to the general requirement. For example:

**250.102(E) Installation.** The equipment bonding jumper *shall be permitted* to be installed inside or outside of a raceway or enclosure.

**(C) Explanatory Material.** References to other standards or sections of the *Code*, or information related to a *Code* rule are included in the form of a Fine Print Note (FPN). Fine Print Notes are for information only and are not enforceable. For example, limiting a circuit's voltage drop is not a requirement, it's a recommendation contained in a fine print note. For example 210.19(A)(4):

FPN No. 4: Conductors for branch circuits as defined in Article 100, sized to prevent a voltage drop exceeding three percent at the farthest outlet of power, heating, and lighting loads, or combinations of such loads, and where the maximum total voltage drop on both feeders and branch circuits to the farthest outlet does not exceed five percent, provide reasonable efficiency of operation.

## 90.6 Formal Interpretations

To promote uniformity of interpretation and application of the provisions of this *Code*, formal interpretation procedures have been established and are found in the NFPA Regulations Governing Committee Projects.

**AUTHOR'S COMMENT:** This is rarely done because it's a very time-consuming process. Not only that, but formal interpretations from the NFPA are not legally binding on the AHJ!

## 90.7 Examination of Equipment for Product Safety

Product evaluation for safety is typically performed by an approved testing laboratory, which publishes a list of equipment that meets a nationally recognized test standard. Products and materials listed, labeled, or identified by an approved testing laboratory are generally approved for use by the AHJ.

Product evaluation by an approved testing laboratory avoids the necessity for repetition of examinations by different persons, frequently with inadequate facilities, and the confusion that could result from conflicting reports.

In addition, listed factory-installed internal wiring and construction of equipment need not be inspected at the time of installation, except to detect alterations or damage [300.1(B)]. Figure 90-11

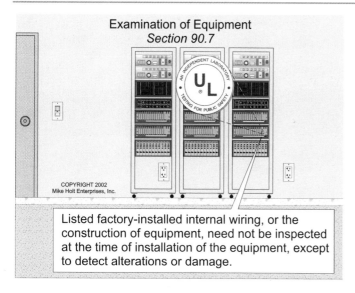

Examination of Equipment
Section 90.7

COPYRIGHT 2002
Mike Holt Enterprises, Inc.

Listed factory-installed internal wiring, or the construction of equipment, need not be inspected at the time of installation of the equipment, except to detect alterations or damage.

**Figure 90-11**

## 90.9 Units of Measurement

**(A) Measurement System.** The metric unit is the preferred measurement system.

**(B) Dual Systems of Units.** Both the metric and inch-pound measurement systems shall be shown in the *NEC*, with the metric units to appear first and the inch-pound system to immediately follow in parentheses.

**(D) Compliance.** Installing electrical systems in accordance with the metric system or the inch-pound system is considered to comply with the *Code*.

**AUTHOR'S COMMENT:** Since compliance with either the metric or the inch-pound unit of measurement system constitutes compliance with this Code, I will only use the inch-pound unit.

## Article 90

1.  Compliance with the provisions of the *Code* will result in _____.
    (a) good electrical service                    (b) an efficient electrical system
    (c) an electrical system essentially free from hazard    (d) all of these

2.  The *Code* contains provisions considered necessary for safety regardless of _____.
    (a) efficient use                              (b) convenience
    (c) good service or future expansion of electrical use    (d) all of these

3.  Hazards often occur because of _____.
    (a) overloading of wiring systems by methods or usage not in conformity with this *Code*
    (b) initial wiring not providing for increases in the use of electricity
    (c) a and b
    (d) none of these

4.  Installation of communications equipment that is under the exclusive control of communications utilities, and located outdoors or in building spaces used exclusively for such installations, _____ covered by the *Code*.
    (a) is            (b) is sometimes        (c) is not            (d) might be

5.  The requirements in "Annexes" must be complied with.
    (a) True          (b) False

6.  The authority having jurisdiction for enforcement of the *Code* has the responsibility _____.
    (a) for making interpretations of the rules of the *Code*
    (b) for deciding upon the approval of equipment and materials
    (c) for waiving specific requirements in the *Code* and allowing alternate methods and material if safety is maintained
    (d) all of these

7.  A *Code* rule may be waived or alternative methods of installation approved that may be contrary to the *NEC* if the AHJ gives verbal or written consent.
    (a) True          (b) False

8.  In the event the *Code* requires new products, constructions, or materials that are not yet available at the time the *Code* is adopted, the _____ may permit the use of the products, constructions, or materials that comply with the most recent previous edition of this *Code* adopted by the jurisdiction.
    (a) architect                              (b) master electrician
    (c) authority having jurisdiction          (d) supply house

9.  Explanatory material such as references to other standards, references to related sections of this *Code*, or information related to a *Code* rule, is included in this *Code* in the form of Fine Print Notes (FPNs).
    (a) True          (b) False

10. Equipment that is listed by a qualified electrical testing laboratory is not required to have the factory-installed _____ wiring inspected at the time of installation, except to detect alterations or damage.
    (a) external        (b) associated        (c) internal        (d) all of these

# Grounding and Bonding Library

Grounding and bonding problems are at epidemic levels. Surveys repeatedly show 90% of power quality problems are due to poor grounding and bonding. Mike has applied electrical theory to this very difficult to understand Article, making it easier for students to grasp the concepts of Grounding & Bonding. Additionally, Mike has color coded the text so you can easily differentiate between Grounding and Bonding.

Order the entire Grounding and Bonding Library including the textbook, two videos, two DVDs and the CD-Rom

## Call Today 1.888.NEC.CODE or visit us online at www.NECcode.com for the latest information and pricing.

# Chapter 1 – General

## Article 100. Definitions

Article 100 contains the definitions of many terms used throughout the *Code*. In general, only those terms used in two or more Articles are contained in Article 100.

**Article Definitions.** Definitions at the beginning of an Article apply only to that Article. For example, the definition of "swimming pool" is in 680.2, because this term applies only to the requirements of Article 680 – Swimming Pools.

**Part Definitions.** Terms defined in a Part of an article apply only to that Part of the Article. For example, the definition of "motor control circuit" at 430.71 applies only to the rules in Part VI Motor Control Circuits of Article 430.

**Section Definitions.** Definitions in a Section apply only to that Section, such as the definition of "hallway" in 210.52(H) or the definition of "AFCI device" in 210.12(A).

Part I of Article 100 contains the definitions of terms used throughout the *Code*, and the definitions of terms in Part II of Article 100 are applicable only to systems that operate at over 600V.

Definitions of standard terms, such as volt, voltage drop, ampere, impedance, or resistance are not listed in the *Code*. The official electrical dictionary of the NFPA is the IEEE Standard Dictionary of Electrical and Electronic Terms, www.ieee.org. If the *NEC* or the IEEE dictionary does not define a term, then a standard dictionary suitable to the AHJ may be consulted. For example, the definition of a kitchen is not contained in the *NEC* or the IEEE. A building code glossary might provide a better definition than your home or school edition dictionary.

## Article 110. Requirements for Electrical Installations

This Article contains the general requirements for electrical installations for the following:

Part I. General

Part II. 600V, Nominal, or less

Part III. Over 600V, Nominal

Part IV. Tunnel Installations over 600V, Nominal

# Article 100
## Definitions

**Accessible as it Applies to Equipment.** Admitting close approach and not guarded by locked doors, elevation, or other effective means. For example:

**210.52(E) Outdoor Outlets.** For a one-family dwelling and each unit of a two-family dwelling that is at grade level, at least one receptacle outlet *accessible* at grade level and not more than $6^1/_2$ ft above grade shall be installed at the front and back of the dwelling. See 210.8(A)(3).

**Accessible as it Applies to Wiring Methods.** Not permanently closed in by the building structure or finish and capable of being removed or exposed without damaging the building structure or finish. For example, raceways, cables and enclosures installed above a suspended ceiling or within a raised floor are considered accessible, because the wiring methods can be accessed without damaging the building structure. See the definitions for Concealed and Exposed. For example: Figure 100-1

**300.15(A) Wiring Methods with Interior Access.** A box or conduit body shall not be required for each splice, junction, switch, pull, termination, or outlet point in wiring methods with removable covers. The covers shall be *accessible* after installation.

**Accessible, Readily (Readily Accessible).** Readily accessible means capable of being reached quickly without having to climb over or remove obstacles or resort to portable ladders. Figure 100-2

**230.70(A)(1) Readily Accessible Location.** The service disconnecting means shall be installed either inside or outside of the building or structure served or where the conductors pass through the building or structure. The disconnecting means shall be at a *readily accessible* location nearest the point of entrance of the conductors. Figure 100-3

**Ampacity.** The current in amperes a conductor can carry continuously, where the temperature will not be raised in excess of the conductor's insulation temperature rating. See 310.10 and 310.15 for examples. Figure 100-4

**Appliance [Article 424].** Electrical equipment, other than industrial equipment, built in standardized sizes, such as ranges, ovens, cooktops, refrigerators, drinking water coolers, beverage dispensers.

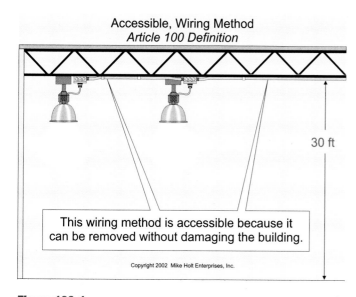

Accessible, Wiring Method
*Article 100 Definition*

30 ft

This wiring method is accessible because it can be removed without damaging the building.

Copyright 2002  Mike Holt Enterprises, Inc.

**Figure 100-1**

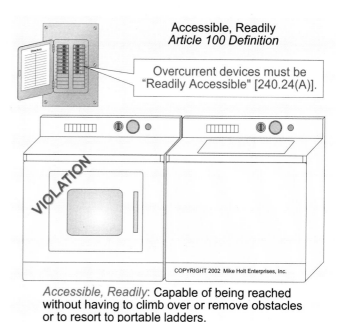

Accessible, Readily
*Article 100 Definition*

Overcurrent devices must be "Readily Accessible" [240.24(A)].

VIOLATION

COPYRIGHT 2002  Mike Holt Enterprises, Inc.

*Accessible, Readily*: Capable of being reached without having to climb over or remove obstacles or to resort to portable ladders.

**Figure 100-2**

Service disconnects located on the second floor still meet the requirements for readily accessible.

**Figure 100-3**

**Approved.** Acceptable to the Authority Having Jurisdiction (AHJ); this is usually the electrical inspector. See 90.4, 90.7, 110.2 and the definitions in Article 100 for Identified, Labeled, and Listed.

> **AUTHOR'S COMMENT:** See the definition for Authority Having Jurisdiction.

**348.30(A) Securely Fastened.** Flexible metal conduit shall be securely fastened in place by an *approved* means within 12 in. of each box, cabinet, conduit body, or other conduit termination and shall be supported and secured at intervals not to exceed $4^1/_2$ ft.

**Attachment Plug (Plug Cap) (Plug).** A wiring device at the end of flexible cord intended to be inserted into a receptacle. Figure 100-5

> **AUTHOR'S COMMENT:** The use of a cord with an attachment plug is limited by 210.50(A), 527.4, 400.7, 410.14, 410.30, 422.33 and 645.5.

**Authority Having Jurisdiction (AHJ).** The AHJ is the organization, office, or individual responsible for approving equipment, materials, an installation, or a procedure.

> FPN: The Authority Having Jurisdiction may be a federal, state or local government, or an individual such as a fire chief, fire marshal, chief of a fire prevention bureau, labor department or health department, building official or electrical inspector, or others having statutory authority. In some circumstances, the property owner or his/her agent assumes the role, and at government installations, the commanding officer or departmental official may be the AHJ.

Typically, the AHJ will be the electrical inspector who has statutory authority. In the absence of federal, state or local regulations, the operator of the facility or his/her agent, such as an architect or engineer of the facility, can assume the role.

> **AUTHOR'S COMMENT:** Many feel that the AHJ should have a strong background in the electrical field such as electrical engineering or an electrical contractor's license, and in some states, this is a legal requirement. Memberships, certifications, and active participation in electrical organizations also speak to an individual's qualifications. See 80.27 for Inspectors' Qualifications.

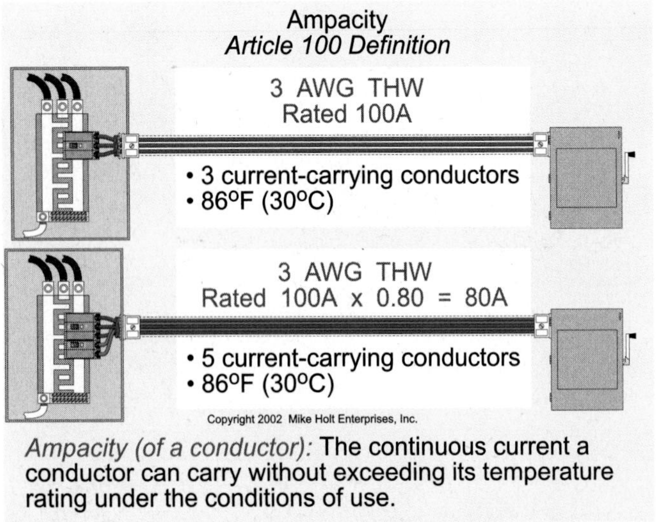

**Ampacity (of a conductor):** The continuous current a conductor can carry without exceeding its temperature rating under the conditions of use.

**Figure 100-4**

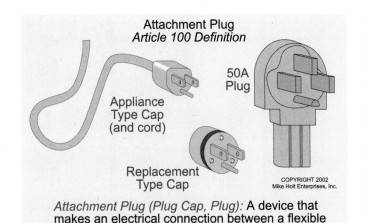

**Attachment Plug (Plug Cap, Plug):** A device that makes an electrical connection between a flexible cord and a receptacle.

**Figure 100-5**

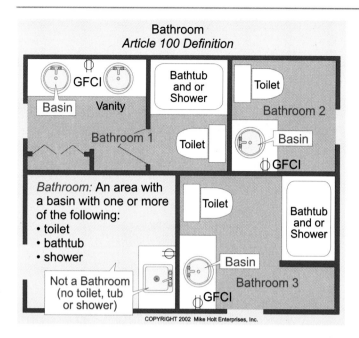

Bathroom
*Article 100 Definition*

GFCI

Basin    Vanity

Bathtub and or Shower

Toilet

Bathroom 2

Bathroom 1

Toilet

Basin

GFCI

*Bathroom:* An area with a basin with one or more of the following:
• toilet
• bathtub
• shower

Toilet

Bathtub and or Shower

Basin

Bathroom 3

Not a Bathroom (no toilet, tub or shower)

GFCI

COPYRIGHT 2002 Mike Holt Enterprises, Inc.

**Figure 100-6**

**Bathroom.** A bathroom is an area including a basin with a toilet, a tub, or a shower. Figure 100-6

**Bonding (Bonded).** The permanent joining of metallic parts to form an electrically conductive path that has the capacity to conduct safely fault current. Figure 100-7

**AUTHOR'S COMMENT:** This could require a separate conductor or it might be accomplished by properly terminating raceways or cables in fittings or boxes as

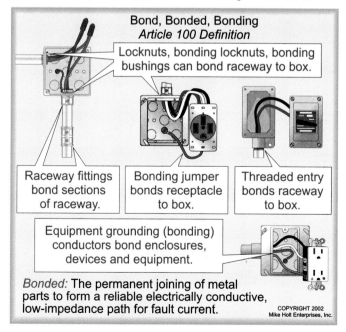

Bond, Bonded, Bonding
*Article 100 Definition*

Locknuts, bonding locknuts, bonding bushings can bond raceway to box.

Raceway fittings bond sections of raceway.

Bonding jumper bonds receptacle to box.

Threaded entry bonds raceway to box.

Equipment grounding (bonding) conductors bond enclosures, devices and equipment.

*Bonded:* The permanent joining of metal parts to form a reliable electrically conductive, low-impedance path for fault current.

COPYRIGHT 2002 Mike Holt Enterprises, Inc.

**Figure 100-7**

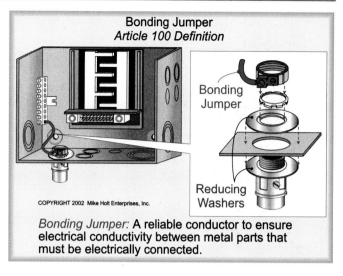

Bonding Jumper
*Article 100 Definition*

Bonding Jumper

Reducing Washers

COPYRIGHT 2002 Mike Holt Enterprises, Inc.

*Bonding Jumper:* A reliable conductor to ensure electrical conductivity between metal parts that must be electrically connected.

**Figure 100-8**

required by Article 250 to form an effective ground-fault current path.

**Bonding Jumper.** A conductor that ensures electrical conductivity between metal parts of the electrical system for metal parts that shall be electrically connected in accordance with Article 250. Figure 100-8

**Bonding Jumper, Main.** The conductor, screw, or strap that bonds the equipment grounding conductor to the grounded (neutral) conductor at the service equipment. For more details see 250.24(A)(4), 250.28, and 408.3(C).

**Branch Circuit [Article 210].** The conductors between the final overcurrent device and the receptacle outlets, lighting outlets, or other outlets as defined in Article 100. Figure 100-9

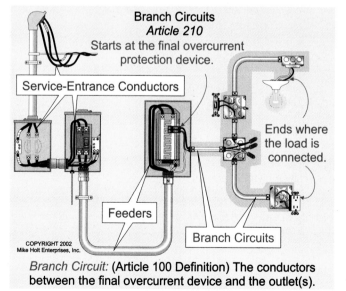

Branch Circuits
*Article 210*
Starts at the final overcurrent protection device.

Service-Entrance Conductors

Ends where the load is connected.

Feeders

Branch Circuits

COPYRIGHT 2002 Mike Holt Enterprises, Inc.

*Branch Circuit:* (Article 100 Definition) The conductors between the final overcurrent device and the outlet(s).

**Figure 100-9**

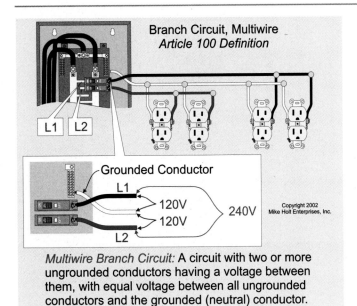

Branch Circuit, Multiwire
*Article 100 Definition*

Grounded Conductor

Copyright 2002
Mike Holt Enterprises, Inc.

*Multiwire Branch Circuit:* A circuit with two or more ungrounded conductors having a voltage between them, with equal voltage between all ungrounded conductors and the grounded (neutral) conductor.

**Figure 100-10**

**Branch Circuit, Multiwire.** A branch circuit that consists of two or more ungrounded conductors with a grounded (neutral) conductor, where there is a voltage potential between the ungrounded conductors and an equal voltage potential from each ungrounded (hot) conductor to the grounded conductor. Figure 100-10

> **AUTHOR'S COMMENT:** Multiwire branch circuits offer the advantage of fewer conductors in a raceway, smaller raceway sizing, and a reduction of material and labor cost. In addition, multiwire branch circuits reduce circuit voltage drop by as much as 50 percent. However, because of the dangers associated with multiwire branch circuits, the NEC contains some additional requirements to ensure a safe installation. See 210.4, 300.13(B) and 408.20 for details.

**Building.** A structure that stands alone or is cut off from other structures by firewalls with all openings protected by approved fire doors. Figure 100-11

**Cabinet [Article 312].** An enclosure for either surface mounting or flush mounting provided with a frame in which a door can be hung. Figure 100-12

**Circuit Breaker [Article 240-VII].** A device designed to be opened and closed manually, and which opens automatically on a predetermined overcurrent without damage to itself. Circuit breakers are available in different configurations such as inverse-time molded case, adjustable (electronically controlled), and instantaneous trip/motor circuit protector.

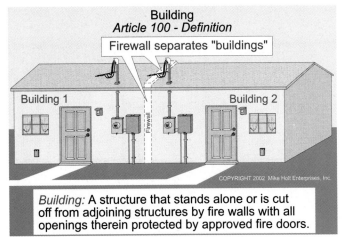

Building
*Article 100 - Definition*

Firewall separates "buildings"

Building 1                          Building 2

COPYRIGHT 2002 Mike Holt Enterprises, Inc.

*Building:* A structure that stands alone or is cut off from adjoining structures by fire walls with all openings therein protected by approved fire doors.

**Figure 100-11**

**Adjustable.** A circuit breaker that can be set to trip at various values of current, time, or both, within a predetermined range.

**Inverse Time.** An inverse-time circuit breaker has a delay in the tripping action, which decreases as the current increases.

**Concealed.** Rendered inaccessible by the structure or finish of the building. Conductors in a concealed raceway are considered concealed, even though they may become accessible by withdrawing them. Figure 100-13

> **AUTHOR'S COMMENT:** Wiring behind panels designed to allow access is considered exposed.

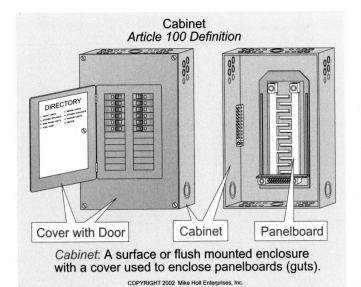

Cabinet
*Article 100 Definition*

DIRECTORY

Cover with Door          Cabinet          Panelboard

*Cabinet:* A surface or flush mounted enclosure with a cover used to enclose panelboards (guts).

COPYRIGHT 2002 Mike Holt Enterprises, Inc.

**Figure 100-12**

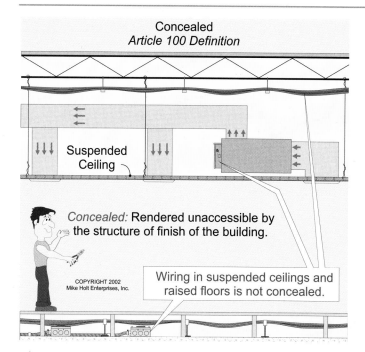

Concealed
*Article 100 Definition*

*Concealed:* Rendered unaccessible by the structure of finish of the building.

COPYRIGHT 2002 Mike Holt Enterprises, Inc.

Wiring in suspended ceilings and raised floors is not concealed.

**Figure 100-13**

**Conduit Body.** A fitting that provides access to conductors through a removable cover. See 314.16(C) for conductor splicing rules. Figure 100-14

**Continuous Load.** A load where the current is expected to run for three hours or more, such as store or parking lot lighting.

**Controller.** A device or several devices that controls, in some predetermined manner, the electric power delivered to electrical equipment. This would include time clocks, lighting contactors, photocells, etc. See 430.71 for the definition of a motor controller. Figure 100-15

**AUTHOR'S COMMENT:** Controls can be operating or protective devices. Thus, a contactor could be considered to be an operating control, a circuit breaker a safety control and a motor starter both an operating control,

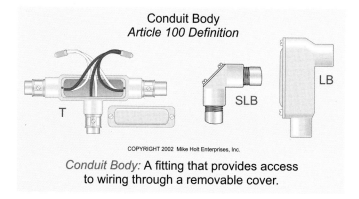

Conduit Body
*Article 100 Definition*

*Conduit Body:* A fitting that provides access to wiring through a removable cover.

**Figure 100-14**

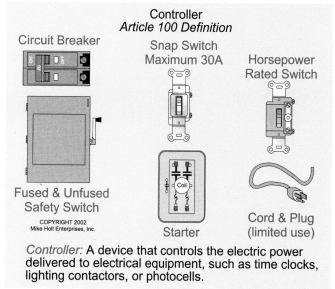

Controller
*Article 100 Definition*

Circuit Breaker

Snap Switch Maximum 30A

Horsepower Rated Switch

Fused & Unfused Safety Switch

Starter

Cord & Plug (limited use)

COPYRIGHT 2002 Mike Holt Enterprises, Inc.

*Controller:* A device that controls the electric power delivered to electrical equipment, such as time clocks, lighting contactors, or photocells.

**Figure 100-15**

and a protective control. A starter starts and stops and also provides overload protection.

**Device.** A unit of an electrical system that is intended to carry, but not utilize (consume) electrical energy. Examples would include receptacles, switches, circuit breakers, fuses, etc., but not locknuts or other mechanical fittings. Figure 100-16

**Disconnecting Means.** A device, or group of devices, or other means by which the conductors of a circuit can be disconnected from their source of supply. Switches, attachment plugs and receptacles, automatic and nonautomatic circuit breakers, and knife and safety switches can be used as a disconnecting means, depending on the conditions. Figure 100-17

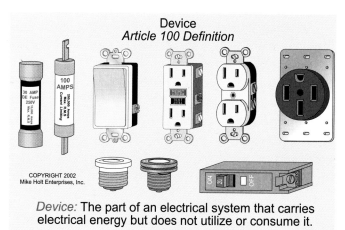

Device
*Article 100 Definition*

COPYRIGHT 2002 Mike Holt Enterprises, Inc.

*Device:* The part of an electrical system that carries electrical energy but does not utilize or consume it.

**Figure 100-16**

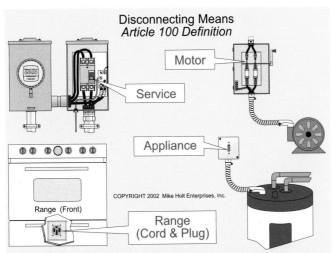

Disconnecting Means
*Article 100 Definition*

Motor

Service

Appliance

Range (Front)

Range
(Cord & Plug)

COPYRIGHT 2002 Mike Holt Enterprises, Inc.

*Disconnecting Means:* A device by which the conductors of a circuit can be disconnected from their source of supply.

**Figure 100-17**

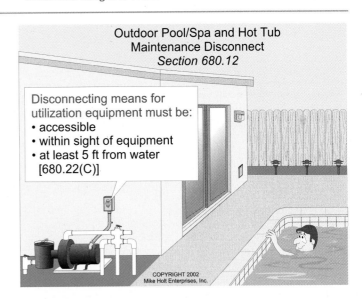

Outdoor Pool/Spa and Hot Tub
Maintenance Disconnect
*Section 680.12*

Disconnecting means for utilization equipment must be:
• accessible
• within sight of equipment
• at least 5 ft from water
[680.22(C)]

COPYRIGHT 2002
Mike Holt Enterprises, Inc.

**Figure 100-18**

For more information on the application of disconnecting means, review the following *Code* rules:

- Air-conditioning, 440.14
- Appliances, 422-Part III
- Building (feeder supplied), 225-Part II
- Electric space heating, 424.19
- Electric duct heaters, 424.65
- Motor control conductors, 430.74
- Motor controllers, 430.102(A)
- Motors, 430.102(B)
- Refrigeration equipment, 440.14
- Services, 230-Part VI
- Swimming pool equipment, 680.12 Figure 100-18

**Dwelling Unit.** One or more rooms for use as a living facility with space for eating, living, and sleeping, and permanent provisions for cooking and sanitation.

**Dwelling, Multifamily.** A multifamily dwelling is a building with three or more dwelling units.

**Energized.** Electrically connected to a source of potential difference.

**Exposed as it Applies to Wiring Methods.** On or attached to the surface of a building or behind panels designed to allow access, such as suspended ceilings or raised floor space. Figure 100-19

**Feeder.** The conductors between the service equipment, the source of a separately derived system, or other power-supply source and the final branch-circuit overcurrent device. Secondary conductors from a transformer and conductors to a remote building are feeders, but conductors from the electric utility are service conductors, not feeder conductors. See the definition of Service Conductors. Figure 100-20

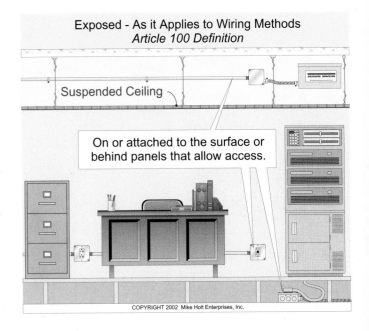

Exposed - As it Applies to Wiring Methods
*Article 100 Definition*

Suspended Ceiling

On or attached to the surface or behind panels that allow access.

COPYRIGHT 2002 Mike Holt Enterprises, Inc.

**Figure 100-19**

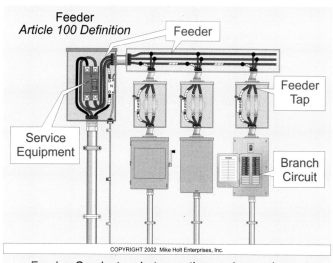

Feeder
*Article 100 Definition*

*Feeder:* Conductors between the service equipment or the source of a separately derived system and the final branch-circuit overcurrent protection device.

**Figure 100–20**

**AUTHOR'S COMMENT:** Other sources of power would include, but not be limited to, a battery, a solar voltaic (photovoltaic or PV) system, a UPS, a generator, or converter windings.

**Fitting.** An accessory such as a locknut that is intended to perform a mechanical function.

**Garage.** A building or portion of a building where self-propelled vehicles can be kept.

FPN: For commercial garages, repair and storage. See Article 511.

**AUTHOR'S COMMENT:** Receptacles can be installed at any height in a dwelling unit garage, but they are typically installed no less than 18 in. above the floor for a commercial garage. See 511.3(B)(1).

NOTE: The definitions for the following terms are located in the beginning of Article 250 in this textbook

• Ground
• Grounded
• Grounded Conductor
• Grounded, Effectively
• Grounding Conductor
• Grounding Conductor, Equipment
• Grounding Electrode Conductor

**Ground-Fault Circuit Interrupter.** A device intended to protect people by de-energizing a circuit within a short time if the

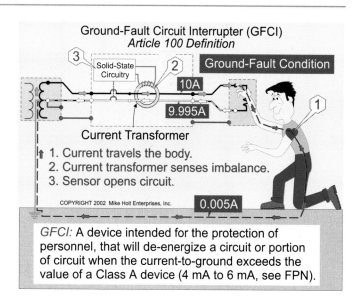

*GFCI:* A device intended for the protection of personnel, that will de-energize a circuit or portion of circuit when the current-to-ground exceeds the value of a Class A device (4 mA to 6 mA, see FPN).

**Figure 100-21**

current-to-ground exceeds the value established for a "Class A" device. Figure 100–21

FPN: A "Class A" ground-fault circuit interrupter opens the circuit when the current to ground has a value between 4 mA and 6 mA.

**AUTHOR'S COMMENT:** GFCI protective devices are commercially available in receptacles, circuit breakers, cord sets, and other types of devices. Figure 100-22

## HOW GFCIs WORK

A ground-fault circuit interrupter (GFCI) is specifically designed to protect people against electric shock from an electrical system. A GFCI protection device operates on the principle of monitoring the imbalance of current between the circuit's ungrounded (hot) and grounded (neutral) conductor. An interest-

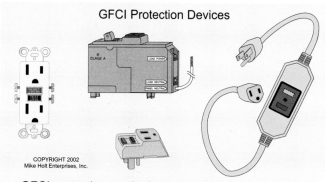

GFCI protection can be incorporated into receptacles, circuit breakers, cord sets and other types of devices.

**Figure 100-22**

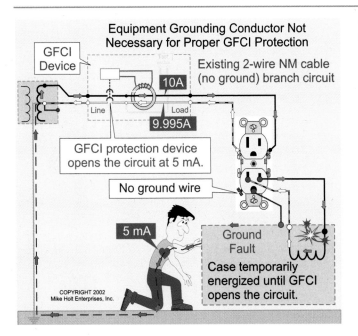

**Figure 100-23**

ing point about these devices is that despite their name, they will operate on a circuit with or without an equipment grounding conductor. Figure 100-23

During the normal operation of a typical 2-wire circuit, the current returning to the power supply will be equal to the current leaving the power supply. If the difference between the current leaving and returning through the current transformer of the GFCI protection device exceeds 5 mA (± 1 mA), the solid-state circuitry opens the switching contacts and de-energizes the circuit.

The mA used above stands for one thousands of an ampere, so 5 mA is equal to $^5/_{1000}$th of an ampere.

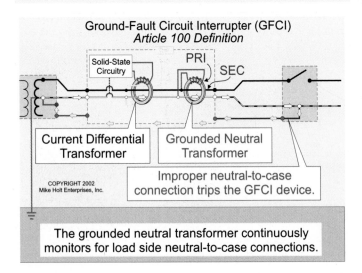

**Figure 100-24**

**AUTHOR'S COMMENT:** Another function of a GFCI device is the detection of downstream neutral-to-case connections. A second current transformer in the GFCI induces a voltage on the circuit conductors. If there is a neutral-to-case connection on the load side of the GFCI protection device. the GFCI will sense the imbalance of the current returning on the equipment grounding conductor and prevent the GFCI device from being turned on. This feature can give the appearance of a "defective" GFCI device because it trips when the circuit is energized, even with no loads on or connected. Figure 100-24

*WARNING: Severe electric shock or death can occur if a person touches the hot and neutral conductor at the same time, even if the circuit is GFCI-protected. This is because the current transformer within the GFCI protection device does not sense an imbalance between the departing and returning current. Figure 100-25*

*Danger: Typically when a GFCI protection device fails, the switching contacts remain closed and the device will continue to provide power without GFCI protection. According to a study by the American Society of Home Inspectors published in the November/ December 1999 issue of the IAEI News (International Association of Electrical Inspectors magazine), 21 percent of GFCI circuit breakers and 19 percent of GFCI receptacles did not provide GFCI protection, yet the circuit remained energized!*

The failures of the GFCI sensing circuits were mostly due to damage to the internal transient voltage surge protectors (metal-oxide varistors) that protect the GFCI sensing circuit. This damage resulted from voltage surges from lightning and other transients. In areas of high lightning activity, such as southwest Florida, the failure rate for GFCI circuit breakers was over 50 percent! Figure 100-26

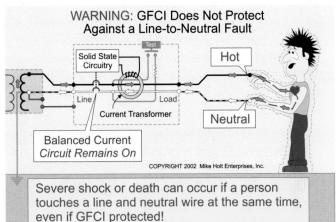

**Figure 100-25**

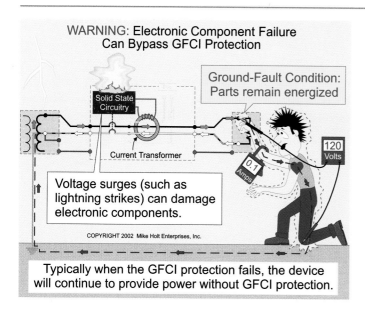

WARNING: Electronic Component Failure Can Bypass GFCI Protection

Ground-Fault Condition: Parts remain energized

Solid State Circuitry

Current Transformer

0.1 Amps

120 Volts

Voltage surges (such as lightning strikes) can damage electronic components.

COPYRIGHT 2002 Mike Holt Enterprises, Inc.

Typically when the GFCI protection fails, the device will continue to provide power without GFCI protection.

**Figure 100-26**

At least one leading manufacturer markets a listed 15A, 125V GFCI receptacle you cannot reset if the GFCI circuit no longer provides ground-fault protection. As an added safety improvement, this particular GFCI receptacle has a built-in line-load reversal feature that prevents the GFCI from resetting if the installer mistakenly reverses the load and line connections.

One final thought on GFCI protection is that you should press the test feature of the GFCI protection device to ensure that it turns the power off to the connected load. Do not assume that a GFCI protection device is operational unless you properly test it!

**Ground-Fault Protection of Equipment.** A system intended to provide protection of equipment from damaging line-to-ground fault currents by opening all ungrounded conductors of the faulted circuit. This protection is provided at current levels less than those required to protect conductors from damage through the operation of a supply circuit overcurrent device. See 215.10, 230.95, and 240.13. This type of protective device is not intended to protect people, only connected utilization equipment.

**Identified (as applied to equipment).** Recognizable as suitable for a specific purpose, function, or environment by listing and labeling. See 90.4, 90.7, 110.3(A)(1) and the definitions in Article 100 for Approved, Labeled, and Listed.

**110.14(A) Terminals.** Terminals for more than one conductor and terminals used to connect aluminum shall be *identified* for the purpose.

**In Sight From (Within Sight From, Within Sight).** The specified equipment is to be visible and not more than 50 ft distant from the other. Figure 100-27

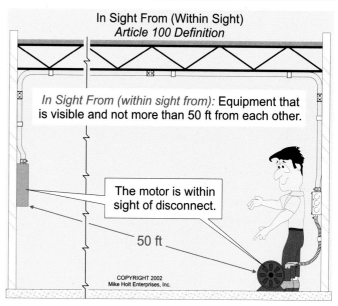

In Sight From (Within Sight)
*Article 100 Definition*

*In Sight From (within sight from):* Equipment that is visible and not more than 50 ft from each other.

The motor is within sight of disconnect.

50 ft

COPYRIGHT 2002 Mike Holt Enterprises, Inc.

**Figure 100-27**

**440.14 Location.** A disconnecting means shall be located *within sight* and readily accessible from A/C or refrigeration equipment.

**Interrupting Rating.** The highest short-circuit current at rated voltage that the device can safely interrupt. For more information, see 110.9.

**Labeled.** Equipment or materials that have a label, symbol, or other identifying mark in the form of a sticker, decal, or printed label, or molded or stamped into the product by an approved testing laboratory. See the definitions of Identified and Listed. Figure 100-28

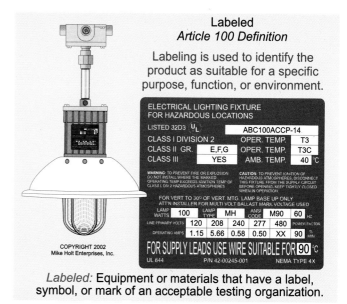

Labeled
*Article 100 Definition*

Labeling is used to identify the product as suitable for a specific purpose, function, or environment.

ELECTRICAL LIGHTING FIXTURE FOR HAZARDOUS LOCATIONS

| LISTED 32D3 UL | ABC100ACCP-14 | |
| CLASS I DIVISION 2 | OPER. TEMP. | T3 |
| CLASS II GR. | E,F,G | OPER. TEMP. | T3C |
| CLASS III | YES | AMB. TEMP. | 40 °C |

FOR SUPPLY LEADS USE WIRE SUITABLE FOR 90 °C

COPYRIGHT 2002 Mike Holt Enterprises, Inc.

*Labeled:* Equipment or materials that have a label, symbol, or mark of an acceptable testing organization.

**Figure 100-28**

**Lighting Outlet.** An outlet for the connection of a lampholder, luminaire, or pendant cord terminating in a lampholder. Figure 100-29

**Listed.** Equipment or materials included in a list published by an approved testing laboratory that is concerned with product evaluation. The listing organization shall maintain periodic inspection of production of listed equipment or materials to ensure that the equipment or material meets appropriate designated standards and is suitable for a specified purpose. See the definitions of Identified and Labeled.

**Location, Damp.** Locations protected from weather and not subject to saturation with water or other liquids, such as partially protected locations under canopies, marquees, roofed open porches, and interior locations subject to moderate degrees of moisture, such as some basements, some barns, and some cold-storage warehouses.

**Location, Wet.** Underground installations or installations in concrete slabs and masonry in direct contact with the earth, locations subject to saturation with water, and unprotected locations exposed to weather.

**Location, Dry.** An area not normally subjected to dampness or wetness, but which may be subject temporarily to dampness or wetness, such as a building under construction.

**Luminaire.** A complete lighting unit that consists of a lamp or lamps together with the parts designed to distribute the light. Figure 100–30

**Nonlinear Load.** A load where the wave shape of the steady state current does not follow the wave shape of the applied voltage. See 210.4(A) FPN, 220.22 FPN 2, 310.15(B)(4)(C), and 450.3 FPN 2. Figure 100-31

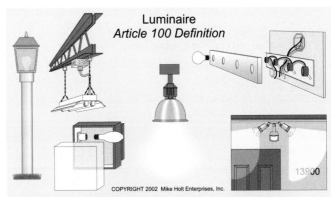

*Luminaire:* A complete lighting unit consisting of a lamp(s) with the parts that distribute the light.

**Figure 100-30**

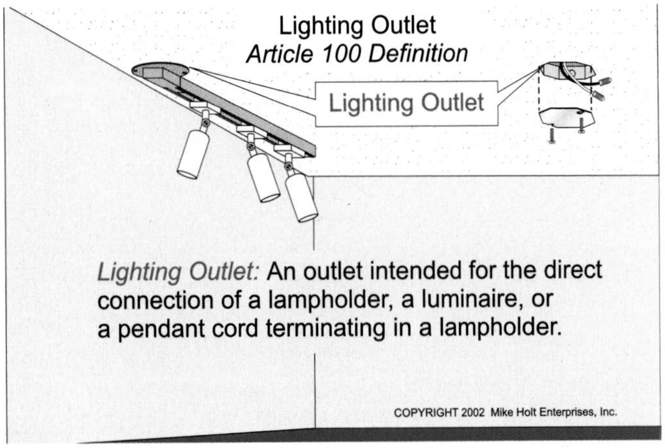

*Lighting Outlet:* An outlet intended for the direct connection of a lampholder, a luminaire, or a pendant cord terminating in a lampholder.

**Figure 100-29**

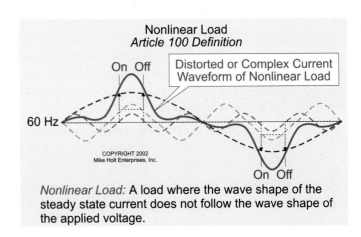

*Nonlinear Load:* A load where the wave shape of the steady state current does not follow the wave shape of the applied voltage.

**Figure 100-31**

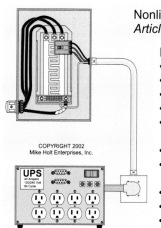

## Nonlinear Load Examples
### Article 100 Definition FPN

Nonlinear loads can include:
- Fluorescent Lighting
- HID Lighting
- Electronic Ballasts
- Electronic Dimmers
- Motors with Variable Frequency Drives
- Computers
- Uninterrupted Power Supplies (UPS)
- Laser Printers
- Data-Processing Equipment
- Other Electronic Equipment

COPYRIGHT 2002
Mike Holt Enterprises, Inc.

**Figure 100-32**

FPN. The NEC does not specifically identify when a load becomes nonlinear, but typical single-phase (1Ø) nonlinear loads include electronic equipment such as copy machines, laser printers, and electric-discharge lighting. Three-phase (3Ø) nonlinear loads would include uninterruptible power supplies (UPS's), induction motors, and electronic switching devices such as variable-frequency speed drives (VFD-VSD). Figure 100-32

**AUTHOR'S COMMENT:** The subject of nonlinear loads is outside the scope of this book. For more information, visit http://www.mikeholt.com/studies/harmonic.htm

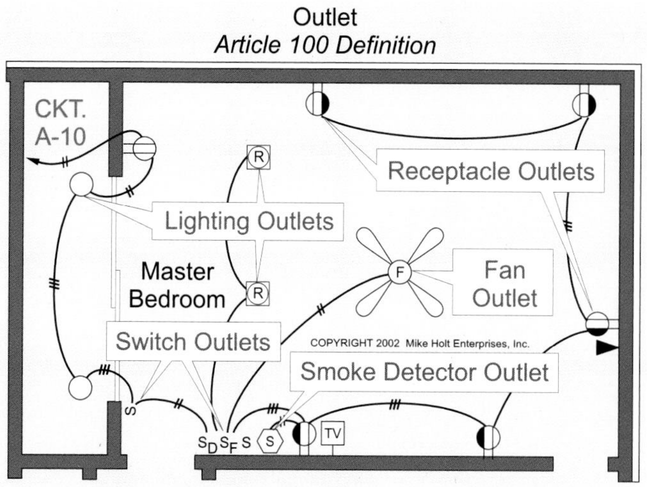

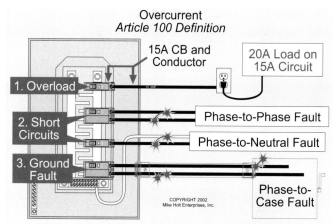

## Overcurrent
### Article 100 Definition

15A CB and Conductor

20A Load on 15A Circuit

1. Overload

2. Short Circuits

Phase-to-Phase Fault

Phase-to-Neutral Fault

3. Ground Fault

COPYRIGHT 2002
Mike Holt Enterprises, Inc.

Phase-to-Case Fault

*Overcurrent:* Current in excess of equipment rating caused from an overload, short circuit, or ground fault.

**Figure 100-34**

**Outlet.** A point in the wiring system where electric current is taken to supply electrical equipment, such as receptacle(s), luminaire(s), and appliance(s). Figure 100-33

**Overcurrent.** Current in amperes that is greater than the rated current of the equipment or conductors, resulting from an overload, short circuit, or ground fault. Figure 100-34

**Overload.** The operation of equipment above its full-load current rating, or current in excess of conductor ampacity. When an overload condition persists for a sufficient length of time, it could cause damaging or dangerous overheating, which could result in equipment failure or a fire. A fault, such as a short circuit or ground fault, is not an overload.

**Panelboard [Article 408].** A distribution point for branch circuit or feeder protection devices designed to be placed in a cabinet and accessible only from the front. The slang term in the electrical field for a panelboard is "the guts." Figure 100-35

## Outlet
### Article 100 Definition

CKT. A-10

Receptacle Outlets

Lighting Outlets

Master Bedroom

Fan Outlet

Switch Outlets

COPYRIGHT 2002 Mike Holt Enterprises, Inc.

Smoke Detector Outlet

TV

*Outlet:* A point on the wiring system where current is taken to supply electrical equipment.

**Figure 100-33**

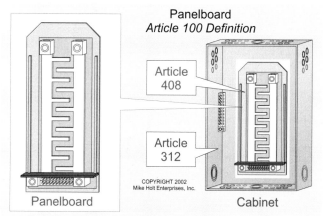

## Panelboard
### Article 100 Definition

Article 408

Article 312

COPYRIGHT 2002
Mike Holt Enterprises, Inc.

Panelboard

Cabinet

*Panelboard:* A distribution point designed to be placed in a cabinet and accessible only from the front.

**Figure 100-35**

**Premises Wiring.** The interior and exterior wiring, including power, lighting, control, and signal circuits and all associated hardware, fittings, and wiring devices, both permanently and temporarily installed. It does not include the internal wiring of electrical equipment and appliances, such as luminaires, dishwashers, water heaters, motors, controllers, motor control centers, A/C equipment, etc. See 90.7 and 300.1(B).

**Qualified Person.** A person who has the skill and knowledge related to the construction and operation of the electrical equipment and its installation. This person shall have received safety training on the hazards involved with electrical systems. See 110.16. Figure 100-36

---

**AUTHOR'S COMMENT:** In many parts of the United States, electricians, electrical contractors, electrical inspectors and electrical engineers must complete between 6 to 24 hours of continuing education per year as a requirement of licensing regulations to maintain their competency.

---

### 110.16 Flash Protection.

Switchboards, panelboards, industrial control panels, and motor control centers that are in other than dwelling occupancies and are likely to require examination, adjustment, servicing, or maintenance while energized shall be field marked to warn qualified persons of potential electric arc flash hazards. The marking shall be located to be clearly visible to *qualified persons* before examination, adjustment, servicing, or maintenance of the equipment.

---

**AUTHOR'S COMMENT:** This term is used over 90 times in the NEC and the definition is in line with OSHA

*Qualified Person:* One who has knowledge and skill related to the construction, operation, and installation of electrical equipment. This person must have safety training on the hazards involved with electrical systems.

**Figure 100-36**

and NFPA 70E *Electrical Safety Requirements for Employee Workplaces* standards.

---

**Raceway.** An enclosure designed for the installation of conductors, cables, or busbars. Raceways in the *NEC* include:

| | |
|---|---|
| 342 | Intermediate Metal Conduit (IMC) |
| 344 | Rigid Metal Conduit (RMC) |
| 348 | Flexible Metal Conduit (FMC) |
| 350 | Liquidtight Flexible Metal Conduit (LFMC) |
| 352 | Rigid Nonmetallic Conduit (RNC) |
| 354 | Nonmetallic Underground Conduit with Conductors (NUCC) |
| 356 | Liquidtight Flexible Nonmetallic Conduit (LFNC) |
| 358 | Electrical Metallic Tubing (EMT) |
| 362 | Electrical Nonmetallic Tubing (ENT) |
| 368 | Busways |
| 370 | Cablebus |
| 372 | Cellular Concrete Raceway |
| 374 | Cellular Metal Floor Raceway |
| 376 | Metal Wireways |
| 378 | Nonmetallic Wireways |
| 380 | Multioutlet Assembly |
| 382 | Nonmetallic Extensions |
| 384 | Strut-Type Channel Raceways |
| 386 | Surface Metal Raceways |
| 388 | Surface Nonmetallic Raceways |
| 390 | Underfloor Raceways |

---

**AUTHOR'S COMMENT:** A cable tray system is not a raceway; it's a support system for cables, raceways and enclosures.

---

**Receptacle.** A contact device installed at an outlet for the connection of an attachment plug. A single receptacle contains one device on a yoke, and a multiple (duplex) receptacle contains two devices on a common yoke. Figure 100-37

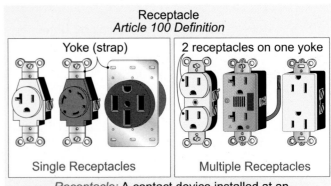

*Receptacle:* A contact device installed at an outlet for the connection of an attachment plug.

**Figure 100-37**

---

**Receptacle Outlet.** An opening in an outlet box where one or more receptacles have been installed.

**Remote-Control Circuit.** An electric circuit that controls another circuit through a relay, solid-state, or equivalent device, installed in accordance with the requirements of Article 725.

**Separately Derived System.** A wiring system from a battery, solar photovoltaic system, transformer, or converter winding, where there is no direct electrical connection to the supply conductors.

> **AUTHOR'S COMMENT:** Understanding a separately derived system is actually a lot more complicated than the above definition. For more information, see 250.20(D) and 250.30.

**Service.** The conductors from the electric utility to the wiring system of the premises.

> **AUTHOR'S COMMENT:** Conductors from a UPS system, battery or solar voltaic system, generator, transformer, or phase converter are not service conductors. See the definitions for Feeder and Service Conductors.

**Service Conductors.** Conductors originating from the "service point" and terminating in "service equipment." To better understand this, review the definitions of Service Point and Service Equipment. Figure 100-38

> **AUTHOR'S COMMENT:** Conductors from a generator, UPS system, or transformer are feeder conductors, not service conductors.

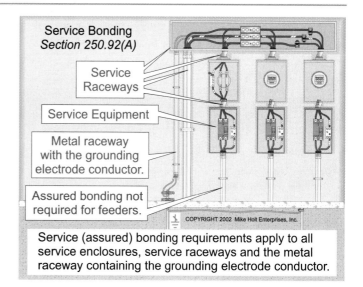

Figure 100-39

**Service Equipment.** The one to six disconnects connected to the load end of service conductors that are intended to control and cut off the supply to the building or structure. See Article 230, Part VI.

> **AUTHOR'S COMMENT:** It is important to know where a service begins and ends to properly apply the bonding requirements contained in 250.92(A) that are necessary for the high fault-current typically available at service equipment. Figure 100-39. In some cases, the service ends before the metering equipment. Figure 100-40

**Service Point.** The point where the electrical utility conductors make contact with premises wiring.

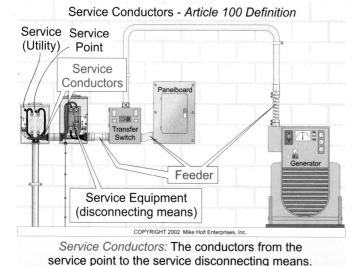

Service Conductors - *Article 100 Definition*

*Service Conductors:* The conductors from the service point to the service disconnecting means.

Figure 100-38

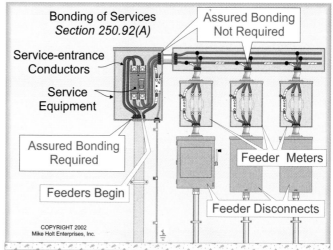

Figure 100-40

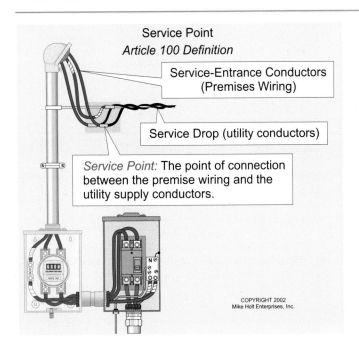

Service Point
*Article 100 Definition*

Service-Entrance Conductors
(Premises Wiring)

Service Drop (utility conductors)

*Service Point:* The point of connection
between the premise wiring and the
utility supply conductors.

COPYRIGHT 2002
Mike Holt Enterprises, Inc.

**Figure 100-41**

**AUTHOR'S COMMENT:** The service point can be at
the pad-mounted transformer (underground residential
distribution or URD), service weatherhead, or meter
enclosure, depending on where the utility conductors
end. Figure 100-41

**Signaling Circuit.** Any electric circuit that energizes signaling equipment.

**Special Permission.** Written consent from the Authority
Having Jurisdiction (AHJ).

**Structure.** That which is built or constructed.

**Switch, General-Use.** A switch used for branch circuits,
rated in amperes, and capable of interrupting its rated current at
its rated voltage.

**Switch, General-Use Snap.** A form of general-use switch
installed in device boxes or on box covers.

**Voltage of a Circuit.** The greatest effective root-mean-square (RMS) difference of potential between any two conductors
of the circuit.

**Voltage, Nominal.** A nominal value assigned to a circuit or
system for the purpose of conveniently designating its voltage
class, such as 120/240V or 480Y/277V. See 220.2(A). Figure
100-42

The actual voltage at which a circuit operates can vary from
the nominal within a range that permits satisfactory operation of
equipment.

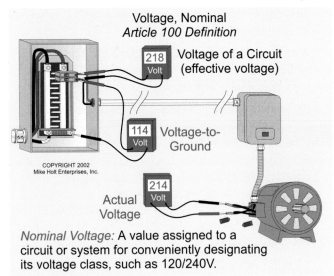

Voltage, Nominal
*Article 100 Definition*

218 Volt  Voltage of a Circuit
(effective voltage)

114 Volt  Voltage-to-
Ground

214 Volt  Actual
Voltage

COPYRIGHT 2002
Mike Holt Enterprises, Inc.

*Nominal Voltage:* A value assigned to a
circuit or system for conveniently designating
its voltage class, such as 120/240V.

**Figure 100-42**

**Voltage to Ground.** The highest voltage between any
ungrounded (hot) conductor and the grounded (neutral) conductor, such as 120 volts-to-ground of a 208Y/120V system. Figure
100-43

**AUTHOR'S COMMENT:** Voltages can be reported in
many ways: peak, peak-to-peak, average, and root-mean-square (RMS). Electrical voltage measuring
equipment is commercially available that provides readings in average and RMS. RMS meters provide the most
accurate reading when used on circuits with nonlinear
and high harmonic contents. The NEC defines the voltage of a circuit using RMS measurements.

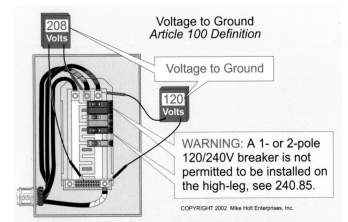

Voltage to Ground
*Article 100 Definition*

208 Volts

Voltage to Ground

120 Volts

WARNING: A 1- or 2-pole
120/240V breaker is not
permitted to be installed on
the high-leg, see 240.85.

COPYRIGHT 2002  Mike Holt Enterprises, Inc.

*Voltage to Ground* (for a grounded circuit). The highest
voltage between any ungrounded conductor and the
grounded conductor, such as 120V-to-ground of a
120/240V system, 208Y/120V system or 208V-to-
ground for a 120/240V high-leg delta system.

**Figure 100-43**

## Article 100

1. Capable of being reached quickly for operation, renewal, or inspections without resorting to portable ladders and such is known as _____.
   (a) accessible (equipment)
   (b) accessible (wiring methods)
   (c) accessible, readily
   (d) all of these

2. Capable of being removed or exposed without damaging the building structure or finish, or not permanently closed in by the structure or finish of the building defines _____.
   (a) accessible (equipment)
   (b) accessible (wiring methods)
   (c) accessible, readily
   (d) all of these

3. Acceptable to the authority having jurisdiction means _____.
   (a) identified
   (b) listed
   (c) approved
   (d) labeled

4. A separate portion of a conduit or tubing system that provides access through a removable cover(s) to the interior of the system at a junction of two or more sections of the system, or at a terminal point of the system, is defined as a(n) _____.
   (a) junction box
   (b) accessible raceway
   (c) conduit body
   (d) pressure connector

5. A device or group of devices that serves to govern, in some predetermined manner, the electric power delivered to the apparatus to which it is connected is a _____.
   (a) relay
   (b) breaker
   (c) transformer
   (d) controller

6. A Class A, GFCI protection device is designed to de-energize the circuit when the ground-fault current is approximately _____.
   (a) 4 mA
   (b) 5 mA
   (c) 6 mA
   (d) any of these

7. Recognized as suitable for the specific purpose, function, use, environment, and application is the definition of _____.
   (a) labeled
   (b) identified (as applied to equipment)
   (c) listed
   (d) approved

8. The environment of a wiring method under the eaves of a house, having a roofed open porch, would be considered a _____ location.
   (a) dry
   (b) damp
   (c) wet
   (d) moist

9. Conduit installed underground or encased in concrete slabs, that are in direct contact with the earth, shall be considered a _____ location.
   (a) dry
   (b) damp
   (c) wet
   (d) moist

10. A(n) _____ is a point on the wiring system at which current is taken to supply utilization equipment.
    (a) box
    (b) receptacle
    (c) outlet
    (d) device

11. The *Code* defines a(n) _____ as one familiar with the construction and operation of the electrical equipment and installations, and who has received safety training on the hazards involved.
    (a) inspector
    (b) master electrician
    (c) journeyman electrician
    (d) qualified person

12. Service conductors only originate from the service point and terminate at the service equipment (disconnect).
    (a) True
    (b) False

13. The _____ is the necessary equipment, usually consisting of a circuit breaker(s) or switch(es) and fuse(s) and their accessories, connected to the load end of service conductors to a building or other structure, or an otherwise designated area, and intended to constitute the main control and cutoff of the supply.
    (a) service equipment
    (b) service
    (c) service disconnect
    (d) service overcurrent protection device

14.  The _____ is the point of connection between the facilities of the serving utility and the premises wiring.
     (a) service entrance                          (b) service point
     (c) overcurrent protection                    (d) beginning of the wiring system

15.  A value assigned to a circuit or system for the purpose of conveniently designating its voltage class such as 120/240V is called _____ voltage.
     (a) root-mean-square     (b) circuit          (c) nominal          (d) source

## Part I. General Requirements

### 110.1 Scope

This article covers the general requirements for the examination and approval, installation and use, access to and spaces about electrical equipment.

### 110.2 Approval of Equipment

Prior to installation, all electrical conductors and equipment shall be approved by the AHJ. Figure 110-1

> **AUTHOR'S COMMENT:** For a better understanding of product approval, review 90.4, 90.7, 110.3 and the definitions for Approved, Identified, Labeled, and Listed in Article 100.

### 110.3 Examination, Identification, Installation, and Use of Equipment

**(A) Guidelines for Approval.** Equipment shall be approved by the AHJ based on the following factors:

(1) Listing or labeling

(2) Mechanical strength and durability

(3) Wire-bending and connection space

(4) Electrical insulation

(5) Heating effects under all conditions likely to arise

(6) Arcing effects

(7) Classification by type, size, voltage, current capacity, and specific use

(8) Other factors contributing to the practical safeguarding of persons using or in contact with the equipment

**(B) Installation and Use.** Equipment shall be installed and used in accordance with any instructions included in the listing or labeling requirements.

> **AUTHOR'S COMMENT:** Equipment is listed for a specific condition of use, operation, or installation and shall be installed and used in accordance with listed instructions. Failure to follow listed instructions, such as torquing of terminals and sizing of conductors, violates this Code rule. Figure 110-2.
>
> When an air-conditioner nameplate specifies "Maximum Fuse Size," one-time or dual-element fuses must be used to protect the equipment. Figure 110-3

### 110.4 Voltages

Electrical equipment shall be installed on a circuit where the system "nominal voltage" does not exceed the voltage rating of

Approval of Equipment
*Section 110.2*

Conductors and equipment can only be installed if approved by the AHJ.

*Approved:* Acceptable to the authority having jurisdiction.

**Figure 110-1**

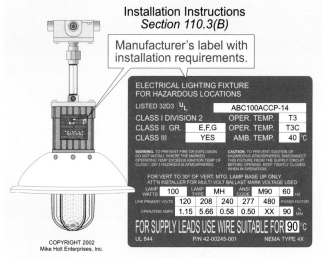

Equipment must be installed and used in accordance with instructions included in the listing or labeling.

**Figure 110-2**

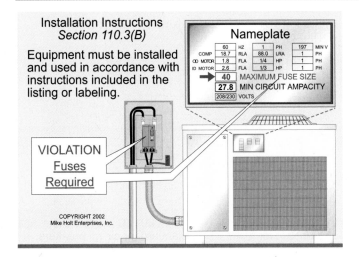

**Figure 110-3**

the equipment. This means that on a 240 nominal voltage system, you cannot install a 208V rated motor. Figure 110-4

> **AUTHOR'S COMMENT:** Electrical equipment must not be connected to a nominal circuit voltage that is rated less than the equipment voltage rating [110.3(B)]. For example, you cannot place a 230V rated motor on a 208 nominal voltage system. Figure 110-5

## 110.5 Copper Conductors

Where conductor material is not specified in a rule, the material and the sizes given are based on copper.

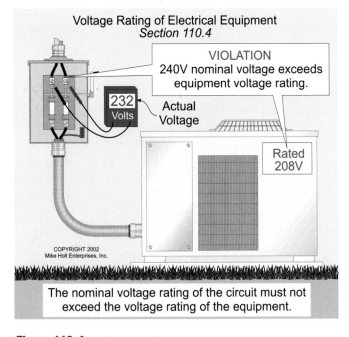

**Figure 110-4**

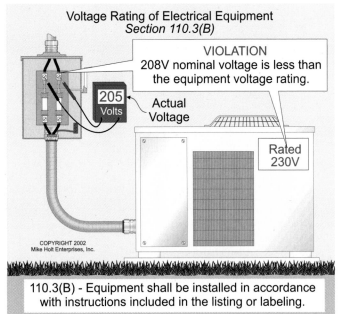

110.3(B) - Equipment shall be installed in accordance with instructions included in the listing or labeling.

**Figure 110-5**

## 110.6 Conductor Sizes

Conductor sizes are expressed in American Wire Gage (AWG) up to 4/0 AWG, and conductor sizes larger than 4/0 AWG are expressed in kcmil (thousand circular mils). Figure 110-6

## 110.7 Conductor Insulation

All wiring shall be installed to be free from short circuits and ground faults.

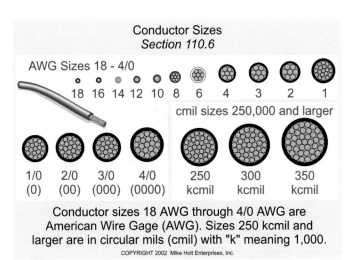

**Figure 110-6**

**AUTHOR'S COMMENT:** Short circuits and ground faults often arise from insulation failure due to mishandling and improper installation. This happens when dragging wire on a sharp edge or around corners, scraping insulation on boxes and enclosures, pulling hard, jerking NM or other cables through rafters and over A/C ducts, nicking the insulation when stripping, or installing cable clamps and staples too tight.

To protect against accidental contact with energized conductors, abandoned conductors must have their non-terminated ends covered with an insulating device identified for the purpose, such as a twist-on or push-on wire connector. See 110.14(B).

## 110.8 Suitable Wiring Methods

Only wiring methods recognized as suitable are included in the *NEC*. The *NEC*-recognized wiring methods shall be installed in accordance with the wiring method requirements as contained in the *Code*. See Chapter 3 for power and lighting wiring methods, Chapter 7 for signaling circuits and Chapter 8 for communications circuits. Figure 110-7

## 110.9 Interrupting Protection Rating

Overcurrent protection devices, such as circuit breakers and fuses, are intended to open the circuit at fault levels. They shall have an interrupting rating, such as 10K, 22K, 65K RMS sufficient for the nominal circuit voltage and the current that is avail-

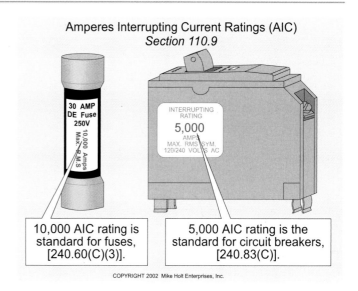

Amperes Interrupting Current Ratings (AIC)
Section 110.9

30 AMP
DE Fuse
250V
10,000
Amps
Max. R.M.S

INTERRUPTING
RATING
5,000
AMPS
MAX. RMS SYM.
120/240 VOLTS AC

10,000 AIC rating is standard for fuses, [240.60(C)(3)].

5,000 AIC rating is the standard for circuit breakers, [240.83(C)].

COPYRIGHT 2002  Mike Holt Enterprises, Inc.

**Figure 110-8**

able at the line terminals of the equipment. See 240.60(C)(3) for the minimum AIC ratings for fuses and 250.83(C) for the minimum AIC rating for circuit breakers. Figure 110-8

## Available Short-Circuit Current

Available short-circuit current (SCA) is the current in amperes that is available at a given point in the electrical system. This available short-circuit current is first determined at the secondary terminals of the utility transformer. Thereafter, the available short-circuit current is calculated at the terminals of the service equipment, branch-circuit and branch-circuit load panelboard.

The available short-circuit current is different at each point of the electrical system, it is highest at the utility transformer and lowest at the branch-circuit load.

The available short-circuit current is dependent on the impedance of the circuit, which increases downstream from the utility transformer. The greater the circuit impedance (utility transformer and the additive impedances of the circuit conductors), the lower the available short-circuit current.

Factors that impact the available short-circuit current at the utility transformer include the system voltage, the transformer kVA rating and its impedance (as expressed in a percentage). Properties that impact the impedance of the circuit include the conductor material (copper versus aluminum), the conductor size, and its length.

The impedance of the circuit increases the further from the utility transformer; therefore, the available short-circuit current is lower downstream from the utility transformer. Figure 110-9

**Interrupting Rating.** Overcurrent protection devices such as circuit breakers and fuses are intended to interrupt the circuit,

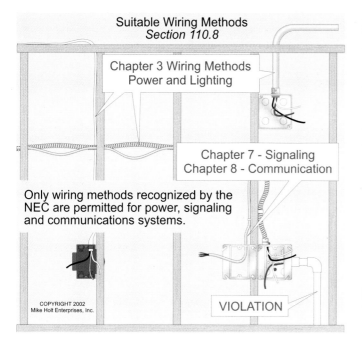

Suitable Wiring Methods
Section 110.8

Chapter 3 Wiring Methods
Power and Lighting

Chapter 7 - Signaling
Chapter 8 - Communication

Only wiring methods recognized by the NEC are permitted for power, signaling and communications systems.

COPYRIGHT 2002
Mike Holt Enterprises, Inc.

VIOLATION

**Figure 110–7**

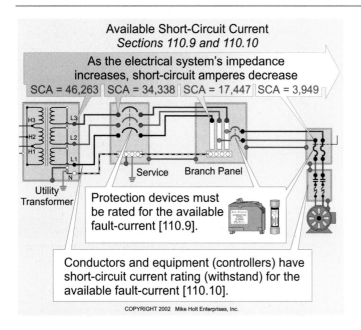

Available Short-Circuit Current
*Sections 110.9 and 110.10*

As the electrical system's impedance increases, short-circuit amperes decrease

SCA = 46,263 | SCA = 34,338 | SCA = 17,447 | SCA = 3,949

Utility Transformer | Service | Branch Panel

Protection devices must be rated for the available fault-current [110.9].

Conductors and equipment (controllers) have short-circuit current rating (withstand) for the available fault-current [110.10].

COPYRIGHT 2002 Mike Holt Enterprises, Inc.

**Figure 110-9**

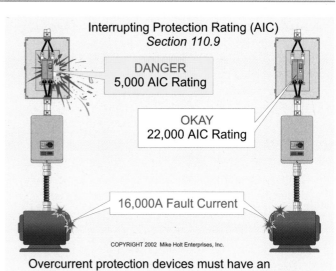

Interrupting Protection Rating (AIC)
*Section 110.9*

DANGER
5,000 AIC Rating

OKAY
22,000 AIC Rating

16,000A Fault Current

COPYRIGHT 2002 Mike Holt Enterprises, Inc.

Overcurrent protection devices must have an interrupting rating that is sufficient for the current that is available at the line terminals of the equipment.

**Figure 110-10**

and they shall have an ampere interrupting rating (AIR) sufficient for the available short-circuit current in accordance with 110.9 and 240.1. Unless marked otherwise, the ampere interrupting rating for branch-circuit circuit breakers is 5,000 amperes [240.83(C)] and 10,000 amperes for branch-circuit fuses [240.60(C)].

> **AUTHOR'S COMMENT:** The terms Ampere Interrupting Rating (AIR) and Amperes Interrupting Current (AIC) are both correct terms meaning the same thing.

Extremely high values of current flow (caused by short circuits or line-to-ground faults) produce tremendously destructive thermal and magnetic forces. If the circuit overcurrent protection device is not rated to interrupt the current at the available fault values, it could explode while attempting to clear the fault. Naturally this can cause serious injury or death, as well as property damage. For information on calculating fault current and sizing equipment for this purpose, visit http://www.mikeholt.com/ Newsletters/Newsletters.htm. Figure 110-10

## 110.10 Short-Circuit Current Rating

Electrical equipment shall have a short-circuit current rating that permits the circuit overcurrent protection device to clear short circuit or ground faults without extensive damage to the electrical components of the circuit. See 110.9, 250.4(A)(5), 250.90, 250.96(A) and Table 250.122 Note.

For example, a motor controller shall have sufficient short-circuit rating for the available fault-current. If the fault exceeds the controller's 5,000A short-circuit current rating, the controller could explode, endangering people and property. To solve this problem, a current-limiting protection device (fast-clearing fuse) can be used to reduce the let-through energy to less than 5,000A. Figure 110-11

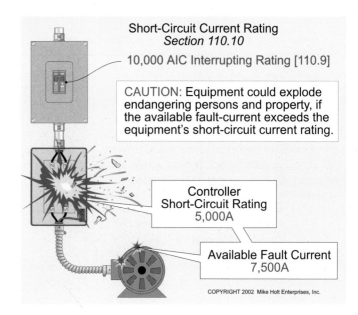

Short-Circuit Current Rating
*Section 110.10*

10,000 AIC Interrupting Rating [110.9]

CAUTION: Equipment could explode endangering persons and property, if the available fault-current exceeds the equipment's short-circuit current rating.

Controller Short-Circuit Rating 5,000A

Available Fault Current 7,500A

COPYRIGHT 2002 Mike Holt Enterprises, Inc.

**Figure 110-11**

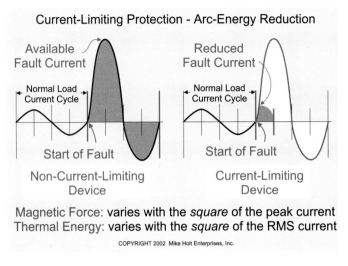

Current-Limiting Protection - Arc-Energy Reduction

Magnetic Force: varies with the *square* of the peak current
Thermal Energy: varies with the *square* of the RMS current

COPYRIGHT 2002 Mike Holt Enterprises, Inc.

**Figure 110–12**

**AUTHOR'S COMMENT:** If the available short-circuit current exceeds the equipment/conductor short-circuit current rating, then the thermal and magnetic forces can cause the equipment circuit conductors as well as grounding conductors to vaporize. The only solution to the problem of excessive available fault current is to:

(1) Install equipment that has a higher short-circuit rating

(2) Protect the components of the circuit by a current-limiting protection device such as a fast-clearing fuse, which can reduce the let-thru energy.

**AUTHOR'S COMMENT:** A breaker or a fuse does limit current, but it may not be listed as a current-limiting device. A thermal-magnetic circuit breaker will typically clear fault current in less than three to five cycles when subjected to a fault of five times its rating. A current-limiting fuse should clear the fault in less than one-half of one cycle. Figure 110-12

*TIP: Both 110.9 and 110.10 deal with ensuring that electrical equipment can either interrupt or withstand the available fault current. 110.9 AIC applies to fuses and circuit breakers, and it requires them to be rated greater than the available short-circuit or ground-fault current. 110.10 applies to equipment and its components, where the short-circuit current rating shall be greater than the available short-circuit or ground-fault. In other words, electrical equipment shall have the ability to withstand the fault current (without extensive damage to the electrical components of the circuit) [110.10] long enough for the overcurrent protection device to open [110.9]. Figure 110-13*

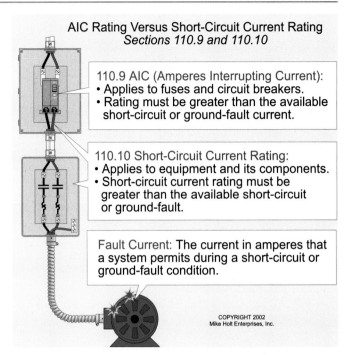

**Figure 110-13**

## 110.11 Deteriorating Agents

Electrical equipment and conductors shall be suitable for the environment or condition of use. Consideration shall be given to the presence of corrosive gases, fumes, vapors, liquids, or chemicals that have a deteriorating effect on conductors or equipment. Figure 110-14

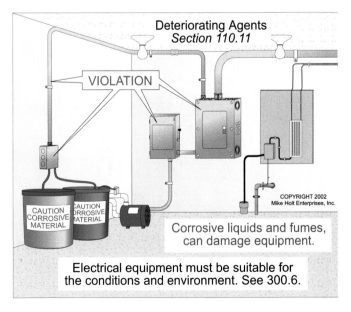

**Figure 110-14**

**AUTHOR'S COMMENT:** Conductors shall not be exposed to ultraviolet rays from the sun or excessive temperature unless identified for the purpose [310.8(D)].

FPN No. 1: See 300.6 for corrosion protection requirements.

FPN No. 2: Some spray cleaning and lubricating compounds contain chemicals that can deteriorate many plastic materials used for insulating and structural applications in equipment.

## 110.12 Mechanical Execution of Work.

Electrical equipment shall be installed in a neat and workmanlike manner.

**AUTHOR'S COMMENT:** What is "a neat and workmanlike manner?" This is a judgment call for the AHJ, but the National Electrical Contractors Association (NECA) has created a series of National Electrical Installation Standards (NEIS), establishing the industry's first quality guidelines for electrical installations. These standards define a benchmark or baseline of quality and workmanship for installing electrical products and systems. They explain what installing electrical products and systems in a "neat and workmanlike manner" means. For more information about these standards, visit http://neca-neis.org/.

**(A) Unused Openings.** Unused cable or raceway openings in electrical equipment shall be effectively closed by fittings that provide protection substantially equivalent to the wall of the equipment. Figure 110–15

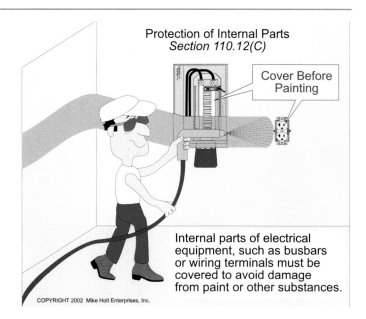

Figure 110-16

**(C) Protection of Internal Parts.** Internal parts of electrical equipment shall not be damaged or contaminated by foreign material, such as paint, plaster, cleaners, etc. For example, precautions shall be taken to provide protection from contaminating the internal parts of panelboards and receptacles during the building construction. Figure 110-16

Electrical equipment that contains damaged parts shall not be installed. Damaged parts would include anything broken, bent or cut, or deteriorated by corrosion, chemical action. Including overheating that may adversely affect the safe operation or the mechanical strength of the equipment, such as cracked insulators, arc shields not in place, overheated fuse clips, as well as damaged or missing switch or circuit-breaker handles. Figure 110-17

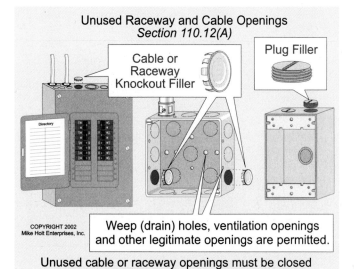

Figure 110-15

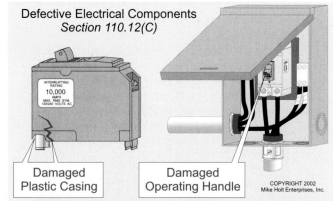

Figure 110-17

## 110.13 Mounting and Cooling of Equipment

**(A) Mounting.** Electrical equipment shall be firmly secured to the surface on which it is mounted. See 314.23.

**(B) Cooling.** Electrical equipment that depends on natural air circulation shall be installed so walls or equipment do not prevent room airflow over such surfaces. The clearance between top surfaces and side surfaces shall be maintained to dissipate rising warm air for equipment designed for floor mounting.

Electrical equipment that is constructed with ventilating openings shall be installed so free air circulation is not inhibited.

## 110.14 Electrical Conductor Termination

**Terminal Conductor Material.** Conductor terminals and splicing shall be properly installed, used, and identified for the conductor material. Devices that are suitable only for aluminum shall be marked AL, and devices that are suitable for both copper and aluminum shall be marked as CU/AL or CO/ALR. See 404.14(C), 406.2(C). Figure 110-18

---

**AUTHOR'S COMMENT:** Terminations of conductors must be in accordance with manufacturer's instructions as required by 110.3(B). For example, if the terminal device states "Suitable for 18-2 AWG Stranded," only stranded conductors can be used with the terminating device. If it states "Suitable for 18-2 AWG Solid," only solid conductors are permitted to be used with the terminating device, and if it states "Suitable for 18-2 AWG," then either solid or stranded conductors can be used with the terminating device.

---

Aluminum: Terminals listed for aluminum conductors are often filled with an antioxidant gel, which is designed to reduce aluminum oxide on the exposed aluminum conductors to reduce the contact resistance between the conductor and the terminal. Split-bolt connectors listed for aluminum-to-aluminum termination will be supplied with instructions that specify how to wire brush the conductors and the requirement for an antioxidant.

Copper: Some terminal manufacturers sell a compound intended to reduce corrosion and heat at terminations, which is especially helpful at high-amperage terminals. This compound is messy, but I have been told that it is effective.

Copper and Aluminum Mixed: Copper and aluminum conductors shall not make contact with each other in a device unless the device is listed and identified for this purpose. Few terminations are listed for use with aluminum wire and copper, but if they are, they will be marked on the product package or terminal device. The reason copper and aluminum cannot be in contact with each other is because corrosion will develop between the conductors, resulting in an increase in contact resistance in the termination or splicing device. This increased resistance can cause overheating of the termination or splice, which could cause a fire. See http://tis-hq.eh.doe.gov/docs/ sn/nsh9001.html for more information on how to properly terminate aluminum and copper conductors together.

---

FPN: Many terminations and equipment are marked with a tightening torque.

---

**AUTHOR'S COMMENT:** All conductors must terminate in devices that have been properly tightened in accordance with the manufacturer's torque specifications that are included with equipment instructions. Failure to torque terminals is a violation of 110.3(B). Figure 110-19

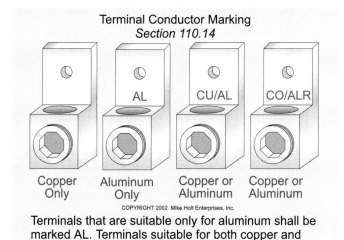

**Terminal Conductor Marking**
*Section 110.14*

| Copper Only | Aluminum Only | Copper or Aluminum | Copper or Aluminum |
|---|---|---|---|
| | AL | CU/AL | CO/ALR |

COPYRIGHT 2002 Mike Holt Enterprises, Inc.

Terminals that are suitable only for aluminum shall be marked AL. Terminals suitable for both copper and aluminum shall be marked as CU/AL or CO/ALR.

*Figure 110-18*

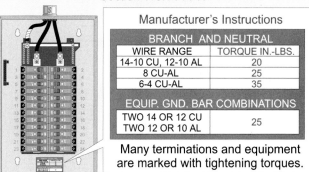

**Manufacturer's Torque Specifications**
*Section 110.14 FPN*

| Manufacturer's Instructions | |
|---|---|
| **BRANCH AND NEUTRAL** | |
| WIRE RANGE | TORQUE IN.-LBS. |
| 14-10 CU, 12-10 AL | 20 |
| 8 CU-AL | 25 |
| 6-4 CU-AL | 35 |
| **EQUIP. GND. BAR COMBINATIONS** | |
| TWO 14 OR 12 CU TWO 12 OR 10 AL | 25 |

Many terminations and equipment are marked with tightening torques.

COPYRIGHT 2002 Mike Holt Enterprises, Inc.

*Figure 110-19*

---

**Question:** What do you do if the torque value is not provided with the device?

**Answer:** Call the manufacturer, visit their web site, or have the supplier make a copy of the installation instructions.

**AUTHOR'S COMMENT:** Terminating conductors without a torque tool can result in an improper and unsafe installation. If you are not using a torquing screwdriver, there is a very good chance that you are not terminating the conductors properly.

**(A) Terminations.** Terminals for conductors shall ensure a good connection without damaging the conductors, and conductors shall terminate to listed pressure connectors.

**Question:** What do you do if the wire you are terminating is larger than the terminal (upsizing conductors for long runs and for nonlinear loads)?

**Answer:** This condition needs to be anticipated in advance and the equipment should be ordered with terminals that will accommodate the larger wire. However, if you are in the field, you should:

• Contact the manufacturer and have them express you the proper terminals, bolts, washers and nuts, or

• Order a terminal device that crimps on the end of the larger conductor and reduces the termination size, or splice the conductors to a smaller wire.

**One Wire Per Terminal.** Terminals are listed to accept no more than one wire per termination, unless marked otherwise. Terminals that are listed for more than one wire shall be identified for this purpose, either within the equipment instructions or on the terminal itself. Figure 110-20

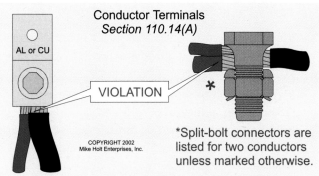

The number of conductors in a terminal is limited to that for which the terminal is designed and listed.

*Figure 110–21*

Most split-bolt connectors are listed for only two conductors, and it is a common industry practice to terminate more than three conductors. However, some split-bolt connectors are listed for three conductors. Figure 110-21

**AUTHOR'S COMMENT:** Some split-bolt connectors are listed for aluminum-to-aluminum or aluminum-to-copper; if so, they will be marked for this application.

**(B) Conductor Splices.** Conductors shall be spliced by a listed splicing device or by exothermic welding. Naturally, using a U-bolt fitting for nonelectrical cables would be an improper and unsafe practice.

**AUTHOR'S COMMENT:** The NEC does not require conductors to be twisted together prior to installing a twist-on wire connector (Wire Nut®), but it is a good practice to twist the grounded (neutral) conductors together for multiwire branch circuits. See 300.13(B). Figure 110-22

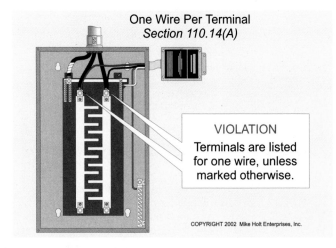

*Figure 110–20*

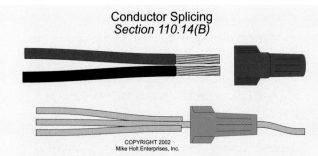

Conductors must be spliced by a listed splicing device or by exothermic welding. Wire connectors must be installed in accordance with manufacturer's instructions and twisting the wires together is not required.

*Figure 110–22*

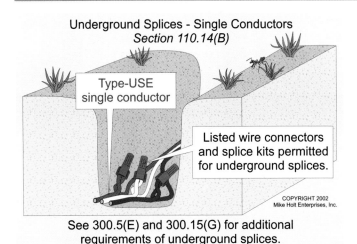

**Underground Splices - Single Conductors**
*Section 110.14(B)*

Type-USE single conductor

Listed wire connectors and splice kits permitted for underground splices.

COPYRIGHT 2002
Mike Holt Enterprises, Inc.

See 300.5(E) and 300.15(G) for additional requirements of underground splices.

*Figure 110–23*

**Insulation.** The free ends of all conductors shall be covered with insulation that is equal to that of the conductor. Spare conductors for future use and conductors that are not currently used can be left, but the free ends of the conductors shall be insulated to prevent the exposed end of the conductor from touching energized parts, which could create an electrical hazard. Both of these requirements can be met by the use of an insulated twist-on or push-on wire connector.

**Underground Splices.**

**Single Conductors.** Single direct-burial conductor of Type UF or USE can be spliced underground without the use of a junction box, but the conductors shall be spliced with a splicing device that is listed for direct-burial installation. See 300.5(E) and 300.15(G). Figure 110-23

**Multiconductor Cable.** Multiconductor Type UF or Type USE cable can be spliced underground with a listed splice kit that encapsulates the conductors and the splice.

**(C) Temperature Limitations.**

**Conductor Size.** Conductors are to be sized in accordance with the lowest temperature rating of any terminal, device, or conductor of the circuit in accordance with (1) and (2) below.

Conductors with temperature ratings higher than specified for terminations shall be permitted to be used for ampacity adjustment, correction, or both. This means that conductor ampacity, when required to be adjusted, is based on the conductor insulation temperature rating in accordance with Table 310.16.

For example, the ampacity of each 12 AWG THHN is 30A based on the values listed in the 90°C column of Table 310.16. If we bundle nine current-carrying 12 AWG THHN conductors, the ampacity for each conductor (as listed in

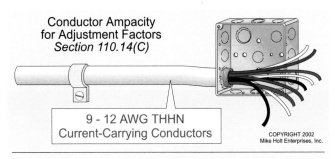

**Conductor Ampacity for Adjustment Factors**
*Section 110.14(C)*

9 - 12 AWG THHN Current-Carrying Conductors

COPYRIGHT 2002
Mike Holt Enterprises, Inc.

Table 310.16: Ampacity of 12 AWG THHN, 90°C = 30A

30A x 0.7 [310.15(B)(2)(a)] = 21A for each 12 AWG

*Figure 110–24*

Table 310.16) needs to be adjusted to 21A (30A × 0.70) as specified by Table 310.15(B)(2)(a). Figure 110-24

**(1) Equipment Provisions.** Unless the equipment is listed and marked otherwise, conductor ampacities used in determining equipment termination provisions shall be based on Table 310.16.

**(a) Conductor Size for Equipment Rated 100A and Less.**

(1) Electrical terminations on equipment rated 100A or less and pressure connector terminals for 14 through 1 AWG conductors are rated for 60°C unless otherwise specified. To keep a 60°C terminal from overheating beyond its rating, the conductor connected to that terminal shall not get hotter than 60°C. This can be done by going to Table 310.16 and selecting a conductor from the 60°C column.

(2) Conductors with temperature ratings greater than 60°C can be used on terminals rated 60°C, but the

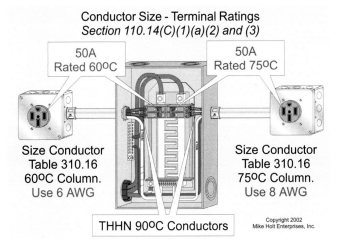

**Conductor Size - Terminal Ratings**
*Section 110.14(C)(1)(a)(2) and (3)*

50A Rated 60°C

50A Rated 75°C

Size Conductor Table 310.16 60°C Column. Use 6 AWG

Size Conductor Table 310.16 75°C Column. Use 8 AWG

THHN 90°C Conductors

Copyright 2002
Mike Holt Enterprises, Inc.

*Figure 110–25*

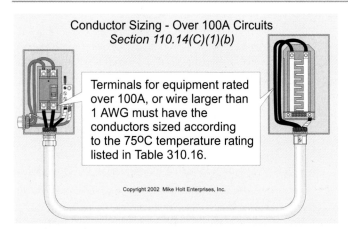

Conductor Sizing - Over 100A Circuits
*Section 110.14(C)(1)(b)*

Terminals for equipment rated over 100A, or wire larger than 1 AWG must have the conductors sized according to the 75ºC temperature rating listed in Table 310.16.

Copyright 2002 Mike Holt Enterprises, Inc.

**Figure 110–26**

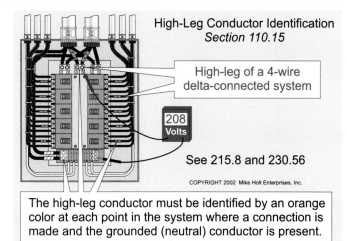

High-Leg Conductor Identification
*Section 110.15*

High-leg of a 4-wire delta-connected system

208 Volts

See 215.8 and 230.56

COPYRIGHT 2002 Mike Holt Enterprises, Inc.

The high-leg conductor must be identified by an orange color at each point in the system where a connection is made and the grounded (neutral) conductor is present.

**Figure 110–27**

conductor shall be sized based on the 60°C temperature column of Table 310.16. Figure 110-25

(3) If the terminals are listed and identified for use with terminals rated 75°C, then the conductors can be sized to the 75°C temperature column of Table 310.16. See Figure 110-25

**(b) Conductor Size for Equipment Rated Over 100A.**

(1) Conductors for equipment rated over 100A and pressure connector terminals for larger than 1 AWG conductors shall be sized to the 75°C temperature column of Table 310.16. Figure 110-26

(2) Conductors with temperature ratings greater than 75°C shall be sized based on the 75°C temperature column of Table 310.16.

**(2) Separate Connector Provisions.** Conductors for separately installed pressure connectors can be sized to the listed temperature rating of the connector. For example, if 90°C pressure connectors are located on busbars, the conductors can be sized to the 90°C temperature column of Table 310.16.

## 110.15 High-Leg Conductor Identification

On a three-phase (3Ø), 4-wire delta-connected system, where the midpoint of one phase winding is grounded, the conductor or busbar having the higher phase voltage-to-ground shall be durably and permanently marked by an outer finish that is orange in color, or by other effective means. Such identification shall be placed at each point on the system where a connection is made if the grounded (neutral) conductor is also present. Figure 110-27

**AUTHOR'S COMMENT:** Similar language is contained in 215.8 for feeders and 230.56 for services.

WARNING: When replacing disconnects, panelboards, meters, switches, or any equipment that contains the high-leg conductor, care must be taken to place the high-leg conductor in the proper location. Failure to install the high-leg properly can result in 120V circuits connected to the 208V high-leg, with disastrous results. See 408.3(E) for proper termination of the high-leg conductor in panelboards and switchboards.

## 110.16 Flash Protection Warning

Switchboards, panelboards, industrial control panels, and motor control centers in commercial and industrial occupancies, that are likely to require examination, adjustment, servicing, or maintenance while energized, shall be field marked to warn qualified persons of the danger of electric arc flash from a line-to-line or line-to-case fault. The marking shall be clearly visible to qualified persons before they examine, adjust, service, or perform maintenance on the equipment.

The "incident energy" for the selection of personal protective equipment under NFPA 70E is not required, just a sign warning that dangerous electric arc flash could be present. Figure 110-28

**AUTHOR'S COMMENT:** This rule is to protect qualified persons who work on energized electrical systems, by ensuring that they are alerted that an arc flash hazard exists so they will select proper personal protective equipment. For more information about flash protection, visit http://www.bussmann.com/safetybasics.

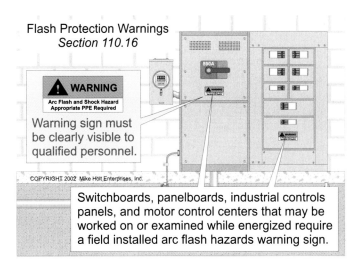

Flash Protection Warnings
Section 110.16

**⚠ WARNING**
Arc Flash and Shock Hazard
Appropriate PPE Required

Warning sign must be clearly visible to qualified personnel.

COPYRIGHT 2002 Mike Holt Enterprises, Inc.

Switchboards, panelboards, industrial controls panels, and motor control centers that may be worked on or examined while energized require a field installed arc flash hazards warning sign.

**Figure 110–28**

**AUTHOR'S COMMENT:** In some installations, the use of current-limiting protection devices may significantly reduce the degree of arc-fault flash hazards. Figure 110-29

FPN No. 1: NFPA 70E-2000, Electrical Safety Requirements for Employee Workplaces, provides assistance in determining the severity of potential exposure, planning safe work practices, and selecting personal protective equipment. For more information, visit www.mikeholt.com/Newsletters/110.16.pdf (66K) or www.mikeholt.com/Newsletters/110.16.doc (293K)

FPN No. 2: ANSI Z535.4-1998, Product Safety Signs and Labels, provides guidelines for the design of safety signs and labels.

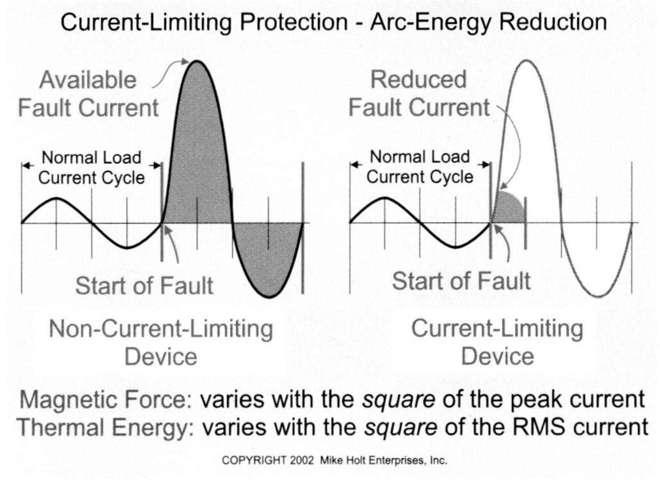

## 110.21 Manufacturer's Markings

The manufacturer's name, trademark, or other descriptive marking shall be placed on all electric equipment. Where required by the *Code*, markings such as voltage, current, wattage, or other ratings shall be provided with sufficient durability to withstand the environment involved.

## 110.22 Identification of Disconnecting Means

All installed disconnecting means shall be legibly marked to indicate their purpose unless they are located and arranged so their purpose is evident. The marking shall be of sufficient durability to withstand the environment involved. Figure 110–30

**AUTHOR'S COMMENT:** A disconnecting means is defined by Article 100 as "A device, or group of devices, or other means by which the conductors of a circuit can be disconnected from their source of supply."

## Part II. 600V, Nominal or Less

## 110.26 Access and Working Space About Electrical Equipment

For the purpose of safe equipment operation and maintenance, all electrical equipment shall have sufficient access and working space. Enclosures housing electrical apparatus that are controlled by lock and key are considered accessible to qualified persons who require access. Figure 110-31

Current-Limiting Protection - Arc-Energy Reduction

Available Fault Current

Normal Load Current Cycle

Start of Fault

Non-Current-Limiting Device

Reduced Fault Current

Normal Load Current Cycle

Start of Fault

Current-Limiting Device

Magnetic Force: varies with the *square* of the peak current
Thermal Energy: varies with the *square* of the RMS current

COPYRIGHT 2002 Mike Holt Enterprises, Inc.

**Figure 110–29**

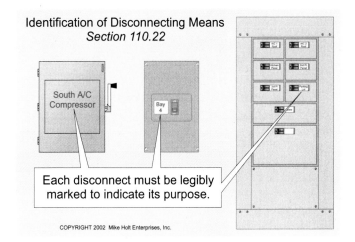

Identification of Disconnecting Means
Section 110.22

South A/C Compressor

Bay 4

Each disconnect must be legibly marked to indicate its purpose.

COPYRIGHT 2002 Mike Holt Enterprises, Inc.

**Figure 110–30**

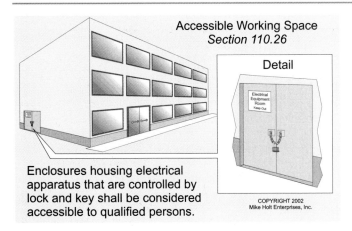

Enclosures housing electrical apparatus that are controlled by lock and key shall be considered accessible to qualified persons.

**Figure 110–31**

| Nominal Voltage to Ground | Condition 1 | Condition 2 | Condition 3 |
|---|---|---|---|
| 0–150V | 3 ft | 3 ft | 3 ft |
| 151–600V | 3 ft | 3½ ft | 4 ft |

Condition 1 – Exposed live parts on one side and no live or grounded parts on the other side of the working space.

Condition 2 – Exposed live parts on one side and grounded parts on the other side of the working space. For the purpose of this table, concrete, brick, or tile walls are considered as grounded.

Condition 3 – Exposed live parts on both sides of the working space.

**(a) Rear and Sides.** Working space is not required in back or sides of assemblies where there are no renewable or adjustable parts on the back or sides and where all connections are accessible from the front. Figure 110-33

**(b) Control and Signal.** The working space requirements apply to control and signal equipment, such as fire alarms, as covered by Chapter 7, but smaller working space is permitted where uninsulated parts operate at not more than 30V ac or 60V dc. Figure 110-34

**AUTHOR'S COMMENT:** Overcurrent protection devices such as fuses and circuit breakers are required to be readily accessible, and each occupant shall have ready access to all overcurrent protection devices protecting the conductors supplying that occupancy [240.24].

**(A) Working Space.** Working space for equipment that may need examination, adjustment, servicing, or maintenance "while energized" shall be installed in a working space in accordance with (1), (2) and (3):

**AUTHOR'S COMMENT:** The phrase "while energized" is the root of many debates. Since electric power to almost all equipment can be turned off, one could argue that working space is never required!

**(1) Depth of Working Space.** The depth of the working space measured from the enclosure front or opening shall not be less than indicated in the following: Figure 110-32

**AUTHOR'S COMMENT:** The working space requirements of 110.26 do not apply to Chapter 8 Communications Equipment. See 90.3.

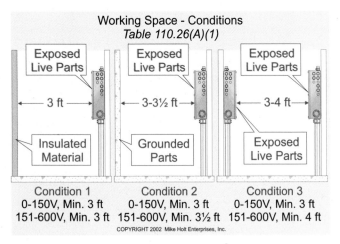

**Figure 110–32**

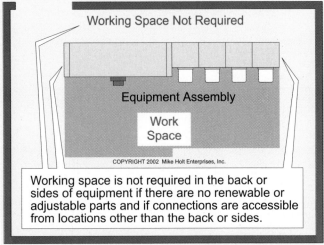

Working space is not required in the back or sides of equipment if there are no renewable or adjustable parts and if connections are accessible from locations other than the back or sides.

**Figure 110–33**

Working Space - 30V and Less Systems
*Section 110.26(A)(1)(b)*

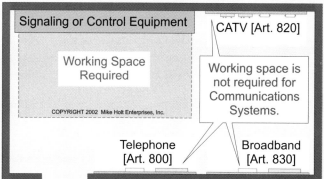

Working space is required for control and signaling equipment, but not communications equipment.

**Figure 110–34**

(2) **Width of Working Space.** The working space shall be a minimum of 30 in. wide, but in no case less than the width of the equipment. Figure 110-35

**AUTHOR'S COMMENT:** The width of the working space is measured from left to right, from right to left, or simply from the centerline of the equipment. In all cases, the working space shall be of sufficient width, depth, and height to permit at least a 90° opening of all equipment doors. Figure 110-36

(3) **Height of Working Space.** The working space in front of equipment shall have a minimum height of not less than 6¹/₂ ft, measured from the grade, floor, or platform for service equipment, switchboards, and motor control equipment [110.26(E)].

Working Space - Width
*Section 110.26(A)(2)*

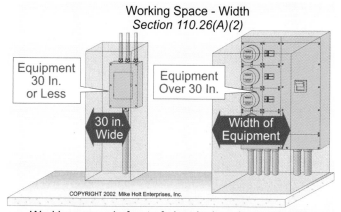

Working space in front of electrical equipment must not be less than 30 in. or the width of the equipment, whichever is greater.

**Figure 110–35**

Working Space - Width
*Section 110.26(A)(2)*

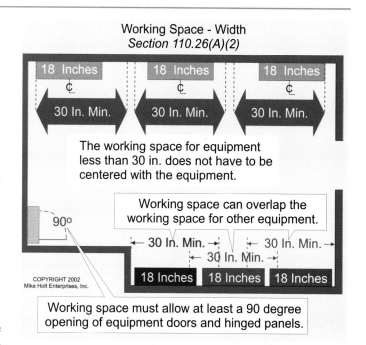

**Figure 110–36**

Equipment such as raceways, cables, wireways, cabinets, panels, etc. can be located above or below electrical equipment, but the equipment shall not extend more than 6 in. into the equipment's working space. For example, a 1-ft-deep wireway can be located above or below a 6-in.-deep panelboard. Figure 110-37

Working Space - Equipment Extending Into
*Section 110.26(A)(3)*

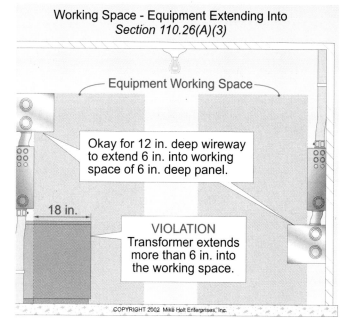

**Figure 110–37**

**(B) Clear Working Space.** The working space required by this section shall be clear at all times, therefore this space shall not be used for storage. It is very dangerous for people who service energized parts to be subjected to additional dangers by working about, around, over or under bicycles, boxes, crates, appliances, and other impediments. Figure 110-38

**AUTHOR'S COMMENT:** Signal and communication equipment as well, and this equipment shall not be installed to encroach on the working space requirements of the power equipment.

**(C) Entrance to Working Space.**

(1) **Minimum Required.** At least one entrance of "sufficient area" shall be provided to give access to the working space. Check with the AHJ for what he or she considers "sufficient area."

(2) **Large Equipment.** For equipment rated 1,200A or more and having a width of over 6 ft, at least one entrance, measuring not less than 24 in. wide and $6^{1}/_{2}$ ft high, shall be provided at each end of the working space. Where the entrance to the working space has a door, the door shall open out and be equipped with panic hardware or other devices so the door can open under simple pressure without the use of hands. Figure 110–39

**AUTHOR'S COMMENT:** Since this requirement is in the NEC, the electrical contractor is responsible to ensure that panic hardware is installed at these locations. Many electrical contractors are offended at being held liable for nonelectrical responsibilities, but this rule should be a little less offensive, given that it is designed to save electricians' lives. For this and other reasons, many

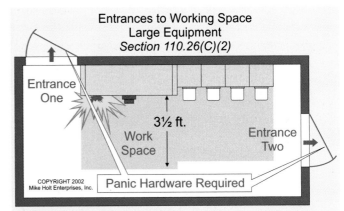

Entrances to Working Space
Large Equipment
*Section 110.26(C)(2)*

Entrance One

$3^{1}/_{2}$ ft.

Work Space

Entrance Two

COPYRIGHT 2002 Mike Holt Enterprises, Inc.

Panic Hardware Required

For equipment rated 1200A or more and over 6 ft. wide, personnel doors must open out and be equipped with panic hardware so as to open without using hands.

*Figure 110–39*

construction professionals routinely hold "pre-construction" or "pre-con" meetings to review potential opportunities for miscommunication – before the work begins.

(a) **Unobstructed Exit.** Only one entrance is required where the location permits a continuous and unobstructed way of exit travel.

(b) **Double Workspace.** Only one entrance is required where the working space is doubled and the equipment is located so the edge of the entrance is no closer than the working space distance. Figure 110-40

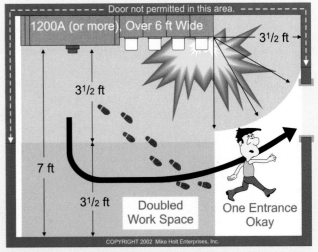

Working Space - Larger Equipment Entrance
*Section 110.26(C)(2)(b)*

Door not permitted in this area.

1200A (or more), Over 6 ft Wide

$3^{1}/_{2}$ ft

$3^{1}/_{2}$ ft

7 ft

$3^{1}/_{2}$ ft

Doubled Work Space

One Entrance Okay

COPYRIGHT 2002 Mike Holt Enterprises, Inc.

For equipment rated 1200A or more and over 6 ft. wide, a single entrance must not be located closer to the equipment than required clearance of Table 110.26(A)(1).

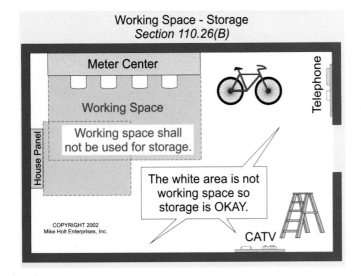

Working Space - Storage
*Section 110.26(B)*

Meter Center

Working Space

Working space shall not be used for storage.

House Panel

Telephone

The white area is not working space so storage is OKAY.

COPYRIGHT 2002 Mike Holt Enterprises, Inc.

CATV

*Figure 110–38*

*Figure 110–40*

**(D) Illumination.** Service equipment, switchboards, panelboards, and motor control centers located indoors shall have the working space illuminated. Illumination can be provided by a lighting source located next to the working space, and it shall not be controlled by automatic means only. Figure 110-41

**(E) Headroom.** The minimum working space headroom about service equipment, panelboards, switchboards or motor control centers shall not be less than $6^1/_2$ ft. When the height of the equipment exceeds $6^1/_2$ ft, the minimum headroom shall not be less than the height of the equipment.

*Exception:* The minimum headroom requirement of $6^1/_2$ ft does not apply to service equipment or panelboards rated 200A or less in an existing dwelling unit.

**(F) Dedicated Equipment Space.** All switchboards, panelboards, distribution boards, and motor control centers shall comply with the following:

**(1) Indoors.**

    **(a) Dedicated Electrical Space.** The space equal to the width and depth of the equipment extending from the floor to a height of 6 ft above the equipment or to the structural ceiling, whichever is lower, shall be dedicated to the electrical installation. No piping, ducts, or other equipment foreign to the electrical installation can be installed in this dedicated space. Figure 110-42

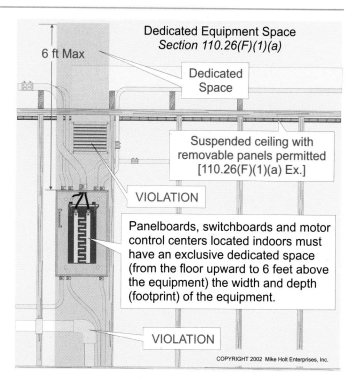

*Figure 110–42*

*Exception:* Suspended ceilings with removable panels can be in the dedicated space.

    **(b) Foreign Systems.** The space above the dedicated space can contain foreign systems, provided protection is installed to avoid damage to the electrical equipment from condensation, leaks, or breaks in such foreign systems. Figure 110-43

    **(c) Sprinkler Protection.** Sprinkler protection can spray water into the dedicated space, but the sprinkler piping itself shall not be located within the dedicated space area.

    **(d) Suspended Ceilings.** A dropped, suspended, or similar ceiling that does not add strength to the building structure is not considered a structural ceiling. This means that a suspended ceiling with lift-out panels can be installed within the dedicated space as defined in 110.26(F)(1)(a).

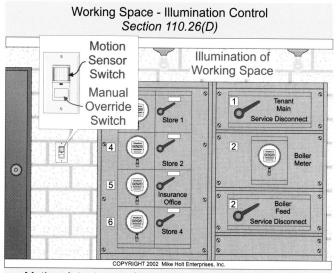

Motion detector can be used to control the illumination of the working space, if a manual override is provided.

*Figure 110–41*

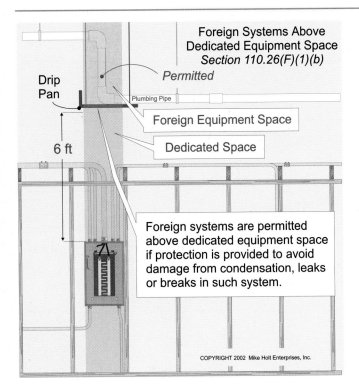

Foreign Systems Above
Dedicated Equipment Space
*Section 110.26(F)(1)(b)*
*Permitted*

Drip
Pan

Plumbing Pipe

Foreign Equipment Space

Dedicated Space

6 ft

Foreign systems are permitted above dedicated equipment space if protection is provided to avoid damage from condensation, leaks or breaks in such system.

COPYRIGHT 2002 Mike Holt Enterprises, Inc.

**Figure 110–43**

## 110.27 Guarding

**(A)** **Guarding Live Parts.** Live parts of electrical equipment operating at 50V or more shall be guarded against accidental contact. Figure 110-44

**(B)** **Prevent Physical Damage.** Electrical equipment shall not be installed where it could be subject to physical damage. Figure 110-45

**AUTHOR'S COMMENT:** What is subject to physical damage? Damage may result from many events, such as automobile accidents and vandalism of public and private property. Check with the AHJ.

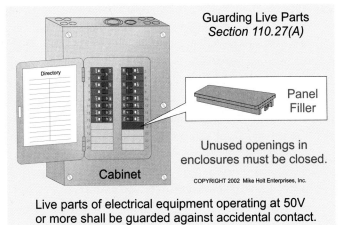

Guarding Live Parts
*Section 110.27(A)*

Directory

Panel
Filler

Cabinet

Unused openings in enclosures must be closed.

COPYRIGHT 2002 Mike Holt Enterprises, Inc.

Live parts of electrical equipment operating at 50V or more shall be guarded against accidental contact.

**Figure 110–44**

Prevention of Physical Damage
*Section 110.27(B)*

CAUTION
Equipment must be protected from physical damage.

Valet
Parking Only

COPYRIGHT 2002 Mike Holt Enterprises, Inc.

**Figure 110–45**

# Article 110

1. To be *Code* compliant, listed or labeled equipment shall be installed and used in accordance with any instructions included in the _____.
   (a) catalog          (b) product          (c) listing or labeling          (d) all of these

2. When referencing the *NEC*, conductors shall be _____ unless otherwise provided in the *Code*.
   (a) bare          (b) stranded          (c) copper          (d) aluminum

3. Only wiring methods recognized as _____ are included in the *Code*.
   (a) identified          (b) efficient          (c) suitable          (d) cost-effective

4. Some spray cleaning and lubricating compounds contain chemicals that cause severe deteriorating reactions with plastics.
   (a) True          (b) False

5. The temperature rating associated with the ampacity of a _____ shall be so selected and coordinated so as to not exceed the lowest temperature rating of any connected termination, conductor, or device.
   (a) terminal          (b) conductor          (c) device          (d) all of these

6. What size THHN conductor is required for a 50A circuit, if the equipment is listed and identified for use with a 75°C conductor? Tip: Table 310.16 lists conductor ampacities.
   (a) 10 AWG          (b) 8 AWG          (c) 6 AWG          (d) all of these

7. Working space distances for enclosed live parts shall be measured from the _____ of equipment or apparatus, if such are enclosed.
   (a) enclosure          (b) opening          (c) a or b          (d) none of these

8. A minimum of _____ feet of working clearance is required to live parts operating at 300V, nominal-to-ground, where there are exposed live parts on one side and no live or grounded parts on the other side.
   (a) 2          (b) 3          (c) 4          (d) 6

9. Concrete, brick, or tile walls shall be considered as _____, when applied to working space requirements.
   (a) inconsequential          (b) in the way          (c) grounded          (d) none of these

10. The working space in front of the electric equipment shall not be less than _____ in. wide or less than the width of the equipment, whichever is greater.
    (a) 15          (b) 30          (c) 40          (d) 60

11. Working space shall not be used for _____.
    (a) storage          (b) raceways          (c) lighting          (d) accessibility

12. The minimum headroom of working spaces for motor control centers shall be _____.
    (a) 3 ft          (b) 5 ft          (c) 6 ft          (d) 6 ft 6 in.

13. The dedicated equipment space for electrical equipment that is required for panelboards is measured from the floor to a height of _____ above the equipment, or to the structural ceiling, whichever is lower.
    (a) 3 ft          (b) 6 ft          (c) 12 ft          (d) 30 ft

14. Ventilated heating or cooling equipment (including ducts) that service the electrical room or space cannot be installed in the dedicated space above a panelboard or switchboard.
    (a) True          (b) False

15. The dedicated space above a panelboard extends to the structural ceiling, which can be a suspended ceiling.
    (a) True          (b) False

# Chapter 2 – Wiring and Protection

Chapter 2 of the *NEC* is a general rules chapter as applied to wiring and protection of conductors. The rules in this chapter apply, except as modified, to Chapters 5, 6, and 7. Along with Chapter 3, this chapter can be considered the heart of the *Code*. Many of the everyday applications of the *NEC* can be found in this chapter. It covers most of the requirements that the electrical industry must deal with on a regular basis.

## Article 200. Use and Identification of Grounded (neutral) Conductors

This article contains the requirements for the use and identification of the grounded (neutral) conductor and its terminals.

## Article 210. Branch Circuits

This article contains the requirements for branch circuits, such as conductor sizing, identification, GFCI receptacle protection, and receptacle and lighting outlet requirements.

Part I. General Provisions
Part II. Branch-Circuit Ratings
Part III. Required Outlets

## Article 215. Feeders

This article covers the rules for installation, minimum size, and ampacity of feeders.

## Article 220. Branch Circuit, Feeder, and Service Calculations

This article provides the requirements for sizing branch circuits, feeders, and services, and for determining the number of receptacles on a circuit and the number of branch circuits required.

Part I. General
Part II. Feeders and Service Calculations
Part III. Optional Calculations

## Article 225. Outside Branch Circuits and Feeders

This article covers installation requirements for equipment, including conductors located outside, on or between buildings, poles, and other structures on the premises.

Part I. General
Part II. More Than One Building or Structure

## Article 230. Services

This article covers the installation requirements for service conductors and equipment. It is very important to know where the service begins and where the service ends when applying Articles 230 and 250.

Conductors and equipment supplied from a battery, uninterruptible power supply system, solar voltaic system, generator, transformer, or phase converters are not considered service conductors; they are feeder conductors.

Part I. General
Part II. Overhead Service–Drop
Part III. Underground Service–Lateral
Part IV. Service-Entrance Conductors
Part V. Service Equipment
Part VI. Disconnecting Means
Part VII. Overcurrent Protection

## Article 240. Overcurrent Protection

This article provides the general requirements for overcurrent protection and overcurrent protective devices not more than 600V, nominal.

Overcurrent protection for conductors and equipment is provided to open the circuit if the current reaches a value that will cause an excessive or dangerous temperature on conductors or conductor insulation.

Part I. General
Part II. Location
Part III. Enclosures
Part IV. Disconnecting and Guarding
Part V. Plug Fuses, Fuseholders, and Adapters
Part VI. Cartridge Fuses and Fuseholders
Part VII. Circuit Breakers

## Article 250. Grounding

Grounding covers the requirements for providing a path(s) to divert high voltage to the earth, requirements for the low-impedance fault current path necessary to facilitate the operation of overcurrent protection devices, and how to remove dangerous voltage potentials between conductive parts of building components and electrical systems.

Part I. General

Part II. Circuit and System Grounding
Part III. Grounding Electrode System and Grounding Electrode Conductor
Part IV. Enclosure and Raceway Grounding (bonding)
Part V. Bonding
Part VI. Equipment Grounding (bonding)
Part VII. Methods of Equipment Grounding (bonding)

## Article 280. Surge Arresters

This article covers general requirements, installation requirements, and connection requirements for surge arresters installed on the line side of service equipment.

## Article 285. Transient Voltage Surge Suppressors (TVSSs)

This article covers general requirements, installation requirements, and connection requirements for transient voltage surge suppressors (TVSSs) permanently installed on the load side of service equipment. It does not apply to cord- and-plug-connected units such as "computer power strips."

## Surge Protection

Protection against voltage surges can take several forms such as preventing the surge at its origin (impossible for lightning and difficult for surges associated with the power system), diverting the surge to ground before it enters the building, and finally, clamping the surge by a Surge Protective Device at the equipment. The industry trend is to call such devices Surge Protective Devices (SPDs), but they are also called Surge Arresters and Transient Voltage Surge Suppressors (TVSSs).

Of these approaches, only an SPD option is available to the end-user and the installation must be done by a professional in accordance with the *NEC*, Articles 280 for Surge Arresters and 285 for Transient Voltage Surge Suppressors (TVSSs).

### Transient Voltage Surges

Transient voltage surges, sometimes called "spikes", are short-term deviations from a desired voltage level or signal, which can cause equipment malfunction or damage. A transient voltage surge on the 60 Hz AC sine wave is very brief in duration (measured in microseconds), and the surge voltage is much higher than the AC line voltage. Equipment driven by microprocessors is especially vulnerable to transient voltage surges.

### Sources of Transients

Transients occur through inductive magnetic coupling and whenever line current is interrupted. Transients outside the facility are often caused by lightning or utility grid switching, but the majority of surges occur within the facility by turning loads on or off, like a variable-speed driven motor.

### What Are Surge Protective Devices?

SPDs are designed to reduce potentially damaging short-duration transients present on utility power lines, data networks, telephone lines, closed circuit and cable TV feeds, and any other power or control lines connected to electronic equipment.

An SPD reduces the magnitude of surges to protect equipment by reducing the surge to a level that can safely be passed through to the load. SPDs can't reduce swells in the AC power, they can't reduce the harmonic conditions produced by non-linear loads, nor can they provide utility bill savings.

Surge arresters can be installed ahead of service equipment accordance with Article 280, and TVSSs shall be installed on the load side of service equipment, in accordance with Article 285.

### How Surge Protective Devices Work

In the simplest terms, SPDs prevent damaging transient voltage surge levels from reaching the devices they protect by providing a shunt path for transients away from the load before they can enter the equipment.

The following analogy should be helpful to understand the action of an SPD. Consider a water mill protected by a pressure relief valve. The pressure relief valve does nothing until an over-pressure pulse occurs in the water supply. When that happens, the valve opens and shunts the extra pressure aside so it won't reach the water wheel. If the relief valve was not present, excessive pressure could damage the water wheel. Even though the relief valve is in place and working properly, some remnant of the pressure pulse will still reach the wheel but pressure will have been reduced enough not to damage the water wheel or disrupt its operation.

### Proper Grounding

For an SPD to function properly, the electrical system and the communications systems must be grounded in accordance with the NEC. Of particular concern is reducing the difference in ground reference between port connections for multi-port appliances. This is accomplished by requiring all communications systems that enter a facility to be grounded to the buildings or structure grounding electrode system. The worst possible mistake, and a violation of the NEC, is to provide separate grounds for the power system and the communications system.

**AUTHOR'S COMMENTS:** For more information, download Leviton' s Applications Manual and Reference Guide for Surge Protection and Line Conditioning Products and EPRI's Developing a Consumer-Oriented Guide on Surge Protection at: http://mikeholt.com/Powerquality/Powerquality.htm.

# Article 200
## Use and Identification of Grounded (neutral) Conductor

## Introduction

Most electrical power supplies are solidly grounded at the power supply by connecting one output terminal to the earth and bonding this terminal to the metal enclosure of the power supply. The purpose of grounding the power supply is to stabilize the output voltage during normal operation. For more information, see 250.4(A)(1).

The conductor that connects the output terminal (typically X0) of the power supply to the earth is called the "grounding electrode conductor." The circuit conductor that is connected to this grounded terminal at the power supply is called a "grounded conductor," which many call the neutral wire. Figure 200-1

According to the IEEE Dictionary, a neutral conductor is the conductor that has an equal potential difference between it and the other output conductors of a 3- or 4-wire system. Therefore, a neutral would be the white/gray wire of a 1Ø, 3-wire, 120/240V or a 3Ø, 4-wire, 208Y/120V system. Figure 200-2

**AUTHOR'S COMMENT:** Since a neutral conductor can only be from a 3- or 4-wire system, the white wire of a 2-wire circuit and the white wire from a 3Ø, 4-wire, 120/240V delta system is not a neutral conductor – it's a grounded conductor. Figure 200-3

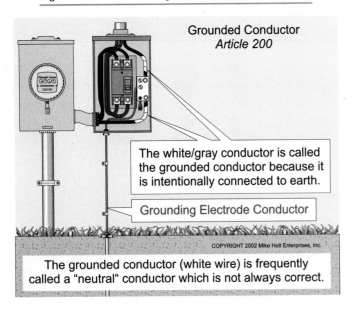

The white/gray conductor is called the grounded conductor because it is intentionally connected to earth.

Grounding Electrode Conductor

COPYRIGHT 2002 Mike Holt Enterprises, Inc.

The grounded conductor (white wire) is frequently called a "neutral" conductor which is not always correct.

**Figure 200-1**

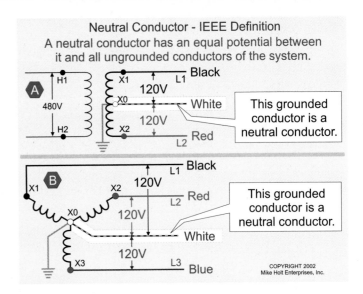

Neutral Conductor - IEEE Definition
A neutral conductor has an equal potential between it and all ungrounded conductors of the system.

This grounded conductor is a neutral conductor.

This grounded conductor is a neutral conductor.

COPYRIGHT 2002
Mike Holt Enterprises, Inc.

**Figure 200-2**

When the electrical trade industry uses the term "neutral," it's always referring to the white wire. However, the proper term for this is "grounded conductor." Technically it's improper to call a "grounded conductor" a "neutral wire" when it's not a neutral, but this is a long-standing industry practice. For the purpose of this book this conductor will be called a grounded (neutral) conductor

### 200.1 Scope

Article 200 contains the requirements for the identification of the grounded (neutral) conductor and its terminals.

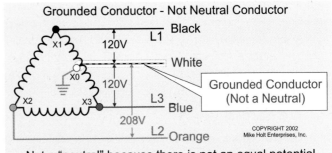

Grounded Conductor - Not Neutral Conductor

Grounded Conductor (Not a Neutral)

COPYRIGHT 2002
Mike Holt Enterprises, Inc.

Not a "neutral" because there is not an equal potential between it and the other ungrounded conductors.

**Figure 200-3**

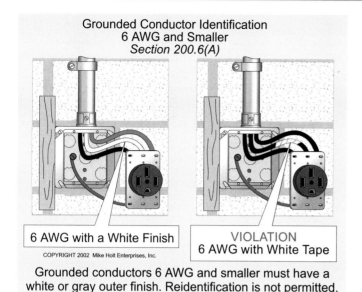

**Grounded Conductor Identification
6 AWG and Smaller
Section 200.6(A)**

6 AWG with a White Finish

VIOLATION
6 AWG with White Tape

COPYRIGHT 2002 Mike Holt Enterprises, Inc.

Grounded conductors 6 AWG and smaller must have a white or gray outer finish. Reidentification is not permitted.

**Figure 200–4**

*CAUTION: A corner-grounded delta system has one output terminal grounded to the earth, and this conductor is also called a grounded conductor, but this topic is beyond the scope of this book.*

## 200.6 Identification of the Grounded Conductor

**(A)  6 AWG or Smaller.** Grounded (neutral) conductors 6 AWG and smaller shall be identified by a continuous white or gray outer finish along its entire length. Figure 200–4.

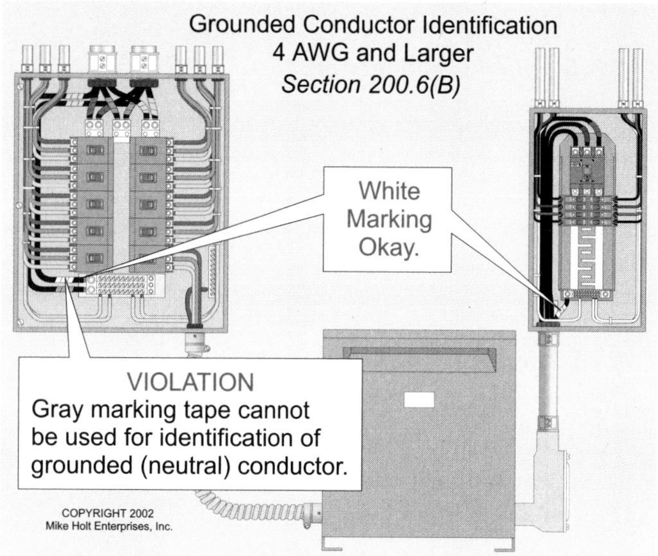

**Grounded Conductor Identification
4 AWG and Larger
Section 200.6(B)**

White Marking Okay.

VIOLATION
Gray marking tape cannot be used for identification of grounded (neutral) conductor.

COPYRIGHT 2002
Mike Holt Enterprises, Inc.

Grounded conductors 4 AWG and larger can be identified by distinctive white markings at terminations.

**Figure 200–5**

**AUTHOR'S COMMENT:** The use of white colored phasing tape, paint, or other methods of re-identification of conductors is not permitted on conductors 6 AWG or smaller.

**(B)  Larger than 6 AWG.** Grounded (neutral) conductors 4 AWG and larger shall be identified by a continuous white or gray outer finish along its entire length, or by distinctive white markings such as tape, paint, or by other effective means at terminations. Figure 200–5

**AUTHOR'S COMMENT:** Re-identification can be with white, but not gray tape or paint according to the NEC.

**(D)  Mixing Conductors from Different Systems.** Where conductors from different systems are installed in the same raceway, cable, or enclosure, one system grounded (neutral) conductor shall have an outer covering of white or gray. The other system grounded (neutral) conductor shall have an outer covering of white with a readily distinguishable different color stripe (not green) running along its entire length. Figure 200–6

FPN: Care should be taken when working on existing systems because the color gray may have been used as an ungrounded (hot) conductor.

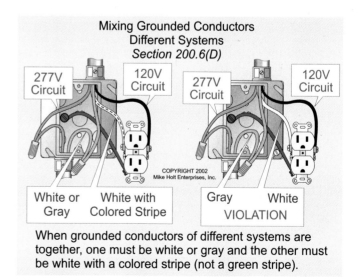

**Mixing Grounded Conductors
Different Systems
Section 200.6(D)**

277V Circuit    120V Circuit    277V Circuit    120V Circuit

COPYRIGHT 2002
Mike Holt Enterprises, Inc.

White or Gray    White with Colored Stripe    Gray    White
VIOLATION

When grounded conductors of different systems are together, one must be white or gray and the other must be white with a colored stripe (not a green stripe).

**Figure 200–6**

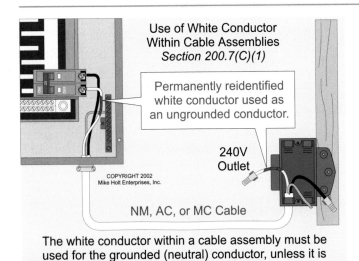

Use of White Conductor
Within Cable Assemblies
*Section 200.7(C)(1)*

Permanently reidentified white conductor used as an ungrounded conductor.

240V Outlet

COPYRIGHT 2002
Mike Holt Enterprises, Inc.

NM, AC, or MC Cable

The white conductor within a cable assembly must be used for the grounded (neutral) conductor, unless it is permanently reidentified where visible.

*Figure 200–7*

## 200.7 Use of White or Natural Gray Color

**(C) Circuits Over 50V.** The use of white or gray insulation for an ungrounded conductor is permitted only for:

**(1) Cable Assembly.** White or gray conductors within a cable can be used for the ungrounded conductor, but the white or gray conductor shall be permanently reidentified to indicate its use as an ungrounded conductor at each location where the conductor is visible. Figure 200–7

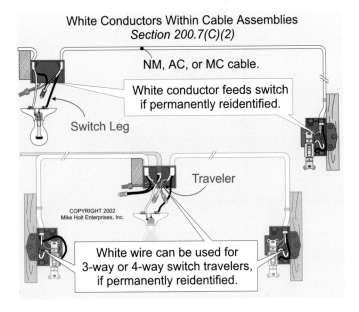

White Conductors Within Cable Assemblies
*Section 200.7(C)(2)*

NM, AC, or MC cable.

White conductor feeds switch if permanently reidentified.

Switch Leg

Traveler

COPYRIGHT 2002
Mike Holt Enterprises, Inc.

White wire can be used for 3-way or 4-way switch travelers, if permanently reidentified.

*Figure 200–8*

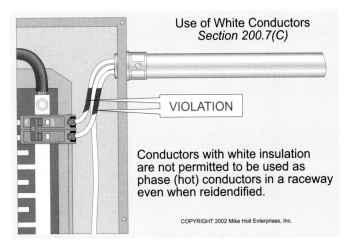

Use of White Conductors
*Section 200.7(C)*

VIOLATION

Conductors with white insulation are not permitted to be used as phase (hot) conductors in a raceway even when reidentified.

COPYRIGHT 2002 Mike Holt Enterprises, Inc.

*Figure 200–9*

**(2) Switches.** White or gray conductors within a cable can be used for single-pole, 3-way or 4-way switch loops; but, the white or gray conductors shall be permanently reidentified to indicate its use as an ungrounded conductor at each location where the conductor is visible. Figure 200–8

FPN: Care should be taken when working on existing systems because the color gray may have been used as an ungrounded (hot) conductor.

**AUTHOR'S COMMENT:** The NEC does not permit the use of white or gray conductor for ungrounded (hot) conductors in a raceway, even if the conductors are permanently re-identified. Figure 200–9

## 200.9 Terminal Identification

The terminal for the grounded (neutral) conductor shall be white in color (actually the color is silver). The terminal for the ungrounded (hot) conductor shall be a color that is readily distinguishable from white (brass or copper in color). Terminals for equipment grounding conductor shall be a green, not readily removable, terminal screw or pressure wire connector [250.126].

## 200.10 Identification of Terminals

**(C) Screw Shell.** The screw shell of a luminaire or lampholder shall be connected to the grounded (neutral) conductor [410.47]. Figure 200-10

**Figure 200-10**

## 200.11 Polarity.

A grounded (neutral) conductor shall not be connected to terminals, or leads, that will cause reversed polarity [410.23]. Figure 200-10

## Article 200

1.  An insulated grounded conductor of _____ or smaller shall be identified by a continuous white or gray outer finish, or by three continuous white stripes on other than green insulation along its entire length.
    (a) 3 AWG          (b) 4 AWG          (c) 6 AWG          (d) 8 AWG

2.  Grounded (neutral) conductors _____ and larger must be identified by a continuous white or gray outer finish along their entire lengths, or by distinctive white markings such as tape, paint, or other effective means at their terminations.
    (a) 10 AWG          (b) 8 AWG          (c) 6 AWG          (d) 4 AWG

3.  A cable containing an insulated conductor with a white outer finish can be used for 3-way or 4-way switch loops, if it is permanently reidentified by painting or other effective means at its termination, and at each location where the conductor is visible and accessible.
    (a) True          (b) False

4.  Receptacles, polarized attachment plugs, and cord connectors for plugs and polarized plugs must have the terminal intended for connection to the grounded conductor identified. Identification shall be by a metal or metal coating that is substantially _____ in color or by the word white or the letter W located adjacent to the identified terminal.
    (a) green          (b) white          (c) gray          (d) b or c

5.  No _____ shall be attached to any terminal or lead so as to reverse designated polarity.
    (a) grounded conductor          (b) grounding conductor
    (c) ungrounded conductor          (d) grounding connector

# Article 210
# Branch Circuits

## Part I. General Provisions

### 210.1 Scope

Article 210 contains the requirements for branch circuits, such as conductor sizing, identification, receptacle GFCI protection, and receptacle and lighting outlet requirements. Figure 210-1

See Article 100 for several branch-circuit definitions.

### 210.2 Other Articles

Other sections that have specific rules for branch circuits include:

Air Conditioning, 440.6, 440.31 and 440.32

Appliances, 422.10

Computer Rooms (Data Processing), 645.5

Electric Space Heating, 424.3(B)

Motors, 430.22

Signs, 600.5

### 210.3 Branch-Circuit Rating

The rating of a branch circuit is determined by the rating of the branch-circuit overcurrent protection device. For example, the

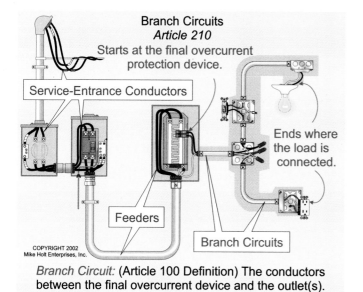

Branch Circuit: (Article 100 Definition) The conductors between the final overcurrent device and the outlet(s).

**Figure 210-1**

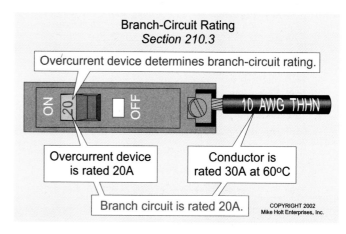

**Figure 210-2**

branch-circuit rating of 10 AWG THHN (rated 30A at 60°C) on a 20A circuit breaker is 20A. Figure 210-2

### 210.4 Multiwire Branch Circuits

(A) **General.** For the purpose of this section a multiwire branch circuit can be considered as a single circuit or as a multiple circuit, depending on its use.

**Single Circuit Example:** Where a building is supplied by one branch circuit, a grounding electrode is not required for the building [250.32(A) Ex.] A multiwire branch circuit to this building could be considered as a single circuit for this rule. [225.30]

**Multiple Circuits Example:** The *NEC* requires two small-appliance circuits for countertop receptacles in dwelling unit kitchens [210.11(C)(1) and 210.52(B)(2)], and one 1Ø, 3-wire 120/240V multiwire branch circuit could be used for this purpose.

**Originate from Same Panel.** To prevent inductive heating and to reduce conductor impedance for fault-currents, all multiwire branch-circuit conductors shall originate from the same panelboard. For more information on inductive heating, see 300.3(B), 300.5(I), and 300.20(A).

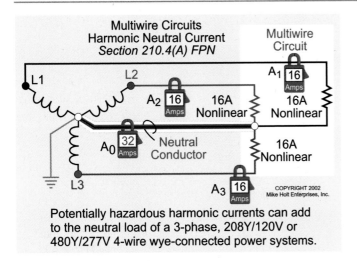

Potentially hazardous harmonic currents can add to the neutral load of a 3-phase, 208Y/120V or 480Y/277V 4-wire wye-connected power systems.

*Figure 210-3*

FPN: Unwanted and potentially hazardous harmonic currents can add on to the neutral conductor of a 3Ø, 4-wire wye-connected power systems, such as 208Y/120V or 480Y/277V, which supplies nonlinear loads. To prevent a fire or equipment damage from excessive neutral current, the designer should consider: (1) increasing the size of the neutral conductor, or (2) installing a separate neutral for each phase. For more information, visit http://www.mike-holt.com/studies/ harmonic.htm. Figure 210-3

**(B)  Dwelling Units.** Multiwire branch circuits that terminate on devices mounted on the same yoke in a dwelling unit shall be provided with a means at the panelboard to disconnect simultaneously all ungrounded circuit conductors. Figure 210-4

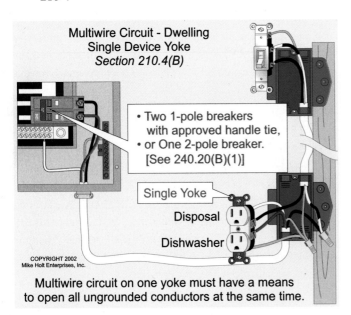

Multiwire circuit on one yoke must have a means to open all ungrounded conductors at the same time.

*Figure 210-4*

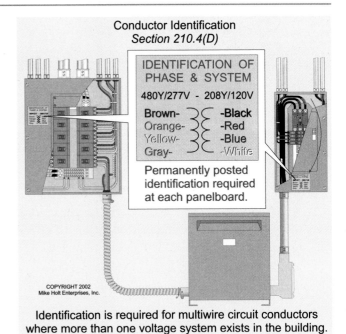

Identification is required for multiwire circuit conductors where more than one voltage system exists in the building.

*Figure 210-5*

---

**AUTHOR'S COMMENT:** Individual single-pole circuit breakers with approved handle ties or a 2-pole breaker with a common internal trip can be used for this purpose [240.20(B)(1)].

---

**(D)  Buildings Containing More than One System Voltage.** Where more than one system voltage exists in a building, the ungrounded conductor of multiwire branch circuits, where accessible, shall be identified by phase and system. The identification can be by color, tagging or other approved means with the identification method posted at each panelboard. Figure 210-5

*WARNING: When different systems are installed in the same raceway or enclosure, the grounded (neutral) conductor shall be properly identified. See 200.6(D) for details.*

### 210.6 Branch-Circuit Voltage Limitation

**(C)  277/480V Circuits.** Circuits not exceeding 277V, nominal-to-ground can supply any of the following luminaires:

(1)  Listed electric-discharge luminaires.

(2)  Listed incandescent luminaires supplied at 120 volts or less from the output of a step-down autotransformer

(3)  Luminaire with a mogul base screw shell.

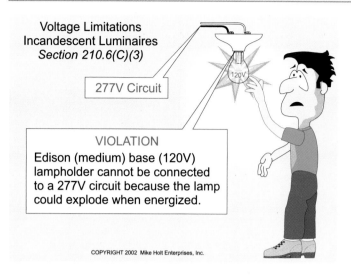

Voltage Limitations
Incandescent Luminaires
*Section 210.6(C)(3)*

277V Circuit

VIOLATION
Edison (medium) base (120V)
lampholder cannot be connected
to a 277V circuit because the lamp
could explode when energized.

COPYRIGHT 2002 Mike Holt Enterprises, Inc.

**Figure 210-6**

*CAUTION: If a 120V rated lampholder is installed on a 277V circuit, the lamp could explode when the lamp is energized. Figure 210-6*

(4)  Lampholders other than the screw shell type.

## 210.7 Receptacles

**(A) Receptacle Outlet Locations.** Receptacle outlets shall be located in accordance with the requirements of 210.50 through 210.70 (Part III of Article 210).

**(B) Receptacle Requirements.** Specific receptacle requirements are contained in Article 406.

**AUTHOR'S COMMENT:** The NEC does not specify the orientation of the ground terminal of a receptacle.

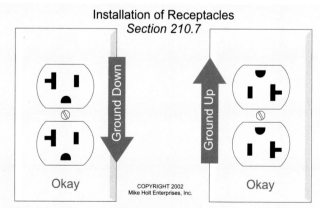

Installation of Receptacles
*Section 210.7*

Ground Down

Ground Up

Okay

Okay

COPYRIGHT 2002
Mike Holt Enterprises, Inc.

The NEC does not specify the orientation of the ground terminal of a receptacle. The ground terminal can be up or down, depending on the customer's needs.

**Figure 210-7**

---

The ground terminal can be up or down, depending on the customer's needs. See http://www.mikeholt.com/Newsletters/9-23-99.htm Figure 210-7

**(C) Multiple Branch Circuits.** Where more than one branch circuit supplies more than one receptacle on the same yoke, a means shall be provided to simultaneously disconnect all the ungrounded circuit conductors at the panelboard. Single-pole breakers with approved handle ties, or a 2-pole circuit breaker can be used for this. Figure 210–8

**AUTHOR'S COMMENT:** This rule is to prevent people from unintentionally working on energized circuits that they thought were disconnected. Technically this rule does not apply to multiwire branch circuits, because a multiwire branch circuit is considered to be one circuit. See Article 100 definition of Branch Circuit, Multiwire.

## 210.8 Ground-Fault Circuit-Interrupter Protection for Personnel.

**(A) Dwelling Units.** GFCI protection shall be provided for all 15 and 20A, 1Ø, 125V receptacles located in the following areas of a dwelling unit:

**(1) Bathroom Area Receptacles.** GFCI protection is required for all 15 and 20A, 1Ø, 125V receptacles in the bathroom area of a dwelling unit. Figure 210-9

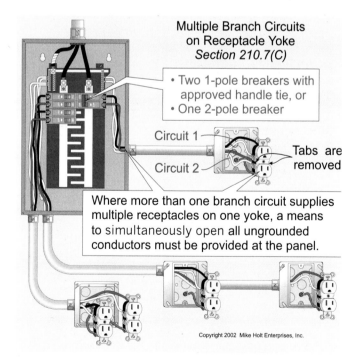

Multiple Branch Circuits
on Receptacle Yoke
*Section 210.7(C)*

• Two 1-pole breakers with approved handle tie, or
• One 2-pole breaker

Circuit 1

Circuit 2

Tabs are removed

Where more than one branch circuit supplies multiple receptacles on one yoke, a means to simultaneously open all ungrounded conductors must be provided at the panel.

Copyright 2002 Mike Holt Enterprises, Inc.

**Figure 210–8**

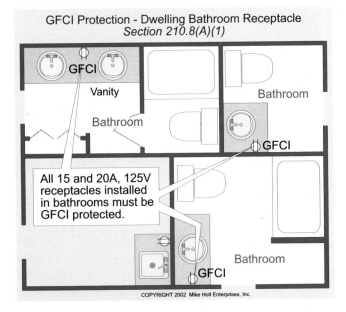

Figure 210-9

**AUTHOR'S COMMENT:** Proposals to allow receptacles for dedicated equipment in the bathroom area to be exempted from the GFCI protection rules were all rejected because it was not in the interest of safety to allow appliances without GFCI protection in this area.

(2) **Garage and Accessory Building Receptacles.** GFCI protection is required for all 15 and 20A, 1Ø, 125V in garages, and in grade-level portions of unfinished or finished accessory buildings used for storage or work areas of a dwelling unit. Figure 210-10

**AUTHOR'S COMMENT:** Receptacles are not required in accessory buildings, but if a 15A or 20A, 1Ø,

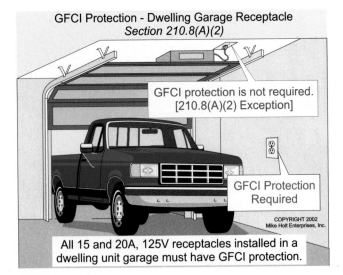

Figure 210-10

All 15 and 20A, 125V receptacles in accessory buildings and similar work areas must be GFCI protected.

Figure 210-11

125V receptacle is installed, it shall be GFCI-protected. Figure 210-11

*Exception No. 1:* GFCI protection is not required for receptacles that are not readily accessible, such as a ceiling-mounted receptacle for the garage door opener.

*Exception No. 2:* GFCI protection is not required for a receptacle on a dedicated branch circuit located and identified for a specific cord-and-plug-connected appliance, such as a refrigerator or freezer. Figure 210-12

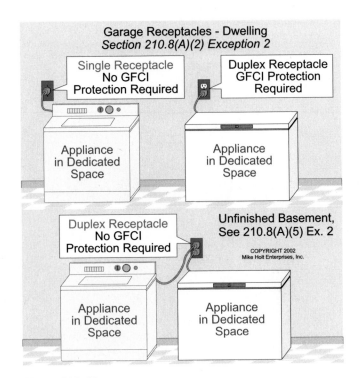

Figure 210-12

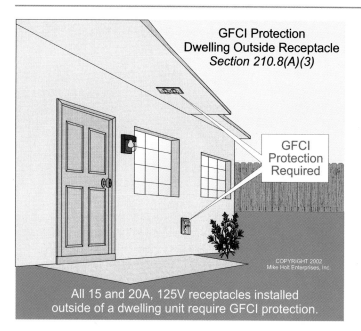

GFCI Protection
Dwelling Outside Receptacle
*Section 210.8(A)(3)*

GFCI
Protection
Required

COPYRIGHT 2002
Mike Holt Enterprises, Inc.

All 15 and 20A, 125V receptacles installed
outside of a dwelling unit require GFCI protection.

**Figure 210-13**

**(3)  Outdoor Receptacles.** All 15 and 20A, 1Ø, 125V receptacles outdoors of dwelling units, including receptacles installed under the eaves of roofs shall be GFCI-protected. Figure 210-13

**AUTHOR'S COMMENT:** The *Code* does not require a receptacle outlet outdoors of a dwelling unit in a multi-family dwelling building (building containing three or more dwelling units). See 210.52(E)

Snow-Melting and Deicing Equipment Receptacle
*Section 210.8(A)(3) Exception*

GFCI

GFCI

GFCI protection
not required.

COPYRIGHT 2002 Mike Holt Enterprises, Inc.

GFCI protection is not required for receptacles that supply snow-melting or deicing equipment, if they are not readily accessible, see 426.28 for GFPE protection.

**Figure 210-14**

*Exception:* GFCI protection is not required for fixed electric snow-melting or deicing equipment receptacles that are not readily accessible and are supplied by a dedicated branch circuit in accordance with 426.28. Figure 210-14

**(4)  Crawl Space Receptacles.** All 15 and 20A, 1Ø, 125V receptacles installed in crawl spaces at or below grade of a dwelling unit shall be GFCI-protected.

**AUTHOR'S COMMENT:** The Code does not require a receptacle to be installed in the crawl space, except when split systems air conditioning units are installed in these spaces. [210-63]

**(5)  Unfinished Basement Receptacles.** GFCI protection is required for all 15 and 20A, 1Ø, 125V receptacles required by 210.52(G) in each unfinished portion of a basement not intended as a habitable room and limited to storage and work areas. Figure 210-15

*Exception No. 1:* GFCI protection is not required for receptacles that are not readily accessible.

*Exception No. 2:* GFCI protection is not required for a receptacle on a dedicated branch circuit located and identified for a specific cord-and-plug-connected appliance, such as a freezer.

**(6)  Kitchen Countertop Surface Receptacles.** GFCI protection is required for all 15 and 20A, 1Ø, 125V receptacles to serve countertop surfaces in a dwelling unit. Figure 210-16

**AUTHOR'S COMMENT:** GFCI protection is not required for receptacles that serve built-in appliances, such as dishwashers or disposals.

GFCI Protection - Dwelling Basement Receptacle
*Section 210.8(A)(5)*

Finished Basement Area:
GFCI protection not required

COPYRIGHT 2002
Mike Holt Enterprises, Inc.

All 15 and 20A, 125V receptacles in unfinished areas of basements must be GFCI protected.

**Figure 210-15**

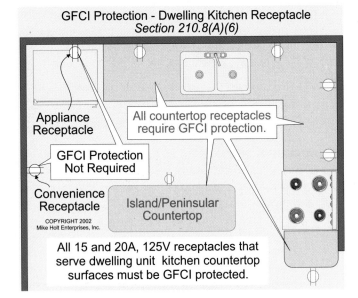

**Figure 210-16**

(7) **Wet Bar Countertop Receptacles.** GFCI protection is required for all 15 and 20A, 1Ø, 125V receptacles installed to serve the countertop surfaces located within 6 ft of the outside edge of a dwelling unit wet bar sink. Figure 210-17

**AUTHOR'S COMMENT:** GFCI protection is not required for receptacles not intended to serve wet bar countertop surfaces, such as refrigerators, ice makers, or water heaters.

(8) **Boathouse Receptacle.** GFCI protection is required for all 15 and 20A, 1Ø, 125V receptacles located in a boathouse. Figure 210–18

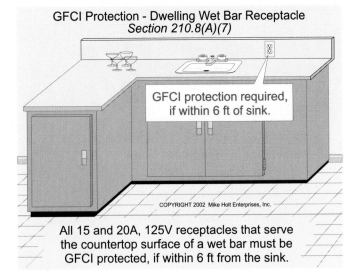

**Figure 210–17**

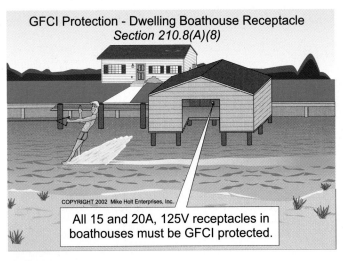

**Figure 210-18**

**AUTHOR'S COMMENT:** The Code does not require a receptacle to be installed in a boathouse.

(B) **Other than Dwelling Units.** GFCI protection is required for all 15 and 20A, 1Ø, 125V receptacles installed in the following commercial/industrial locations:

**AUTHOR'S COMMENT:** GFCI protection is not required for receptacles outside a commercial or industrial occupancy. Figure 210-19

(1) **Bathroom Receptacles.** If a receptacle is installed in a commercial or industrial bathroom, it shall be GFCI-protected. Figure 210-20

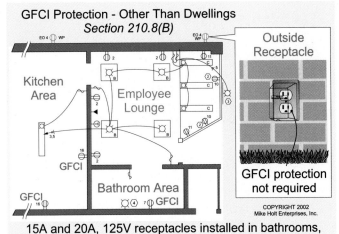

**Figure 210-19**

GFCI Protection - Nondwelling Bathroom Receptacle
Section 210.8(B)(1)

All 15 and 20A, 125V receptacles in nondwelling
unit bathrooms must be GFCI protected.

**Figure 210-20**

**AUTHOR'S COMMENT:** See Article 100 for the definition of a *bathroom*.

**(2) Rooftop Receptacles.** All 15 and 20A, 1Ø, 125V receptacles installed on rooftops shall be GFCI-protected. Figure 210-21

**AUTHOR'S COMMENT:** Section 210.63 requires a 15A or 20A, 1Ø, 125V receptacle outlet be installed within 25 ft of the heating, air-conditioning and refrigeration equipment.

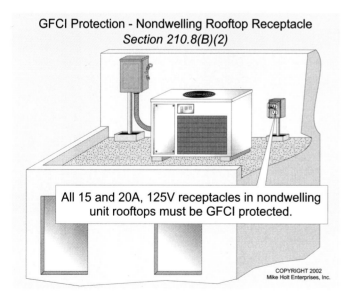

GFCI Protection - Nondwelling Rooftop Receptacle
Section 210.8(B)(2)

All 15 and 20A, 125V receptacles in nondwelling
unit rooftops must be GFCI protected.

**Figure 210-21**

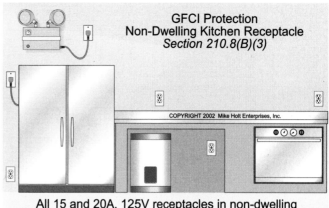

GFCI Protection
Non-Dwelling Kitchen Receptacle
Section 210.8(B)(3)

All 15 and 20A, 125V receptacles in non-dwelling
unit kitchens must be GFCI protected.

**Figure 210-22**

*Exception:* GFCI protection is not required for fixed electric snow-melting or deicing equipment receptacles that are not readily accessible and are supplied by a dedicated branch circuit in accordance with 426.28.

**(3) Kitchens.** All 15 and 20A, 1Ø, 125V receptacles installed in kitchens, even those that do not supply the countertop surface, shall be GFCI-protected. Figure 210-22

**AUTHOR'S COMMENT:** Section 210.8(A)(6) only requires GFCI protection for 15 and 20A, 1Ø, 125V receptacles installed to serve dwelling unit counter surfaces, yet 210.8(B)(3) requires all 15 and 20A, 1Ø, 125V receptacles in a commercial kitchen to be GFCI-protected! I think this rule will be revised during the 2005 Code cycle.

*NOTE: Since 1971, the NEC has been expanding the requirements of GFCI protection to include the following locations:*

- *Agricultural Buildings, 547.5(G), Figure 210-23*
- *Carnivals, Circuses and Fairs, 525.23*
- *Commercial Garages, 511.12, Figure 210-24*
- *Elevator Pits, 620.85*
- *Health Care Facilities, 517.20(A)*
- *Marinas and Boatyards, 555.19(B)(1)*
- *Portable or Mobile Signs, 600.10(C)(2)*
- *Swimming Pools, 680.22(A),*
- *Temporary Installations, 527.6 Figure 210-25*

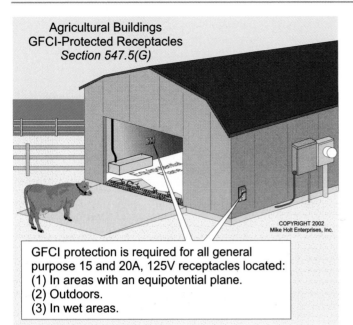

**Figure 210-23**

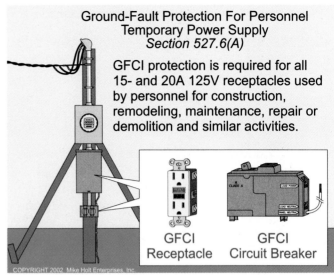

**Figure 210-25**

## Part II. Branch-Circuit Ratings

### 210.11 Branch-Circuit Requirements

**(A) Number of Branch Circuits.** The minimum number of branch circuits required is determined from the total computed load and the rating of the circuits used. In all installations, the number of circuits shall be sufficient to supply the load served. For example, a 2,100 sq ft dwelling requires four 15A or three 20A, 120V circuits for general lighting and receptacles. Figure 210-26

**AUTHOR'S COMMENT:** See Example D1(a) in Annex D.

**(C) Dwelling Unit.**

**(1) Small-Appliance Branch Circuits.** Two or more 20A, 1Ø, 120V small-appliance branch circuits shall be provided to supply power for 15A or 20A receptacle outlets in the dwelling unit kitchen, dining room, breakfast room, pantry, or similar dining areas as required by 210.52(B). Figure 210-27

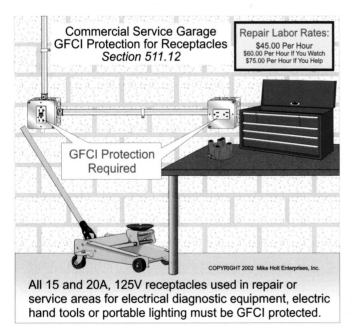

**Figure 210-24**

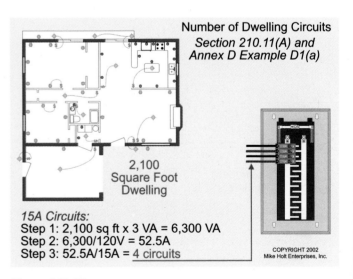

**Figure 210-26**

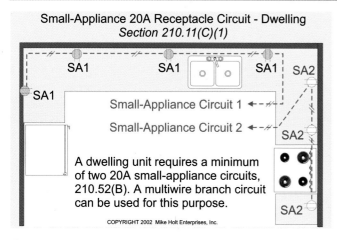

Small-Appliance 20A Receptacle Circuit - Dwelling
*Section 210.11(C)(1)*

**Figure 210-27**

**AUTHOR'S COMMENT:** Lighting outlets or receptacles for other purposes cannot be connected to the small-appliance branch circuit [210.52(B)(2)]. The two small-appliance branch circuits can be supplied by one 3-wire multiwire circuit or by two separate 120V circuits. In addition, there is no Code requirement to supply each counter top with two small-appliance circuits.

**(2) Laundry Branch Circuit.** One 20A 1Ø, 120V branch circuit is required for the dwelling unit laundry room receptacle outlet required by 210.52(F).

**AUTHOR'S COMMENT:** The laundry room receptacle circuit can supply more than one receptacle in the laundry room, but it cannot serve any other outlets, such as the laundry room lighting or receptacles in other rooms. Figure 210-28

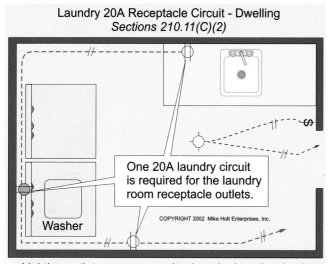

Laundry 20A Receptacle Circuit - Dwelling
*Sections 210.11(C)(2)*

One 20A laundry circuit is required for the laundry room receptacle outlets.

COPYRIGHT 2002 Mike Holt Enterprises, Inc.

Washer

Lighting outlets are not permitted on the laundry circuit.

**Figure 210-28**

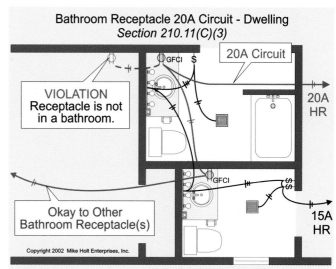

Bathroom Receptacle 20A Circuit - Dwelling
*Section 210.11(C)(3)*

VIOLATION
Receptacle is not in a bathroom.

20A Circuit

20A HR

15A HR

Okay to Other Bathroom Receptacle(s)

Copyright 2002 Mike Holt Enterprises, Inc.

A minimum of one 20A circuit is required to supply the required 15A or 20A bathroom receptacles. Other outlets are not permitted on the bathroom receptacle circuit.

**Figure 210-29**

**(3) Bathroom Branch Circuit.** The bathroom receptacle outlets required in 210.52(D) for a dwelling unit shall be supplied by at least one 20A 1Ø, 120V branch circuit that does not supply any other load. A 15A receptacle is rated for 20A feed-through, it can be used for this purpose [210.21(B)(3)]. Figure 210-29

**AUTHOR'S COMMENT:** A single 20A, 1Ø, 120V branch circuit can be used to supply multiple bathroom receptacles.

*Exception:* Where a single 20A, 1Ø, 120V branch circuit supplies only one bathroom, all of the outlets in the bathroom can be supplied by that circuit if no load fastened in place is rated more than 10A in accordance with 210.23(A). Figure 210-30

**Question:** Can a hydromassage bathtub motor or radiant heater be connected to the 20A, 1Ø, 120V bathroom receptacle circuit?

**Answer:** Yes, if the appliance is not rated more than 10A, and the installation instructions do not require a separate circuit for the equipment.

**Question:** Can a luminaire, ceiling fan, or bath fan be connected to the 20A, 1Ø, 120V receptacle circuit that supplies one bathroom?

**Answer:** Yes.

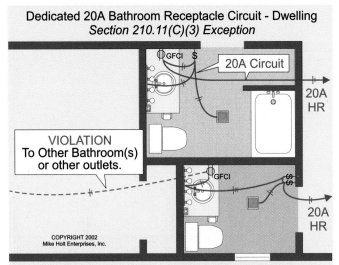

**Dedicated 20A Bathroom Receptacle Circuit - Dwelling**
*Section 210.11(C)(3) Exception*

20A Circuit

20A HR

**VIOLATION**
To Other Bathroom(s)
or other outlets.

20A HR

COPYRIGHT 2002
Mike Holt Enterprises, Inc.

Where a dedicated 20A circuit supplies a single bathroom, other equipment within the bathroom can be connected to the circuit if the equipment rating does not exceed 10A.

**Figure 210-30**

## 210.12 Arc-Fault Circuit-Interrupter (AFCI) Protection

**(A) AFCI Definition.** An AFCI protection device provides protection from an arcing fault by recognizing the characteristics unique to an arcing fault and by functioning to de-energize the circuit when an arc fault is detected.

## Arc-Fault Protection

Arcing is defined as a luminous discharge of electricity across an insulating medium. Electric arcs operate at temperatures between 5,000 and 15,000°F and expel small particles of very hot molten materials. Higher current arcs are more likely to cause a fire because of the higher thermal energy contained in the arc. Greater current will melt more of the conductor metal and therefore expel more hot molten particles.

In electrical circuits, unsafe arcing faults can occur in one of two ways, as series arcing faults or as parallel arcing faults. The most dangerous is the parallel arcing fault. A series arc can occur when the conductor in series with the load is unintentionally broken. Examples might be a frayed conductor in a cord that has pulled apart. A series arc fault current is load limited; the arc's current cannot be greater than the load current the conductor serves. Typically, series arcs do not develop sufficient thermal energy to create a fire.

Parallel arcing faults occur in one of two ways, as a short-circuit or as a ground fault. A short-circuit arc might occur if the wire's insulation is damaged by an excessively or incorrect-

ly driven staple in NM cable, or the insulation of the cord is damaged by a conductive object such as a metal table leg placed on the cord. The result is a decrease in the dielectric strength of insulation separating the conductors allowing a high-impedance low current arcing fault to develop. This arcing fault carbonizes the conductor's insulation further decreasing the dielectric of the insulation separating the conductors, resulting in an increase in current, an exponential increase in thermal energy heat, and the likelihood of a fire.

The current flow in a short-circuit type arc is limited by the system impedance and the impedance of the arcing fault itself. Typically, at a receptacle, fault current will be above 75A, but not likely above 450A. This short-circuit arc is reported as being more common in older homes where the appliance cords and the building wiring have deteriorated due to the negative effects of aging.

A ground-fault type parallel arc fault can only occur when a ground path is present. This type of arcing fault can be cleared by either a GFCI or AFCI circuit protective device.

The RMS current value for parallel arc faults will be considerably less than that of a solid bolted type fault; therefore, a typical 15A circuit protective device might not clear this fault before a fire is ignited.

**AFCI Protection Devices.** To help reduce the hazard of electrical fires from a parallel arcing fault in the branch-circuit wiring, the *NEC* requires a listed AFCI protection device be installed to protect the branch-circuit wiring in dwelling unit bedrooms. UL 1699 contains the listing requirement for 15 and 20A Arc-Fault Circuit-Interrupter devices. Each type of device is intended to protect different aspects of the branch circuit and extension wiring. However, the only AFCI device that currently meets the *NEC* requirements is the Branch/Feeder Type. Figure 210-31

**AUTHOR'S COMMENT:** Visit http://www.ul.com/regulators/afci/categories.html for more information of the six AFCI types and UL product categories.

**Branch/Feeder Type –** This AFCI circuit breaker device is to be installed at the origin of a circuit, such as at a panelboard. It provides parallel arc fault protection for branch-circuit wiring. This device is not listed to provide series arc fault protection, but it is listed to protect against series arc faults in NM cable with ground, but not NM cable without ground.

**Combination Type –** This device provides both parallel and series arc fault protection for cord sets and power-supply cords downstream of the device. Typically, this device is a

## Types of AFCI Devices

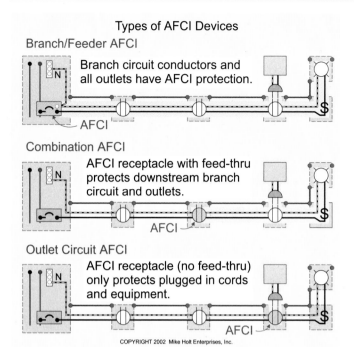

Branch/Feeder AFCI

Branch circuit conductors and all outlets have AFCI protection.

AFCI

Combination AFCI

AFCI receptacle with feed-thru protects downstream branch circuit and outlets.

AFCI

Outlet Circuit AFCI

AFCI receptacle (no feed-thru) only protects plugged in cords and equipment.

AFCI

COPYRIGHT 2002  Mike Holt Enterprises, Inc.

**Figure 210-31**

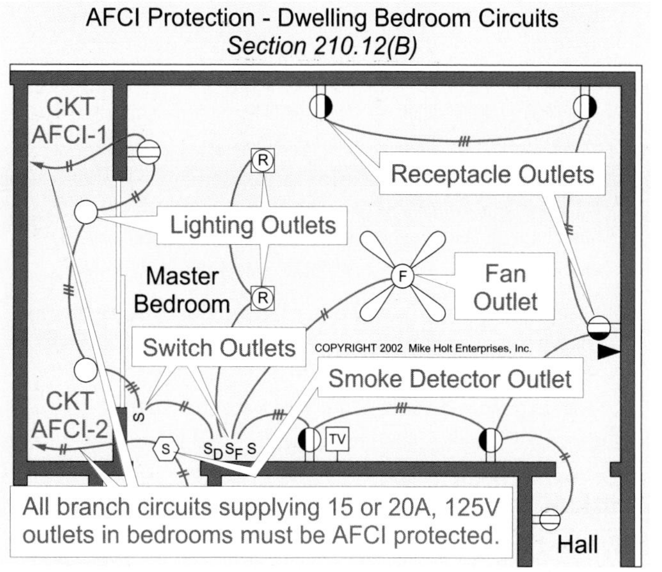

receptacle. This device is not listed to provide parallel arc fault protection upstream of the device.

**Outlet Branch Circuit Type** – A device intended to be installed as the first outlet in a branch circuit. It is intended to provide parallel and series protection to downstream branch-circuit wiring, cord sets, and power-supply cords. This device also provides series protection to upstream branch-circuit wiring.

**Outlet Circuit Type** – This AFCI device, typically an outlet, is installed at a branch-circuit outlet and provides for both parallel and series arcing fault protection of the cord sets and power-supply cords plugged into the outlet. This device is intended to protect extension wiring plugged into the device as well as extensions plugged into downstream receptacles. It does not provide arc fault protection on feed-thru branch-circuit conductors, nor is it listed and tested to provide parallel or series arc fault protection upstream of the device.

UL 1699 requires testing of the AFCI devices through a rigorous set of tests for arc detection ability, unwanted operation (to avoid nuisance operation) and operation inhibition.

While all listed electrical devices are subjected to various types of tests intended to confirm safe operation, AFCI devices are subject to an enhanced level of testing. This is due to the daily occurrences of "safe" arcing events that occur in our homes. Arcing is produced from many safely operating electrical appliances, such as universal motors in

## AFCI Protection - Dwelling Bedroom Circuits
### Section 210.12(B)

CKT AFCI-1

Lighting Outlets

Master Bedroom

Switch Outlets

Receptacle Outlets

Fan Outlet

COPYRIGHT 2002  Mike Holt Enterprises, Inc.

Smoke Detector Outlet

CKT AFCI-2

Hall

All branch circuits supplying 15 or 20A, 125V outlets in bedrooms must be AFCI protected.

**Figure 210-32**

vacuum cleaners, motor starting relays in refrigerators and portable drills.

*WARNING: AFCI protection devices are not designed to prevent fires caused by series arcing at loose connections to a receptacle, switch, device or in a splice.*

**(B) Dwelling Unit Bedrooms.** All branch circuits supplying 15A or 20A, 1Ø, 125V outlets installed in dwelling unit bedrooms shall be AFCI-protected by a device that is a listed device to protect the entire circuit from an arc fault. Figure 210–32

**AUTHOR'S COMMENT:** The 125V limitation to the requirement means that AFCI protection would not be required for a 240V baseboard heater or room air conditioner. For more information, visit http://www.mikeholt.com/Newsletters/Newsletters.htm, go to the "Miscellaneous" section and visit my "AFCI" links.

The intent of this rule is that arc fault protection be provided for all branch-circuit conductors that supply 115A and 20A, 1Ø, 120V outlets. In addition, the AFCI shall be listed so it will protect the 'entire branch circuit' by de-energizing the circuit when an arc fault is detected.

Controversy exists between circuit breaker manufacturers and wiring device manufacturers. The circuit breaker manufacturers insist that the only acceptable device is the AFCI circuit breaker. Wiring device manufacturers take the position that AFCI receptacles, listed to protect the entire branch circuit, should be suitable and Code compliant.

Both make excellent arguments, but 210.12(A) is very specific. It requires the AFCI protection device to de-energize the circuit and protect the entire circuit from an arc fault. The only device that can de-energize the circuit when an arc fault is detected is the AFCI circuit breaker. AFCI receptacles of the type listed to detect upstream series arc faults will not de-energize the circuit from parallel type arcing faults that may occur upstream of the device. Therefore, they cannot be used to meet the NEC requirement of 210.12.

## AFCI CONSIDERATIONS

At the time a dwelling unit is wired, it is hard to tell from looking at the bare walls whether a room will be used as a home office or a bedroom. Also, if you are looking at an efficiency apartment, a room may well be furnished with a foldout couch that is used for sleeping, making it look as much like a bedroom as a living room.

If you are in the practice of using one branch circuit for both lighting and receptacles, the 2002 changes will have no effect to the wiring as compared to the 1999 NEC. But, the practice of separating the lighting from the receptacle circuits in dwelling unit bedrooms will now require two AFCI circuit breakers, or you'll place them all on the same circuit due to the cost of AFCI protection.

The following proposals for the 2002 NEC were all rejected:

• The U. S. Consumer Product Safety Commission's request that existing bedroom branch circuits be protected by an AFCI when the service equipment is replaced.

• Omit AFCI protection for the lighting outlets, because light may be needed when the AFCI device operates.

• Extend AFCI protection for guest room branch circuits of motels and hotels.

• Permit the AFCI receptacle outlet to provide the required protection.

• Omit AFCI protection for the smoke detector circuit conductors.

• Delete the AFCI requirement completely.

## COMMON QUESTIONS ABOUT AFCIs

**Q1.** Why is AFCI protection only required for dwelling unit branch circuits?

**A1.** The NEC wanted the industry to gain experience with these devices in bedroom circuits so in the future their usage might be expanded to other rooms and facilities that could benefit by the added protection. Studies have shown that over 60 per-

cent of fires are from causes in the fixed wiring, switches, receptacle outlets and lighting fixtures that are part of the fixed electrical system of a residence.

**Q2.** What happens when a bedroom appliance has a locked-rotor condition (bedroom window air conditioner)? Will the AFCI breaker respond?

**A2.** No and Yes. The waveform signature of locked-rotor current is not typical of an arc fault, so the AFCI will not respond. Under locked-rotor conditions, either the magnetic function, or, in time, the thermal trip element of the protection device should open the circuit. Many small air-conditioner motors will be equipped with a hermetically encased thermal limit device, which will open before the branch-circuit protection device. However, if the locked-rotor current does not open a protection device, the motor winding will ultimately short-out and the circuit breaker or fuse will open to safely clear the fault.

**Q3.** Will there be lots of nuisance tripping of these devices?

**A3.** Honestly, there has not been sufficient experience in the field to answer this question, but what might appear to be a nuisance tripping condition might be an actual arc fault. At some point, we will all need to learn how to trace down arcing-type faults, perhaps with some type of sensor, perhaps by taking the circuit apart, isolating individual sections. I am not aware of any commercially available meter or instrument that could be used for this purpose.

**Q4.** Will an AFCI prevent fires from loose connections at terminals or splicing devices?

**A4.** No, this product is not listed to protect against a series (glowing) fault.

**Q5.** Are there any AFCI/GFCI combination breakers?

**A5.** Yes

**Q6.** If you have a dedicated line for computer usage, marked Computer Only, would this also need to be on AFCI in the bedroom?

**A6.** Yep, if it's in the bedroom.

**Q7.** Can an AFCI circuit breaker be placed on a multiwire branch circuit?

**A7.** No, not unless it is a 2-pole breaker listed for multiwire branch circuits.

**Q8.** I have heard that when a light bulb burns out, it draws 150-200 amps for one-half cycle and this could cause nuisance tripping of an AFCI.

**A8.** Not true.

## 210.19 Conductor Sizing

### (A) Branch Circuits

**(1) Continuous and Noncontinuous Loads.** Conductors that supply continuous loads shall be sized no less than 125 percent of the continuous loads, plus 100 percent of the noncontinuous loads based on the conductor ampacities as listed in Table 310.16, before any ampacity adjustment in accordance with the terminal temperature rating [110.14(C)].

**AUTHOR'S COMMENT:** Circuit conductors shall have sufficient ampacity, after ampacity adjustment, to carry the load, and they shall be protected against overcurrent in accordance with 210.20(A) and 240.4.

**Question:** What size branch-circuit conductor is required for a 44A continuous load if the terminals are rated for 75°C? Figure 210-33

**Answer:** 6 AWG

44A load × 1.25 = 55A, 6 AWG THHN rated 65A at 75°C, Table 310.16

**AUTHOR'S COMMENT:** Circuit protection devices shall be sized no smaller than 125 percent of the continuous load. In this case, 44A load x 1.25 = 55A, 60A [210.20(A) and 240.6(A)].

*Exception:* Where the assembly and the overcurrent protection device are both listed for 100 percent continuous load operation, the branch-circuit conductors can be sized at 100 percent of the continuous load.

**AUTHOR'S COMMENT:** Equipment suitable for 100 percent continuous loading is rarely available in ratings under 400A.

FPN No. 4: Voltage drop should be considered when sizing branch-circuit conductors. Figure 210-34

**AUTHOR'S COMMENT:** The purpose of the *National Electrical Code* is the practical safeguarding of persons and property from hazards caused by the use of electricity. The *NEC* does not generally consider voltage drop to be a safety issue. As a result, the NEC contains seven recommendations (Fine Print Notes) that circuit conductors be sized sufficiently large enough so reasonable efficiency of equipment operation can be provided. In addition, the *NEC* has three rules that require conductors be sized to accommodate the voltage drop of the circuit conductors.

### Ohm's Law Method – Single-Phase Only

Voltage drop of the circuit conductors can be determined by multiplying the current of the circuit by the total resistance of the circuit conductors: VD = I × R. "I" is equal to the load in amperes and "R" is equal to the resistance of the conductor as listed in Chapter 9, Table 8 for direct current circuit, or in Chapter 9, Table 9 for alternating current circuits. The Ohm's law method cannot be used for 3Ø circuits.

**120V Question:** What is the voltage drop of two 12 AWG conductors that supply a 16A, 120V load which is located 100 ft from the power supply (200 ft of wire)? Figure 210-35

(a) 3.2V          (b) 6.4V

(c) 9.6V          (d) 12.8V

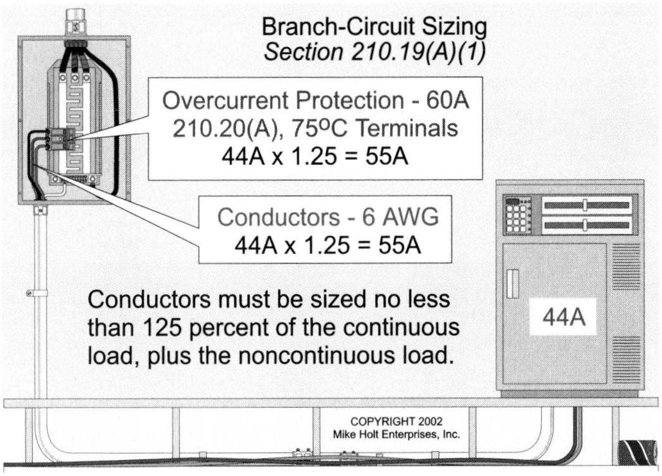

**Branch-Circuit Sizing**
*Section 210.19(A)(1)*

Overcurrent Protection - 60A
210.20(A), 75°C Terminals
44A x 1.25 = 55A

Conductors - 6 AWG
44A x 1.25 = 55A

44A

Conductors must be sized no less than 125 percent of the continuous load, plus the noncontinuous load.

COPYRIGHT 2002
Mike Holt Enterprises, Inc.

**Figure 210-33**

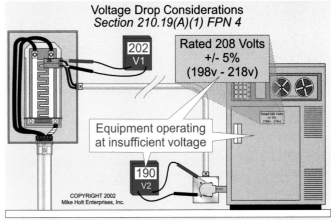

**Voltage Drop Considerations**
*Section 210.19(A)(1) FPN 4*

202
V1

Rated 208 Volts
+/- 5%
(198v - 218v)

Equipment operating
at insufficient voltage

190
V2

COPYRIGHT 2002
Mike Holt Enterprises, Inc.

Voltage drop should be considered when sizing branch-circuit conductors, but this is not a requirement.

**Figure 210-34**

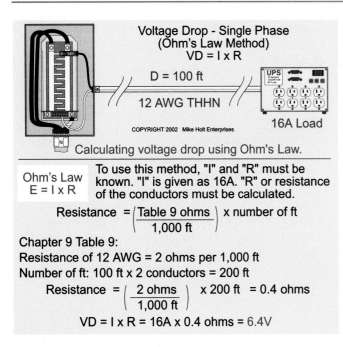

Figure 210-35

**Answer:** (b) 6.4V

Voltage Drop = I × R

"I" is equal to 16A

"R" is equal to 0.4Ω (Chapter 9, Table 9: (2Ω/1,000 ft) × 200 ft

Voltage Drop = 16A × 0.4Ω

Voltage Drop = 6.4V, (6.4 V/120V = 5.3% voltage drop)

Operating Voltage = 120V – 6.4V = 113.6V

The 5.3% voltage drop for the above branch circuit exceeds the *NEC*'s recommendations of 3%, but it does not violate the *NEC* unless the 16A load is rated less than 113.6V [110.3(B)].

**240V Single-Phase Question:** What is the voltage drop of a 24 A, 240 V, 1Ø load located 160 ft from the panelboard, if it is wired with 10 AWG conductors? Figure 210-36

(a) 4.53V                    (b) 9.22V

(c) 3.64V                    (d) 5.54V

**Answer:** (b) 9.22V

Voltage Drop = I × R

"I" is equal to 24A

"R" is equal to 0.384Ω (Chapter 9, Table 9: (1.2Ω/1,000 ft) × 320 ft

Voltage Drop = 24 A × 0.384Ω = 9.216 VD

Voltage Drop = 9.22V, (9.22V/240V = 3.8% voltage drop)

Operating Voltage = 240V – 9.22V = 230.8V

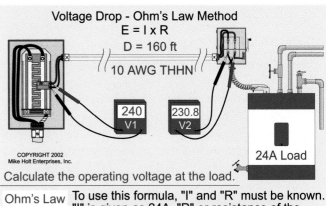

Figure 210-36

## Voltage Drop Using the Formula Method

When the circuit conductors have already been installed, the voltage drop of the conductors can be determined by using one of the following formulas:

$$VD (1Ø) = 2 × K × Q × I × D/CM$$

$$VD (3Ø) = 1.732 × K × Q × I × D/CM$$

"VD" = Volts Dropped: The voltage drop of the circuit conductors as expressed in volts.

"K" = Direct-Current Constant: This is a constant that represents the direct-current resistance for a one thousand circular mils conductor that is one thousand feet long, at an operating temperature of 75ºC. The direct current constant valuefor copper is 12.9 ohms and 21.2 ohms is for aluminum. The "K" constant is suitable for alternating current circuits, where the conductor size is not larger than 1/0 AWG.

"Q" = Alternating Current Adjustment Factor: Alternating current circuits 2/0 AWG and larger should be adjusted for the effects of self-induction (skin effect). The "Q" adjustment factor is determined by dividing alternating current resistance as listed in *NEC* Chapter 9, Table 9, by the direct current resistance as listed in Chapter 9, Table 8.

"I" = Amperes: The load in amperes at 100 percent, not 125 percent for motors or continuous loads.

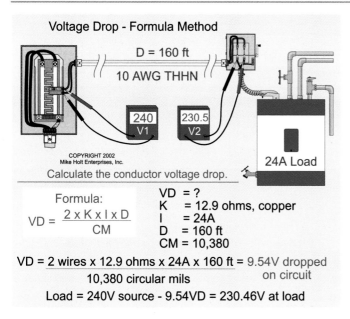

Figure 210-37

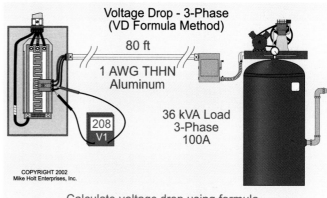

Figure 210-38

"D" = Distance: The distance the load is located from the power supply, not the total length of the circuit conductors.

"CM" = Circular-Mils: The circular mils of the circuit conductor as listed in Chapter 9, Table 8.

**Single-Phase Question:** What is the voltage drop for a 10 AWG copper conductor that supplies a 24A, 240V, 1Ø load located 160 ft from the panelboard? Figure 210-37

(a) 4.25V                          (b) 9.5V

(c) 3 percent                      (d) 5 percent

**Answer:** (b) 9.5V is the closest

$VD = 2 \times K \times I \times D/CM$

K = 12.9 ohms, copper

I = 24A

D = 160 ft

CM = 10,380 circular mils, Chapter 9, Table 8 (10 AWG)

VD = 2 wires × 12.9 ohms × 24A × 160 ft/26,240 circular mils

VD = 9.54V, (9.54V/240V = 4% voltage drop)

Operating Voltage = 240V − 9.54V = 230.46V

**Three-Phase Question:** A 36 kVA, 208V, 1Ø load is located 80 ft from the panelboard and it is wired with 1 AWG THHN aluminum conductors. What is the voltage drop of the conductors to the equipment disconnect? Figure 210-38

(a) 3.5V                          (b) 7V

(c) 3 percent                      (d) 5 percent

**Answer:** (a) 3.5 V

$VD = 1.732 \times K \times I \times D/CM$

K = 21.2Ω, aluminum

I = 100 A, 36,000VA/(208V × 1.732)

D = 80 ft

CM = 83,690 circular mils, Chapter 9, Table 8 (1 AWG)

VD = 1.732 × 21.2Ω × 100A × 80 ft/83,690 circular mils

VD = 3.5V (3.5 V/208V = 1.7%)

Operating Voltage = 208V − 3.5V = 204.5V

**(4) Minimum Conductor Size.** Branch-circuit conductors shall have an ampacity sufficient for the loads served and shall not be smaller than 14 AWG.

**AUTHOR'S COMMENT:** Branch-circuit tap conductors are not permitted for receptacle outlets [210.19(A)(4) Ex. 1(c)]. Figure 210-39

### Voltage Drop Considerations

Fine Print Notes in the *NEC* are for informational purposes only and are not enforceable by the inspection authority [90.5(C)]. However, 110.3(B) requires equipment to be installed in accordance with the equipment instructions. Therefore, electrical equipment shall be installed so that it operates within its voltage rating as specified by the manufacturer. Figure 210-40

Due to voltage drop within the circuit conductors, the operating voltage of electrical equipment will be less than the output

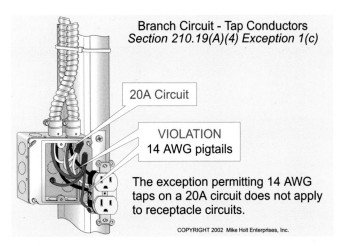

**Figure 210-39**

voltage of the power supply. Inductive loads (i.e., motors, ballasts, etc.) that operate at voltage below their rating can overheat resulting in shorter equipment operating life and increased cost, as well as inconvenience for the customer. Undervoltage for sensitive electronic equipment such as computers, laser printers, copy machines, etc., can cause the equipment to lock-up or suddenly power down resulting in data loss, increased cost and possible equipment failure. Resistive loads (heaters, incandescent lighting) that operate at undervoltages simply will not provide the expected rated power output.

Voltage drop on the conductors can cause incandescent lighting to flicker when other appliances, office equipment, or heating and cooling systems cycle on. Though this might be

annoying for some, it's not dangerous and it does not violate the *NEC*.

## NEC Recommendations

The National Electrical *Code* contains Fine Print Notes to alert the *Code* user that equipment can have improved efficiency of operation if conductor voltage drop is taken into consideration.

**Branch Circuits** – This FPN recomments that branch-circuit conductors be sized to prevent a maximum voltage drop of three percent. The maximum total voltage drop for a combination of both branch circuit and feeder should not exceed five percent. [210.19(A)(1) FPN], Figure 210-41

**Feeders** – This FPN recommends that feeder conductors be sized to prevent a maximum voltage drop of three percent. The maximum total voltage drop for a combination of both branch circuit and feeder should not exceed five percent. [215.2(A)(1) FPN 2]

**Question:** What is the minimum *NEC* recommended operating voltage for a 120V load that is connected to a 120/240V source? Figure 210-37(42)

(a) 120V          (b) 115V

(c) 114V          (d) 116V

**Answer:** (c) 114V. The maximum conductor voltage drop recommended for both the feeder and branch circuit is five percent of the voltage source: 120V × 5% = 6V. The operating voltage at the load is determined by subtracting the conductor's voltage drop from the voltage source: 120V - 6V = 114V.

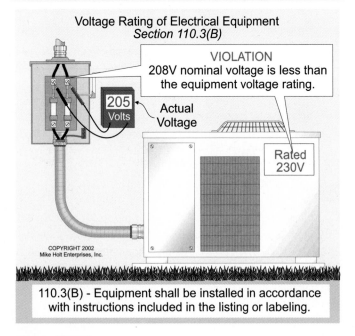

**Figure 210-40**

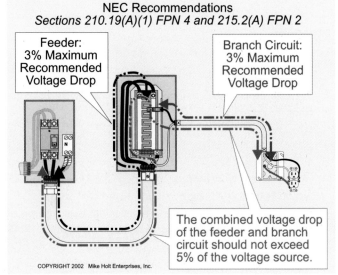

**Figure 210.41**

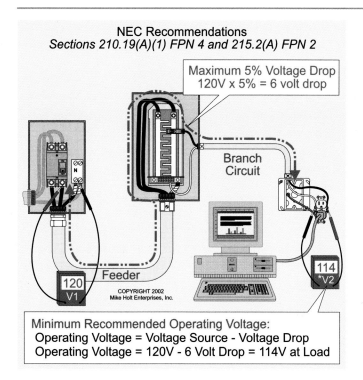

NEC Recommendations
*Sections 210.19(A)(1) FPN 4 and 215.2(A) FPN 2*

Maximum 5% Voltage Drop
120V x 5% = 6 volt drop

Branch
Circuit

Feeder

120
V1

114
*V2

COPYRIGHT 2002
Mike Holt Enterprises, Inc.

Minimum Recommended Operating Voltage:
Operating Voltage = Voltage Source - Voltage Drop
Operating Voltage = 120V - 6 Volt Drop = 114V at Load

**Figure 210-42**

## Determining Circuit Voltage Drop

When the circuit conductors have already been installed, the voltage drop of the conductors can be determined by one of two methods: Ohm's law or the VD formula.

**AUTHOR'S COMMENT:** A free voltage drop calculator can be downloaded at: www.mikeholt.com./free/free.htm

## 210.20 Overcurrent Protection

**(A) Continuous and Noncontinuous Loads.** Branch-circuit overcurrent protection devices shall have an ampacity of no less than 125 percent of the continuous loads, plus 100 percent of the noncontinuous loads. See 210.19(A)(1) for sizing branch-circuit conductors for continuous loads.

*Exception:* Where the assembly and the overcurrent protection device are both listed for 100 percent continuous load operation, the branch-circuit protection device can be sized at 100 percent of the continuous load.

**AUTHOR'S COMMENT:** Equipment suitable for 100 percent continuous loading is rarely available in ratings under 400A.

**(B) Conductor Protection.** Branch-circuit conductors shall be protected against overcurrent in accordance with 240.4.

**(C) Equipment Protection.** Branch-circuit equipment shall be protected in accordance with 240.3.

### 210.21 Outlet Device Rating

**(A) Lampholder Ratings.** Lampholders connected to a branch circuit with a rating over 20A shall be of the heavy-duty type.

*WARNING: Lampholders for fluorescent lamps are not rated heavy duty, so they cannot be installed on circuits rated over 20A. See 210.3 for the definition of Branch-Circuit Rating.*

**(B) Receptacle Ratings and Loadings.**

**(1) Single Receptacle.** A single receptacle installed on an individual branch circuit shall have an ampacity of not less than the rating of the overcurrent protection device.

**AUTHOR'S COMMENT:** A single receptacle is one contact device on a yoke and a duplex receptacle is two or more contact devices on the same yoke [Article 100].

**(2) Multiple Receptacle Loading.** Where connected to a branch circuit supplying two or more receptacles, a receptacle shall not supply a total cord-and-plug-connected load rated in excess of 80 percent of the receptacle rating.

**(3) Multiple Receptacle Rating.** Where connected to a branch circuit supplying two or more receptacles, receptacles shall be rated and installed on circuits as follow: Figure 210-43

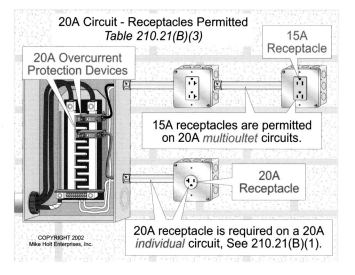

20A Circuit - Receptacles Permitted
*Table 210.21(B)(3)*

20A Overcurrent
Protection Devices

15A
Receptacle

15A receptacles are permitted
on 20A *multioutlet* circuits.

20A
Receptacle

20A receptacle is required on a 20A
*individual* circuit, See 210.21(B)(1).

COPYRIGHT 2002
Mike Holt Enterprises, Inc.

**Figure 210-43**

| Receptacle Rating | Circuit Rating |
|---|---|
| 15A | 15A |
| 15A or 20A | 20A |
| 30A | 30A |
| 40A or 50A | 40A |
| 50A | 50A |

## 210.23 Permissible Loads

An individual branch circuit can supply any load for which it is rated. A multioutlet branch circuit shall supply loads in accordance with (A) below.

**(A) 15A and 20A Circuit.** A single 15A or 20A branch circuit can supply lighting, equipment, or any combination of both.

**AUTHOR'S COMMENT:** 15A or 20A circuits can supply both lighting and receptacles on the same circuit.

**(1) Portable Equipment.** Cord-and-plug-connected equipment shall not be rated more than 80 percent of the branch-circuit rating.

**AUTHOR'S COMMENT:** UL and other approved testing laboratories list portable equipment up to 100 percent of the circuit rating. The NEC is an installation standard, not a product standard.

**(2) Fixed Equipment.** Equipment fastened in place, other than luminaires, shall not be rated more than 50 percent of the branch-circuit ampere rating if this circuit supplies both luminaires and receptacles.

**Question:** Can a 9.8A 120V central vacuum system, which is fastened in place, be supplied by a 15A, 120V, 1Ø receptacle circuit? Figure 210-44

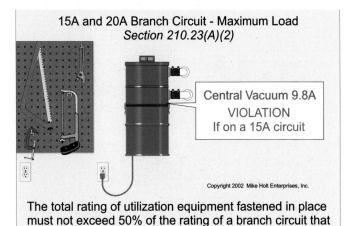

**15A and 20A Branch Circuit - Maximum Load**
*Section 210.23(A)(2)*

Central Vacuum 9.8A
VIOLATION
If on a 15A circuit

The total rating of utilization equipment fastened in place must not exceed 50% of the rating of a branch circuit that supplies other cord-and-plug-connected equipment.

**Figure 210-44**

**Answer:** No, because the rating of the equipment, 9.8A, is more than 50 percent of the branch-circuit ampere rating.

## 210.25 Common Area Branch Circuits

Dwelling unit branch circuits can supply loads only within or associated with the dwelling unit. Branch circuits for house lighting, central alarm, signal, fire alarm, communications, or other needs for public safety shall not originate from a dwelling unit panelboard.

**AUTHOR'S COMMENT:** This rule reduces the likelihood of common area branch circuits of two-family or multifamily dwellings being supplied from an individual dwelling unit for the supply of safety equipment. This prevents common area circuits from being turned off by tenants or because of nonpayment of electric bills.

## Part III. Required Outlets

### 210.50 General

Receptacle outlets shall be installed in accordance with the requirements in 210.52 through 210.63.

**(A) Receptacle Outlet – Cord Pendant.** Permanently installed cord pendant receptacles are considered receptacle outlets, and they shall be installed in accordance with the requirements of Article 400. Figure 210-45

**(C) Appliance Receptacle Outlet Location.** Receptacle outlets installed for a specific appliance, such as a clothes washer, dryer, range, or refrigerator, shall be within 6 ft of the intended location of the appliance.

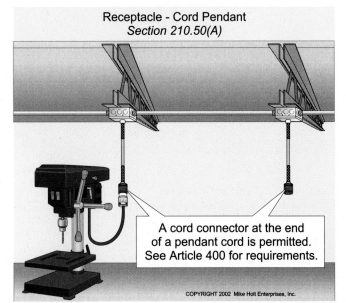

Receptacle - Cord Pendant
*Section 210.50(A)*

A cord connector at the end of a pendant cord is permitted. See Article 400 for requirements.

**Figure 210-45**

## 210.52 Dwelling Unit Receptacle Outlet Requirements

This section contains the requirements for 15 and 20A, 1Ø, 125V receptacle outlets for dwelling units. Receptacle outlets that are: (1) part of a luminaire or appliance, (2) located within cabinets or cupboards, or (3) more than 5¹/₂ ft above the floor shall not be used to meet the requirements of this section.

**(A) General Requirements – Dwelling Unit.** A receptacle outlet shall be installed in every kitchen, family room, dining room, living room, parlor, library, den, bedroom, recreation rooms, and similar rooms or areas in accordance with the following requirements. Figure 210-46

**(1) Receptacle Placement.** A receptacle outlet shall be installed so no point along the wall space will be more than 6 ft, measured horizontally, from a receptacle outlet.

---

**AUTHOR'S COMMENT:** The purpose of this rule is to ensure that a general-purpose receptacle is conveniently located to reduce the likelihood that an extension cord will travel across openings such as doorways or fireplaces.

---

**(2) Wall Space Definition.**

(1) Any space 2 ft or more in width, unbroken along the floor line by doorways, fireplaces, and similar openings.

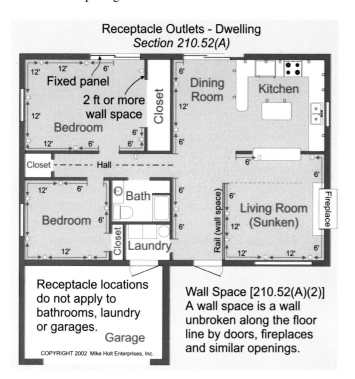

**Figure 210-46**

**Figure 210–47**

(2) The space occupied by fixed panels in exterior walls.

(3) The space occupied by fixed room dividers, such as freestanding bar-type counters or railings.

**(3) Floor Outlets.** Floor outlets shall be counted as the required receptacle outlets if located within 18 in. of the wall. Figure 210-47

**(B) Small-Appliance Circuit – Dwelling Unit.**

**(1) Receptacle Outlets.** Receptacle outlets in the kitchen, pantry, breakfast room, and dining room area of a dwelling unit, including the receptacle outlet for refrigeration equipment, shall be supplied by a 20A, 1Ø, 120V small-appliance circuit [210.11(C)(1)]. Figure 210-48

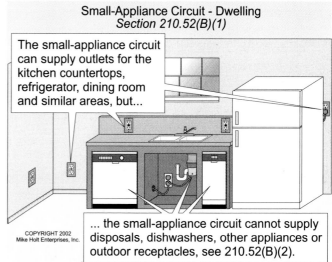

**Figure 210-48**

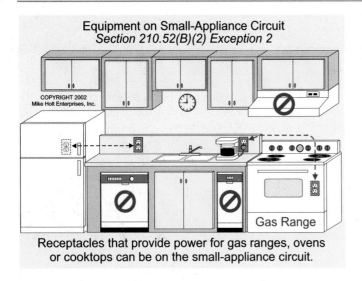

**Equipment on Small-Appliance Circuit**
*Section 210.52(B)(2) Exception 2*

Receptacles that provide power for gas ranges, ovens or cooktops can be on the small-appliance circuit.

*Figure 210-49*

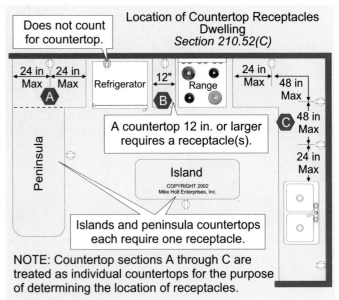

**Location of Countertop Receptacles Dwelling**
*Section 210.52(C)*

Does not count for countertop.

A countertop 12 in. or larger requires a receptacle(s).

Islands and peninsula countertops each require one receptacle.

NOTE: Countertop sections A through C are treated as individual countertops for the purpose of determining the location of receptacles.

*Figure 210-50*

**(2) Not Supply Other Outlets.** The 20A small-appliance circuits shall not supply any other outlet, including outlets for luminaires or appliances.

*Exception No. 1:* A receptacle connected to the small-appliance circuit can be used to supply an electric clock.

*Exception No. 2:* A receptacle connected to the small-appliance circuit can supply gas-fired ranges, ovens, or counter-mounted cooking units. Figure 210-49

**(3) Kitchen Countertop Receptacles.** Kitchen countertop receptacles, as required by 210.52(C), shall be supplied by at least two, 20A, 1Ø, 120V small-appliance, branch circuits [210.11(C)(1)]. Either or both of these circuits can supply receptacle outlets in the same kitchen, pantry, breakfast room, or dining room of the dwelling unit [210.52(B)(1)]. See 210.11(C)(1).

**(C) Countertop Receptacle – Dwelling Unit.** In kitchens and dining rooms of dwelling units, receptacle outlets for countertop spaces shall be installed according to (1) through (5) below. Figure 210-50

**AUTHOR'S COMMENT:** GFCI protection is required for all 15 and 20A 1Ø, 125V receptacles that supply kitchen countertop surfaces appliances [210.8(A)(6)].

**(1) Wall Counter Space.** A receptacle outlet shall be installed for each kitchen and dining area countertop wall space that is 1 ft or wider, and receptacles shall be placed so no point along the countertop wall space is more than 2 ft, measured horizontally from a receptacle outlet.

**(2) Island Countertop Space.** One receptacle outlet shall be installed at each island countertop space that has a long dimension of 2 ft or greater, and a short dimension of 1 ft or greater.

**(3) Peninsular Countertop Space.** One receptacle outlet shall be installed at each peninsular countertop that has a long dimension of 2 ft or greater, and a short dimension of 1 ft or greater, measured from the connecting edge.

**(4) Separate Spaces.** When breaks occur in countertop spaces for appliances, sinks, etc., each countertop space is considered as separate for determining receptacle placement.

**(5) Receptacle Location.** Receptacle outlets required for the countertop space shall be located above, but not more than 20 in. above, the counter surface. Receptacle outlets that are not accessible by appliances fastened in place, that are located in appliance garages, or that supply appliances that occupy dedicated space shall not be used as the required counter surface receptacles. Figure 210-51

**AUTHOR'S COMMENT:** An "appliance garage" is an enclosed area on the counter surface where an appliance can be stored and hidden from view when not in use. If a receptacle is installed inside an appliance garage, it cannot count as a kitchen counter surface outlet required for dwelling units.

Kitchen Countertop
Receptacles – Dwelling
*Section 210.52(C)(5)*

20 in.

Must be located within 20 in. above countertop.

COPYRIGHT 2002 Mike Holt Enterprises, Inc.

**Figure 210-51**

*Exception:* The receptacle outlet for the countertop space can be installed below the countertop where necessary for the physically impaired or where there is no wall space above an island or peninsular counter. The receptacle shall be not more than 1 ft below the countertop surface and no more than 6 in., measured horizontally, from the counter's edge. Figure 210-52

**(D)** **Bathrooms – Dwelling Unit.** In dwelling units, at least one 15A or 20A, 1Ø, 125V receptacle outlet shall be installed within 3 ft of the outside edge of each bathroom basin. The receptacle outlet shall be located on a wall or partition that is next to the basin or basin counter surface. See 210.11(C)(3). Figure 210-53

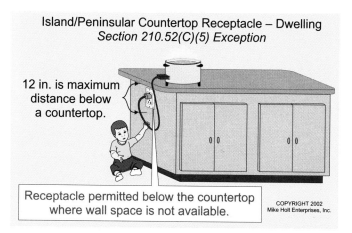

Island/Peninsular Countertop Receptacle – Dwelling
*Section 210.52(C)(5) Exception*

12 in. is maximum distance below a countertop.

Receptacle permitted below the countertop where wall space is not available.

COPYRIGHT 2002 Mike Holt Enterprises, Inc.

**Figure 210-52**

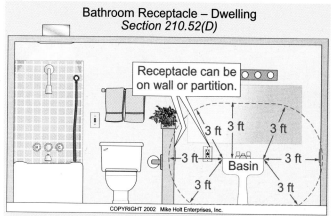

Bathroom Receptacle – Dwelling
*Section 210.52(D)*

Receptacle can be on wall or partition.

3 ft  3 ft  3 ft

3 ft                3 ft

Basin

3 ft  3 ft

COPYRIGHT 2002 Mike Holt Enterprises, Inc.

A 15 or 20A, 125V single-phase receptacle outlet must be installed within 3 ft of the outside edge of each basin.

**Figure 210-53**

**AUTHOR'S COMMENT:** One receptacle outlet could be located between two basins, if the outlet is within 3 ft of the outside edge of each basin. Figure 210-54

**(E)** **Outdoor Receptacle – Dwelling Units.**

**One-Family Dwelling Unit.** Two receptacle outlets shall be installed outdoors not more than $6^1/_2$ ft above grade, one at the front and one at the back of a one-family dwelling. These receptacles shall be GFCI-protected. [210.8(A)(3)].

**Two-Family Dwelling Unit.** A two-family dwelling shall have two receptacle outlets installed outdoors not more than $6^1/_2$ ft above grade, one at the front and one at the back of each dwelling unit at grade level. These receptacles shall be GFCI-protected. [210.8(A)(3)].

Bathroom Receptacle – Dwelling
*Section 210.52(D)*

Okay if within 3 ft of each.

Basin    Basin

COPYRIGHT 2002 Mike Holt Enterprises, Inc.

A 15 or 20A, 125V single-phase receptacle outlet must be installed within 3 ft of the outside edge of each basin.

**Figure 210-54**

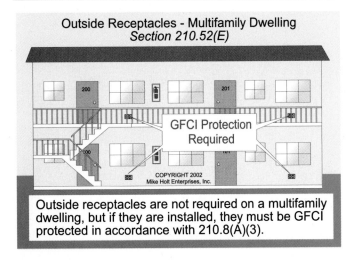

**Outside Receptacles - Multifamily Dwelling**
*Section 210.52(E)*

Outside receptacles are not required on a multifamily dwelling, but if they are installed, they must be GFCI protected in accordance with 210.8(A)(3).

**Figure 210-55**

**Basement Receptacle - Dwelling**
*Section 210.52(G)*

Each area of an unfinished basement in a one-family dwelling must have a 15 or 20A, 125V single-phase receptacle outlet. GFCI protection required, 210.8(A)(5).

**Figure 210-56**

**AUTHOR'S COMMENT:** This rule does not require a receptacle outlet to be outdoors of a dwelling unit in a multifamily dwelling (building containing three or more dwelling units), but where receptacles are installed, they shall be GFCI-protected [210.8(A)(3)]. Figure 210-55

(F) **Laundry Area Receptacle – Dwelling Unit.** Each dwelling unit shall have at least one 15A or 20A, 1Ø, 125V receptacle installed for the laundry area, supplied by the 20A, 1Ø, 125V 2laundry circuit. See 210.11(C)(2).

*Exception No. 1:* A laundry receptacle outlet is not required in a dwelling unit of a multifamily building if laundry facilities are available to all building occupants.

(G) **Basement and Garage Receptacles – Dwelling Unit.** For a one-family dwelling, at least one 15A or 20A, 1Ø, 125V receptacle outlet, in addition to any provided for laundry equipment, shall be installed in each basement and in each attached garage, and in each detached garage with electric power. See 210.8(A) and (A)(5) for GFCI protection requirements.

Where a portion of the basement is finished into one or more habitable rooms, each separate unfinished portion shall have a 15A or 20A, 1Ø, 125V receptacle outlet installed. The purpose of this rule is to prevent an extension cord from a non-GFCI-protected receptacle to supply power to loads in the unfinished portion of the basement. Figure 210-56

(H) **Hallway Receptacle – Dwelling Unit.** One 15A or 20A, 1Ø, 125V receptacle outlet shall be installed for each corridor that is at least 10 ft long, measured along the centerline of the hall, without passing through a doorway.

### 210.60 Receptacles in Guest Rooms for Hotels and Motels

(A) **General Requirements.** Receptacle outlets for guest rooms in hotels, motels, and similar occupancies shall be installed in accordance with the requirements of 210.52, such as:

**210.52(A)** Receptacle outlets shall be installed so no point along the floor line in any wall space is more than 6 ft, measured horizontally from an outlet in that space, including any wall space 2 ft or more in width.

**210.52(D)** At least one 15A or 20A, 1Ø, 125V GFCI-protected receptacle outlet shall be installed within 3 ft of the outside edge of each bathroom basin [210.8(B)(1)].

(B) **Receptacle Placement.** The number of receptacle outlets required for guest rooms shall not be less than that required for a dwelling unit, in accordance with 210.52(A).

To eliminate the need for extension cords by guests for ironing, computers, refrigerators, etc., receptacles can be located to be convenient for permanent furniture layout, but at least two receptacle outlets shall be readily accessible.

Receptacle outlets behind a bed shall be located so the bed will not make contact with an attachment plug, or the receptacle shall be provided with a suitable guard. Figure 210-57

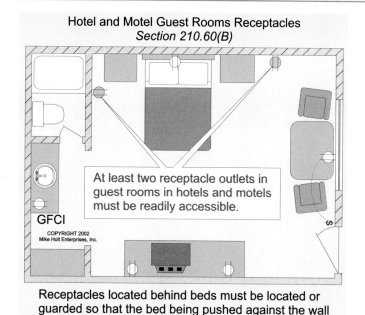

Hotel and Motel Guest Rooms Receptacles
Section 210.60(B)

At least two receptacle outlets in guest rooms in hotels and motels must be readily accessible.

GFCI

COPYRIGHT 2002
Mike Holt Enterprises, Inc.

Receptacles located behind beds must be located or guarded so that the bed being pushed against the wall will not damage the attachment plug.

**Figure 210-57**

## 210.63 Heating, Air-Conditioning, and Refrigeration Equipment.

A 15A or 20A, 1Ø, 125V receptacle outlet shall be installed at an accessible location for the servicing of heating, air-conditioning, and refrigeration equipment. The receptacle shall be located within 25 ft of, and on the same level as, the heating, air-conditioning, and refrigeration equipment. This receptacle shall not be connected to the load side of the equipment disconnecting means. Figure 210–58

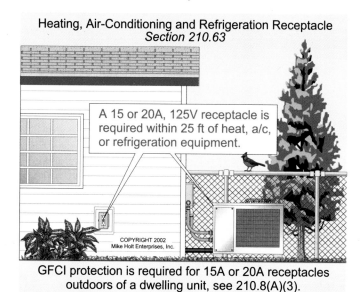

Heating, Air-Conditioning and Refrigeration Receptacle
Section 210.63

A 15 or 20A, 125V receptacle is required within 25 ft of heat, a/c, or refrigeration equipment.

COPYRIGHT 2002
Mike Holt Enterprises, Inc.

GFCI protection is required for 15A or 20A receptacles outdoors of a dwelling unit, see 210.8(A)(3).

**Figure 210-58**

**AUTHORS COMMENT:** Note that ventilation-only equipment is not included in the preceding requirement. It's HACR, not HVACR.

This receptacle shall be GFCI-protected if it is located on a rooftop [210.8(B)(2)], outdoors of a dwelling unit [210.8(A) (3)] or in the crawl space of a dwelling unit [210.8(A)(4)].

**AUTHOR'S COMMENT:** The outdoor 15A or 20A, 1Ø, 125V receptacle outlet required for dwelling units [210.52(E)] could be used to satisfy this requirement, if located within 25 ft of the HACR equipment.

## 210.70 Lighting Outlet Requirements

**(A)** **Dwelling Unit Lighting Outlet.** Figure 210-59

**(1)** **Habitable Rooms.** At least one wall switch-controlled lighting outlet shall be installed in every habitable room and bathroom of a dwelling unit. Figure 210-60

*Exception No. 1:* In other than kitchens and bathrooms, a receptacle controlled by a wall switch can be used instead of a lighting outlet.

Lighting Outlets - Dwelling
Section 210.70(A)

Bedroom 1

Closet

Dining Room

Kitchen

Hall

Bedroom 2

Bath

Living Room

Fireplace

Laundry

Garage

COPYRIGHT 2002  Mike Holt Enterprises, Inc.

A lighting outlet must be installed at all:
(1) Habitable rooms and bathrooms
(2) Hallways, stairways, attached garages and outdoor entrances
(3) Storage or equipment spaces

**Figure 210-59**

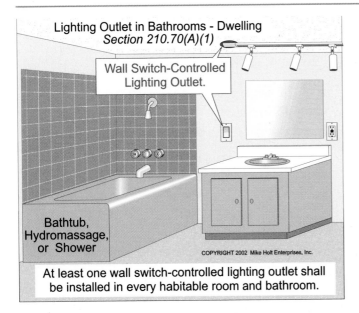

Lighting Outlet in Bathrooms - Dwelling
*Section 210.70(A)(1)*

Wall Switch-Controlled Lighting Outlet.

Bathtub, Hydromassage, or Shower

COPYRIGHT 2002 Mike Holt Enterprises, Inc.

At least one wall switch-controlled lighting outlet shall be installed in every habitable room and bathroom.

**Figure 210-60**

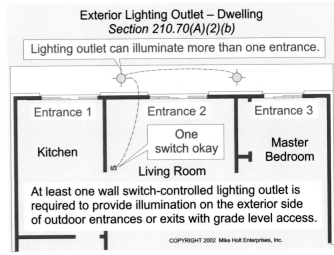

Exterior Lighting Outlet – Dwelling
*Section 210.70(A)(2)(b)*

Lighting outlet can illuminate more than one entrance.

Entrance 1    Entrance 2    Entrance 3

Kitchen    One switch okay    Master Bedroom

Living Room

At least one wall switch-controlled lighting outlet is required to provide illumination on the exterior side of outdoor entrances or exits with grade level access.

COPYRIGHT 2002 Mike Holt Enterprises, Inc.

**Figure 210-62**

*Exception No. 2:* Lighting outlets can be controlled by occupancy sensors equipped with a manual override that will allow the sensor to function as a wall switch. Figure 210-61

**(2) Other Areas.**

    **(a) Hallways, Stairways, Garages.** In dwelling units, at least one wall switch-controlled lighting outlet shall be installed in hallways, stairways, attached garages, and detached garages with electric power.

    **(b) Exterior Entrances.** At least one wall switch-controlled lighting outlet shall provide illumination on the exterior side of outdoor entrances or exits with

grade level access. A vehicle door is not considered an outdoor entrance or exit for the purpose of this rule. Figure 210-62

**AUTHOR'S COMMENT:** The Code contains the requirement for the location of the lighting outlet, but does not specify the location for the switch. Naturally, you would not want to install a switch behind a door or other inconvenient location, but the Code does not require you to relocate the switch to suit the swing of the door. When in doubt as to the best location to place a light switch, consult the job plans or ask your boss. If you're the boss and you don't know, check with the AHJ. Figure 210-63

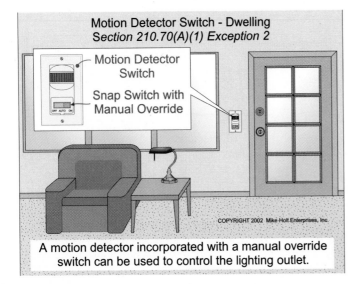

Motion Detector Switch - Dwelling
*Section 210.70(A)(1) Exception 2*

Motion Detector Switch

Snap Switch with Manual Override

COPYRIGHT 2002 Mike Holt Enterprises, Inc.

A motion detector incorporated with a manual override switch can be used to control the lighting outlet.

**Figure 210-61**

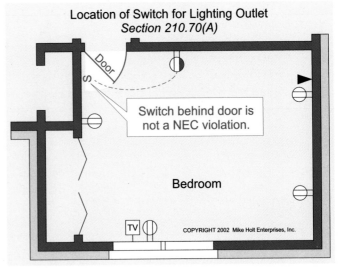

Location of Switch for Lighting Outlet
*Section 210.70(A)*

Door

Switch behind door is not a NEC violation.

Bedroom

TV

COPYRIGHT 2002 Mike Holt Enterprises, Inc.

**Figure 210-63**

Outdoor Entrance Lighting Outlet – Dwelling
*Section 210.70(A)(2) Exception*

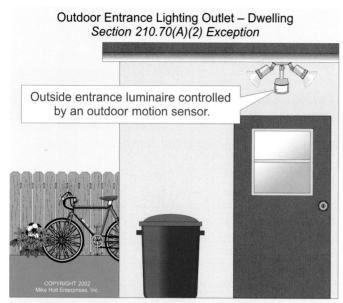

Outside entrance luminaire controlled by an outdoor motion sensor.

At outdoor entrances of a dwelling unit, remote, central or automatic control of lighting is permitted in lieu of a switch.

**Figure 210-64**

Storage and Equipment Space Lighting Outlet – Dwelling
*Section 210.70(A)(3)*

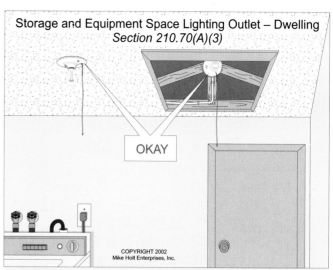

OKAY

For attics, underfloor spaces, utility rooms and basements, at least one lighting outlet containing a switch or controlled by a wall switch must be installed where these spaces are used for storage or contain equipment for servicing.

**Figure 210-65**

(c) **Stairway.** Where the stairway between floor levels has six risers or more, lighting outlets for interior stairways shall be controlled by a wall switch that is located at each floor and landing that includes an entryway.

*Exception to (a), (b), and (c):* Lighting outlets for hallways, stairways, and outdoor entrances can be switched on and off by remote, central, or automatic control. Figure 210-64

(3) **Storage and Equipment Rooms.** At least one lighting outlet containing a switch or controlled by a wall switch shall be installed in attics, underfloor spaces, utility rooms, and basements used for storage or containing equipment requiring service. The switch shall be located at the usual point of entry to these spaces, and the lighting outlet shall be located at or near the equipment requiring service. Figure 210-65

(B) **Guest Rooms.** At least one wall switch-controlled lighting outlet, or wall switch-controlled receptacle, shall be installed in guest rooms of hotels, motels, and similar occupancies.

(C) **Commercial Spaces.** At least one lighting outlet containing a switch or controlled by a wall switch shall be installed in attics and underfloor spaces if these spaces contain equipment that requires service. The switch shall be located at the usual point of entry to these spaces, and the lighting outlet shall be located at or near the equipment requiring service. In addition, a 15A or 20A, 1Ø, 125V receptacle shall be installed within 25 ft of HACR equipment. [210.63]

# Article 210

1.  When more than one nominal voltage system exists in a building, each ungrounded system conductor shall be identified by phase and system. The means of identification shall be permanently posted at each branch-circuit panelboard.
    (a) True           (b) False

2.  All 115And 20A, 1Ø, 125V receptacles on _____ of commercial occupancies must be GFCI-protected.
    (a) bathrooms     (b) rooftops     (c) kitchens     (d) all of these

3.  A dedicated 20A circuit is permitted to supply power to a dwelling unit bathroom for receptacle outlet(s) and other equipment within the same bathroom.
    (a) True           (b) False

4.  All branch circuits that supply 125V, 1Ø, 15 and 20A outlets installed in dwelling unit bedrooms shall be protected by a(n) _____ listed to provide protection of the entire branch circuit.
    (a) AFCI     (b) GFCI     (c) a and b     (d) none of these

5.  Branch-circuit conductors that supply a continuous load or any combination of continuous and noncontinuous loads shall have an ampacity of not less than 125 percent of the continuous load, plus 100 percent of the noncontinuous load.
    (a) True           (b) False

6.  Where a branch circuit supplies continuous loads or any combination of continuous and noncontinuous loads, the rating of the overcurrent device shall not be less than the noncontinuous load plus 125 percent of the continuous load.
    (a) True           (b) False

7.  If a 20A branch circuit supplies multiple 125V receptacles, the receptacles must have an ampere rating of no less than _____.
    (a) 10A     (b) 15A     (c) 20A     (d) 30A

8.  _____ in dwelling units shall supply only loads within that dwelling unit or loads associated only with that dwelling unit.
    (a) Service-entrance conductors     (b) Ground-fault protection
    (c) Branch circuits     (d) none of these

9.  In a dwelling unit, each wall space of _____ or wider requires a receptacle.
    (a) 2 ft     (b) 3 ft     (c) 4 ft     (d) 5 ft

10. Receptacle outlets shall, insofar as practicable, be spaced equal distances apart in a dwelling unit. Receptacle outlets in floors shall not be counted as part of the required number of receptacle outlets unless they are located within _____ of the wall.
    (a) 6 in.     (b) 12 in.     (c) 18 in.     (d) close to the wall

11. In dwelling units, outdoor receptacles can be connected to the 20A small-appliance branch circuit.
    (a) True           (b) False

12. A receptacle outlet shall be installed at each wall counter space that is 12 in. or wider so that no point along the wall line is more than _____, measured horizontally from a receptacle outlet in that space.
    (a) 10 in.     (b) 12 in.     (c) 16 in.     (d) 24 in.

13. One receptacle outlet shall be installed at each island or peninsular countertop having a long dimension of _____ or greater and a short dimension of _____ or greater.
    (a) 12 in., 24 in.     (b) 24 in., 12 in.     (c) 24 in., 48 in.     (d) 48 in., 24 in.

14. When breaks occur in dwelling unit kitchen-countertop spaces for ranges, refrigerators, sinks, etc., each countertop surface is considered a separate counter space for determining receptacle placement.
    (a) True           (b) False

15. Where a portion of the dwelling unit basement is finished into one or more habitable rooms, each separate unfinished portion shall have a receptacle outlet installed.
    (a) True            (b) False

16. A 15A or 20A, 1Ø, 125V receptacle outlet must be located within 25 ft of heating, air-conditioning, and refrigeration equipment for _____ occupancies.
    (a) dwelling         (b) commercial        (c) industrial        (d) all of these

17. Which rooms in a dwelling unit must have a switch-controlled lighting outlet?
    (a) Every habitable room.                (b) Bathrooms.
    (c) Hallways and stairways.             (d) all of these

18. In _____ rooms other than kitchens and bathrooms of dwelling units, one or more receptacles controlled by a wall switch shall be permitted in lieu of lighting outlets.
    (a) habitable         (b) finished        (c) all        (d) a and b

19. In a dwelling unit, illumination from a lighting outlet shall be provided at the exterior side of each outdoor entrance or exit that has grade-level access.
    (a) True            (b) False

20. For other than dwelling units, a wall-switch lighting outlet is required near equipment requiring servicing in attics or underfloor spaces, and the switch must be located at the point of entrance to the attic or underfloor space.
    (a) True            (b) False

# Article 215
# Feeders

## 215.1 Scope

Article 215 covers the installation and minimum conductor sizing requirements for feeders. Figure 215-1

## 215.2 Minimum Rating and Size

### (A) Not Over 600V

**(1) Continuous and Noncontinuous Loads.** Feeders that supply continuous loads shall be sized no less than 125 percent of the continuous loads, plus 100 percent of the noncontinuous loads based on the conductor ampacities as listed in Table 310.16, before any ampacity adjustment in accordance with the terminal temperature rating [110.14(C)].

**AUTHOR'S COMMENT:** Feeder conductors shall have sufficient ampacity, after ampacity adjustment, to carry the load; and, they shall be protected against overcurrent in accordance with 215.13 and 240.4.

**Question:** What size feeder conductor is required for a 184A continuous load if the terminals are rated for 75°C? Figure 215-2

(a) 1/0 AWG  (b) 2/0 AWG
(c) 3/0 AWG  (d) 4/0 AWG

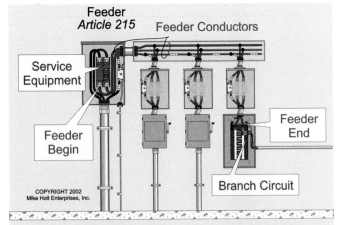

*Feeder:* The circuit conductors between the service equipment or the source of a separately derived system and the final branch-circuit overcurrent device.

**Figure 215-1**

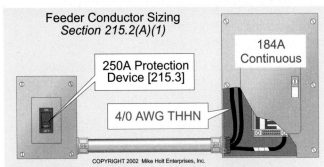

Feeder must be sized no less than 125% of the continuous loads, plus 100% of the noncontinuous loads, based on the ampacity listed in Table 310.16.

**Figure 215-2**

**Answer:** (d) 4/0 AWG

Size the conductors at 125 percent of the load [215.2(A)(1)].

184A load × 1.25 = 230A, 4/0 AWG THHN is rated 230A at 75°C, Table 310.16

**AUTHOR'S COMMENT:** A feeder protection device shall be sized no smaller than 125 percent of the continuous load, in this case 184A load x 1.25 = 230A, next size up 250A [215.3, 240.4(B) and 240.6(A)].

**(4) Dwelling Unit and Mobile Home Feeder Sizing.** Feeder conductors for individual dwelling units or mobile homes need not be larger than service-entrance conductors sized in accordance with 310.15(B)(6).

FPN No. 2: Voltage drop should be considered when sizing feeder conductors, but this is not a Code requirement. See 210.19 for examples.

**AUTHOR'S COMMENT:** For more information on this topic, visit http://www.mikeholt.com/studies/vd.htm.

## 215.3 Overcurrent Protection

The feeder overcurrent protection device shall have an ampere rating of no less than 125 percent of the continuous loads, plus 100 percent of the noncontinuous loads.

**AUTHOR'S COMMENT:** See 215.2(A)(1) for sizing feeder circuit conductors for continuous loads.

*Exception No. 1:* Where the assembly and the overcurrent protection device are both listed for 100 percent continuous load operation, the branch-circuit load can be sized at 100 percent of the continuous load.

**AUTHOR'S COMMENT:** Equipment suitable for 100 percent continuous loading is rarely available in ratings under 400A.

## 215.8 High-Leg Conductor Identification

On a 3Ø, 4-wire delta-connected system where the midpoint of one phase winding is grounded, the conductor or busbar having the higher phase voltage-to-ground shall be durably and permanently marked by an outer finish that is orange in color, or by other effective means. Such identification shall be placed at each point on the system where a connection is made if the grounded (neutral) conductor is also present. Figure 215-3

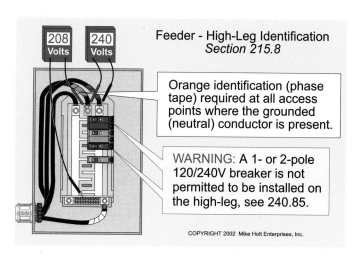

Feeder - High-Leg Identification
*Section 215.8*

208 Volts    240 Volts

Orange identification (phase tape) required at all access points where the grounded (neutral) conductor is present.

WARNING: A 1- or 2-pole 120/240V breaker is not permitted to be installed on the high-leg, see 240.85.

COPYRIGHT 2002 Mike Holt Enterprises, Inc.

**Figure 215-3**

**AUTHOR'S COMMENT:** Similar language is contained in 110.15 for branch circuits, 230.56 for services and 408.3(E) for panelboards.

*WARNING: When replacing disconnects, panelboards, meters, switches, or any equipment that contains the high-leg conductor, care shall be taken to ensure that the high-leg conductor terminates to the proper terminal. Failure to install the high-leg properly can result in 120V circuits connected to the 208V high-leg, with disastrous consequences. It's very important that you see 240.85 for the proper selection of circuit breakers installed on 3Ø, 4-wire delta-connected systems.*

High-leg conductors are required by the National Electrical Safety *Code* (utility code) to be terminated to the "C" phase, and not the "B" phase. So take the extra time when working with these types of systems to confirm just where the utility company connected the high-leg.

## 215.10 Ground-Fault Protection of Equipment

Each 480Y/277V feeder disconnect rated 1,000A or more shall be protected against ground fault in accordance with the requirements in 230.95. See 240.13.

*Exception No. 3.* Equipment ground-fault protection is not required if ground-fault protection is provided on the supply side of the feeder.

**AUTHOR'S COMMENT:** See Article 100 for the definition of *Ground-Fault Protection of Equipment.*

## Article 215

1.  Dwelling unit or mobile home feeder conductors not over 400A can be sized according to 310.15(B)(6).
    (a) True            (b) False

2.  If required by the authority having jurisdiction, a diagram showing feeder details shall be provided _____ the installation of the feeders.
    (a) after
    (c) before the final inspection of
    (b) prior to
    (d) diagrams are not required

3.  When a feeder supplies _____ in which equipment grounding conductors are required, the feeder shall include or provide a grounding means to which the equipment grounding conductors of the branch circuits shall be connected.
    (a) equipment disconnecting means
    (c) branch circuits
    (b) electrical systems
    (d) electric discharge lighting equipment

4.  Ground-fault protection is required for the feeder disconnect if _____.
    (a) the feeder is rated 1,000A or more
    (b) it is a solidly grounded wye system
    (c) more than 150 volts-to-ground, but not exceeding 600 volts phase-to-phase
    (d) all of these

5.  Ground-fault protection of equipment shall not be required if ground-fault protection of equipment is provided on the _____ side of the feeder.
    (a) load       (b) supply       (c) service       (d) none of these

# Article 220
## Branch-Circuit, Feeder, and Service Calculations

**AUTHOR'S COMMENT:** The text and graphics contained in this article of the book only cover some of the basic and important requirements of the NEC as it relates to Electrical Calculations. To properly understand how to perform load calculations, you should consider purchasing our Electrical Calculations products. Visit: www.MikeHolt.com for more information.

## Part I. General

### 220.1 Scope

This article contains the requirements necessary for sizing branch circuits, feeders, and services. In addition, this article can be used to determine the number of receptacles on a circuit, and the number of branch circuits required.

**AUTHOR'S COMMENT:** See the following Code rules for specific conductor sizing and overcurrent protection requirements:

Air-Conditioning, 440.6, 440.21, 440.22, 440.31, 440.32 and 440.62

Appliances, 422.10 and 422.11

Branch Circuits, 210.19 and 210.20(A)

Computers (Data Processing), 645.2 and 645.5(A)

Conductors, 310.15

Feeders, 215.2(A) and 215.3

Marinas, 555.12, 555.19(A)(4), and 555.19(B)

Mobile Homes and Manufactured Homes, 550.12 and 550.18

Motors, 430.6(A), 430.22(A), 430.24, 430.52 and 430.62

Overcurrent Protection Rules, 240.4 and 240.20

Refrigeration (Hermetic), 440.6 and Part IV

Services, 230.42(A) and 230.79

Signs, 600.5

Space Heating, 424.3(B)

Transformers, 450.3

### 220.2 Voltage for Calculations

**(A) Voltage Used for Calculations.** Unless other voltages are specified, branch circuit, feeder, and service loads shall be computed using the nominal system voltage such as 120, 120/240, 208Y/120, 240, 347, 480Y/277, 480, 600Y/347, and 600V. See Article 100 for the definition of Voltage, Nominal. Figure 220-1

**AUTHOR'S COMMENT:** Systems of 600Y/347V are not common in the United States, but they are often utilized in Canada.

**(B) Fractions of an Ampere (Rounding Amperes).** Calculations that result in a fraction of less than one-half of an ampere can be dropped. For example: 44.49A would rounded to 44A, and 44.50A would be rounded to 45A.

**AUTHOR'S COMMENT:** When do you round – after each calculation, or at the final calculation? The NEC is not specific on this issue, but I guess it all depends on the answer you want to see!

**Question:** According to 424.3(B) the conductors and overcurrent protection device for electric space-heating equipment shall be sized no less than 125 percent of the total load. What size conductor is required to supply a 9 kW (37.5A), 240V, 1Ø fixed space heater that has a 3A

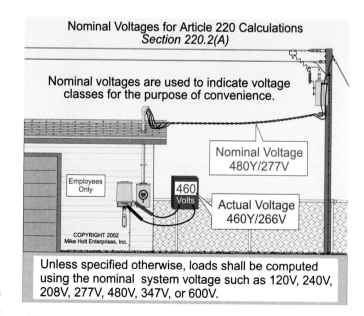

Nominal Voltages for Article 220 Calculations
*Section 220.2(A)*

Nominal voltages are used to indicate voltage classes for the purpose of convenience.

Employees Only

460 Volts

Nominal Voltage 480Y/277V

Actual Voltage 460Y/266V

COPYRIGHT 2002
Mike Holt Enterprises, Inc.

Unless specified otherwise, loads shall be computed using the nominal system voltage such as 120V, 240V, 208V, 277V, 480V, 347V, or 600V.

*Figure 220-1*

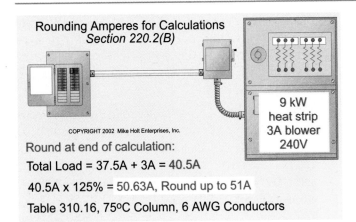

Rounding Amperes for Calculations
*Section 220.2(B)*

9 kW
heat strip
3A blower
240V

COPYRIGHT 2002 Mike Holt Enterprises, Inc.

Round at end of calculation:

Total Load = 37.5A + 3A = 40.5A

40.5A x 125% = 50.63A, Round up to 51A

Table 310.16, 75ºC Column, 6 AWG Conductors

**Figure 220-2**

blower motor if equipment terminals are rated 75°C?
Figure 220-2

(a) 10 AWG                    (b) 8 AWG
(c) 6 AWG                     (d) 4 AWG

**Answer:** (c) 6 AWG

Conductor size = 37.5 + 3A = 40.5A × 1.25 = 50.63,
round up to 51A, 6 AWG rated 65A at 75°C

## 220.3 Computation of Loads

**(A) General Lighting.** The general lighting load specified in Table 220.3(A) shall be computed from the outside dimensions of the building or area involved.

| Occupancy | VA per ft² |
|---|---|
| Armories and auditoriums | 1 |
| Assembly halls and auditoriums | 1 |
| Banks | 3[b] |
| Barber shops and beauty parlors | 3 |
| Churches | 1 |
| Clubs | 2 |
| Court rooms | 2 |
| Dwelling units | 3[a] |
| Garages — commercial (storage) | 1/2 |
| Halls, corridors, closets, stairways | 1/2 |
| Hospitals | 2 |
| Hotels and motels without cooking facilities | 2 |
| Industrial commercial (loft) buildings | 2 |
| Lodge rooms | 1 1/2 |
| Office buildings | 3 1/2[b] |
| Restaurants | 2 |
| Schools | 3 |
| Storage spaces | 1/4 |
| Stores | 3 |
| Warehouses (storage) | 1/4 |

Table 220.3(A) [a] Dwelling Units. For dwelling units, the computed floor area shall not include open porches, garages, or unfinished

spaces not adaptable for future use. Receptacle and lighting outlets are included in the 3 VA ft² general lighting and no additional load calculations shall be required, see 220.3(B)(10).

**Question:** What is the general lighting load for a 40 ft × 50 ft (2,000 ft²) dwelling unit? Figure 220-3

(a) 2,000 VA                    (b) 3,000 VA
(c) 5,000 VA                    (d) 6,000 VA

**Answer:** (d) 6,000 VA

General Lighting Load = 40 × 50 ft = 2,000 ft²×3 VA per ft² = 6,000 VA

Table 220.3(A) [b] Banks and Offices. If the number of receptacles for a bank or office is unknown, an additional 1 VA ft² shall be included for the feeder and/or service load.

**Question:** What is the general lighting and receptacle load for a 10,000 ft² office building if the number of receptacles is unknown? Figure 220-4

(a) 43,750 VA                    (b) 10,000 VA
(c) 53,750 VA                    (d) none of these

**Answer:** (c) 53,750 VA

Lighting Load = 10,000 ft² × 3.5 VA × 1.25* = 43,750 VA

Receptacle Load = 10,000 ft² × 1 VA =                     10,000 VA

Total General Lighting and Receptacle Load = 53,750 VA

*215.2(A)(1) and 230.42(A) require the conductors to be sized at 125 percent of the continuous load.

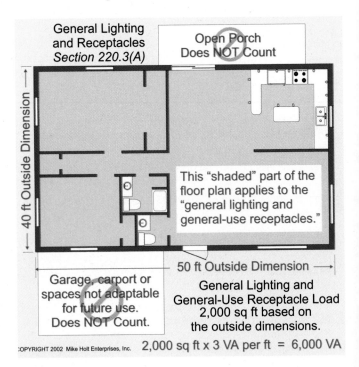

General Lighting and Receptacles
*Section 220.3(A)*

Open Porch
Does NOT Count

40 ft Outside Dimension

This "shaded" part of the floor plan applies to the "general lighting and general-use receptacles."

Garage, carport or spaces not adaptable for future use. Does NOT Count.

50 ft Outside Dimension

General Lighting and General-Use Receptacle Load 2,000 sq ft based on the outside dimensions.

COPYRIGHT 2002 Mike Holt Enterprises, Inc.     2,000 sq ft x 3 VA per ft = 6,000 VA

**Figure 220-3**

Bank - General Lighting and Receptacle Demand Load
*Section 220.3(A) Note*[b]

Bank - 10,000 sq ft
Number of Receptacles Unknown

Determine demand load for lighting and receptacles.

Table 220.3(A), Lighting load for a bank is 3½ VA per sq ft

10,000 sq ft x 3½ VA per sq ft = 35,000 VA lighting load

35,000 VA lighting load x 1.25 = 43,750 VA lighting demand load, but Table 220.3(A) Note "b" requires an additional unit load of 1 VA per sq ft when the number of receptacles is unknown.

43,750 VA lighting + 10,000 VA receptacle = 53,750 VA demand load

**Figure 220-4**

**AUTHOR'S COMMENT:** For commercial and industrial occupancies, the lighting load shall be assumed to be continuous and the total value shall be increased 25 percent; see 210.19(A)(1) for branch circuits, 215.2(A)(1) for feeders and 230.42(A) for services. Figure 220-5

**(B) Other Loads — All Occupancies.** The minimum feeder/service VA load for each outlet shall not be less than that computed in 220.3(B)(1) through (11).

**(1) Specific Appliances or Loads.** The VA load for an outlet not covered in (2) through (11) shall be computed based on the ampere and nominal voltage rating of the equipment.

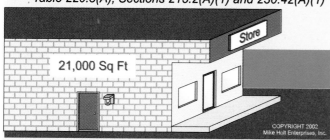

General Lighting Demand at 125%
*Table 220.3(A), Sections 215.2(A)(1) and 230.42(A)(1)*

21,000 Sq Ft

Store

Determine the general lighting demand load.

Table 220.3(A), store lighting is 3 VA per sq ft.
21,000 sq ft x 3 VA per sq ft = 63,000 VA lighting load
Store lighting is a continuous load at 125%
63,000 VA lighting load x 1.25 = 78,750 VA demand load

**Figure 220-5**

**(2) Electric Dryers and Household Electric Cooking Appliances.** The net computed load shall be in accordance with 220.18 for household electric dryers, and 220.19 for household electric ranges and other cooking appliances.

**(3) Motor Loads.** The net computed load for motors shall be in accordance with 430.22 or 430.24, see 220.4(A) and 220.14.

**AUTHOR'S COMMENT:** The net computed load for air-conditioning and refrigeration equipment shall be in accordance with 440.32, 440.33, and 440.34.

**(4) Recessed Luminaires.** The net computed load for recessed luminaire(s) shall be computed based on the maximum VA rating for which the luminaire is rated.

**AUTHOR'S COMMENT:** This value shall be increased by 25 percent for continuous loads; see 210.19(A)(1) for branch circuits, 215.2(A)(1) for feeders and 230.42(A) for services.

**(5) Heavy-Duty Lampholders.** The VA load for heavy-duty lampholders shall be computed at a minimum of 600 VA.

**AUTHOR'S COMMENT:** This value shall be increased by 25 percent for continuous loads; see 210.19(A)(1) for branch circuits, 215.2(A)(1) for feeders and 230.42(A) for services.

**(6) Sign and Outline Lighting.** Each commercial building and each commercial occupancy accessible to pedestrians shall have at least one outlet for a sign or outline lighting located at an accessible location outside the entrance to each tenant space. The outlet(s) shall be supplied by a 20A branch circuit that supplies no other load [600.5(A)]. The VA load for the sign outlet shall be no less than 1,200 VA. Figure 220-6

**AUTHOR'S COMMENT:** This value shall be increased by 25 percent for continuous loads; see 210.19(A)(1) for branch circuits, 215.2(A)(1) for feeders and 230.42(A) for services.

**(7) Show Windows.** The VA load for show-window lighting:

(1) 180 VA per show-window receptacle outlet in accordance with 220.3(B)(9), or

(2) 200 VA per linear foot of show-window lighting, see 220.12. Figure 220-7

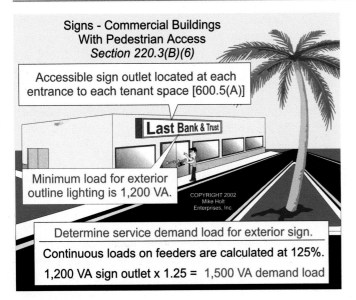

Figure 220-6

**AUTHOR'S COMMENT:** This value shall be increased by 25 percent for continuous loads; see 210.19(A)(1) for branch circuits, 215.2(A)(1) for feeders and 230.42(A) for services.

(8) **Fixed Multioutlet Assemblies.** Fixed multioutlet assemblies in commercial occupancies shall be:

   (1) Where appliances are unlikely to be used simultaneously, each 5 ft, or fraction of 5 ft, of multioutlet assembly shall be 180 VA. Figure 220-8

   (2) Where appliances are likely to be used simultaneously, each 1 ft or fraction of a foot of multioutlet assembly shall be 180VA.

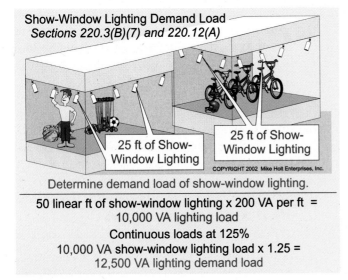

Figure 220-7

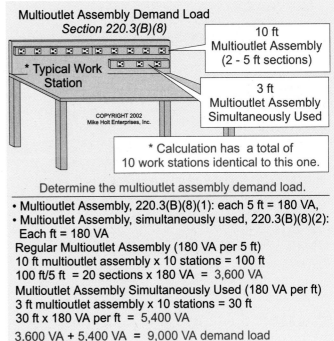

Figure 220-8

(9) **Commercial Receptacle Load.** Each 15A or 20A, 1Ø, 125V general-use receptacle outlet shall be considered to have a load rating of 180 VA per mounting strap. Figure 220-9

A single piece of equipment consisting of a multiple receptacle comprised of four or more receptacles shall be computed at not less than 90 VA per receptacle. Figure 220-10

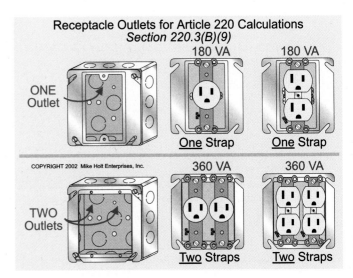

Figure 220-9

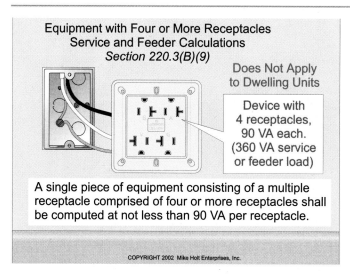

**Equipment with Four or More Receptacles Service and Feeder Calculations**
*Section 220.3(B)(9)*

**Does Not Apply to Dwelling Units**

Device with 4 receptacles, 90 VA each. (360 VA service or feeder load)

A single piece of equipment consisting of a multiple receptacle comprised of four or more receptacles shall be computed at not less than 90 VA per receptacle.

COPYRIGHT 2002 Mike Holt Enterprises, Inc.

**Figure 220-10**

**Question:** What is the maximum number of 15A , 1Ø, 125V receptacle outlets permitted on a 15A, 1Ø, 120V circuit in a commercial occupancy? Figure 220-11

(a) 4             (b) 6
(c) 10            (d) 13

**Answer:** (c) 10

Step 1. Determine the VA rating for the 120V, 15A circuit.

Circuit VA = 120V × 15A = 1,800 VA

Step 2. Determine the number of receptacles permitted on the circuit.

Number of Receptacles = 1,800 VA/180 VA* = 10

*It is a generally accepted practice that receptacles are not considered a continuous load.

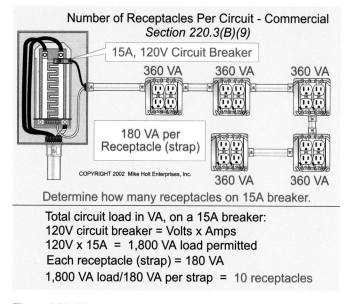

**Number of Receptacles Per Circuit - Commercial**
*Section 220.3(B)(9)*

15A, 120V Circuit Breaker

360 VA    360 VA    360 VA

180 VA per Receptacle (strap)

COPYRIGHT 2002 Mike Holt Enterprises, Inc.

360 VA    360 VA

Determine how many receptacles on 15A breaker.

Total circuit load in VA, on a 15A breaker:
120V circuit breaker = Volts x Amps
120V x 15A = 1,800 VA load permitted
Each receptacle (strap) = 180 VA
1,800 VA load/180 VA per strap = 10 receptacles

**Figure 220-11**

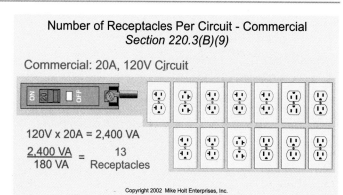

**Number of Receptacles Per Circuit - Commercial**
*Section 220.3(B)(9)*

Commercial: 20A, 120V Circuit

120V x 20A = 2,400 VA

$\frac{2,400\ VA}{180\ VA} = 13$ Receptacles

Copyright 2002 Mike Holt Enterprises, Inc.

**Figure 220-12**

**Question:** What is the maximum number of 15A or 20A, 1Ø, 125V receptacle outlets permitted on a 20A, 120V circuit in a commercial occupancy? Figure 220-12

(a) 4             (b) 6
(c) 8             (d) 13

**Answer:** (d) 13

Step 1. Determine the VA rating for the 120V, 20A circuit.

Circuit VA = 120V × 20A = 2,400 VA

Step 2. Determine the number of receptacles permitted on the circuit.

Number of Receptacles = 2,400 VA/180 VA* = 13

*It is a generally accepted practice that receptacles are not considered a continuous load.

**(10) Residential Receptacle Load.** There is no VA load rating for 15A and 20A, 125V general-use dwelling unit receptacle and lighting outlets, because the load for these devices are part of the general lighting (3 VA per ft$^2$) load [Table 220.3(A)].

**Question:** What is the maximum number of 15A or 20A, 1Ø, 125V receptacle and lighting outlets permitted on a 20A, 120V circuit in a dwelling unit?

(a) 4             (b) 6
(c) 8             (d) No limit

**Answer:** (d) No limit. The *NEC* does not specify the maximum number of 15A or 20A general-use receptacle outlets permitted on a circuit in a dwelling unit. However, the *NEC* Handbook (published by the NFPA) clarifies that there is no limit as to the number of receptacle outlets on a dwelling unit circuit, because residential receptacles are lightly used. Figure 220-13

*CAUTION: Check with the AHJ to find out if there is any local requirement that places a limitation on the number of receptacles and lighting outlets on a circuit.*

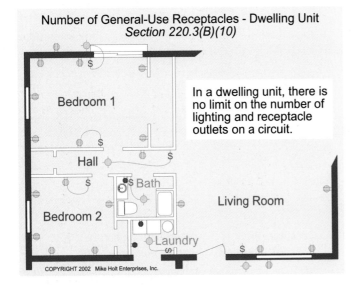

Number of General-Use Receptacles - Dwelling Unit
Section 220.3(B)(10)

In a dwelling unit, there is no limit on the number of lighting and receptacle outlets on a circuit.

**Figure 220-13**

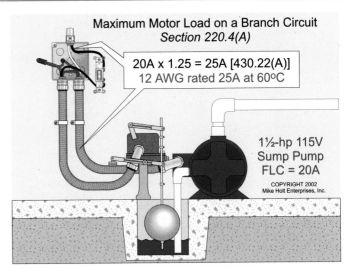

Maximum Motor Load on a Branch Circuit
Section 220.4(A)

20A x 1.25 = 25A [430.22(A)]
12 AWG rated 25A at 60ºC

1½-hp 115V
Sump Pump
FLC = 20A

COPYRIGHT 2002
Mike Holt Enterprises, Inc.

**Figure 220-14**

---

**AUTHOR'S COMMENT:** Though there is no limit on the number of lighting and/or receptacle outlets on dwelling general lighting branch circuits, the NEC does require a minimum number of circuits to be installed for general purpose receptacles and lighting outlets [210.11(A)]. In addition, the receptacle and lighting loads shall be evenly distributed among the required circuits [210.11(B)].

---

**(11) Other Outlets.** Receptacle and lighting outlets not covered in (1) through (10) shall have a net computed load based on 180 VA per outlet.

---

**AUTHOR'S COMMENT:** This value shall be increased by 25 percent for continuous loads; see 210.19(A)(1) for branch circuits, 215.2(A)(1) for feeders and 230.42(A) for services.

---

## 220.4 Maximum Load on a Branch Circuit

**(A) Motors, Air Conditioners and Refrigeration Equipment.** Branch circuits supplying motors, air-conditioning, or refrigeration equipment shall be sized no smaller than 125 percent of the current rating of the motor in accordance with 430.6(A) and 430.22(A) for motors, and 440.32 for air-conditioning and refrigeration equipment.

**Question:** What is the minimum size conductor (terminals rated 75°C) for a $1^1/_2$-hp, 115V motor? Figure 220-14
(a) 14 AWG           (b) 12 AWG
(c) 10 AWG           (d) 8 AWG

**Answer:** (b) 12 AWG

Step 1.  Determine motor full-load current [Table 430.148].
$1^1/_2$-hp FLC = 20A

Step 2. Size the branch-circuit conductors in accordance with Table 310.16 [430.22(A)].
Branch-Circuit Conductors = 20A × 1.25 = 25A, 12 AWG THHN is rated 25A at 75°C.

**(B) Inductive Lighting Loads.** Branch circuits that supply inductive luminaires (fluorescent and HID) shall have the conductors sized to the ampere rating of the ballast, not to the wattage of the lamps. Figure 220-15

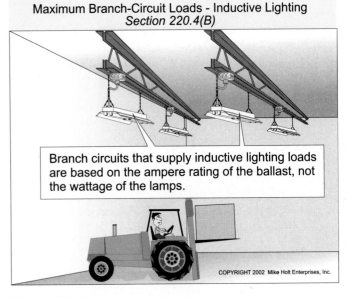

Maximum Branch-Circuit Loads - Inductive Lighting
Section 220.4(B)

Branch circuits that supply inductive lighting loads are based on the ampere rating of the ballast, not the wattage of the lamps.

COPYRIGHT 2002  Mike Holt Enterprises, Inc.

**Figure 220-15**

**Question:** What is the maximum number of 4-lamp, 34W fluorescent luminaires (each rated 1.34A) permitted on a 20A circuit, if they are on for 3 hours or more?

(a) 8        (b) 11
(c) 13       (d) 15

**Answer:** (b) 11

Step 1. Determine the maximum continuous load permitted on a 20A circuit, [210.19(A)(1)]. The circuit protection device and conductors shall be sized no smaller than 125% of the continuous load. Another way of saying this is that the maximum continuous load on a circuit protection device is limited to 80% of the circuit rating. Maximum Load = 20A × 0.80 = 16A

Step 2. Determine the maximum number of luminaires on a 20A circuit.

Maximum Number of Luminaires on the Circuit = 16A/1.34A = 11.94 or 11 luminaires

**AUTHOR'S COMMENT:** Inductive luminaires (fluorescent, high-pressure sodium, etc.) have a ballast which causes the voltage and the current of the circuit to be out-of-phase with each other (power factor). As a result, the input, 162 VA to the ballast (120V × 1.34A), will be greater than the 136W output (34W × 4 lamps) of the lamps. I know this is getting complicated, but just remember: size all circuits supplying inductive loads to the ampere rating of the load, not the wattage of the load.

**(C) Household Cooking Appliances.** Branch-circuit conductors for household cooking appliances can be sized in accordance with the requirements of Table 220.19, specifically Note 4.

**AUTHOR'S COMMENT:** For ranges of $8^3/_4$ kW or more rating, the minimum branch-circuit rating shall be 40 amperes. See 210.19(A)(3).

## Part II. Feeder and Service Calculations

### 220.10 General

The computed load of a feeder or service shall not be less than the sum of the loads on the branch circuits supplied, as determined by Part I of this article, after applying demand factors contained in Parts II, III, or IV.

FPN: See Examples D1(A) through D10 in Annex D.

### 220.11 General Lighting Demand Factors

The *Code* recognizes that not all luminaires will be on at the same time and it permits the following demand factors to be applied:

| Type of Occupancy | Lighting VA Load | Demand Factor (Percent) |
|---|---|---|
| Dwelling units[a] | First 3,000 VA | 100 |
| | Next 117,000 VA | 35 |
| | Remainder | 25 |
| Hotels/motels without provision for cooking | First 20,000 VA | 50 |
| | Remainder | 30 |
| Warehouses (storage) | First 12,500 VA | 100 |
| | Remainder | 50 |
| All others | Total VA | 100[b] |

[a] The dwelling unit demand factors apply to the general lighting load, 3 VA per ft² [Table 220.3(A)], the 3,000 VA for the two small-appliance circuits [210.11(C)(1), 220.16(A)], and the 1,500 VA for the laundry circuit [210.11(C)(2), 220.16(B)]. Figure 220-16
[b] The general lighting demand factor shall be increased by 25 percent for continuous loads. See 210.19(A)(1) for branch circuits, 215.2(A)(1) for feeders and 230.42(A) for services.

**AUTHOR'S COMMENT:** For commercial occupancies, the VA load for receptacles [220.3(B)(9)] and fixed multioutlet assemblies [220.3(B)(8)] can be added to the general lighting load and subjected to the demand factors. See 220.13.

**AUTHOR'S COMMENT:** See 210.11(A) in this book for an example on how to calculate the number of general lighting circuits required for a dwelling unit.

General Lighting, Small Appliance and Laundry Demand
*Table 220.11*

*Figure 220-16*

## 220.12 Commercial - Show Window and Track Lighting Load

**(A) Show Windows.** For show-window lighting, a load of not less than 200 VA per linear foot shall be included for a show window, measured horizontally along its base. See 220.3(B)(7)(2).

**(B) Track Lighting.** The feeder/service net computed load for track lighting in a commercial occupancy shall be 150 VA for every 2 ft of track lighting or fraction thereof. Figure 220-17

**AUTHOR'S COMMENT:** This value shall be increased by 25 percent for continuous loads; see 215.2(A)(1) for feeders and 230.42(A) for services.

**Question:** The approximate feeder/service net computed load for 150 ft of track lighting in a commercial occupancy is _____.

(a) 10,000 VA            (b) 12,000 VA
(c) 14,063 VA            (d) 16,000 VA

**Answer:** (c) 14,063 VA

150 ft/2 ft = 75 units × 150 VA × 1.25 = 14,063 VA

**AUTHOR'S COMMENT:** This rule does not apply to branch circuits, therefore the maximum number of lampholders permitted on a track lighting system is based on the wattage rating of the lamps, and the voltage and ampere rating of the circuit.

**Question:** How many 75 W lampholders can be installed on a 20A, 120V track lighting system in a commercial occupancy?

(a) 10            (b) 15
(c) 20            (d) 25

**Answer:** (d) 25

Step 1. Determine the maximum continuous VA load permitted on a 20A circuit, see 210.19(A)(1).
Maximum Load =
20A × 0.80 = 16A × 120V = 1,920 VA

Step 2. Determine the maximum number of lampholders on a 20A circuit.
Maximum Number of Lampholders on Circuit =
1,920 VA/75 VA = 25.6, or 25 lampholders

## 220.13 Commercial - Receptacle Load

The net computed load for receptacles [220.3(B)(9)] and fixed multioutlet assemblies [220.3(B)(8)] shall be determined by:

Adding the receptacle and fixed multioutlet assembly load to the general lighting load [Table 220.3(A)] and adjusting this total by the demand factors contained in Table 220.11, or

Applying a 50 percent demand factor to that portion of the receptacle and fixed multioutlet receptacle load in excess of 10 kVA.

**Question:** What is the net computed load for 150 receptacles installed in a commercial occupancy? Figure 220-18

(a) 8,500 VA            (b) 10,000 VA
(c) 18,500 VA            (d) 27,000 VA

**Answer:** (c) 18,500 VA

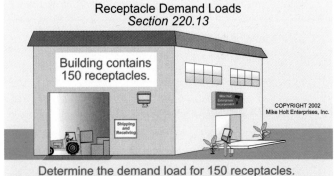

Receptacle Demand Loads
Section 220.13

Building contains 150 receptacles.

COPYRIGHT 2002 Mike Holt Enterprises, Inc.

Determine the demand load for 150 receptacles.

220.3(B)(9), each receptacle = 180 VA
Table 220.13: First 10 kVA at 100%, remainder at 50%
150 receptacles x 180 VA = 27,000 VA
1st 10,000 VA at 100%   = -10,000 VA x 1.00 = 10,000 VA
Remainder at 50%   =      17,000 VA x 0.50 = +8,500 VA
Receptacle Demand Load =                    18,500 VA

**Figure 220-18**

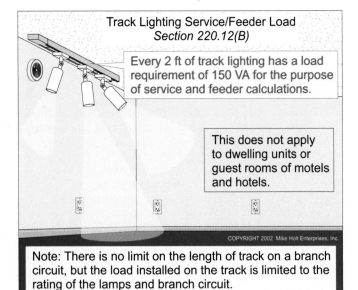

Track Lighting Service/Feeder Load
Section 220.12(B)

Every 2 ft of track lighting has a load requirement of 150 VA for the purpose of service and feeder calculations.

This does not apply to dwelling units or guest rooms of motels and hotels.

COPYRIGHT 2002 Mike Holt Enterprises, Inc.

Note: There is no limit on the length of track on a branch circuit, but the load installed on the track is limited to the rating of the lamps and branch circuit.

**Figure 220-17**

Step 1. Determine the total connected receptacle load at 180 VA per receptacle.

Total Connected Load = 150 receptacles × 180 VA = 27,000 VA

Step 2. Determine the net computed load for the 150 receptacles.

| | | |
|---|---|---|
| Total Connected Load | 27,000 VA | |
| First 10,000 VA at 100% | - 10,000 VA @ 100% | = 10,000 VA |
| Remainder at 50% | 17,000 VA @ 50% | = 8,500 VA |
| Total Receptacle Demand Load | | 18,500 VA |

## 220.14 Motor Load

The feeder/service conductor for motors shall be sized no smaller than 125 percent of the largest motor FLC, plus the sum of the other motor FLC's. See 430.24.

## 220.15 Fixed Electric Space Heating Load

The feeder/service load for fixed electric space heating equipment shall be computed at 100 percent of the total connected load.

## 220.16 Dwelling Unit - Small-appliance and Laundry Load

**(A) Small-appliance Circuit Load.** Each dwelling unit shall have a minimum of two, 20A, 1Ø, 120V small-appliance branch circuits for the kitchen and dining room receptacles as required by 210.52(B)(1) [210.11(C)(1)]. The feeder/service load for each small-appliance circuit shall be 1,500 VA, and this load can be subjected to the general lighting demand factors contained in Table 220.11.

> **AUTHOR'S COMMENT:** Receptacles rated 15A or 20A, 125V can be installed on the 20A small-appliance branch circuit. See Table 210.21(B)(3).

**(B) Laundry Circuit Load.** Each dwelling unit shall have a 20A laundry circuit to supply power for the laundry room receptacles as required by 210.52(F) [210.11(C)(2)]. The feeder/service load for the laundry circuit shall be 1,500 VA and this load can be subjected to the general lighting demand factors contained in Table 220.11.

**Question:** What is the general lighting and receptacle net computed load for a 2,700 ft² dwelling unit, including the small-appliance and laundry circuits? Figure 220-19

(a) 1,500 VA          (b) 3,000 VA
(c) 4,500 VA          (d) 6,360 VA

**Answer:** (d) 6,360 VA

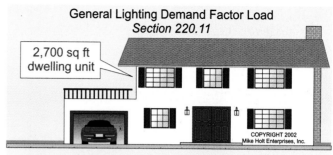

General Lighting Demand Factor Load
Section 220.11

2,700 sq ft dwelling unit

COPYRIGHT 2002
Mike Holt Enterprises, Inc.

Step 1. Determine the total connected load
General lighting   2,700 sq ft x 3 VA =        8,100 VA
Small-appliance   1,500 VA x 2 circuits =   3,000 VA
Laundry circuit    1,500 VA x 1 circuit  =   1,500 VA
Total connected load                             12,600 VA

Step 2. Determine the general lighting demand load
Total connected load   12,600 VA
1st 3,000 VA at 100%   - 3,000 VA at 100%   3,000 VA
                            9,600 VA at 35%     3,360 VA
Total demand load                                6,360 VA

**Figure 220-19**

Step 1. Determine the Total Connected Load

| | | |
|---|---|---|
| General lighting | 2,700 ft² x 3 VA per ft² | 8,100 VA |
| Small-appliance | 1,500 VA x 2 circuits | 3,000 VA |
| Laundry circuit | 1,500 VA x 1 circuit | 1,500 VA |
| Total Connected Load | | 12,600 VA |

Step 2. Determine the General Lighting Demand Load

| | | |
|---|---|---|
| Total Connected Load | 12,600 VA | |
| First 3,000 VA at 100% | - 3,000 VA at 100% | 3,000 VA |
| Remainder at 35% | 9,600 VA at 35% | 3,360 VA |
| Total Demand Load | | 6,360 VA |

> **AUTHOR'S COMMENT:** A laundry circuit is not required in a dwelling unit of a multifamily building, if laundry facilities are provided on the premises for all building occupants. See 210.52(F) Ex. 1.

**Question:** What is the general lighting and receptacle net computed load for a 12-unit multifamily dwelling if each dwelling is 2,700 ft², and if laundry facilities are provided on the premises for all building occupants? Figure 220-20

(a) 11,500 VA          (b) 33,000 VA
(c) 41,500 VA          (d) 47,250 VA

**Answer:** (d) 47,250 VA

Step 1. Determine the Total Connected Load

| | | |
|---|---|---|
| General lighting | 2,700 ft² x 3 VA per ft² | 8,100 VA |
| Small-appliance | 1,500 VA x 2 circuits | 3,000 VA |
| Laundry circuit | | 0,000 VA * |
| Total Connected Load | 11,100 VA x 12 units = | 133,200 VA |

* see 210.52(F) Ex. 1

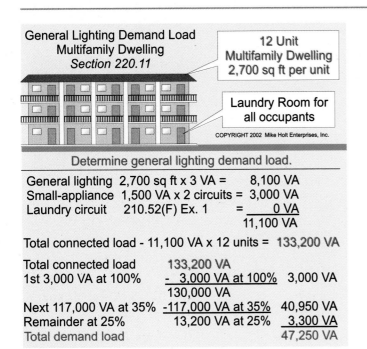

General Lighting Demand Load
Multifamily Dwelling
*Section 220.11*

12 Unit
Multifamily Dwelling
2,700 sq ft per unit

Laundry Room for
all occupants

COPYRIGHT 2002 Mike Holt Enterprises, Inc.

Determine general lighting demand load.

General lighting  2,700 sq ft x 3 VA =      8,100 VA
Small-appliance  1,500 VA x 2 circuits =  3,000 VA
Laundry circuit    210.52(F) Ex. 1       =      0 VA
                                                       11,100 VA

Total connected load - 11,100 VA x 12 units =  133,200 VA

Total connected load       133,200 VA
1st 3,000 VA at 100%      -  3,000 VA at 100%   3,000 VA
                                    130,000 VA
Next 117,000 VA at 35%  -117,000 VA at 35%   40,950 VA
Remainder at 25%            13,200 VA at 25%     3,300 VA
Total demand load                                    47,250 VA

**Figure 220-20**

Step 2. Determine the General Lighting Demand Load

Total Connected Load    133,200 VA
First 3,000 VA at 100% -   3,000 VA at 100% 3,000 VA
                                    130,200 VA
Next 117,000 VA at 35%  117,000 VA at 35% 40,950 VA
Remainder at 25%            13,200 VA at 25%  3,300 VA
Total Demand Load                                 47,250 VA

## 220.17 Dwelling Unit - Appliance Load

A demand factor of 75 percent can be applied to the total connected load of four or more appliances to determine the appliance feeder/service net computed load. This demand factor does not apply to electric space-heating equipment [220.15], electric clothes dryers [220.18], electric ranges [220.19], electric air-conditioning equipment [Article 440, Part IV], or motors [220.14].

**Question:** What is the feeder/service appliance net computed load for a dwelling unit that contains a 1,000 VA disposal, a 1,500 VA dishwasher, and a 4,500 VA water heater? Figure 220-21

(a) 3,000 VA                    (b) 4,500 VA
(c) 6,000 VA                    (d) 7,000 VA

**Answer:** (d) 7,000 VA. No demand factor for three appliances.

**Question:** What is the feeder/service appliance net computed load for a 12-unit multifamily dwelling if each unit contained a 1,000 VA disposal, a 1,500 VA dishwasher and a 4,500 VA water heater?

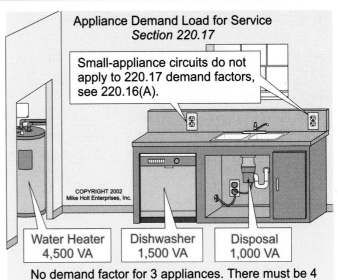

Appliance Demand Load for Service
*Section 220.17*

Small-appliance circuits do not apply to 220.17 demand factors, see 220.16(A).

COPYRIGHT 2002
Mike Holt Enterprises, Inc.

Water Heater        Dishwasher        Disposal
4,500 VA            1,500 VA          1,000 VA

No demand factor for 3 appliances. There must be 4 or more appliances to apply 220.17 demand factors.

**Figure 220-21**

(a) 23,000 VA                  (b) 43,500 VA
(c) 63,000 VA                  (d) 71,000 VA

**Answer:** (c) 63,000 VA

Demand Load = 7,000 VA × 12 units × 0.75 = 63,000 VA

## 220.18 Dwelling Unit - Electric Clothes Dryer Load

The feeder/service load for electric clothes dryers located in a dwelling unit shall not be less than 5,000W, or the nameplate rating if greater than 5,000W. When a building contains five or more dryers, it shall be permissible to apply the demand factors listed in Table 220.18 to the total connected dryer load. Figure 220-22

**AUTHOR'S COMMENT:** A clothes dryer load is not required if the dwelling unit does not have an electric clothes dryer circuit.

**Question:** What is the feeder/service net computed load for a 10-unit multifamily building that contains a 5kW dryer in each unit?

(a) 25,000W                    (b) 43,500W
(c) 63,000W                    (d) 71,000W

**Answer:** (a) 25,000W

Demand Load = 10 units × 5,000W × 0.50 = 25,000W

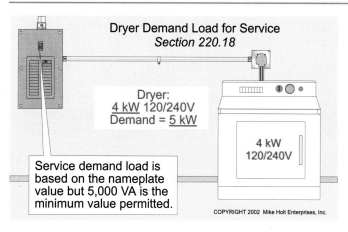

**Dryer Demand Load for Service**
*Section 220.18*

Dryer:
4 kW 120/240V
Demand = 5 kW

4 kW
120/240V

Service demand load is based on the nameplate value but 5,000 VA is the minimum value permitted.

COPYRIGHT 2002 Mike Holt Enterprises, Inc.

**Figure 220-22**

## 220.19 Dwelling Unit - Electric Ranges and Cooking Appliance Load

Household cooking appliances rated over $1^3/_4$ kW can have the feeder and service loads calculated according to the demand factors of Table 220.19. Table 220.19 Note 4 is used to calculate the branch circuit load for household cooking appliances, such as electric ranges, counter-mounted cooktops, and ovens. See 220.20 for commercial cooking equipment.

### Table 220.19 Column A (Less than $3^1/_2$ kW rating)

**Question:** According to Table 220.19 Column A, what is the net computed service or feeder load for ten 3 kW cooktops?

(a) 13,000W                (b) 14,700W
(c) 16,000W                (d) 17,000W

**Answer:** (b) 14,700W

Demand Load = 3 kW × 10 units × 0.49 = 14.7 kW

### Table 220.19 Column B ($3^1/_2$ kW to $8^3/_4$ kW Rating)

**Question:** According to Table 220.19 Column B, what is the net computed service or feeder load for ten 8 kW ranges?

(a) 27,200W                (b) 28,400W
(c) 29,200W                (d) 31,000W

**Answer:** (a) 27,200W

Demand Load = 8 kW × 10 units × 0.34 = 27.2 kW

### Table 220.19 Column A and B, Note 3

When the rating of the cooking appliances falls under both Column A and B, the demand factors for each column shall be applied to the appliances for that column, and the results added together.

**Question:** According to Table 220.19, what is the net computed load for ten 3kW ovens and ten 6kW cooktops?

(a) 41,900W                (b) 58,400W
(c) 59,200W                (d) 61,000W

**Answer:** (a) 41,900W

Demand Load = 3 kW × 10 units × 0.49 = 14.7 kW

Demand Load = 8 kW × 10 units × 0.34 = 27.2 kW

Total Demand Load =                           41.9 kW

**AUTHOR'S COMMENT:** If you check Column C (12 kW ranges), the maximum demand load for 20 units would be 35 kW.

### Table 220.19 Column C (Not over 12 kW Rating)

**Question:** According to Table 220.19 Column C, what is the net computed load for nine 12 kW ranges?

(a) 13,000W                (b) 14,700W
(c) 16,000W                (d) 24,000W

**Answer:** (d) 24,000W

**AUTHOR'S COMMENT:** Column C lists maximum demands (KW); Columns A and B list demand factors (percentage).

### Table 220.19 Note 1

For identically sized ranges individually rated more than 12 kW, the maximum demand in Column C shall be increased 5 percent for each additional kilowatt of rating or major fraction thereof by which the rating of individual ranges exceeds 12 kW.

**Question:** According to Table 220.19 Note 1, what is the net computed load for nine 16 kW ranges?

(a) 15,000W                (b) 20,000W
(c) 26,000W                (d) 28,800W

**Answer:** (d) 28,800W

Step 1. Determine Column C net computed load for nine 12 kW ranges. (24 kW)

Step 2. Increase Column C net computed load 5 percent for each kW over 12 kW by which the rating of individual ranges exceeds 12 kW.

Total Demand Load = 24 kW × 1.20 (16 kW – 12 kW = 4 × 5%) = 28.8 kW

## Table 220.19 Note 2

For ranges individually rated more than $8^3/_4$ kW, but none exceeding 27 kW, and of different ratings, an average rating shall be computed by adding together the ratings of all ranges to obtain the total connected load (using 12 kW for any range rated less than 12 kW) and dividing this total by the number of ranges. Then the maximum demand in Column C shall be increased 5 percent for each kilowatt or major fraction thereof by which this average value exceeds 12 kW.

**Question:** According to Table 220.19 Note 2, what is the net computed service or feeder load for three 14 kW ranges, three 16 kW ranges and three 18 kW ranges?

(a) 15,000W
(b) 20,000W
(c) 26,000W
(d) 28,800W

**Answer:** (d) 28,800W

Step 1. Determine Column C net computed load for nine 12 kW ranges. (24 kW)

Step 2. Determine the average rating of the ranges.

3 ranges at 14 kW =   42 kW
3 ranges at 16 kW =   48 kW
3 ranges at 18 kW =   54 kW
Total Rating          144 kW

Average Rating = 144 kW/9 ranges = 16 kW

Step 3. Increase Column C net computed load 5 percent for each kW over 12 kW by which the rating of average range exceeds 12 kW.

Total Demand Load = 24 kW × 1.20 (16 kW – 12 kW = 4 × 5%) = 28.8 kW

## Table 220.19 Note 4 Branch-Circuit Calculations

**Branch-Circuit Load for One Range** – It shall be permissible to compute the branch-circuit load for one range in accordance with Table 220.19.

**Question:** What is the branch-circuit net load in amperes for a 12 kW range that is rated 120/240V? Figure 220-23

(a) 20A
(b) 33A
(c) 41A
(d) 50A

**Answer:** (b) 33A

Step 1. Determine Column C net computed load for one 12 kW range. One unit = 8 kW

Step 2. Determine branch-circuit load in amperes.

I = P/E
P = 8,000W, E = 240V
I = 8,000W/240V = 33.33A

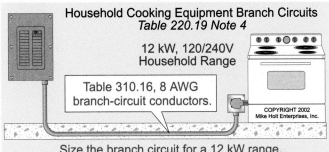

**Household Cooking Equipment Branch Circuits**
*Table 220.19 Note 4*

12 kW, 120/240V Household Range

Table 310.16, 8 AWG branch-circuit conductors.

COPYRIGHT 2002 Mike Holt Enterprises, Inc.

Size the branch circuit for a 12 kW range.

Step 1: Column C, one unit = 8 kW demand
Step 2: Convert the demand load into amperes
I = P/E  = 8,000W/240V = 33.33A

**Figure 220-23**

**Branch-Circuit Load for One Wall Mounted Oven or One Counter-Mounted Cooking Unit** – The branch-circuit load for one wall-mounted oven or one counter-mounted cooking unit shall be the nameplate rating of the appliance.

**Question:** What size branch circuit is required for a 6 kW wall-mounted oven? Figure 220-24

(a) 14 AWG
(b) 12 AWG
(c) 10 AWG
(d) 8 AWG

**Answer:** (c) 10 AWG

I = P/E
P = 6,000W, E = 240V
I = 6,000W/240V = 25A, 10 AWG rated 30A at 60°C [Table 310.16 and 110.14(C)(1)(a)(1)]

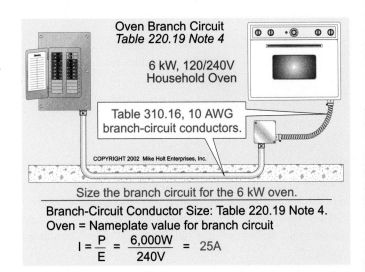

**Oven Branch Circuit**
*Table 220.19 Note 4*

6 kW, 120/240V Household Oven

Table 310.16, 10 AWG branch-circuit conductors.

COPYRIGHT 2002 Mike Holt Enterprises, Inc.

Size the branch circuit for the 6 kW oven.

Branch-Circuit Conductor Size: Table 220.19 Note 4.
Oven = Nameplate value for branch circuit

$I = \dfrac{P}{E} = \dfrac{6,000W}{240V} = 25A$

**Figure 220-24**

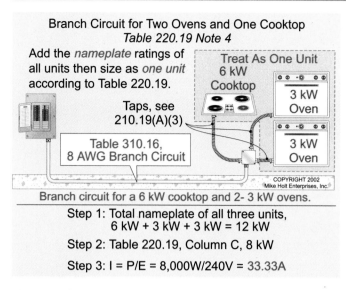

Branch Circuit for Two Ovens and One Cooktop
*Table 220.19 Note 4*

Add the *nameplate* ratings of all units then size as *one unit* according to Table 220.19.

Taps, see 210.19(A)(3)

Treat As One Unit
6 kW Cooktop
3 kW Oven
3 kW Oven

Table 310.16, 8 AWG Branch Circuit

COPYRIGHT 2002 Mike Holt Enterprises, Inc.

Branch circuit for a 6 kW cooktop and 2- 3 kW ovens.

Step 1: Total nameplate of all three units, 6 kW + 3 kW + 3 kW = 12 kW

Step 2: Table 220.19, Column C, 8 kW

Step 3: I = P/E = 8,000W/240V = 33.33A

**Figure 220-25**

**Branch-Circuit Load for One Counter-Mounted Cooking Unit and Up to Two Wall-Mounted Ovens** – The branch-circuit load for one counter-mounted cooking unit and up to two wall-mounted ovens is determined by adding the nameplate ratings together and treating this value as a single range.

**Question:** What size branch circuit is required for one 6 kW counter-mounted cooking unit, and two 3 kW wall-mounted ovens? Figure 220-25

(a) 14 AWG  (b) 12 AWG
(c) 10 AWG  (d) 8 AWG

**Answer:** (d) 8 AWG

Step 1. Determine the total load. 6 kW + 3 kW + 3 kW = 12 kW

Step 2. Determine net computed load as a 12 kW range.

Table 220.19 Column C = 8 kW

Step 3. Size the branch-circuit conductors.

I = P/E

P = 8,000W, E = 240V

I = 8,000W/240V = 33.33A, 8 AWG rated 40A at 60°C [Table 310.16 and 110.14(C)(1)(a)(1)]

## 220.20 Commercial - Kitchen Equipment Load

Table 220.20 is used to calculate the net computed load for thermostat controlled or intermittently used commercial electric cooking equipment, such as dishwasher booster heaters, water heaters, and other kitchen loads. The kitchen equipment feeder/service net computed load shall not be less than the sum of the two largest kitchen equipment loads.

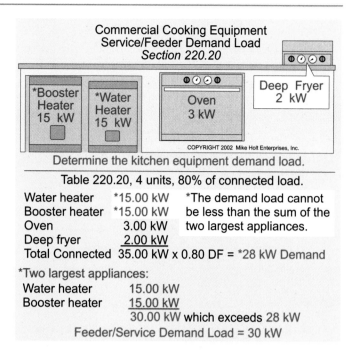

Commercial Cooking Equipment Service/Feeder Demand Load
*Section 220.20*

*Booster Heater 15 kW
*Water Heater 15 kW
Oven 3 kW
Deep Fryer 2 kW

COPYRIGHT 2002 Mike Holt Enterprises, Inc.

Determine the kitchen equipment demand load.

Table 220.20, 4 units, 80% of connected load.

| | | |
|---|---|---|
| Water heater | *15.00 kW | *The demand load cannot |
| Booster heater | *15.00 kW | be less than the sum of the |
| Oven | 3.00 kW | two largest appliances. |
| Deep fryer | 2.00 kW | |
| Total Connected | 35.00 kW x 0.80 DF = *28 kW Demand | |

*Two largest appliances:
Water heater　15.00 kW
Booster heater　15.00 kW
30.00 kW which exceeds 28 kW
Feeder/Service Demand Load = 30 kW

**Figure 220-26**

Table 220.22 demand factors do not apply to space-heating, ventilating, or air-conditioning equipment.

**Question:** What is the commercial kitchen equipment net computed load for one 15 kW booster water heater, one 15 kW water heater, one 3 kW oven, and one 2 kW deep fryer? Figure 220-26

(a) 15,000W  (b) 20,000W
(c) 26,000W  (d) 30,000W

**Answer:** (d) 30,000W

Step 1. Determine the total connected load.

Total Load = 15 kW + 15 kW + 3 kW + 2 kW = 35 kW

Step 2. Determine the total net computed load.

35 kW × 0.8 = 28 kW, but it shall not be less than the sum of the two largest appliances, or 30 kW.

## 220.21 Noncoincident Loads

Where it is unlikely that two or more loads will be used at the same time, it shall be permissible to only include the largest load(s) that will be used at one time when determining the feeder/service load. Figure 220-27

**Question:** What is the feeder/service net computed load for a 5-hp, 230V air conditioner that has a current rating of 28A versus 9 kW heating? Figure 220-28

(a) 5,000W  (b) 6,000W
(c) 7,500W  (d) 9,000W

**Answer:** (d) 9,000W

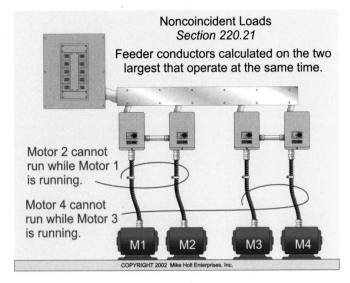

**Figure 220-27**

Air-conditioner Load = 230V × 28A = 6,440 VA (omit)

Heat Load = 9,000W

## 220.22 Feeder/Service Neutral Load

The neutral load shall be the maximum net computed load between the neutral and any one ungrounded (hot) conductor. Line-to-line loads do not place any load on the neutral, therefore they are not considered when sizing the feeder or service neutral conductor. Figure 220-29

**Dwelling Unit Range and Cooking Appliances Load.** The feeder/service neutral load for household electric ranges, wall-mounted ovens, or counter-mounted cooking units shall be calculated at 70 percent of the cooking equipment net computed load in accordance with Table 220.19.

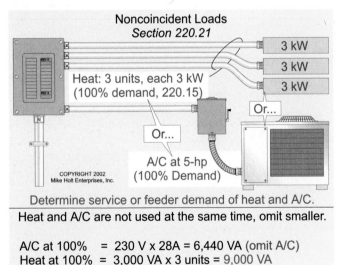

**Figure 220-28**

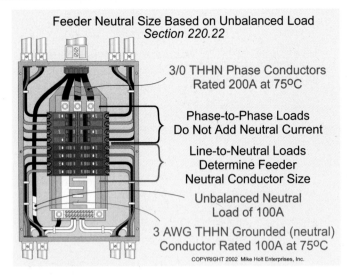

**Figure 220-29**

**Question:** What is the feeder/service neutral load for nine 12 kW household ranges?

(a) 13,000W                    (b) 14,700W
(c) 16,800W                    (d) 24,000W

**Answer:** (c) 16,800W

Step 1. Table 220.19 Column C = 24 kW

Step 2. Neutral Load = 24,000W × 0.70 = 16,800W

**Dwelling Unit Dryer Load.** The feeder/service neutral load for household electric dryers shall be calculated at 70 percent of the dryer net computed load in accordance with Table 220.18.

**Question:** A 10-unit multifamily building has a 5 kW electric clothes dryer in each unit. What is the feeder/service neutral load for these dryers?

(a) 17,500W                    (b) 23,500W
(c) 33,000W                    (d) 41,000W

**Answer:** (a) 17,500W

Step 1. Table 220.18 = 10 units × 5,000W × 0.50 = 25,000W

Step 2. Neutral Load = 25,000W × 0.70  = 17,500W

**Over 200A Neutral Reduction.** The feeder/service load for 3-wire, 1Ø, or 4–wire, 3Ø systems supplying linear loads can be reduced for that portion of the unbalanced load over 200A by a multiplier of 70 percent.

**Question:** What is the neutral load for a 600A feeder/service, if the 600A feeder/service load (two parallel raceways) consists of 100A of 240V loads, 100A of household ranges, 50A of household dryers, and 350A of 120V loads? Figure 220-30

(a) 200A                       (b) 379A
(c) 455A                       (d) 600A

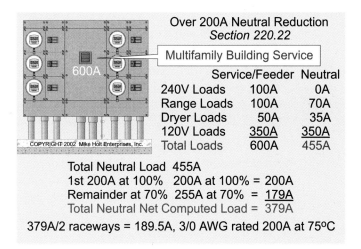

Over 200A Neutral Reduction
*Section 220.22*

Multifamily Building Service

| | Service/Feeder | Neutral |
|---|---|---|
| 240V Loads | 100A | 0A |
| Range Loads | 100A | 70A |
| Dryer Loads | 50A | 35A |
| 120V Loads | 350A | 350A |
| Total Loads | 600A | 455A |

Total Neutral Load 455A
1st 200A at 100% 200A at 100% = 200A
Remainder at 70% 255A at 70% = 179A
Total Neutral Net Computed Load = 379A
379A/2 raceways = 189.5A, 3/0 AWG rated 200A at 75ºC

**Figure 220-30**

**Answer:** (b) 379A

Step 1. Determine Neutral Load

| | Feeder/Service | Neutral Load |
|---|---|---|
| 240V Loads | 100A | 0A |
| Range Load | 100A | 70A |
| Dryer Load | 50A | 35A |
| 120V Loads | 350A | 350A |
| Total Load | 600A | 455A |

Step 2. Determine the Reduced Neutral Load

| Total Neutral Load | 455A | |
|---|---|---|
| First 200A at 100% | 200A at 100% = | 200A |
| Remainder at 70% | 255A at 70% = | 179A |
| Total neutral net computed load | | 379A |

379A/2 raceways = 189.5A, 3/0 AWG rated 200A at 75°C [Table 310.16]

**Harmonic (Nonlinear) Loads.** The feeder/service neutral net computed load cannot be reduced for nonlinear loads supplied from a 4–wire, wye-connected, 3-phase system.

**AUTHOR'S COMMENT:** Not only is the neutral conductor not permitted to be reduced when it supplies nonlinear loads, it might have to be increased in size! Nonlinear loads of the line-to-neutral type (120V and 277V) generate odd tripling harmonic neutral currents that add (instead of cancel) on the neutral conductor. The result is that the load on the neutral conductor could be almost twice the maximum neutral load on any ungrounded (hot) conductor, which can cause overheating of the neutral conductor unless it is sized to accommodate this condition. It is extremely rare that the neutral would ever have to be sized larger than the ungrounded (hot) conductors. Figure 220-31

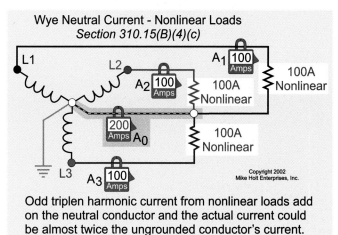

Wye Neutral Current - Nonlinear Loads
*Section 310.15(B)(4)(c)*

Odd triplen harmonic current from nonlinear loads add on the neutral conductor and the actual current could be almost twice the ungrounded conductor's current.

**Figure 220-31**

**Question:** How many amperes could flow on the neutral conductor for a commercial office building if the 600A, 208Y/120V feeder/service load (two parallel raceways) consists of 200A of 208V loads, 200A of 120V nonlinear load loads, and 200A of linear loads?

(a) 200A       (b) 379A
(c) 455A       (d) 600A

**Answer:** (d) 600A. The current on the neutral conductor for nonlinear loads could be as much as twice the maximum neutral load.

| Nonlinear Load | 200A × 2 = | 400A |
|---|---|---|
| Linear Load | 200A × 1 = | 200A |

Total Adjusted Neutral Load 600A/2 raceways = 300A, 350 kcmil rated 310A at 75°C [Table 310.16]

**3-Wire Wye Systems.** The feeder/service neutral net computed load shall not be permitted to be reduced for 3-wire circuits consisting of 2-phase wires and a neutral supplied from a 4–wire, wye-connected, 3-phase system. This is because the neutral of a 3-wire circuit connected to a 4–wire, 3-phase system carries approximately the same current as the phase conductors [310.15(B)(4)(c)].

**Question:** How many amperes could flow on the neutral conductor for a building if the 400A, 3-wire 208Y/120V feeder load (from a 3Ø, 4-wire service) consists of 200A of 240V loads and 200A of 120V linear loads? Figure 220-32

(a) 200A       (b) 379A
(c) 455A       (d) 600A

**Answer:** (a) 200A

*WARNING: The grounded (neutral) service conductor shall be brought to each service disconnect [250.24(B)] and it must be sized no smaller than required by Table 250.66.*

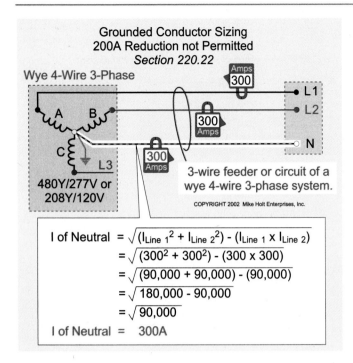

**Grounded Conductor Sizing**
**200A Reduction not Permitted**
*Section 220.22*

Wye 4-Wire 3-Phase

480Y/277V or
208Y/120V

3-wire feeder or circuit of a
wye 4-wire 3-phase system.

COPYRIGHT 2002 Mike Holt Enterprises, Inc.

$$I \text{ of Neutral} = \sqrt{(I_{Line\,1}^2 + I_{Line\,2}^2) - (I_{Line\,1} \times I_{Line\,2})}$$
$$= \sqrt{(300^2 + 300^2) - (300 \times 300)}$$
$$= \sqrt{(90,000 + 90,000) - (90,000)}$$
$$= \sqrt{180,000 - 90,000}$$
$$= \sqrt{90,000}$$
$$I \text{ of Neutral} = 300A$$

**Figure 220-32**

**Question:** What size grounded neutral service conductor is required for a structure that has a 400A service supplied with 500 kcmil conductors, if the maximum unbalanced neutral load is no more than 100A?

(a) 3 AWG          (b) 2 AWG
(c) 1 AWG          (d) 1/0 AWG

**Answer:** (d) 1/0 AWG

According to Table 310.16, a 3 AWG rated 100A at 75°C [110.14(C)] would be of sufficient size to carry the 100A unbalanced load. However, the grounded (neutral) conductor for a service shall be sized in accordance with Table 250.66, and this Table requires a 1/0 AWG, based on the 500 kcmil service conductors, see 250.24(B) for details.

## PART III. OPTIONAL CALCULATIONS FOR COMPUTING FEEDER AND SERVICE LOADS

### 220.30 Dwelling Unit - Optional Load Calculation

**(A) Feeder/Service Load.** A 3-wire feeder/service load for a dwelling unit can be calculated by adding the loads from 220.30(B) and (C), instead of the standard method specified in Part II of Article 220 for a dwelling unit having an ampacity of 100A or more. The neutral load is determined by 220.22 (standard calculation method).

**(B) General Loads.** The net computed load shall not be less than 100 percent of the first 10 kVA plus 40 percent of the remainder of the following loads.

**(1) Small-Appliance and Laundry Circuits.** A load of 1,500 VA shall apply for each 20A small-appliance and laundry branch circuit. Since it is required there be two small-appliance circuits and a laundry circuit, the minimum will be 4,500 VA.

**(2) General Lighting.** 3 VA per ft$^2$ for general lighting and general-use receptacles. The floor area shall be computed from the outside dimensions of the dwelling unit not including open porches, garages, or unused or unfinished spaces not adaptable for future use.

**(3) Appliances.** The nameplate rating of all appliances that are fastened in place, permanently connected, or located to be on a specific circuit.

**(4) Motor VA.** The nameplate VA rating of all motors.

**(C) Heating and Air Conditioning.** Include the largest of the following:

(1) Air-Conditioning – 100 percent

(2) Heat Pump and Supplemental Heating – 100 percent

(3) Thermal Storage Heating – 100 percent

Thermal storage heating is the process of heating bricks or water at night when the electric rates are lower. Then during the day, the building uses the thermally stored heat.

(4) Central Space Heating – 65 percent

(5) Space Heating (three or less units) – 65 percent

(6) Space Heating (four or more units) – 40 percent

**Question:** What size 120/240V, 3-wire feeder/service is required for a 1,500 ft$^2$ dwelling unit that contains the following loads:

| | |
|---|---|
| Dishwasher, 1,200 VA | Water heater, 4,500 VA |
| Disposal, 900 VA | Dryer, 4,000 VA |
| Cooktop, 6,000 VA | Oven, 3,000 VA |
| Heat, 7,000 VA | A/C, 5-hp |

(a) 100A          (b) 110A
(c) 125A          (d) 150A

**Answer:** (a) 100A

Step 1.  Determine net computed load:

    (1)  Small-appliance 1,500 x 2   3,000 VA
          Laundry                           1,500 VA
    (2)  General lighting 1,500 x 3   4,500 VA
    (3)  Dishwasher                    1,200 VA
          Water heater                  4,500 VA
          Disposal                      900 VA
          Dryer                       4,000 VA
          Cooktop                    6,000 VA
          Oven 3,000               <u>3,000 VA</u>
                                     28,600 VA

First 10,000 at 100%    <u>10,000 VA</u>  at 100% 10,000 VA
Remainder at 40%       18,600 VA  at 40%   7,440 VA
Heat (7,000 x 0.65 = 4,550 VA) (omit)
A/C 5-hp (230V x 28A)                <u>6,440 VA</u>
                                     23,880 VA

Step 2.  Determine feeder/service size:

$$I = VA/E = 23{,}880 \text{ VA}/240\text{V} = 99.5\text{A, 4 AWG}$$
[215.2(A)(4) and 310.15(B)(6)]

## 220.32 Multifamily - Optional Load Calculation

**(A)  Feeder or Service Load.** It shall be permissible to compute the load of a feeder or service for a building that has more than two dwelling units in accordance with Table 220.32 instead of Part II of Article 220, if each dwelling unit is equipped with electric cooking equipment and either electric space heating, air conditioning, or both. The neutral load is determined by 220.22.

**Question:** What size 208Y/120V, 3-phase service is required for a multifamily building that has twenty 1,500 ft$^2$ dwelling units each containing the following loads:

Dishwasher, 1,200 VA      Water heater, 4,500 VA
Disposal, 900 VA            Dryer, 4,000 VA
Cooktop, 6,000 VA         Oven, 3,000 VA
A/C, 5-hp (6,440 VA)       Heat, 7,000 VA

    (a) 400A                  (b) 600A
    (c) 800A                 (d) 1,200A

   **Answer:** (c) 800A

Step 1.  Determine net computed load:

  (1)  Small-appliance 1,500 x 2   3,000 VA
        Laundry                         1,500 VA
  (2)  General lighting 1,500 x 3   4,500 VA
  (3)  Dishwasher                   1,200 VA
        Water heater                 4,500 VA
        Disposal                    900 VA
        Dryer                      4,000 VA
        Cooktop                  6,000 VA
        Oven 3,000             3,000 VA
        A/C 5-hp 6,440 VA (omit)
        Heat                     <u>7,000 VA</u>
            35,600 VA x 20 x 0.38 = 270,560 VA

Step 2.  Determine feeder/service size:

$$I = VA/(E \times 1.732) =$$
$$270{,}560 \text{ VA}/(208\text{V} \times 1.732) = 751\text{A}$$

Two parallel sets of 500 kcmil, each rated
$380\text{A} \times 2 = 760\text{A}$ [310.15(B)(6)]

# Article 220

1.  The 3 VA per square foot general lighting load for dwelling units shall not include _____.
    (a) open porches
    (b) garages
    (c) unused or unfinished spaces not adaptable for future use
    (d) all of these

2.  For other than dwelling units, the feeder and service load calculation for track lighting is to be determined at 150 VA for every _____ of track installed.
    (a) 4 ft                (b) 6 ft                (c) 2 ft                (d) none of these

3.  Receptacle loads for nondwelling units, computed at not more than 180 VA per outlet in accordance with 220.3(B)(9), shall be permitted to be _____.
    (a) added to the lighting loads and made subject to the demand factors of Table 220.11
    (b) made subject to the demand factors of Table 220.13
    (c) made subject to the lighting demand loads of Table 220.3(B)
    (d) a or b

4.  Loads that are computed for dwelling unit small-appliance branch circuits can be included with the _____ load and subject to the demand factors permitted in Table 220.11 for the general lighting load.
    (a) general lighting        (b) feeder              (c) appliance           (d) receptacle

5.  Using standard load calculations, the feeder demand factor for five household clothes dryers is _____ percent.
    (a) 70                  (b) 85                  (c) 50                  (d) 100

6.  To determine the feeder demand load for ten 3 kW household cooking appliances, use _____ of Table 220.19.
    (a) Column A            (b) Column B            (c) Column C            (d) none of these

7.  The feeder demand load for four 6 kW cooktops is _____.
    (a) 17 kW               (b) 4 kW                (c) 12 kW               (d) 24 kW

8.  Where it is unlikely that two or more loads will be in use simultaneously, it shall be permissible to use only the _____ loads at any given time in computing the total load to a feeder.
    (a) smaller                              (b) larger
    (c) difference between the               (d) none of these

9.  The maximum unbalanced feeder load for household electric ranges, wall-mounted ovens, counter-mounted cooking units, and electric dryers shall be considered as _____ percent of the load on the ungrounded conductors, as determined in accordance with Table 220.19 for ranges and Table 220.18 for dryers.
    (a) 50                  (b) 70                  (c) 85                  (d) 115

10. Under the optional method for calculating a single-family dwelling, general loads beyond the initial 10 kW are to be assessed at a _____ percent demand factor.
    (a) 40                  (b) 50                  (c) 60                  (d) 75

# Article 225
# Outside Wiring

## Part I General Requirements

### 225.1 Scope

This article contains the installation requirements for outside branch circuits and feeders run on or between buildings, structures, or poles. Figure 225-1

### 225.2 Other Articles

Other articles that contain requirements for outside wiring include:

Class 1, Class 2, and Class 3 remote-control, signaling, and power-limited circuits, Article 725

Communications circuits, Article 800

Community antenna television and radio distribution systems, Article 820

Conductors for general wiring, Article 310

Electric signs and outline lighting, Article 600

Floating buildings, Article 553

Grounding, Article 250

Marinas and boatyards, Article 555

Messenger supported wiring, Article 396

Open wiring on insulators, Article 398

Radio and television equipment, Article 810

Services, Article 230

Solar photovoltaic systems, Article 690

Swimming pools, fountains, and similar installations, Article 680

### 225.6 Minimum Size Conductors

**(A) Overhead Conductor.**

**(1) Spans.** The minimum size overhead conductor for spans up to 50 ft is 10 AWG; and for spans over 50 ft, the minimum size permitted is 8 AWG. Figure 225-2

**(B) Festoon Lighting.** The minimum size conductor for festoon lighting is 12 AWG copper. Figure 225-3

---

**AUTHOR'S COMMENT:** Festoon lighting is a string of outdoor lights suspended between two points [Article 100]. This type of wiring is commonly installed at fairs. See 525.20(C) for the installation requirement of outdoor lampholders.

---

### Outside Branch Circuits and Feeders
### Article 225

Overhead Branch Circuit or Feeder Conductor Spans

COPYRIGHT 2002
Mike Holt Enterprises, Inc.

Underground Branch Circuit or Feeder

Festoon Lighting

Article 225 applies to outside branch circuits and feeders on or between buildings or structures and wiring for utilization equipment located outside.

**Figure 225-1**

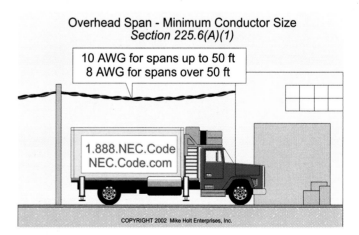

Overhead Span - Minimum Conductor Size
*Section 225.6(A)(1)*

10 AWG for spans up to 50 ft
8 AWG for spans over 50 ft

1.888.NEC.Code
NEC.Code.com

COPYRIGHT 2002 Mike Holt Enterprises, Inc.

**Figure 225-2**

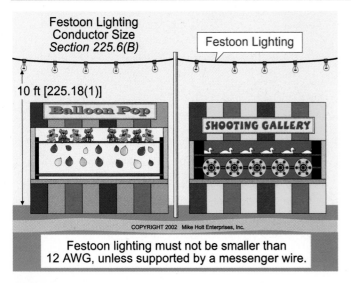

**Festoon lighting must not be smaller than 12 AWG, unless supported by a messenger wire.**

**Figure 225-3**

## 225.7 Luminaires Installed Outdoors

**(C) 277V-to-Ground.** Luminaires connected on 277V or 480V circuits are permitted outdoors, but not within 3 ft of windows that open, platforms, fire escapes, and the like. Figure 225-4

---

**AUTHOR'S COMMENT:** See 210.6(C) for the types of luminaires permitted on 277V or 480V branch circuits.

---

## 225.15 Supports over Buildings

Supports over a building shall be in accordance with 230.29.

## 225.16 Point of Attachment to Buildings

The point of attachment to a building shall be in accordance with 230.26.

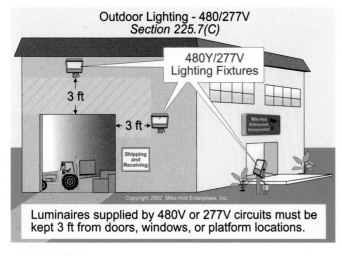

**Luminaires supplied by 480V or 277V circuits must be kept 3 ft from doors, windows, or platform locations.**

**Figure 225-4**

## 225.17 Means of Attachment to Buildings

The means of attachment to a building shall be in accordance with 230.27.

## 225.18 Clearances

Overhead conductor spans not over 600V, nominal, shall maintain a clearance of:

(1)  10 ft above finished grade, sidewalks, or platforms or projections from which they might be accessible to pedestrians, where the voltage is not in excess of 150 volts-to-ground.

(2)  12 ft above residential property and driveways, and those commercial areas not subject to truck traffic, where the voltage does not exceed 300 volts-to-ground.

(3)  15 ft above those areas listed in the 12 ft classification, where the voltage exceeds 300 volts-to-ground.

(4)  18 ft over public streets, alleys, roads, parking areas subject to truck traffic, driveways on other than residential property, and other areas traversed by vehicles such as those used for cultivation, grazing, forestry, and orchards.

---

**AUTHOR'S COMMENT:** See 680.8 for clearances over pools.

---

## 225.19 Clearances From Building

**(A) Above Roofs.** Overhead conductors shall maintain a minimum of 8 ft above the surface of a roof for a minimum distance of 3 ft in all directions from the edge of the roof. Figure 225-5

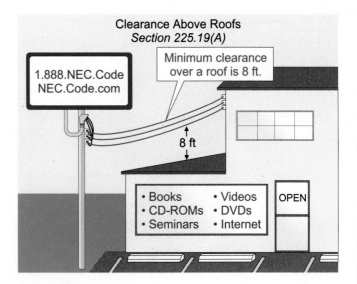

**Figure 225-5**

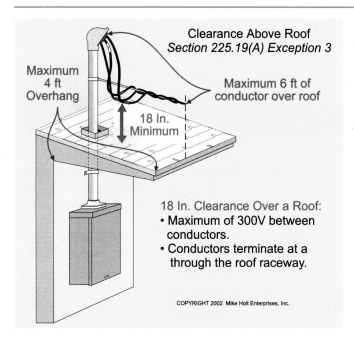

**Figure 225-6**

*Exception No. 2:* Where the voltage does not exceed 300V between conductors, overhead conductor clearances from the roof can be reduced from 8 ft to 3 ft, if the slope of the roof exceeds 4 in. in 12 in.

*Exception No. 3:* If the voltage between conductors does not exceed 300V, the conductor clearance over the roof overhang can be reduced from 8 ft to 18 in., if no more than 6 ft of overhead conductors pass over no more than 4 ft of roof overhang, and the conductors terminate at a through-the-roof raceway or approved support. Figure 225-6

*Exception No. 4:* The 3 ft vertical clearance that extends from the roof shall not apply when the point of attachment is on the side of the building below the roof.

**(B)  From Other Structures.** Overhead conductors not over 600V, nominal, shall maintain a vertical, diagonal, and horizontal clearance of not less than 3 ft from signs, chimneys, radio and television antennas, tanks, and other nonbuilding or nonbridge structures.

**(D)  Final Span Clearance.**

    **(1)  Clearance From Windows.** An overhead conductor to a building shall maintain a clearance of 3 ft from windows that are designed to be opened, doors, porches, balconies, ladders, stairs, fire escapes, or similar locations. Figure 225-7

*Exception:* Overhead conductors that run above a window are not required to maintain the 3 ft distance.

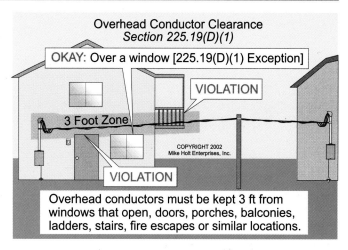

**Figure 225-7**

    **(2)  Vertical Clearance.** Overhead conductors shall maintain a vertical clearance of not less than 10 ft above platforms, projections or surfaces from which they might be reached. This vertical clearance shall be maintained for 3 ft, measured horizontally, from the platforms, projections, or surfaces from which they might be reached.

    **(3)  Below Opening.** Overhead conductors shall not be installed under an opening through which materials might pass, and they shall not be installed where they will obstruct an entrance to building openings. Figure 225-8

### 225.26 Trees for Conductor Support

Trees or other vegetation shall not be used for the support of overhead conductor spans, but they can be used to support electrical equipment and luminaires. See 410.16(H). Figure 225-9

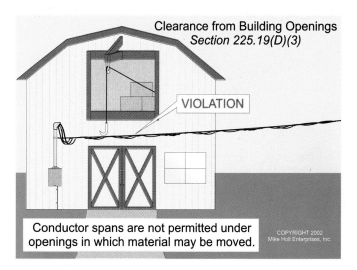

**Figure 225-8**

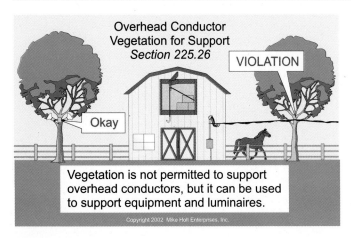

**Figure 225-9**

---

**AUTHOR'S COMMENT:** Overhead conductor spans cannot be supported to trees. This includes services [230.10] and temporary wiring [527.4(J)], as well as branch circuits and feeders. Figure 225-10

---

## Part II. More Than One Building or Structure.

### 225.30. Number of Supplies

No more than one feeder or branch circuit shall supply each building or other structure, except as permitted in (A) through (E). For the purpose of this section, a multiwire branch circuit shall be considered a single circuit.

---

**AUTHOR'S COMMENT:** If you need more than one branch circuit (or more than one multiwire branch circuit), a single feeder shall be run to the building or structure to feed distribution equipment that can provide multiple branch circuits.

---

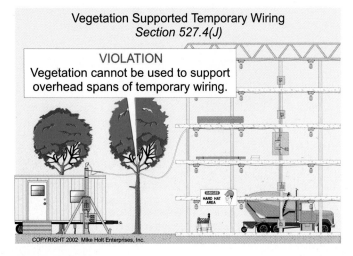

**Figure 225-10**

**(A) Special Conditions.** Additional circuits are permitted for:

(1) Fire pumps

(2) Emergency systems

(3) Legally required standby systems

(4) Optional standby systems

(5) Parallel power production systems

---

**AUTHOR'S COMMENT:** To minimize the possibility of accidental interruption, the disconnecting means for the fire pump or standby power shall be located remotely away from the normal power disconnect [225.34(B)].

---

**(B) Special Occupancies.** By special permission, additional circuits are permitted for:

(1) Multiple-occupancy buildings where there is no available space for supply equipment accessible to all occupants, or

(2) A building or structure that is so large that two or more supplies are necessary.

**(C) Capacity Requirements.** Additional supplies are permitted where the capacity requirements are in excess of 2,000A at a supply voltage of 600V or less.

**(D) Different Characteristics.** Additional supplies are permitted for different voltages, frequencies, or uses, such as control of outside lighting from multiple locations.

**(E) Documented Switching Procedures.** Additional supplies are permitted where documented safe switching procedures are established and maintained for disconnection.

### 225.31 Disconnecting Means

A disconnect shall be provided to disconnect all conductors that enter or pass through a building or structure.

### 25.32 Disconnect Location

The disconnecting means for a building or structure shall be installed at a readily accessible location, either outside the building or structure, or inside nearest the point of entrance of the conductors. Figure 225-11

If the disconnecting means is outdoors, the *NEC* does not require it to be located on the building or structure. How far away can the disconnecting means be from the building? This is up to the AHJ.

Supply conductors are considered outside of a building or other structure where encased or installed under not less than 2 in. of concrete or brick [230.6]. Figure 225-12

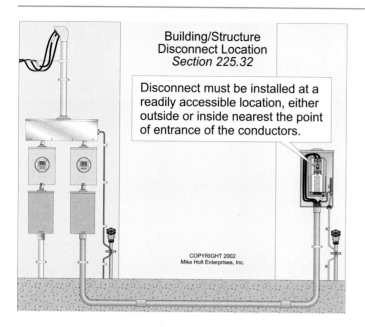

Building/Structure
Disconnect Location
*Section 225.32*

Disconnect must be installed at a readily accessible location, either outside or inside nearest the point of entrance of the conductors.

COPYRIGHT 2002
Mike Holt Enterprises, Inc.

**Figure 225-11**

*Exception No. 1.* Where documented safe switching procedures are established and maintained, the building/structure disconnecting means can be located elsewhere on the premises if monitored by qualified persons.

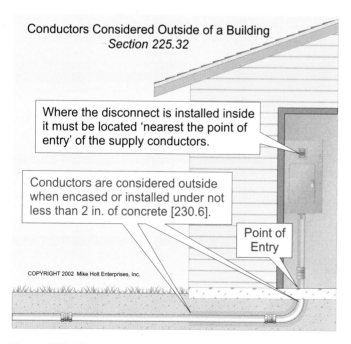

Conductors Considered Outside of a Building
*Section 225.32*

Where the disconnect is installed inside it must be located 'nearest the point of entry' of the supply conductors.

Conductors are considered outside when encased or installed under not less than 2 in. of concrete [230.6].

Point of Entry

COPYRIGHT 2002 Mike Holt Enterprises, Inc.

**Figure 225-12**

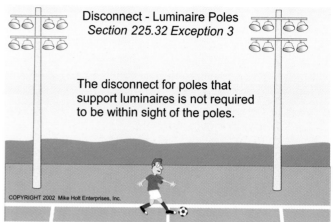

Disconnect - Luminaire Poles
*Section 225.32 Exception 3*

The disconnect for poles that support luminaires is not required to be within sight of the poles.

COPYRIGHT 2002 Mike Holt Enterprises, Inc.

**Figure 225-13**

**AUTHOR'S COMMENT:** The definition of "Qualified Persons" in Article 100 ensures that a person has the skills and knowledge related to the construction and operation of the electrical equipment and installation, and that they have received safety training on the hazards involved with electrical systems.

*Exception No. 3:* The disconnecting means for poles that support luminaires may be located remotely from the pole. Figure 225-13

*Exception No. 4:* The disconnecting means for a sign is not required to be readily accessible if it is installed in accordance with the requirements for signs. Figure 225-14

**AUTHOR'S COMMENT:** Each sign shall be controlled by an externally operable switch or circuit breaker that opens all ungrounded conductors to the sign. The sign disconnecting means shall be within sight of the sign, or the disconnecting means shall be capable of being locked in the open position [600.6(A)].

Disconnect - Sign Structure
*Section 225.32 Exception 4*

Mike Holt Enterprises
888-NEC-CODE

Disconnect not required to be readily accessible.

COPYRIGHT 2002 Mike Holt Enterprises, Inc.

The sign disconnecting means must be installed in accordance with 600.6.

**Figure 225-14**

## 225.33 Maximum Number of Disconnects

**(A) Six.** The building or structure disconnecting means must consist of no more than six switches or six circuit breakers in a single enclosure, or separate enclosures for each supply permitted in 225.30.

## 225.34 Grouping of Disconnects

**(A) Two to Six Disconnects.** Building or structure disconnection means shall be grouped in one location, and they shall be marked to indicate the load served [110.22].

**(B) Additional Disconnects.** To minimize the possibility of accidental interruption of the critical power systems, the disconnecting means for a fire pump, or standby power, as permitted in 225.30(A), shall be located remotely away from the normal power disconnect.

## 225.35 Access to Occupants

In a multiple-occupancy building, each occupant shall have access to the disconnecting means for their occupancy.

*Exception:* The occupant disconnect can be accessible to building management only, if electrical maintenance under continuous supervision is provided by the building management.

## 225.36 Identified As Suitable for Service Equipment

The building or structure disconnecting means shall be identified as suitable for use as service equipment.

**AUTHOR'S COMMENT:** This means that the disconnect shall be supplied with a main bonding jumper so that a neutral-to-ground connection can be made as permitted in 250.32(B)(2) and 250.142(A). Figure 225-15

*Exception:* A snap switch, or a set of 3-way or 4-way snap switches, can be used as the disconnecting means for garages and outbuildings on residential property, without having a "suitable for use as service equipment" rating. Figure 225-16

## 225.37 Identification of Multiple Supplies

Where a building or structure is supplied by more than one service, or a combination of branch circuits, feeders, and services, a permanent plaque or directory shall be installed at each service, feeder, or branch-circuit disconnect location denoting all other services, feeders, and branch circuits supplying that building or structure and the area served by each.

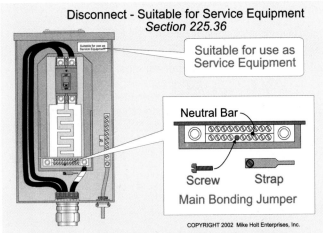

Disconnect - Suitable for Service Equipment
*Section 225.36*

Suitable for use as service equipment means the equipment is supplied with a main bonding jumper that can be used to bond the neutral to the case.

**Figure 225-15**

## 225.38 Disconnect Construction

**(A) Manual or Power-Operated Circuit Breakers.** The building or structure disconnecting means can consist of either a manually operable switch or circuit breaker, or a power-operated switch or circuit breaker. If power-operated, the switch or circuit breaker shall be capable of being manually operated.

**AUTHOR'S COMMENT:** A shunt-trip pushbutton can be used to open a power-operated circuit breaker; the circuit breaker is the disconnecting means, not the pushbutton.

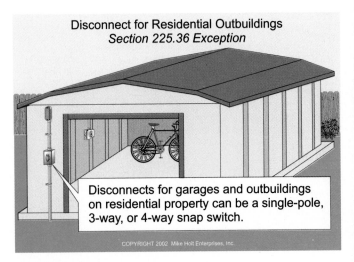

Disconnect for Residential Outbuildings
*Section 225.36 Exception*

Disconnects for garages and outbuildings on residential property can be a single-pole, 3-way, or 4-way snap switch.

**Figure 225-16**

## 225.39 Rating of Disconnecting Means

The disconnecting means for a building or structure shall have an ampere rating of not less than the computed load according to Article 220 and in no case less than:

**(A)** **15A.** For a single branch circuit.

**(B)** **30A.** For an installation consisting of not more than two, 2-wire branch circuits.

**(C)** **100A.** For a one-family dwelling where the initial computed load is 10 kVA or more, or where the initial installation consists of six or more 2-wire branch circuits.

**(D)** **60A.** For all other installations.

# Article 225

1. Open individual conductors shall not be smaller than _____ AWG copper for spans up to 50 ft in length and _____ AWG copper for a longer span, unless supported by a messenger wire.
   (a) 10, 8      (b) 6, 8      (c) 6, 6      (d) 8, 8

2. The minimum point of attachment of overhead conductors to a building shall in no case be less than _____ above finished grade.
   (a) 8 ft      (b) 10 ft      (c) 12 ft      (d) 15 ft

3. Overhead conductors shall have a minimum of _____ vertical clearance from final grade over residential property and driveways, as well as those commercial areas not subject to truck traffic where the voltage is limited to 300 volts-to-ground.
   (a) 10 ft      (b) 12 ft      (c) 15 ft      (d) 18 ft

4. Overhead conductors installed over roofs shall have a vertical clearance of _____ above the roof surface.
   (a) 8 ft      (b) 12 ft      (c) 15 ft      (d) 3 ft

5. Overhead conductors to a building shall maintain a vertical clearance of final spans above, or within, _____ measured horizontally from the platforms, projections or surfaces from which they might be reached.
   (a) 3 ft      (b) 6 ft      (c) 8 ft      (d) 10 ft

6. A building or structure shall be supplied by a maximum of _____ feeder(s) or branch circuit.
   (a) one      (b) two      (c) three      (d) as many as desired

7. The building disconnecting means shall be installed at a(n) _____location.
   (a) accessible      (b) readily accessible      (c) outdoor      (d) indoor

8. There shall be no more than _____ disconnects installed for each supply.
   (a) two      (b) four      (c) six      (d) none of these

9. The two to six disconnects as permitted in 225.33 shall be _____. Each disconnect shall be marked to indicate the load served.
   (a) the same size      (b) grouped      (c) in the same enclosure      (d) none of these

10. In a multiple-occupancy building, each occupant shall have access to his or her own _____.
    (a) disconnecting means      (b) building drops
    (c) building-entrance assembly      (d) lateral conductors

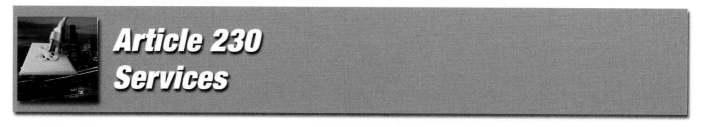

## Article 230
## Services

### Part I. General

Understanding where the service begins and where it ends is critical in the proper application of many Code rules. To understand how to apply these rules, let's review the following definitions from Article 100.

Service point - The point of connection between the facilities of the serving utility and the premises wiring.

Service conductors - The conductors from the service point to the service disconnecting means (service equipment, not meter). Service conductors would include service-entrance conductors for both overhead (service drop) and underground (service lateral).

Service equipment - The necessary equipment, usually consisting of circuit breakers or switches and fuses and their accessories, connected to the load end of service conductors to a building or other structure, or an otherwise designated area, and intended to constitute the main control and cutoff of the supply. Service equipment does not include the metering equipment, such as the meter and/or meter enclosures [230.66].

After reviewing these three definitions, you should understand that service conductors originate at the serving utility (service point) and terminate on the line side of the service disconnecting means (service equipment). Conductors and equipment on the load side of service equipment are considered feeder conductors and this would include: Figure 230-1

- Secondary conductors from customer-owned transformers.
- Conductors from generators, UPS systems, or photovoltaic systems (separately derived systems).
- Conductors to remote buildings or structures.
- Feeder conductors shall be installed in accordance with the requirements contained in Articles 215 and 225.

### 230.1 Scope

Article 230 covers the installation requirements for service conductors and equipment.

### 230.2 Number of Services

A building or structure can be served by only one service (service drop or service-lateral), except as permitted by (A) through (D). Figure 230-2

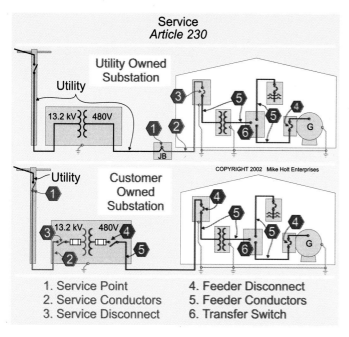

1. Service Point
2. Service Conductors
3. Service Disconnect
4. Feeder Disconnect
5. Feeder Conductors
6. Transfer Switch

**Figure 230-1**

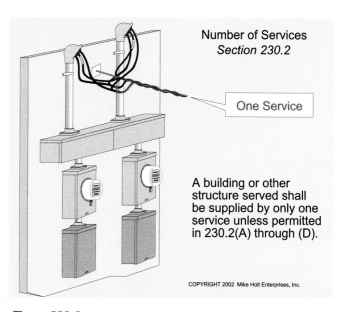

Number of Services
*Section 230.2*

One Service

A building or other structure served shall be supplied by only one service unless permitted in 230.2(A) through (D).

COPYRIGHT 2002 Mike Holt Enterprises, Inc.

**Figure 230-2**

**(A) Special Conditions.**

(1) Fire pumps

(2) Emergency power

(3) Legally required standby power

(4) Optional standby power

(5) Parallel power production systems

**(B) Special Occupancies.** By special permission, additional services are permitted for:

(1) Multiple-occupancy buildings where there is no available space for supply equipment accessible to all occupants, or

(2) A building or other structure so large that two or more supplies are necessary.

**(C) Capacity Requirements.** Additional services are permitted:

(1) Where the capacity requirements are in excess of 2,000A at a supply voltage of 600V or less.

(2) Where the load requirements of a single-phase installation exceed the utility's capacity.

(3) By special permission.

**(D) Different Characteristics.** An additional service is permitted for different voltages, frequencies, or phases, or for different uses, such as for different electricity rate schedules.

**(E) Identification of Multiple Services.** Where a building or structure is supplied by more than one service, or a combination of branch circuits, feeders, and services, a permanent plaque or directory shall be installed at each service, feeder, or branch-circuit disconnect location denoting all other services, feeders, and branch circuits supplying that building or structure and the area served by each. Figure 230-3

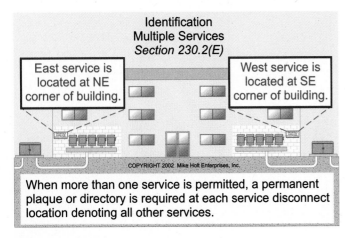

Identification
Multiple Services
Section 230.2(E)

East service is located at NE corner of building.

West service is located at SE corner of building.

When more than one service is permitted, a permanent plaque or directory is required at each service disconnect location denoting all other services.

**Figure 230-3**

## 230.3 Pass Through a Building or Structure

Service conductors shall not pass through the interior of another building or other structure.

## 230.6 Conductors Considered Outside a Building

Conductors are considered outside a building when they are installed:

(1) Under not less than 2 in. of concrete beneath a building or structure. Figure 230-4

(2) Within a building or structure in a raceway that is encased in not less than 2 in. thick of concrete or brick.

(3) Installed in a vault that meets the construction requirements of Article 450, Part III.

(4) In conduit under not less than 18 in. of earth beneath a building or structure.

## 230.7 Service Conductors Separate From Other Conductors

Service conductors shall not be installed in the same raceway or cable with feeder or branch-circuit conductors. Figure 230-5

**AUTHOR'S COMMENT:** This rule does not prohibit the mixing of service, feeder, and branch-circuit conductors in the same "service equipment enclosure."

*WARNING: Overcurrent protection for the feeder conductors could be bypassed if we mixed service conductors with other conductors in the same raceway and a fault occurred between the service and feeder conductors.*

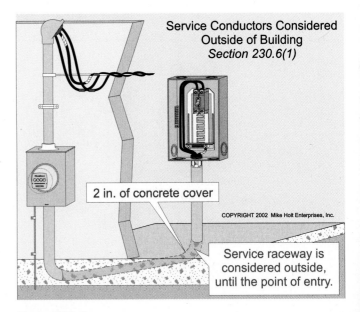

Service Conductors Considered
Outside of Building
Section 230.6(1)

2 in. of concrete cover

Service raceway is considered outside, until the point of entry.

**Figure 230-4**

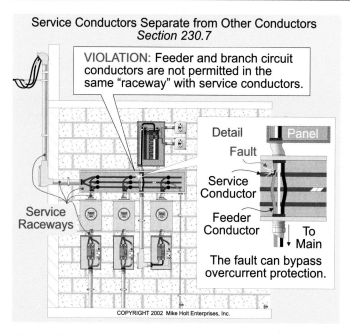

Figure 230-5

**AUTHOR'S COMMENT:** This requirement may be the root of the misconception that "line" and "load" conductors are not permitted in the same raceway. It is true that service conductors are not permitted in the same raceway with feeder or branch-circuit conductors, but line and load conductors of feeders and branch circuits can be in the same raceway, cable, or enclosure. Figure 230-6

## 230.8 Raceway Seals

Used or unused underground raceways shall be sealed or plugged at either or both ends [300.5(G)] to prevent moisture from contacting energized live parts.

> **AUTHOR'S COMMENT:** This can be accomplished with the use of a putty-like material called duct seal or a fitting identified for the purpose. A seal of the type required in Chapter 5 for hazardous (classified) locations is not required.

## 230.9 Clearance From Building Openings

**(A) Clearance From Windows.** Overhead service conductors shall maintain a clearance of 3 ft from windows that are designed to be opened, doors, porches, balconies, ladders, stairs, fire escapes, or similar locations. Figure 230-7

*Exception:* Overhead conductors run above a window are not required to maintain the 3 ft distance.

**(B) Vertical Clearance.** Overhead service conductors shall maintain a vertical clearance of not less than 10 ft above platforms, projections or surfaces from which they might be reached [230.24(B)]. This vertical clearance shall be maintained for 3 ft, measured horizontally, from the platform, projections or surfaces from which they might be reached.

**(C) Below Opening.** Service conductors shall not be installed under an opening through which materials might pass, and they shall not be installed where they will obstruct entrance to building openings. For example, the upper opening in a barn loft is often used to move hay in or out of the loft storage area. Figure 230-8

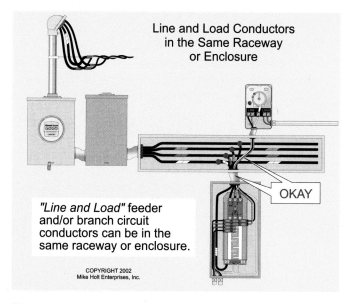

Figure 230-6

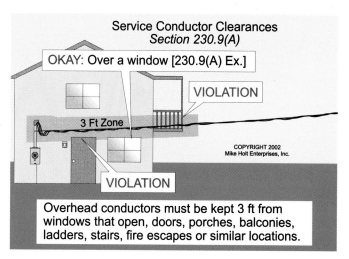

Figure 230-7

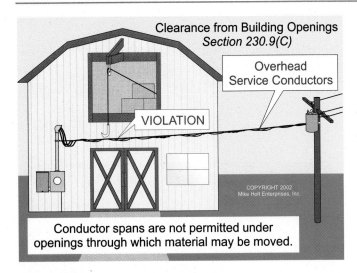

**Figure 230-8**

## 230.10 Vegetation as Support.

Vegetation such as trees shall not be used for the support of overhead service conductors. Figure 230-9

## Part II. Overhead Service-Drop Conductors

**AUTHOR'S COMMENT:** Overhead service-drop conductors installed by the electric utility shall be in accordance with the National Electric Safety Code (NESC), not the NEC [90.2(B)(5)], but overhead service conductors not under the exclusive control of the electric utility shall be installed in accordance with the NEC.

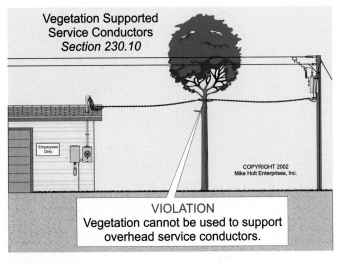

**Figure 230-9**

## 230.23 Size and Rating

(A) **Ampacity of Service-Drop Conductors.** Service-drop conductors shall have adequate mechanical strength and sufficient ampacity in accordance with Article 220 calculations.

(B) **Ungrounded Conductor Size.** Service-drop conductors shall not be smaller than 8 AWG copper or 6 AWG aluminum.

*Exception:* Service-drop conductors can be as small as 12 AWG for limited-load installations.

(C) **Grounded (neutral) Conductor Size.** The grounded (neutral) service drop conductor shall be sized to carry the maximum unbalanced load in accordance with 220.22 and shall not be sized smaller than required by 250.24(B).

**AUTHOR'S COMMENT:** Section 250.24(B) requires that a grounded (neutral) conductor be bonded to the service equipment enclosure and this conductor shall be sized in accordance with Table 250.66.

## 230.24 Clearances

Service-drop conductors shall be located so they are not readily accessible, and they shall comply with the following clearance requirements:

(A) **Above Roofs.** Overhead service conductors must maintain a minimum clearance of 8 ft above the surface of a roof for a minimum distance of 3 ft in all directions from the edge of the roof.

*Exception No. 2:* Where the voltage does not exceed 300V between conductors, overhead conductor clearances from the roof can be reduced from 8 ft to 3 ft, if the slope of the roof exceeds 4 in. in 12 in.

*Exception No. 3:* If the voltage between conductors does not exceed 300V, the conductor clearance over the roof overhang can be reduced from 8 ft to 1 1/2 ft, if no more than 6 ft of overhead conductors pass over no more than 4 ft of roof overhang, and the conductors terminate at a through-the-roof raceway or approved support. Figure 230-10

*Exception No. 4:* The 3 ft vertical clearance that extends from the roof does not apply when the point of attachment is on the side of the building below the roof.

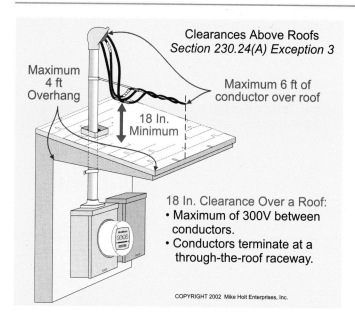

**Figure 230-10**

**(B) Clearances.** Overhead conductor spans shall maintain the following clearances: Figure 230-11

(1) 10 ft at the electric service entrance to buildings, at the lowest point of the drip loop of the building electric entrance, above finished grade, sidewalks, or platform or projection from which they might be accessible to pedestrians, where the voltage is not in excess of 150 volts-to-ground.

(2) 12 ft above residential property and driveways, and those commercial areas not subject to truck traffic, where the voltage does not exceed 300 volts-to-ground.

(3) 15 ft above those areas listed in the 12 ft classification, where the voltage exceeds 300 volts-to-ground.

(4) 18 ft over public streets, alleys, roads, parking areas subject to truck traffic, driveways on other than residential property and other areas such as cultivated fields, grazing areas, forests and orchards traversed by vehicles. Department of Transportation (DOT) type right-of-ways in rural areas are often used by slow-moving and tall farming machinery to avoid impeding traffic flow.

**(D) Swimming Pools.** Service conductors above pools, diving structures, observation stands, towers, or platforms shall comply with 680.8.

## 230.26 Point of Attachment

The point of attachment for service-drop conductors shall not be less than 10 ft above the finish grade and shall be located so the minimum service conductor clearance required by 230.24(B) can be maintained.

*CAUTION: Conductors might need to have the point of attachment raised so the overhead conductors will comply with the clearances required by 230.24. Figure 230-12*

## 230.28 Service Masts Used as Supports

The service mast shall have adequate mechanical strength, or braces or guy wires shall support it, to withstand the strain caused by the service-drop conductors as determined by the AHJ.

**AUTHOR'S COMMENT:** Some local building codes require a minimum 2 in. rigid metal conduit to be used for the service mast. In addition, many electric utilities contain specific requirements for the service mast.

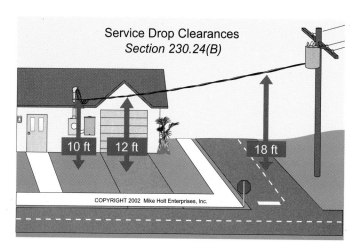

**Figure 230-11**

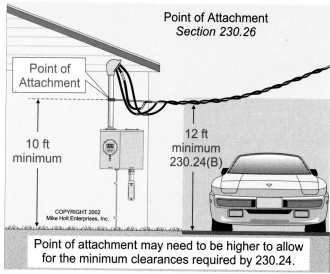

**Figure 230-12**

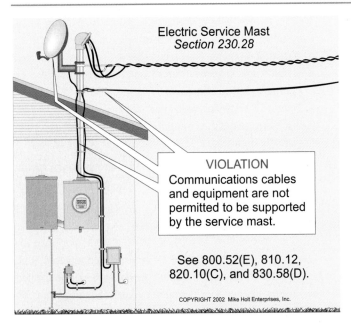

Electric Service Mast
*Section 230.28*

VIOLATION
Communications cables
and equipment are not
permitted to be supported
by the service mast.

See 800.52(E), 810.12,
820.10(C), and 830.58(D).

COPYRIGHT 2002 Mike Holt Enterprises, Inc.

**Figure 230-13**

Only electric utility service-drop conductors can be attached to a service mast, and then only with listed devices.

> **AUTHOR'S COMMENT:** Sections 810.12 and 820.10(C) specify that aerial cables for radio, TV, or CATV cannot be attached to the electric service mast, and 810.12 prohibits antennas from being attached to the service mast. In addition, 800.52(E) and 830.58(D) prohibit communications cables from being attached to raceways, including a service mast for power conductors. Figure 230-13

## Part III. Underground Service Lateral Conductors

> **AUTHOR'S COMMENT:** Underground service-lateral conductors installed by the electric utility shall be in accordance with the National Electric Safety Code (NESC), not the NEC [90.2(B)(5)], but underground conductors not under the exclusive control of the electric utility shall be installed in accordance with the NEC.

### 230.31 Size and Rating

**(A)** **Service-Laterals.** Service-lateral conductors shall have adequate mechanical strength and sufficient ampacity in accordance with Article 220.

**(B)** **Ungrounded Conductor Size.** Service-lateral conductors shall not be smaller than 8 AWG copper or 6 AWG aluminum.

*Exception:* Service-lateral conductors can be as small as 12 AWG for limited-load installations.

**(C)** **Grounded (neutral) Conductor Size.** The grounded (neutral) service-lateral conductor shall be sized to carry the maximum unbalanced load in accordance with 220.22 and shall not be sized smaller than required by 250.24(B).

> **AUTHOR'S COMMENT:** Section 250.24(B) for services requires that a grounded (neutral) conductor be bonded to the service equipment enclosure and this conductor shall be sized in accordance with Table 250.66.

### 230.32 Protection Against Damage

Underground service conductors shall be protected against physical damage and be installed in accordance with the minimum burial depths listed in Table 300.5. Service-lateral conductors entering a building shall be encased in 2 in. of concrete or brick in accordance with 230.6, or they shall be protected by a raceway identified in 230.43.

## Part IV. Service–Entrance Conductors

### 230.40 Number of Service-Entrance Conductor Sets

Each service drop or lateral shall supply only one set of service-entrance conductors.

*Exception No. 1:* Buildings with more than one occupancy can have one set of service-entrance conductors for each service of different characteristics [230.2(D)] run to each occupancy.

*Exception No. 2:* One set of service-entrance conductors can supply two to six service disconnecting means as permitted in 230.71(A).

*Exception No. 3:* A single-family dwelling unit with a separate structure can have one set of service- entrance conductors run to each structure from a single service drop or lateral.

### 230.42 Size and Rating

**(A)** **Load Calculations.** Service-entrance conductors shall have sufficient ampacity for the loads to be served in accordance with Article 220.

**(1)** **Continuous Loads.** Service conductors that supply continuous loads shall be sized no less than 125 percent of the continuous loads, plus 100 percent of the non-continuous loads. The conductor is selected based on the conductor ampacities as listed in Table 310.16, before any ampacity adjustment in accordance with the terminal rating [110.14(C)].

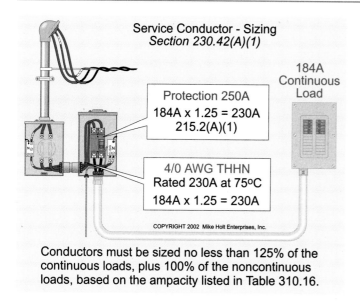

Service Conductor - Sizing
Section 230.42(A)(1)

184A Continuous Load

Protection 250A
184A x 1.25 = 230A
215.2(A)(1)

4/0 AWG THHN
Rated 230A at 75ºC
184A x 1.25 = 230A

COPYRIGHT 2002 Mike Holt Enterprises, Inc.

Conductors must be sized no less than 125% of the continuous loads, plus 100% of the noncontinuous loads, based on the ampacity listed in Table 310.16.

**Figure 230-14**

**Question.** What size service conductor is required for a 184A continuous load if the terminals are rated for 75ºC? Figure 230-14

**Answer:** 4/0 AWG

Step 1. Size the conductors at 125 percent of the load.

184A load × 1.25 = 230A, 4/0 AWG THHN is rated 230A at 75ºC, Table 310.16

**AUTHOR'S COMMENT:** Protection devices shall be sized no smaller than 125 percent of the continuous load 184A load x 1.25 = 230A, next size up 250A [215.3 and 240.6(A)].

**(C) Grounded (neutral) Conductor Size.** The grounded (neutral) service conductor shall be sized to carry the maximum unbalanced load in accordance with 220.22 and shall not be sized smaller than required by 250.24(B).

## 230.43 Wiring Methods

Service conductors shall be installed in one of the following wiring methods:

(1) Open wiring on insulators
(3) Rigid metal conduit, RMC
(4) Intermediate metal conduit, IMC
(5) Electrical metallic tubing, EMT
(6) Electrical nonmetallic tubing, ENT
(7) Service-entrance cables, SE or USE
(8) Wireways

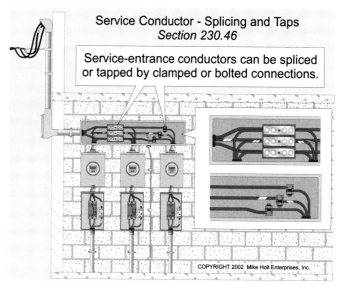

Service Conductor - Splicing and Taps
Section 230.46

Service-entrance conductors can be spliced or tapped by clamped or bolted connections.

COPYRIGHT 2002 Mike Holt Enterprises, Inc.

**Figure 230-15**

(9) Busways
(11) Rigid nonmetallic conduit, RNC
(13) Type MC cable
(15) Flexible metal conduit or liquidtight flexible metal conduit not over 6 ft long
(16) Liquidtight flexible nonmetallic conduit

## 230.46 Spliced Conductors

Service-entrance conductors can be spliced or tapped. Figure 230-15

## 230.50 Protection Against Physical Damage – Aboveground

**(A) Service Cables.** Service cables subject to physical damage shall be protected by any of the following methods: Figure 230-16

(1). Rigid metal conduit, RMC
(2). Intermediate metal conduit, IMC
(3). Schedule 80 rigid nonmetallic conduit, RNC
(4). Electrical metallic tubing, EMT
(5). Other approved means acceptable to the AHJ

**AUTHOR'S COMMENT:** If the AHJ determines that the raceway is not subject to physical damage, Schedule 40 rigid nonmetallic conduit can be used. See 300.5(D).

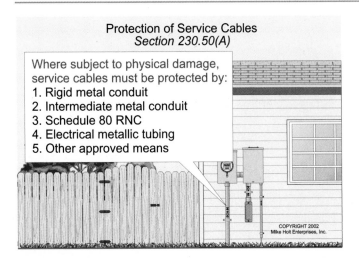

**Figure 230-16**

## 230.51 Service Cable Supports

**(A) Service Cable Supports.** Service-entrance cable shall be supported within 1 ft of service heads and raceway connections and at intervals not exceeding 30 in.

## 230.54 Connections at Service Head (Weatherheads)

**(A) Raintight.** Raceways for overhead service drops shall have a raintight service head.

**(C) Above the Point of Attachment.** Service heads shall be located above the point of attachment. See 230.26.

*Exception:* Where it is impractical to locate the service head above the point of attachment, it shall be located within 2 ft of the point of attachment.

**(E) Opposite Polarity Through Separately Bushed Holes.** Service heads shall provide a bushed opening, and phase conductors shall be in separate openings.

**(F) Drip Loops.** Drip loop conductors shall be below the service head or below the termination of the service-entrance cable sheath.

**(G) Arranged so Water will not Enter.** Service drops and service-entrance conductors shall be arranged to prevent water from entering the service equipment. This is accomplished by installing the point of attachment below the weatherhead in accordance with 230.54(C).

## 230.56 High-Leg Identification

On a 3Ø, 4-wire delta-connected system, where the midpoint of one phase winding is grounded, the conductor or busbar having the higher-phase voltage-to-ground shall be durably and permanently marked by an outer finish that is orange in color, or

by other effective means. Such identification shall be placed at each point on the system where a connection is made if the grounded (neutral) conductor is also present.

**AUTHOR'S COMMENT:** Similar language is contained in 110.15 for branch circuits and 215.8 for feeders.

*WARNING: When replacing disconnects, panelboards, meters, switches, or any equipment that contains the high-leg conductor, care shall be taken to connect the high-leg conductor to the proper terminal. Failure to install the high-leg properly can result in 120V circuits connected to the 208V high-leg, with disastrous consequences. See Figures 110-27 and 215-3.*

*CAUTION: Electric utilities require the high-leg conductor in metering equipment to terminate on the "C" phase, whereas 408.3(E) requires the high-leg conductor to terminate on the "B" or center phase of panelboards and switchboards.*

## Part V. Service Equipment – General

## 230.66 Identified as Suitable for Service Equipment

The service disconnecting means shall be identified as suitable for use as service equipment. This means that the disconnect shall be supplied with a main bonding jumper [250.28] so a neutral-to-ground connection can be made as required in 250.24(B). Figure 230-17.

Individual meter cans are not considered service equipment.

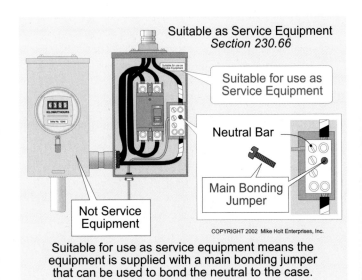

Suitable for use as service equipment means the equipment is supplied with a main bonding jumper that can be used to bond the neutral to the case.

**Figure 230-17**

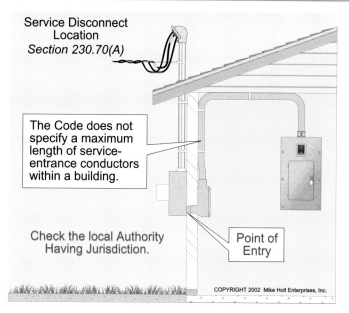

The Code does not specify a maximum length of service-entrance conductors within a building.

Service Disconnect Location
Section 230.70(A)

Check the local Authority Having Jurisdiction.

Point of Entry

COPYRIGHT 2002 Mike Holt Enterprises, Inc.

**Figure 230-18**

## Part VI. Service Equipment – Disconnecting Means

### 230.70 General

The service disconnect shall disconnect all service-entrance conductors from the building or structure premises wiring.

**(A) Location.**

**(1) Readily Accessible.** The service disconnect shall be at a readily accessible location either outside the building or structure, or inside nearest the point of entry of the service conductors.

*WARNING: Service conductors do not have short-circuit or ground-fault protection, so they must be limited in length when installed inside a building. Some local jurisdictions have a specific requirement as to the maximum length permitted within a building. Figure 230-18*

**AUTHOR'S COMMENT:** If the service disconnect is outdoors, the NEC does not require it to be located on the building or structure. Check with the AHJ on how far the disconnecting means can be from the building.

**(2) Bathrooms.** The service disconnecting means cannot be installed in a bathroom. Figure 230-19

**(3) Remote Control of Service Disconnect.** Where a remote-control device is used to actuate the service disconnecting means, the service disconnecting means shall be at a readily accessible location either outside the building or structure, or nearest the point of entry of

Service Disconnect - In Bathroom
Section 230.70(A)(2)

VIOLATION
Service disconnect is not permitted in a bathroom.

Bathroom

COPYRIGHT 2002 Mike Holt Enterprises, Inc.

240.24(E) prohibits overcurrent protection devices in dwelling unit or guest room bathrooms.

**Figure 230-19**

the service conductors as required by 230.70(A)(1). Figure 230-20

**AUTHOR'S COMMENT:** The construction of the disconnecting means shall be in accordance with 230.76, and a pushbutton that activates the electromagnetic coil of a shunt-trip circuit breaker does not meet this requirement. See 230.71(A) and 230.76.

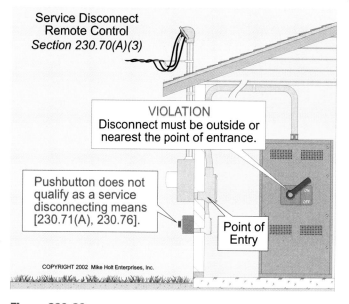

Service Disconnect Remote Control
Section 230.70(A)(3)

VIOLATION
Disconnect must be outside or nearest the point of entrance.

Pushbutton does not qualify as a service disconnecting means [230.71(A), 230.76].

Point of Entry

COPYRIGHT 2002 Mike Holt Enterprises, Inc.

**Figure 230-20**

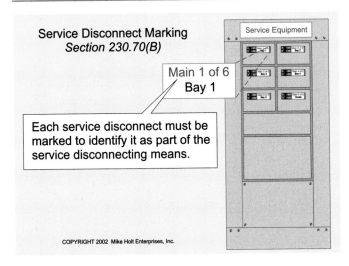

**Figure 230-21**

**(B) Disconnect Identification.** Each service disconnect shall be permanently marked to identify it as part of the service disconnecting means. All disconnecting means to be legibly marked to indicate their purpose. In addition, the marking shall be of sufficient durability to withstand the environment involved [110.22]. Figure 230-21

---

**AUTHOR'S COMMENT:** When a building or structure has two or more services, a plaque is required at each service location to show the location of the other service. See 230.2(E).

---

**(C) Suitable for the Conditions.** Service disconnecting means shall be suitable for the prevailing conditions.

## 230.71 Number of Disconnects

**(A) Maximum.** There shall be no more than six disconnects for each service permitted by 230.2, or each set of service-entrance conductors permitted by 230.40, Ex. 1, 3, 4 or 5.

The service disconnecting means can consist of up to six switches or six circuit breakers mounted in a single enclosure, or mounted in a single enclosure, in a group of separate enclosures, or in or on a switchboard. Figure 230-22

*CAUTION: The rule is six disconnects for each service, not each building. If the building has two services, then there can be a total of twelve disconnects: two groups of six. Figure 230-23*

## 230.72 Grouping of Disconnects

**(A) Two to Six Disconnects.** The disconnecting means for each service shall be grouped.

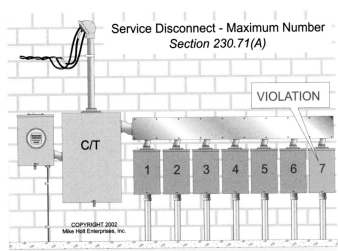

The service disconnect can consist of up to six switches or six circuit breakers mounted in a single enclosure, in a group of separate enclosures.

**Figure 230-22**

**(B) Fire Pump and Standby Power Service.** To minimize the possibility of accidental interruption of power, the disconnecting means for fire pumps or standby power services, as permitted in 230.2(A)(1), shall be located remotely away from the two to six disconnects for normal service.

**(C) Access to Occupants.** In a multiple-occupancy building, each occupant shall have access to his or her disconnecting means.

*Exception:* In multiple-occupancy buildings where electrical maintenance is provided by continuous building management, the service disconnecting means can be accessible only to building management personnel.

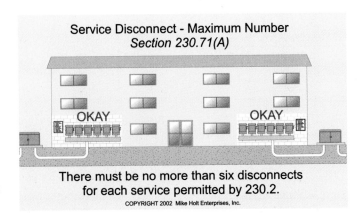

There must be no more than six disconnects for each service permitted by 230.2.

**Figure 230-23**

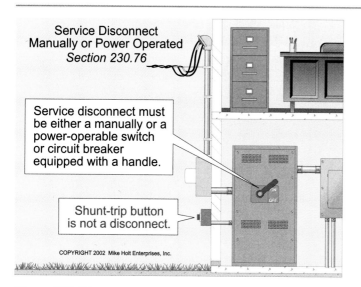

Service Disconnect
Manually or Power Operated
*Section 230.76*

Service disconnect must be either a manually or a power-operable switch or circuit breaker equipped with a handle.

Shunt-trip button is not a disconnect.

COPYRIGHT 2002 Mike Holt Enterprises, Inc.

**Figure 230-24**

## 230.76 Manual or Power-Operated Circuit Breakers

The disconnecting means can consist of either a manually operable switch or circuit breaker, or a power-operated switch or circuit breaker. If power operated, the switch or circuit breaker shall be capable of being operated manually. Figure 230-24

## 230.79 Rating of Disconnect

The disconnecting means for the building or structure shall have an ampere rating of not less than the computed load according to Article 220 and in no case less than:

**(A) 15A.** For a single branch circuit.

**(B) 30A.** For an installation consisting of not more than two 2-wire branch circuits.

**(C) 100A.** For a one-family dwelling where the initial computed load is 10 kVA or more, or where the initial installation consists of six or more 2-wire branch circuits.

**(D) 60A.** For all other installations.

## 230.82 Equipment on the Supply Side

Electrical equipment shall not be connected to the supply side of the service disconnect enclosure, except:

(2) Meters.

(4) Tap conductors for legal and optional standby power systems, fire-pump equipment, fire and sprinkler alarms, and load (energy) management devices.

**AUTHOR'S COMMENT:** Emergency standby power cannot be connected ahead of service equipment. Figure 230-25

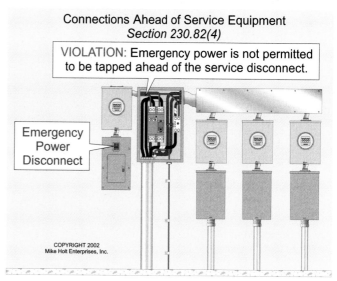

Connections Ahead of Service Equipment
*Section 230.82(4)*

VIOLATION: Emergency power is not permitted to be tapped ahead of the service disconnect.

Emergency Power Disconnect

COPYRIGHT 2002
Mike Holt Enterprises, Inc.

**Figure 230-25**

(5) Solar photovoltaic systems, fuel-cell systems, or interconnected electric power production sources.

## Part VII. Service Equipment Overcurrent Protection

**AUTHOR'S COMMENT:** The NEC does not require service conductors to be provided with short-circuit or ground-fault protection, but overload protection is provided by the feeder protection device.

## 230.90 Overload Protection Required

Each ungrounded service conductor shall have overload protection at the point where the service conductors terminate. See 240.21(D). Figure 230-26

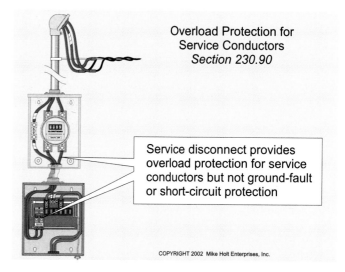

Overload Protection for Service Conductors
*Section 230.90*

Service disconnect provides overload protection for service conductors but not ground-fault or short-circuit protection

COPYRIGHT 2002 Mike Holt Enterprises, Inc.

**Figure 230-26**

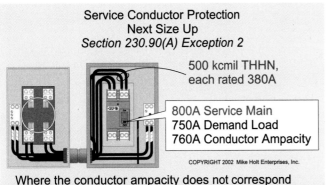

Service Conductor Protection
Next Size Up
*Section 230.90(A) Exception 2*

500 kcmil THHN,
each rated 380A

800A Service Main
750A Demand Load
760A Conductor Ampacity

COPYRIGHT 2002 Mike Holt Enterprises, Inc.

Where the conductor ampacity does not correspond
with the rating of overcurrent protection device, the next
higher device can be used, if not over 800A [240.4(B)].

**Figure 230-27**

**(A) Overcurrent Protection Size.** The rating of the protection device shall not be greater than the ampacity of the conductors.

*Exception No. 2:* Where the ampacity of the conductors does not correspond with the standard rating of overcurrent protection devices as listed in 240.6(A), the next higher protection device can be used if it does not exceed 800A in accordance with 240.4(B).

For example, two sets of 500 kcmil THHN conductors (each rated 380A at 75°C) can be protected by an 800A overcurrent protection device. Figure 230-27

**AUTHOR'S COMMENT:** Typically, conductors are sized to the 75°C ampacity listed in Table 310.16. See 110.14(C) for more information.

*Exception No. 3:* The sum of the ratings of two to six circuit service disconnecting means can exceed the ampacity of the service conductors, provided the calculated load in accordance with Article 220 does not exceed the ampacity of the service conductors. Figure 230-28

*Exception No. 5:* Overload protection for 3-wire, 120/240V dwelling service conductors can be in accordance with the requirements of 310.15(B)(6). Figure 230-29

## 230.95 Ground-Fault Protection of Equipment

Ground-fault protection of equipment shall be provided for each service disconnect rated 480Y/277V, three-phase, 1,000A or more.

The rating of the service disconnect shall be considered to be the rating of the largest fuse that can be installed or the highest continuous current trip setting for which the actual overcurrent device installed in a circuit breaker is rated or can be adjusted.

**AUTHOR'S COMMENT:** See Article 100 for the definition of Ground-Fault Protection of Equipment.

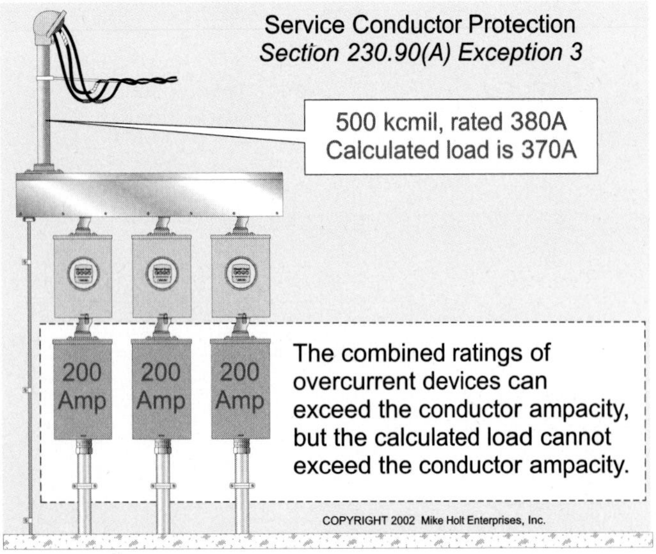

Service Conductor Protection
*Section 230.90(A) Exception 3*

500 kcmil, rated 380A
Calculated load is 370A

200 Amp   200 Amp   200 Amp

The combined ratings of overcurrent devices can exceed the conductor ampacity, but the calculated load cannot exceed the conductor ampacity.

COPYRIGHT 2002 Mike Holt Enterprises, Inc.

**Figure 230-28**

Service Conductor Protection - Dwelling Unit
*Section 230.90(A) Exception 5*

200A

2/0 AWG THHN
Rated 175A

COPYRIGHT 2002
Mike Holt Enterprises, Inc.

Overload protection for 120/240V dwelling unit service conductors can be in accordance with 310.15(B)(6).

**Figure 230-29**

## Article 230

1.  A building or structure shall be supplied by a maximum of _____ service(s).
    (a) one        (b) two        (c) three        (d) as many as desired

2.  Service conductors supplying a building or other structure shall not _____ of another building or other structure.
    (a) be installed on the exterior walls        (b) pass through the interior
    (c) a and b        (d) none of these

3.  Conductors other than service conductors shall not be installed in the same _____.
    (a) service raceway     (b) service cable     (c) enclosure     (d) a or b

4.  Overhead service-drop conductors shall have a horizontal clearance of _____ from a pool.
    (a) 6 ft        (b) 10 ft        (c) 8 ft        (d) 4 ft

5.  When two to six service disconnecting means in separate enclosures are grouped at one location and supply separate loads from one service drop or lateral, _____ set(s) of service-entrance conductors shall be permitted to supply each or several such service equipment enclosures.
    (a) one        (b) two        (c) three        (d) four

6.  Service conductors shall be sized no less than _____ percent of the continuous load, plus 100 percent of the noncontinuous load.
    (a) 100        (b) 115        (c) 125        (d) 150

7.  Wiring methods permitted for service conductors include _____.
    (a) rigid metal conduit        (b) electrical metallic tubing
    (c) rigid nonmetallic conduit        (d) all of these

8.  Service-entrance conductors can be spliced or tapped by clamped or bolted connections at any time as long as _____.
    (a) the free ends of conductors are covered with an insulation that is equivalent to that of the conductors or with an insulating device identified for the purpose [110.14]
    (b) wire connectors or other splicing means installed on conductors that are buried in the earth shall be listed for direct burial [300.5(E)]
    (c) no splice is made in a raceway [300.13(A)]
    (d) all of these

9.  Service cables, where subject to physical damage, shall be protected.
    (a) True        (b) False

10. Service heads must be located _____.
    (a) above the point of attachment        (b) below the point of attachment
    (c) even with the point of attachment        (d) none of these

11. The service disconnecting means shall be installed at a(n) _____ location.
    (a) dry     (b) readily accessible     (c) outdoor     (d) indoor

12. There shall be no more than _____ disconnects installed for each service, or for each set of service-entrance conductors as permitted in 230.2 and 230.40.
    (a) two        (b) four        (c) six        (d) none of these

13. When the service contains two to six service disconnecting means, they shall be _____.
    (a) the same size        (b) grouped at one location
    (c) in the same enclosure        (d) none of these

14.    _____ for power-operable service disconnects can be connected on the supply side of the service disconnecting means if suitable overcurrent protection and disconnecting means are provided.
       (a) Control circuits                          (b) Distribution panels
       (c) Grounding conductors                      (d) none of these

15.    Circuits used only for the operation of fire alarms, other protective signaling systems, or the supply to fire pump equipment shall be permitted to be connected on the _____ of the service overcurrent protection device where separately provided with overcurrent protection.
       (a) base              (b) load side           (c) supply side           (d) top

## Article 240
## Overcurrent Protection

## Part I. General

### 240.1 Scope

Article 240 covers the general requirements for overcurrent protection and the installation requirements of overcurrent protection devices. Figure 240-1 This article is divided into seven parts:

Part I. General

Part II. Location

Part III. Enclosures

Part IV. Disconnecting and Grounding

Part V. Plug Fuses, Fuseholders, and Adapters

Part VI. Cartridge Fuses and Fuseholders

Part VII. Circuit Breakers

Overcurrent is a condition where the current exceeds the rating of conductors or equipment due to overload, short circuit or ground fault [Article 100]. Figure 240-2

FPN: An overcurrent protection device protects the circuit by opening the device when the current reaches a value that will cause an excessive or dangerous temperature rise in conductors. Overcurrent protection devices shall have an interrupting rating sufficient for the maximum possible fault current available on the line-side terminals of the equipment [110.9]. Electrical equipment shall have a short-circuit cur-

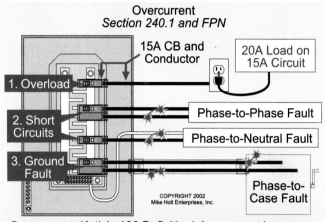

Overcurrent: (Article 100 Definition) Any current in excess of the rated current of equipment or materials. Causes of overcurrent are overloads, short circuits and ground faults.

**Figure 240-2**

rent rating that permits the circuit's overcurrent protection device to clear short circuit or ground faults without extensive damage to the circuit's electrical components [110.10].

### 240.2 Definitions

**Current-Limiting Overcurrent Protective Device.** An overcurrent protective device reduces the fault current to a

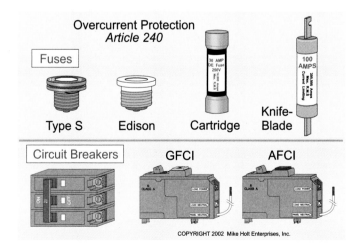

**Figure 240-1**

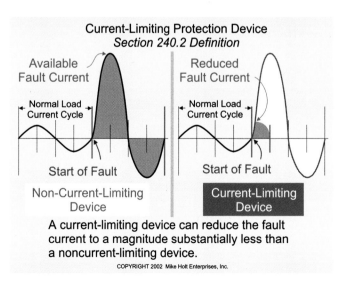

A current-limiting device can reduce the fault current to a magnitude substantially less than a noncurrent-limiting device.

**Figure 240-3**

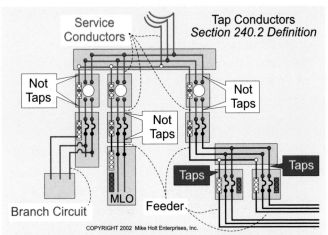

Figure 240-4

*Tap:* Conductors, other than service conductors, that have overcurrent protection ahead of its point of supply that exceeds the value permitted for similar conductors.

magnitude substantially less than that obtainable in the same circuit if the current-limiting device was not used. See 240.40 and 240.60(B). Figure 240-3

**Tap Conductors.** As used in this article, a tap conductor is defined as a conductor, other than a service conductor, that has overcurrent protection ahead of its point of supply that exceeds the value permitted for similar conductors. Figure 240-4

**AUTHOR'S COMMENT:** This definition is needed in the application of branch-circuit taps [210.19(D)] and feeder taps [240.21(B)].

## 240.3 Protection of Equipment

Equipment and their conductors shall be protected against overcurrent in accordance with the following:

Air-Conditioning, 440.22

Branch Circuit, 210.20

Feeder Conductors, 215.3

Flexible Cords, 240.5(B)(1)

Fixed Electric Heating, 424.3(B)

Fixture Wire, 240.5(B)(2)

Panelboards, 408.16(A)

Service Conductors, 230.90(A)

Transformers, 450.3

## 240.4 Protection of Conductors

Except as permitted by (A) through (G), conductors shall be protected against overcurrent in accordance with their ampacity after ampacity adjustment. See 310.15.

**(A)** **Power Loss Hazard.** Conductor overload protection shall not be required, but short-circuit protection shall be provide where the interruption of the circuit would create a hazard, such as in a material-handling magnet circuit or fire-pump circuit.

**(B)** **Overcurrent Protection Not Over 800A.** The next higher standard overcurrent device rating (above the ampacity of the conductors being protected) is permitted, provided all of the following conditions are met:

(1) The conductors do not supply multioutlet receptacle branch circuits.

(2) The ampacity of a conductor, after ampacity adjustment, does not correspond with the standard rating of a fuse or circuit breaker in 240.6(A).

(3) The protection device rating does not exceed 800A. For example, two sets of 500 kcmil THHN conductors (each rated 380A at 75°C) can be protected by an 800A overcurrent protection device. Figure 240-5

**AUTHOR'S COMMENT:** Typically, conductors are sized to the 75°C ampacity listed in Table 310.16. See 110.14(C) for more information.

**(C)** **Overcurrent Protection Over 800A.** If the circuit's overcurrent protection device exceeds 800A, the circuit conductor ampacity, after ampacity adjustment, shall have a rating not less than the rating of the overcurrent device.

Overcurrent Protection - Not Over 800A
Next Size Up
*Section 240.4(B)(3)*

Where the conductor ampacity does not correspond with the standard rating of an overcurrent protection device, the next higher device rating can be used if it is not over 800A.

400A

500 kcmil THHN
Rated 380A

COPYRIGHT 2002 Mike Holt Enterprises, Inc.

Figure 240-5

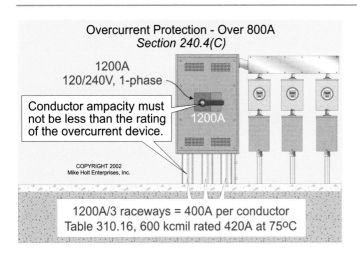

Overcurrent Protection - Over 800A
*Section 240.4(C)*

1200A
120/240V, 1-phase

Conductor ampacity must not be less than the rating of the overcurrent device.

1200A

COPYRIGHT 2002
Mike Holt Enterprises, Inc.

1200A/3 raceways = 400A per conductor
Table 310.16, 600 kcmil rated 420A at 75ºC

*Figure 240-6*

For example, a 1,200A, 120/240V, 1Ø feeder requires three sets of 600 kcmil conductors per phase, where each conductor has an ampacity of 420A at 75°C in accordance with Table 310.16. Figure 240-6

**(D) Small Conductors.** Unless specifically permitted in 240.4(E), (F) and (G), overcurrent protection shall not exceed 15A for 14 AWG, 20A for 12 AWG, and 30A for 10 AWG copper, or 15A for 12 AWG and 25A for 10 AWG aluminum. Figure 240-7

**(E) Tap Conductors.** The overcurrent protection requirements for tap conductors are contained in 210.19(A)(3) and (4) for branch circuits, 240.5(B)(2) for fixture wires, 240.21(B) for feeders, 240.21(C) for secondary conductors, and 430.53(D) for motors.

**(F) Transformer Secondary Conductors.** The primary overcurrent protection device can protect the secondary conductors of a 2-wire system, provided the protective device is sized in accordance with 450.3, it does not exceed the value determined by multiplying the secondary conductor ampacity by the secondary-to-primary transformer voltage ratio.

**AUTHOR'S COMMENT:** The primary protection device does not protect the secondary conductors of a 3- or 4-wire system.

**(G) Overcurrent for Specific Applications.** Overcurrent protection for equipment and conductors shall be in accordance with the equipment article. For example:

**Air-Conditioning or Refrigeration [Article 440].** Air-conditioning and refrigeration equipment and circuit conductors shall be protected against overcurrent in accordance with 440.22.

**Question:** What size conductor and protection device is required for an air conditioner, when the nameplate indicates that the minimum circuit ampacity is 23A and the maximum overcurrent protection setting is 40A fuses? Figure 240-8

(a) 12 AWG, 40A breaker      (b) 12 AWG, 40A fuses

(c) a or b                              (d) none of these

**Answer:** (b) 12 AWG, 40A fuses

Size of Conductor [440.32]

18A × 1.25 = 22.5A, 12 AWG THHN is rated 25A at 60°C

Size of Overcurrent Protection [440.22(A)]

18A × 2.25 = 40.5A, 40A fuses in accordance with the manufacturer's instructions [110.3(B) and 240.6(A)]

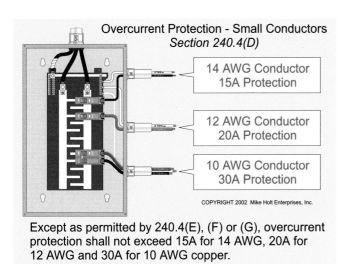

Overcurrent Protection - Small Conductors
*Section 240.4(D)*

14 AWG Conductor
15A Protection

12 AWG Conductor
20A Protection

10 AWG Conductor
30A Protection

COPYRIGHT 2002 Mike Holt Enterprises, Inc.

Except as permitted by 240.4(E), (F) or (G), overcurrent protection shall not exceed 15A for 14 AWG, 20A for 12 AWG and 30A for 10 AWG copper.

*Figure 240-7*

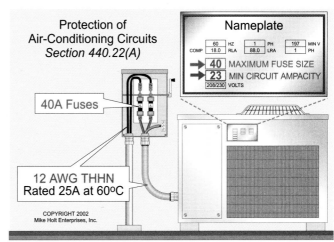

Protection of
Air-Conditioning Circuits
*Section 440.22(A)*

Nameplate

| COMP | 60 HZ | 1 PH | 197 MIN V |
| | 18.0 RLA | 88.0 LRA | 1 PH |

**40** MAXIMUM FUSE SIZE
**23** MIN CIRCUIT AMPACITY
208/230 VOLTS

40A Fuses

12 AWG THHN
Rated 25A at 60ºC

COPYRIGHT 2002
Mike Holt Enterprises, Inc.

*Figure 240-8*

**Motors [Article 430].** Motor circuit conductors shall be protected against short circuits and ground faults in accordance with the requirements of 430.52 and 430.62 [430.51].

**Question:** What size branch-circuit conductor and protection device (circuit breaker) is required for a 7$^1/_2$-hp, 230V, 3Ø motor? Figure 240-9

(a) 10 AWG, 50A breaker      (b) 10 AWG, 60A breaker

(c) a or b                   (d) none of these

**Answer:** (c) 10 AWG, 50A or 60A breaker
Branch-Circuit Conductor [Table 310.16, 430.22, and Table 430.150]
22A × 1.25 = 28A, 10 AWG, rated 30A at 60°C
Branch-Circuit Protection [240.6(A), 430.52(C)(1) Ex. 1, Table 430.150]
Inverse-time breaker: 22A × 2.5 = 55A, next size up = 60A

**Motor Control [Article 430].** Motor control circuit conductors shall be sized and protected in accordance with 430.72.

**Remote-Control, Signaling, and Power-Limited Circuits [Article 725].** Remote-control, signaling, and power-limited circuit conductors shall be protected against overcurrent according to 725.23 and 725.41.

### 240.5 Protection of Flexible Cords and Fixture Wires

**(B) Branch-Circuit Overcurrent Protection**

**(1) Cord for Listed Appliance or Portable Lamps.** Where flexible cord is approved for and used with a listed appliance or portable lamp, it can be supplied by a branch circuit in accordance with the following:

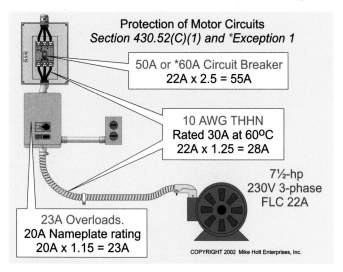

Figure 240-9

(1) 20A circuits – 18 AWG cord and larger

(2) 30A circuits – 16 AWG cord and larger

(3) 40A circuits – cord of 20A capacity and over

(4) 50A circuits – cord of 20A capacity and over

**(2) Fixture Wire.** Fixture wire can be tapped to branch-circuits in accordance with the following:

(1) 20A circuit – 18 AWG, up to 50 ft of run length

(2) 20A circuit – 16 AWG, up to 100 ft of run length

(3) 20A circuit – 14 AWG and larger

(4) 30A circuit – 14 AWG and larger

(5) 40A circuit – 12 AWG and larger

(6) 50A circuit – 12 AWG and larger

### 240.6 Standard Ampere Ratings

**(A) Fuses and Fixed-Trip Circuit Breakers.** The standard ratings in amperes for fuses and inverse-time breakers are as follows: 1, 3, 6, 10, 15, 20, 25, 30, 35, 40, 45, 50, 60, 70, 80, 90, 100, 110, 125, 150, 175, 200, 225, 250, 300, 350, 400, 450, 500, 600, 601, 700, 800, 1000, 1200, 1600, 2000, 2500, 3000, 4000, 5000 and 6000. Figure 240-10

**AUTHOR'S COMMENT:** Fuses rated less than 15A are sometimes required for the protection of fractional horsepower motor circuits [430.52], motor control circuits [430.72], and remote-control circuit conductors [725.23].

**(B) Adjustable Circuit Breakers.** The ampere rating of an adjustable circuit breaker is the maximum possible long-time pickup current setting that is permitted.

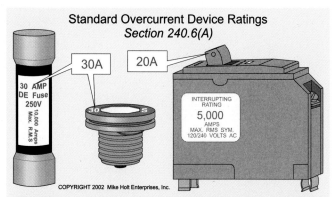

The standard ratings for fuses and inverse time breakers include: 1, 3, 6, 10, 15, 20, 25, 30, 35, 40, 45, 50, 60,70, 80, 90, 100, 110, 125, 150, 175, 200, 225, 250, 300, 350, 400, 450, 500, 600, 601, 700, 800, 1000, 1200 amperes.

Figure 240-10

**(C) Restricted Access Adjustable-Trip Circuit Breakers.** A circuit breaker that has restricted access to the adjusting means can have an ampere rating equal to the long-time pickup current setting.

## 240.10 Supplementary Overcurrent Protection.

Supplementary overcurrent protection devices often used for luminaires, appliances, and other equipment or for internal circuits and components of equipment are not required to be readily accessible.

## 240.13 Ground Fault Protection of Equipment

Each circuit rated 1,000A or more installed on solidly grounded 4-wire, 480Y/277V, 3Ø systems shall be protected against ground-fault protection in accordance with the requirements contained in 230.95. See 215.10.

> **AUTHOR'S COMMENT:** See Article 100 for the definition of *Ground-Fault Protection of Equipment*.

Ground-Fault Protection of Equipment does not apply to:

(1) Continuous industrial processes where a nonorderly shutdown would introduce additional or increased hazards.

(2) Installations where ground-fault protection of equipment is already provided.

(3) Fire pumps installed in accordance with Article 695.

## Part II. Location

## 240.20 Ungrounded Conductors

**(A) Overcurrent Protection Device Required.** Overcurrent protection devices shall be installed in each ungrounded conductor.

**(B) Circuit Breaker as an Overcurrent Protection Device.** Circuit breakers shall open all ungrounded conductors of the circuit, except as permitted by (1), (2), or (3):

(1) **Multiwire Branch Circuit.** Except where limited by 210.4(B) for dwelling units, individual single-pole breakers can be installed on each ungrounded conductor of a multiwire branch circuit that supplies single-phase line-to-neutral loads. Figure 240-11

> **AUTHOR'S COMMENT:** Multiwire branch circuits that terminate on devices mounted on the same yoke in a dwelling unit shall be provided with a means to disconnect simultaneously all ungrounded circuit conductors. This can be accomplished by single-pole circuit

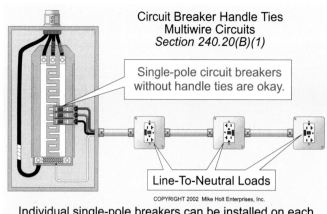

Circuit Breaker Handle Ties
Multiwire Circuits
*Section 240.20(B)(1)*

Single-pole circuit breakers without handle ties are okay.

Line-To-Neutral Loads

COPYRIGHT 2002 Mike Holt Enterprises, Inc.

Individual single-pole breakers can be installed on each ungrounded (hot) conductor of a multiwire branch circuit if it only supplies single-phase line-to-neutral loads.

**Figure 240-11**

breakers with approved handle ties or a 2-pole breaker with common internal trip [210.4(B)]. Figure 240-12

(2) **Single-Phase Line-to-Line Loads.** Individual single-pole circuit breakers with an approved handle tie can be used for each ungrounded conductor of a branch circuit that supplies single-phase line-to-line loads. Figure 240-13

(3) **Three-Phase Line-to-Line Loads.** Individual single-pole breakers with an approved handle tie can be used for each ungrounded conductor of a branch circuit that serves three-phase line-to-line loads. Figure 240-14

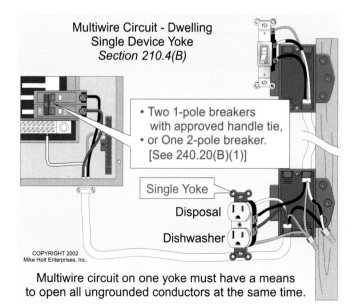

Multiwire Circuit - Dwelling
Single Device Yoke
*Section 210.4(B)*

• Two 1-pole breakers with approved handle tie,
• or One 2-pole breaker.
[See 240.20(B)(1)]

Single Yoke

Disposal

Dishwasher

COPYRIGHT 2002
Mike Holt Enterprises, Inc.

Multiwire circuit on one yoke must have a means to open all ungrounded conductors at the same time.

**Figure 240-12**

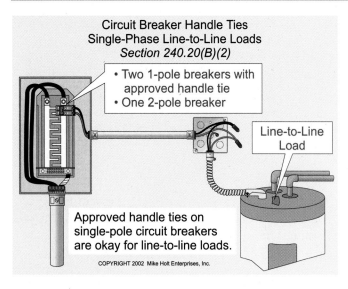

**Figure 240-13**

---

**AUTHOR'S COMMENT:** Handle ties made from nails, screws, wires or other nonconforming methods are generally not approved by the AHJ. Figure 240-15

---

## 240.21 Location in Circuit

Except as permitted by (A) through (G), overcurrent protection devices shall be placed at the point where the branch or feeder conductors receive their power.

A tap conductor is not permitted to supply another tap conductor. In other words, you can not make a tap from a tap.

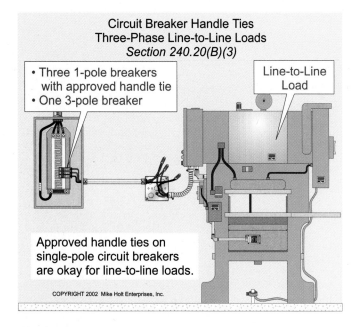

**Figure 240-14**

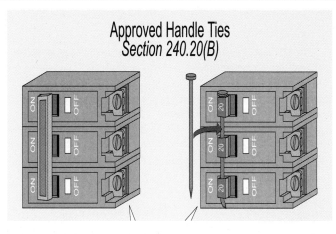

**Figure 240-15**

**(A) Branch-Circuit Taps.** For branch-circuit taps, see 210.19(A)(4) Ex. No.1.

**(B) Feeder Taps.** Conductors can be tapped from a feeder if they are installed in accordance with the following:

**(1) 10-Foot Feeder Tap Rule**

Feeder tap conductors not over 10 ft without overcurrent protection at the point they receive their supply are permitted, but they shall be installed in accordance with the following: Figure 240-16

(1) The ampacity of the tap conductor shall not be less than:

• The computed load in accordance with Article 220, and

• The rating of the device supplied by the tap conductors or the overcurrent protective device at the termination of the tap conductors.

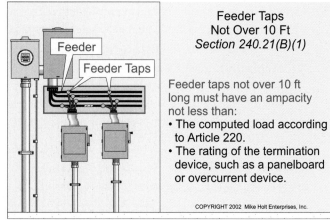

**Figure 240-16**

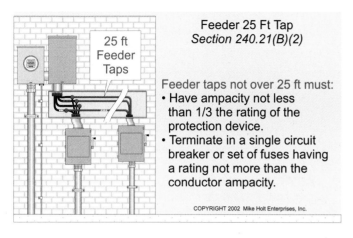

Feeder 25 Ft Tap
*Section 240.21(B)(2)*

Feeder taps not over 25 ft must:
• Have ampacity not less than 1/3 the rating of the protection device.
• Terminate in a single circuit breaker or set of fuses having a rating not more than the conductor ampacity.

COPYRIGHT 2002 Mike Holt Enterprises, Inc.

**Figure 240-17**

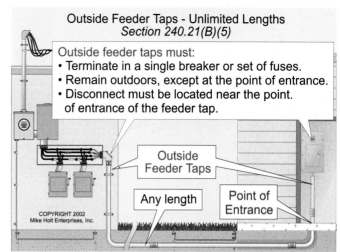

Outside Feeder Taps - Unlimited Lengths
*Section 240.21(B)(5)*

Outside feeder taps must:
• Terminate in a single breaker or set of fuses.
• Remain outdoors, except at the point of entrance.
• Disconnect must be located near the point. of entrance of the feeder tap.

Outside Feeder Taps

Any length

Point of Entrance

COPYRIGHT 2002 Mike Holt Enterprises, Inc.

**Figure 240-18**

(2) The tap conductors shall not extend beyond the equipment they supply.

(3) The tap conductors shall be installed in a raceway if they leave the enclosure.

(4) The tap conductors shall have an ampacity of not less than 10 percent of the ampacity of the overcurrent protection device from which the conductors are tapped.

**(2) 25-Foot Feeder Tap Rule**

Feeder tap conductors not over 25 ft without overcurrent protection at the point they receive their supply are permitted, but they shall be installed in accordance with the following: Figure 240-17

(1) The ampacity of the tap conductors shall not be less than one-third the ampacity of the overcurrent protection device protecting the feeder.

(2) The tap conductors shall terminate in a single circuit breaker, or set of fuses having a rating no greater than the tap conductor ampacity as listed in Table 310.16.

(3) The tap conductors shall be suitably protected from physical damage or they shall be enclosed in a raceway.

**(5) Outside Feeder Tap of Unlimited Length Rule**

Outside feeder tap conductors can be of unlimited length without overcurrent protection at the point they receive their supply, but they shall be installed in accordance with the following: Figure 240-18

(1) The tap conductors shall be suitably protected from physical damage.

(2) The tap conductors shall terminate at a single circuit breaker or a single set of fuses that limit the load to the ampacity of the conductors.

(3) The overcurrent device for the tap conductors shall be an integral part of the disconnecting means or it shall be located immediately adjacent thereto.

(4) The disconnect shall be located at a readily accessible location either outside the building or structure, or nearest the point of entry of the conductors.

**(C) Transformer Secondary Conductors.** Secondary conductors can be run without secondary overcurrent protection when installed in accordance with the following:

**(2) 10-Foot Secondary Conductor Rule**

Secondary conductors can be run up to 10 ft without overcurrent protection, but they shall be installed in accordance with the following: Figure 240-19

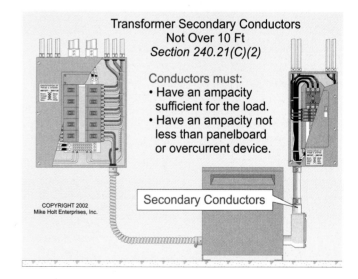

Transformer Secondary Conductors
Not Over 10 Ft
*Section 240.21(C)(2)*

Conductors must:
• Have an ampacity sufficient for the load.
• Have an ampacity not less than panelboard or overcurrent device.

Secondary Conductors

COPYRIGHT 2002 Mike Holt Enterprises, Inc.

**Figure 240-19**

(1) The ampacity of the secondary conductor shall not be less than:

- The computed load in accordance with Article 220, and

- The rating of the device supplied by the tap conductors or the overcurrent protective device at the termination of the tap conductors.

(2) The secondary conductors shall not extend beyond the switchboard, panelboard, disconnecting means, or control devices they supply.

(3) The secondary conductors shall be enclosed in a raceway.

**AUTHOR'S COMMENT:** Lighting and appliance branch-circuit panelboards require overcurrent protection to be located on the secondary side of the transformer. See 408.16(D).

### (4) Outside Secondary Conductors of Unlimited Length

Outside secondary conductors can be run of unlimited length without overcurrent protection at the point they receive their supply, but they shall be installed in accordance with the following: Figure 240-20

(1) The conductors shall be suitably protected from physical damage.

(2) The conductors shall terminate at a single circuit breaker or a single set of fuses that limit the load to the ampacity of the conductors.

(3) The overcurrent device for the conductors shall be an integral part of a disconnecting means or it shall be located immediately adjacent thereto.

(4) The disconnecting means shall be located at a readily accessible location complying with one of the following:

a. Outside of a building or structure.

b. Inside, nearest the point of entrance of the conductors.

c. Where installed in accordance with 230.6, nearest the point of entrance of the conductors,

**AUTHOR'S COMMENT:** This rule requires outside secondary conductors from customer owned transformers to terminate in a single overcurrent protection device. However, outside secondary conductors from transformers under the exclusive control of the electric utility can terminate in up to six overcurrent protection devices in accordance with 230.71(A).

### (6) 25-Foot Secondary Conductor Rule

Secondary conductors run not over 25 ft without overcurrent protection at the point they receive their supply are permitted, but they shall be installed in accordance with the following: Figure 240-21

(1) The secondary conductors shall have an ampacity that (when multiplied by the ratio of the secondary-to-primary voltage) is at least one-third the rating of the overcurrent device protecting the primary of the transformer.

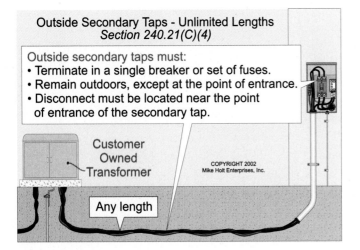

**Outside Secondary Taps - Unlimited Lengths**
*Section 240.21(C)(4)*

Outside secondary taps must:
- Terminate in a single breaker or set of fuses.
- Remain outdoors, except at the point of entrance.
- Disconnect must be located near the point of entrance of the secondary tap.

Customer Owned Transformer

COPYRIGHT 2002 Mike Holt Enterprises, Inc.

Any length

**Figure 240-20**

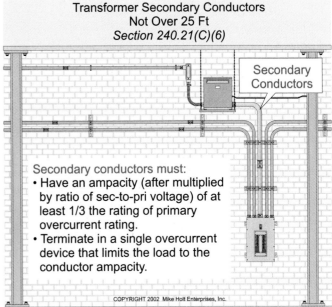

**Transformer Secondary Conductors Not Over 25 Ft**
*Section 240.21(C)(6)*

Secondary Conductors

Secondary conductors must:
- Have an ampacity (after multiplied by ratio of sec-to-pri voltage) of at least 1/3 the rating of primary overcurrent rating.
- Terminate in a single overcurrent device that limits the load to the conductor ampacity.

COPYRIGHT 2002 Mike Holt Enterprises, Inc.

**Figure 240-21**

(2) The secondary conductors shall terminate in a single circuit breaker or set of fuses that has a rating not greater than the secondary conductor ampacity.

(3) The secondary conductors shall be protected from physical damage.

**AUTHOR'S COMMENT:** The secondary conductor shall be sized no smaller than one-third the ampacity of the primary protection device based on the secondary/primary ratio. For all practical purposes, this can be ignored, because the secondary conductors are typically sized at 100 percent of the primary protection device rating.

**(D) Service Conductors.** Service-entrance conductors shall be protected against overload in accordance with 230.91.

## 240.24 Location in Premises

**(A) Readily Accessible.** Overcurrent protection devices shall be readily accessible, except for the following:

**AUTHOR'S COMMENT:** "Readily accessible" means located so it is capable of being reached quickly without having to climb over or remove obstacles. Figure 240-22

(2) Supplementary overcurrent protection devices are not required to be readily accessible. See 240.10. Figure 240-23

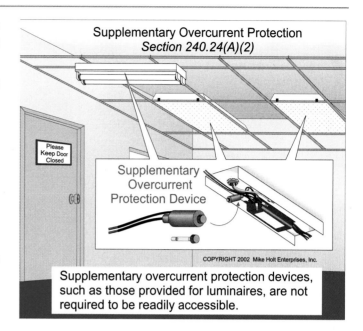

Supplementary overcurrent protection devices, such as those provided for luminaires, are not required to be readily accessible.

**Figure 240-23**

(4) Overcurrent protection devices located next to equipment they supply can be accessible by portable means. See 404.8(A) Ex. 2. Figure 240-24

**(B) Occupancy.** Each occupant shall have ready access to all overcurrent protection devices protecting the conductors supplying that occupancy.

*Exception No. 1:* Service and feeder overcurrent protection devices are not required to be accessible to occupants of multiple-occupancy buildings or guest rooms of hotels and motels, if electric maintenance is provided under continuous building management.

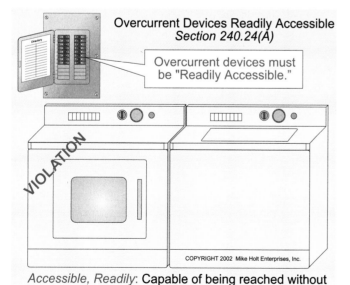

Accessible, Readily: Capable of being reached without the use of ladders or climbing over or moving obstacles.

**Figure 240-22**

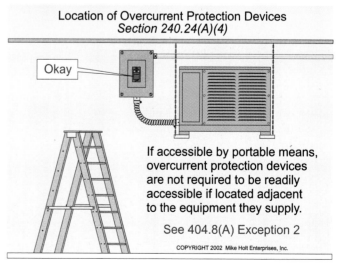

If accessible by portable means, overcurrent protection devices are not required to be readily accessible if located adjacent to the equipment they supply.

See 404.8(A) Exception 2

**Figure 240-24**

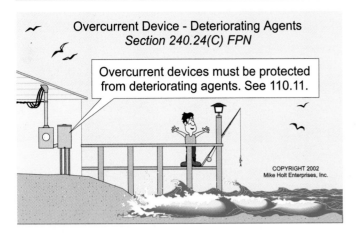

Figure 240-25

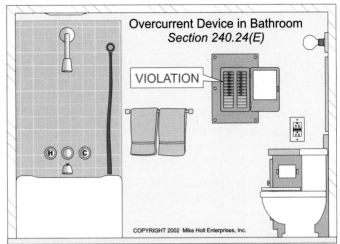

Overcurrent devices are not permitted in dwelling unit
bathrooms or guest room bathrooms of hotels and motels.

Figure 240-27

*Exception No. 2:* Branch-circuit overcurrent protection devices are not required to be accessible to occupants of guest rooms of hotels and motels, if electric maintenance is provided in a facility that is under continuous building management.

**(C) Exposed to Physical Damage.** Overcurrent protection devices shall not be exposed to physical damage. See 110.27(B).

> FPN: See 110.11, Deteriorating Agents. Figure 240-25

**(D) Vicinity of Easily Ignitable Material.** Overcurrent protection devices shall not be located near easily ignitable material, such as in clothes closets. The purpose of keeping overcurrent protection devices away from easily ignitable material is to prevent fires, not to keep them out of clothes closets. Figure 240-26

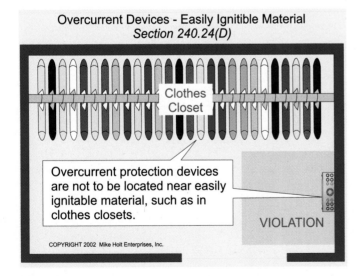

Figure 240-26

**(E) Not in Bathrooms.** Overcurrent protection devices shall not be located in the bathrooms of dwelling units or guest rooms of hotels or motels. Figure 240-27

**AUTHOR'S COMMENT:** The service disconnect is not permitted to be located in any bathroom. See 230.70(A)(2).

## Part III. Enclosures

### 240.32 Damp or Wet Locations

Enclosures in damp or wet locations containing overcurrent protection devices shall prevent moisture or water from entering or accumulating within the enclosure. When the enclosure is surface-mounted in a wet location, the enclosure shall be mounted with at least a $1/4$ in. air space between it and the mounting surface. See 312.2(A).

### 240.33 Vertical Position

Enclosures containing overcurrent protection devices shall be mounted in a vertical position unless this is impracticable. Circuit breaker enclosures can be installed horizontally, if the circuit breaker is installed in accordance with 240.81. Figure 240-28

**AUTHOR'S COMMENT:** Section 240.81 specifies that where circuit breaker handles are operated vertically, the "up" position of the handle shall be in the "on" position. So in effect, an enclosure that contains one circuit breaker can be mounted horizontally, but an enclosure containing a panelboard/loadcenter with multiple

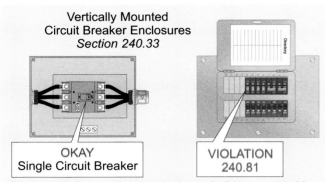

**Figure 240-28**

circuit breakers would have to be mounted vertically. Figure 240-28

## Part V. Plug Fuses, Fuseholders, and Adapters

### 240.51 Edison-base Fuse

(A) **Classification.** Edison-base fuses are classified to operate at not more than 125V and have an ampere rating of not more than 30A.

(B) **Replacement Only.** Edison-base fuses can be used only for replacement in an existing installation where there is no evidence of tampering or overfusing.

*WARNING: Edison-base screw shells do not restrict the installation of a 30A fuse on 14 AWG!*

### 240.53 Type S Fuses

(A) **Classification.** Type S fuses are classified to operate at not more than 125V and have an ampere rating of 0-15A, 16-20A and 21-30A. Figure 240-29

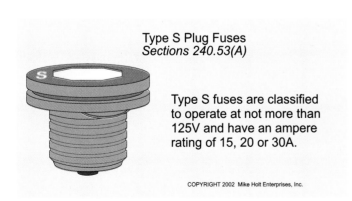

**Figure 240-29**

(B) **Not Interchangeable.** Type S fuses are made so ampere classifications shall not be interchanged.

### 240.54 Type S Fuses, Adapters, and Fuseholders

(A) **Type S Adapters.** Type S adapters are designed to fit Edison-base fuseholders.

(B) **Prevent Edison-base Fuses.** Type S fuseholders and adapters are designed for Type S fuses only.

(C) **Nonremovable Adapters.** Type S adapters are designed so that they cannot be removed once installed.

## Part VI. Cartridge Fuses and Fuseholders

### 240.60 General

**AUTHOR'S COMMENT:** There are two basic shapes of cartridge fuses, the ferrule type with a maximum rating of 60A and the knife-blade type with an ampere rating over 60A. The physical size of the fuse, such as length and diameter, depends on the fuse voltage and current rating. Figure 240-30

(A) **Maximum Voltage – 300V Type.** Cartridge fuses and fuseholders of the 300V type can only be used for:

Circuits not exceeding 300V between conductors.

Circuits not exceeding 300V from any ungrounded conductor to the grounded (neutral) conductor.

(B) **Noninterchangeable Fuseholders.** Fuseholders shall be designed to make interchanging of fuses for different voltages and current classifications difficult. Fuseholders for current-limiting fuses shall be designed so that only current-limiting fuses can be inserted. See 240.2. Figure 240-31

(C) **Marking.** Cartridge fuses shall have an interrupting rating of 10,000A, unless marked otherwise.

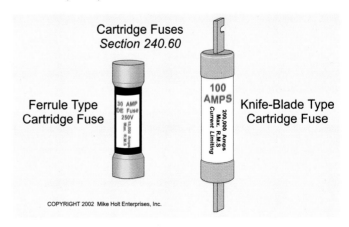

**Figure 240-30**

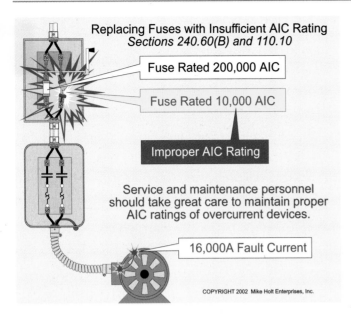

**Replacing Fuses with Insufficient AIC Rating**
*Sections 240.60(B) and 110.10*

Fuse Rated 200,000 AIC

Fuse Rated 10,000 AIC

Improper AIC Rating

Service and maintenance personnel should take great care to maintain proper AIC ratings of overcurrent devices.

16,000A Fault Current

COPYRIGHT 2002 Mike Holt Enterprises, Inc.

**Figure 240-31**

*WARNING: Care should be taken to ensure that fuses have sufficient interrupting circuit rating for the available fault current. Using a fuse with inadequate interrupting current rating could cause equipment to be destroyed from a line-to-line or line-to-ground fault, as well as death or serious injury. See 110.9 for more details. See Figure 240-31*

## 240.61 Classification

Cartridge fuses and fuseholders shall be classified according to voltage and amperage ranges. Fuses rated 600V, nominal, or less can be used for voltages at or below their ratings.

## Part VII. Circuit Breakers

### 240.80 Method of Operation

Circuit breakers shall be capable of being opened and closed by hand. Nonmanual means of operating a circuit breaker, such as electrical (shunt trip) or pneumatic operation, are permitted as long as the circuit breaker can also be operated manually.

### 240.81 Indicating

Circuit breakers shall clearly indicate whether they are in the open "off" or closed "on" position. When the handle of a circuit breaker is operated vertically, the "up" position of the handle shall be the "on" position. See 240.33 and 404.6(C). Figure 240-32

### 240.83 Markings

**(C)  Interrupting Rating.** Circuit breakers shall have an interrupting rating of 5,000A, unless marked otherwise.

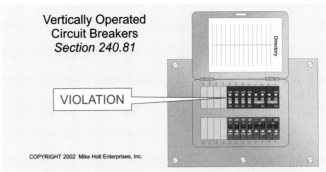

**Vertically Operated Circuit Breakers**
*Section 240.81*

Directory

VIOLATION

COPYRIGHT 2002 Mike Holt Enterprises, Inc.

Where the circuit breaker is operated vertically, the "up" position of the handle must be the "on" position.

**Figure 240-32**

*WARNING: Care shall be taken to ensure that the circuit breaker has sufficient interrupting circuit rating for the available fault current. Using a circuit breaker with inadequate interrupting current rating could cause equipment to be destroyed from a line-to-line or line-to-ground fault, as well as death or serious injury. See 110.9 for more details. Figure 240-33*

**(D)  Used as Switches.** Circuit breakers used to switch 120V or 277V fluorescent lighting circuits shall be listed and marked "SWD" or "HID." Circuit breakers used to switch high-intensity discharge lighting circuits shall be listed and marked "HID." Figure 240-34

**AUTHOR'S COMMENT:** UL 489 Standard on Molded Case Circuit Breakers, permits "HID" breakers to be rated up to 50A, whereas an "SWD" breaker is rated up to 20A. The tests for "HID" breakers include an

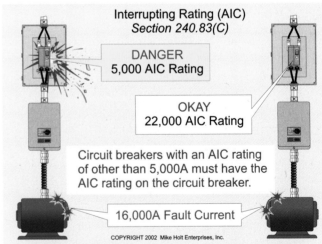

**Interrupting Rating (AIC)**
*Section 240.83(C)*

DANGER
5,000 AIC Rating

OKAY
22,000 AIC Rating

Circuit breakers with an AIC rating of other than 5,000A must have the AIC rating on the circuit breaker.

16,000A Fault Current

COPYRIGHT 2002 Mike Holt Enterprises, Inc.

CAUTION: Overcurrent protection devices must have an interrupting rating that is sufficient for the current that is available at the line terminals of the equipment [110.9].

**Figure 240-33**

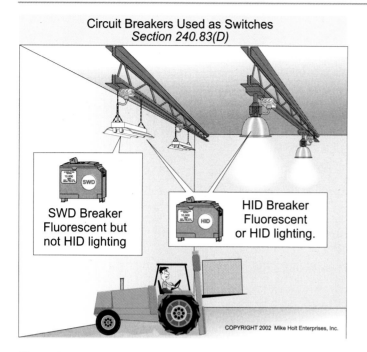

Circuit Breakers Used as Switches
*Section 240.83(D)*

SWD Breaker
Fluorescent but
not HID lighting

HID Breaker
Fluorescent
or HID lighting.

COPYRIGHT 2002 Mike Holt Enterprises, Inc.

**Figure 240-34**

endurance test at 75 percent power factor, whereas "SWD" breakers are endurance tested at 100 percent power factor. The contacts and the spring of an "HID" breaker are a heavier duty to dissipate the increased heat due to greater current flow in the circuit generated from the "HID" luminaire taking a minute or two for the lamp to ignite.

**(E)  Voltage Markings.** Circuit breakers shall be marked with a voltage rating that corresponds with their interrupting rating. See 240.85.

## 240.85 Applications

A circuit breaker with a straight voltage rating such as 240V or 480V, can be used on a circuit where the nominal voltage between any two conductors does not exceed the circuit breaker's voltage rating. Figure 240-35

A circuit breaker with a slash rating, such as 120/240V or 480Y/277V, can be used on a solidly grounded circuit where the nominal voltage of any one conductor to ground does not exceed the lower of the two values and the nominal voltage between any two conductors does not exceed the higher value.

**AUTHOR'S COMMENT:** A 120/240V slash circuit breaker cannot be used on the high-leg of a solidly grounded 120/240V delta system, because the phase-to-ground voltage of the high-leg is 208V, which exceeds the 120 line-to-ground voltage rating. Figure 240-36

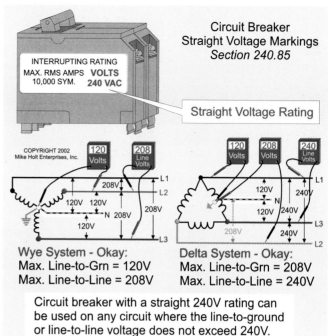

Circuit Breaker
Straight Voltage Markings
*Section 240.85*

INTERRUPTING RATING
MAX. RMS AMPS   **VOLTS**
10,000 SYM.   **240 VAC**

Straight Voltage Rating

COPYRIGHT 2002
Mike Holt Enterprises, Inc.

Wye System - Okay:
Max. Line-to-Grn = 120V
Max. Line-to-Line = 208V

Delta System - Okay:
Max. Line-to-Grn = 208V
Max. Line-to-Line = 240V

Circuit breaker with a straight 240V rating can be used on any circuit where the line-to-ground or line-to-line voltage does not exceed 240V.

**Figure 240-35**

FPN: Circuit breakers on corner-grounded delta systems, should consider the individual pole interrupting capability. For more information on this subject, visit www.mikeholt.com/Newsletters/ 240.85.pdf (73K), or www.mikeholt.com/Newsletters/240.85.doc (137K)

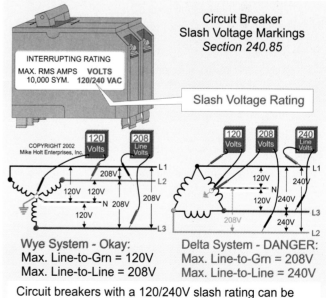

Circuit Breaker
Slash Voltage Markings
*Section 240.85*

INTERRUPTING RATING
MAX. RMS AMPS   **VOLTS**
10,000 SYM.   **120/240 VAC**

Slash Voltage Rating

COPYRIGHT 2002
Mike Holt Enterprises, Inc.

Wye System - Okay:
Max. Line-to-Grn = 120V
Max. Line-to-Line = 208V

Delta System - DANGER:
Max. Line-to-Grn = 208V
Max. Line-to-Line = 240V

Circuit breakers with a 120/240V slash rating can be used where the line-to-ground voltage does not exceed 120V and the line-to-line voltage does not exceed 240V.

**Figure 240-36**

# Article 240

1. Overcurrent protection for conductors and equipment is designed to _____ the circuit if the current reaches a value that will cause an excessive or dangerous temperature in conductors or conductor insulation.
   (a) open     (b) close     (c) monitor     (d) record

2. Unless specifically permitted in 240.4(E) through 240.4(G), the overcurrent protection shall not exceed _____ after any correction factors for ambient temperature and the number of conductors have been applied.
   (a) 15A for 14 AWG copper     (b) 20A for 12 AWG copper
   (c) 30A for 10 AWG copper     (d) all of these

3. Which of the following is not a standard size for fuses or inverse-time circuit breakers?
   (a) 45     (b) 70     (c) 75     (d) 80

4. Which of the following statements about supplementary overcurrent protection is correct?
   (a) Shall not be used in luminaires.
   (b) May be used as a substitute for a branch-circuit overcurrent protection device.
   (c) May be used to protect internal circuits of equipment.
   (d) Shall be readily accessible.

5. Single-pole breakers with approved handle ties can be used for the protection of each ungrounded conductor for line-to-line connected loads.
   (a) True     (b) False

6. Overcurrent protection for tap conductors not over 25 ft is not required at the point where the conductors receive their supply providing the _____.
   (a) ampacity of the tap conductors is not less than one-third the rating of the overcurrent device protecting the feeder conductors being tapped
   (b) tap conductors terminate in a single circuit breaker or set of fuses that limit the load to the ampacity of the tap conductors
   (c) tap conductors are suitably protected from physical damage
   (d) all of these

7. Overcurrent protection devices shall be _____.
   (a) accessible (as applied to wiring methods)     (b) accessible (as applied to equipment)
   (c) readily accessible     (d) inaccessible to unauthorized personnel

8. Enclosures for overcurrent protection devices must be mounted in a _____ position unless that is shown to be impracticable.
   (a) vertical     (b) horizontal     (c) vertical or horizontal     (d) there are no requirements

9. Plug fuses of the Edison-base type shall be used _____.
   (a) where overfusing is necessary     (b) only as replacement in existing installations
   (c) as a replacement for Type S fuses     (d) only for 50A and above

10. Type S fuses, fuseholders, and adapters are required to be designed so that _____ would be difficult.
    (a) installation     (b) tampering     (c) shunting     (d) b or c

11. Fuseholders for cartridge fuses shall be so designed that it is difficult to put a fuse of any given class into a fuseholder that is designed for a lower _____ or a higher _____ than that of the class to which the fuse belongs.
    (a) voltage, wattage     (b) wattage, voltage     (c) voltage, current     (d) current, voltage

12. Circuit breakers, used as switches in 120V or 277V fluorescent-lighting circuits, shall be listed and marked _____.
    (a) UL     (b) SWD or HID     (c) Amps     (d) VA

13. Circuit breakers used to switch high-intensity discharge lighting circuits must be listed and must be marked as _____.
    (a) SWD     (b) HID     (c) a or b     (d) a and b

14. A circuit breaker with a straight voltage rating (240V or 480V) can be used on a circuit where the nominal voltage between any two conductors does not exceed the circuit breaker's voltage rating.
    (a) True             (b) False

15. A circuit breaker with a slash rating (120/240V or 480Y/277V) can be used for a solidly grounded circuit where the nominal voltage of any one conductor to _____ does not exceed the lower of the two values, and the nominal voltage between any two conductors does not exceed the higher value.
    (a) another conductor      (b) an enclosure      (c) earth      (d) ground

# *Introduction to Article 250 – Grounding and Bonding*

The purpose of the *National Electrical Code* is the practical safeguarding of persons and property from hazards arising from the use of electricity [90.1(A)]. In addition, the *NEC* contains provisions that are considered necessary for safety. Compliance with the *NEC* combined with proper maintenance shall result in an installation that is essentially free from hazard [90.1(B)].

## Understanding the Basics of Electrical Systems

Contrasting the electric utility system wiring, which is governed by the National Electrical Safety Code (NESC), with those of premises wiring which are covered by the *NEC*, should be helpful to enhance an understanding of some basic electrical principles such as current flow, neutral and fault-current paths.

**Utility Current Flow.** Electrons leaving a power supply are always trying to return to the same power supply; they are not trying to go into the earth. When alternating current is applied to the primary of a transformer, it induces a voltage in the secondary. This induced secondary voltage causes electrons to leave one end of the transformer's secondary, travel over the circuit's conductors through the load and return over the remaining circuit's conductors to the other end of the transformer's secondary. Figure 250-1

**Utility Neutral Current Path.** The electric utility grounds the primary and secondary neutral conductor to the earth at multiple locations to create a parallel path so as to reduce the impedance of the return neutral current path. This multipoint grounded

utility neutral helps in reducing primary utility neutral voltage drop, the clearing of utility line-to-neutral faults and in reducing elevated line-to-ground voltages caused by ground faults. Figure 250-2

**Utility Ground-Fault Current Path.** Metal parts of the electric utility equipment (transformer and capacitor cases, guy wires, luminaires, etc.) are grounded to the earth and bonded to the grounded (neutral) conductor to provide a low-impedance parallel path for the purpose of clearing a line-to-case ground fault. If the utility grounded (neutral) conductor is inadvertently opened, the earth itself should still have sufficiently low impedance to permit sufficient fault current to flow to blow the fuse, thereby clearing the high-voltage ground fault.

For example, a 7,200V line is typically protected by a 60A to 100A fuse (depending on wire size). The earth, having an impedance of 25Ω, would have no problem carrying sufficient fault current to blow a 100A fuse. (I = E/Z, I = 7,200V/25Ω, I = 288A). Figure 250-3

**Premises Neutral Current Path.** Neutral current should only flow on the grounded (neutral) conductor, not on metal parts of the electrical installation [250.6]. Figure 250-4

**Premises Ground-Fault Current Path.** Metal parts of premises wiring are bonded to a low-impedance path designed and intended to carry fault current from the point of a line-to-case fault on a wiring system to the grounded (neutral) conductor at the electrical supply source. This low-impedance fault-current path

Electrons leaving a voltage source
must return to that voltage source.

**Figure 250–1**

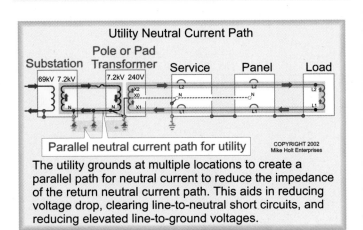

The utility grounds at multiple locations to create a parallel path for neutral current to reduce the impedance of the return neutral current path. This aids in reducing voltage drop, clearing line-to-neutral short circuits, and reducing elevated line-to-ground voltages.

**Figure 250–2**

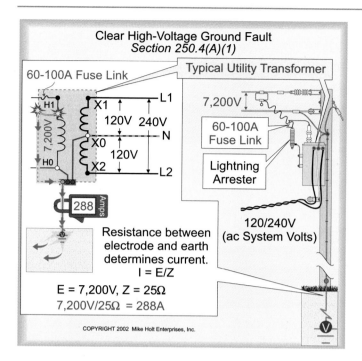

**Clear High-Voltage Ground Fault**
*Section 250.4(A)(1)*

60-100A Fuse Link

Typical Utility Transformer

7,200V

60-100A Fuse Link

Lightning Arrester

X1
H1
7,200V
X0
H0
X2

120V   240V
120V

L1
N
L2

288 Amps

120/240V
(ac System Volts)

Resistance between electrode and earth determines current.
I = E/Z
E = 7,200V, Z = 25Ω
7,200V/25Ω = 288A

COPYRIGHT 2002 Mike Holt Enterprises, Inc.

**Figure 250-3**

ensures that the ground fault will be quickly cleared by the opening of the circuit protection device. Figure 250-5

For systems operating at 600V or less, the earth will not carry sufficient fault current to clear a line-to-case ground fault. For example, a 120V fault to the earth of 25Ω will only draw 4.8A (I = E/Z, I = 120V/25Ω, I = 4.8A), not enough to open a 15A protection device [250.4(A)(5)]. Figure 250-6

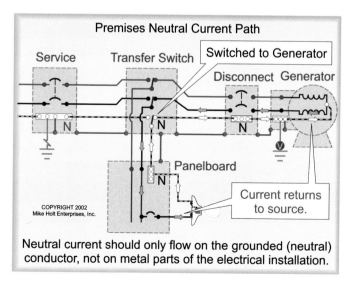

**Premises Neutral Current Path**

Service    Transfer Switch    Switched to Generator

Disconnect  Generator

N          N          N

N
Panelboard

Current returns to source.

COPYRIGHT 2002
Mike Holt Enterprises, Inc.

Neutral current should only flow on the grounded (neutral) conductor, not on metal parts of the electrical installation.

**Figure 250-4**

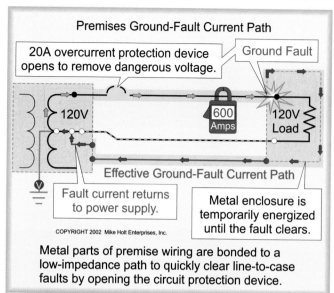

**Premises Ground-Fault Current Path**

20A overcurrent protection device opens to remove dangerous voltage.

Ground Fault

120V

600 Amps

120V Load

Effective Ground-Fault Current Path

Fault current returns to power supply.

Metal enclosure is temporarily energized until the fault clears.

COPYRIGHT 2002 Mike Holt Enterprises, Inc.

Metal parts of premise wiring are bonded to a low-impedance path to quickly clear line-to-case faults by opening the circuit protection device.

**Figure 250-5**

## Understanding Electrical Shock Hazard

If an electrical system is not properly wired to remove dangerous voltage from a ground fault, persons can be subjected to electric shock, which can result in injury or death. The National Safety Council estimates that approximately 300 people in the United States die each year because of an electric shock from

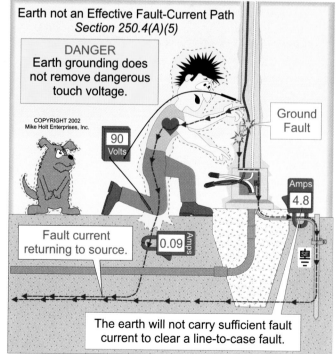

**Earth not an Effective Fault-Current Path**
*Section 250.4(A)(5)*

DANGER
Earth grounding does not remove dangerous touch voltage.

COPYRIGHT 2002
Mike Holt Enterprises, Inc.

Ground Fault

90 Volts

Amps 4.8

Fault current returning to source.

0.09 Amps

The earth will not carry sufficient fault current to clear a line-to-case fault.

**Figure 250-6**

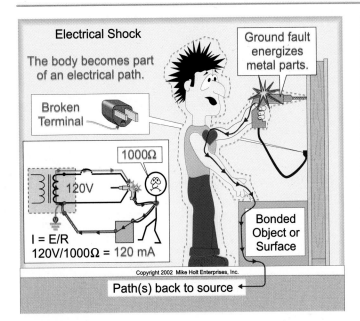

**Figure 250-7**

120V and 277V circuits. People become injured and death occurs when voltage pushes electrons through the human body, particularly through the heart. An electrical shock from as little as 30V alternating current for as little as one second can disrupt the heart's electrical circuitry, causing it to go into ventricular fibrillation. Ventricular fibrillation prevents the blood from circulating through the brain, resulting in death in a matter of minutes. Figure 250-7

**AUTHOR'S COMMENT:** According to the American Heart Association, ventricular fibrillation (VF) is a life-threatening condition in which the heart no longer beats but "quivers" or fibrillates very rapidly — 350 times per minute or more. To avoid sudden cardiac death, the person must be treated with a defibrillator immediately. Cardiopulmonary resuscitation (CPR) provides some extra time, but defibrillation is essential for surviving ventricular fibrillation.

## What Determines the Severity of Electric Shock?

The severity of an electric shock is dependent on the current flowing through the body, which is impacted by the electromotive force (E), measured in volts, and the contact resistance (R), measured in ohms. Current can be determined by the formula I = E/R.

The typical resistances of individual elements of human circuits include:

|  | Dry | Wet |
|---|---|---|
| Foot Immersed in Water |  | 100Ω |
| Hand Immersed in Water |  | 300Ω |
| Hand Around 1$^1/_2$ in. Pipe | 1,000Ω | 500Ω |
| Hand Holding Pliers | 8,000Ω | 1,000Ω |
| Finger-Thumb Grasp | 30,000Ω | 8,000Ω |
| Finger Touch | 100,000Ω | 12,000Ω |

The effects of 60 Hz alternating current on an average human includes: Figure 250-8

• Electrical Sensation. Tingle sensation occurs at about 0.3 mA to 0.4 mA for an adult female and 0.5 mA for an adult male.

• Perception Let-Go. Current over 0.7 - 1.1 mA is very uncomfortable to both sexes.

• Maximum Let-Go Level. The maximum let-go threshold level for a female is approximately 10 mA and for a male it is about 16 mA.

*The "let-go threshold" is the current level where we lose control of our muscles and the electricity causes muscles to contract until the current is removed.*

• Fibrillation Level – 50 mA for 0.2 seconds (female) and 75 mA for 0.5 seconds (male)

According to IEEE Std. 80, *IEEE Guide for Safety in AC Substations* the maximum safe shock duration can be determined by the formula:

Seconds = 0.116/(E/R), where "R" (the resistance of a person) is assumed to be 1,000Ω.

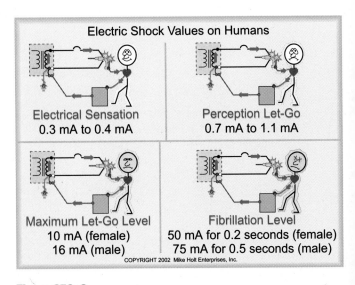

**Figure 250-8**

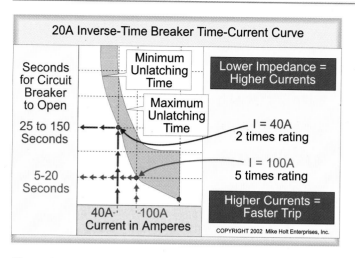

**Figure 250–9**

## Example:

For a 120V circuit, the maximum shock duration = 0.116/(120V/1,000Ω) = 1 Second.

For a 277V circuit, the maximum shock duration = 0.116/(277V/1,000Ω) = 0.43 Second.

## Clearing a Ground Fault

To protect against electric shock from dangerous voltages on metal parts of electrical equipment, a ground fault must quickly be removed by opening the circuit's overcurrent protection device. The time it takes for an overcurrent protection device to open is inversely proportionate to the magnitude of the fault current. Thus, the higher the ground-fault current value, the less time it will take for the protection device to open and clear the fault. For example, a 20A circuit with an overload of 40A (two times the rating) would trip a breaker in 25 to 150 seconds. At 100A (five times the rating) the breaker would trip in 5 to 20 seconds. Figure 250-9

To remove dangerous touch voltage on metal parts from a ground fault, the fault-current path must have sufficiently low impedance to allow the fault current to quickly rise to facilitate the opening of the branch-circuit overcurrent protection device.

**Example:** Approximately how much ground-fault current can flow in a 100A circuit, which consists of:

Ungrounded Conductors – 200 ft of 3 AWG at 0.05Ω

Equipment Grounding (bonding) Conductor – 200 ft of 8 AWG at 0.156Ω, Figure 250-10

(a) 100A        (b) 200A

(c) 600A        (d) 800A

**Answer:** (c) 600A

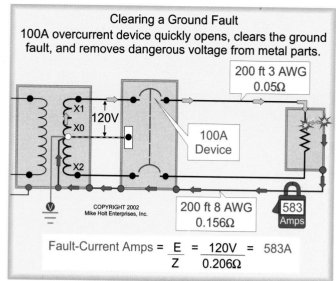

**Figure 250–10**

Fault Current = E/Z

E = 120V

Z = 0.05Ω + 0.156Ω = 0.206Ω

Fault Current = 120V/0.206Ω = 583A

## Why Grounding is Often Difficult to Understand

The reason it's difficult to understand the rules contained in Article 250 – Grounding is because many do not understand that this article applies to both grounding and bonding. In addition, the proper definitions of many important terms such as "bond, bonded, bonding, ground, grounded, grounding, and effectively grounded," and their intended application is not understood or improperly used. So before we get too deep into this subject, let's review the differences between grounding and bonding.

**Bond, Bonded or Bonding (Article 100).** The permanent joining of metallic parts to form an electrically conductive path, that ensures electrical continuity and the capacity to conduct safely any fault current that is likely to be imposed. Bonding is intended to create a low-impedance path for the purpose of removing dangerous touch voltage from metal parts from a ground fault by quickly opening the overcurrent protection device.

Bonding is generally accomplished by properly mechanically terminating metal raceways and cables to enclosures. It is also accomplished by properly bonding electrical devices to enclosures. Figure 250-11

**Ground (100 Definition).** The earth. Figure 250-12

**Grounded (100 Definition).** The connection of metal parts to earth for the purpose of directing lightning and other high-voltage surges into the ground. See Figure 250-12

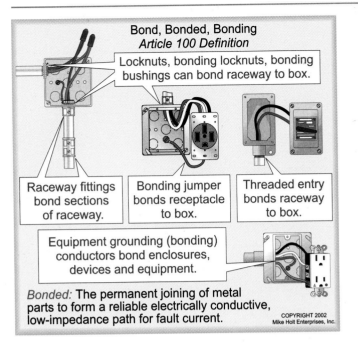

**Figure 250–11**

Grounded, Effectively (100 Definition). Intentionally connected to earth. See Figure 250-12

**AUTHOR'S COMMENT:** In addition, electrical power-supply systems over 1,000 VA are grounded to earth to stabilize the system voltage [250.4(A)(1)].

Grounded Conductor (100 Definition). The conductor that is intentionally grounded to earth at a power supply.

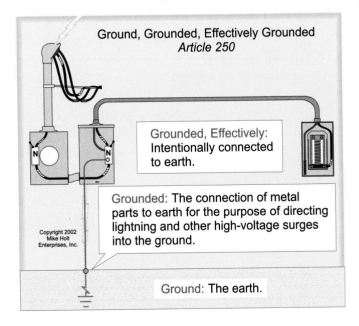

**Figure 250–12**

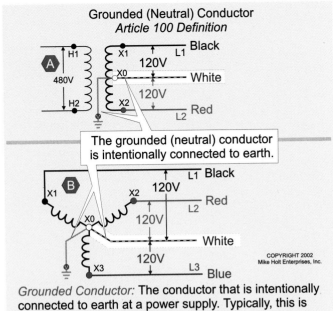

**Figure 250–13**

Typically, this conductor is called the neutral wire and it is identified with the color white or gray [200.6]. Figure 250-13

**AUTHOR'S COMMENT:** In this textbook, the "grounded conductor" will be identified as the "grounded (neutral) conductor." For more information on the differences between the grounded conductor and the neutral conductor, see the introduction to Article 200 in this textbook.

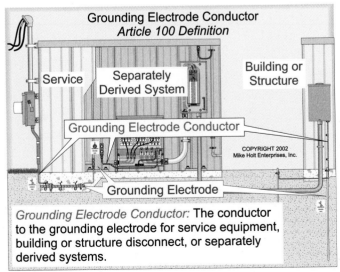

**Figure 250–14**

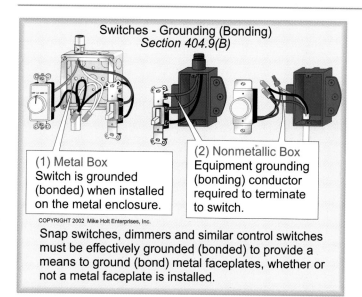

**Switches - Grounding (Bonding)**
*Section 404.9(B)*

**(1) Metal Box**
Switch is grounded (bonded) when installed on the metal enclosure.

**(2) Nonmetallic Box**
Equipment grounding (bonding) conductor required to terminate to switch.

COPYRIGHT 2002 Mike Holt Enterprises, Inc.

Snap switches, dimmers and similar control switches must be effectively grounded (bonded) to provide a means to ground (bond) metal faceplates, whether or not a metal faceplate is installed.

**Figure 250-15**

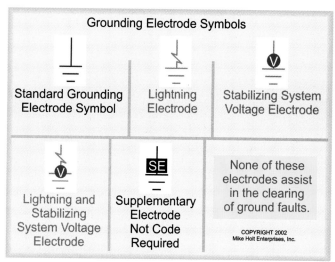

**Grounding Electrode Symbols**

| Standard Grounding Electrode Symbol | Lightning Electrode | Stabilizing System Voltage Electrode |
| --- | --- | --- |
| Lightning and Stabilizing System Voltage Electrode | Supplementary Electrode Not Code Required | None of these electrodes assist in the clearing of ground faults. |

COPYRIGHT 2002
Mike Holt Enterprises, Inc.

**Figure 250-16**

**Grounding Electrode Conductor (100 Definition).** The conductor that connects the equipment grounding (bonding) conductor, the grounded (neutral) conductor, or both, to the grounding electrode at the service equipment, at separately derived systems, and at each building or structure in accordance with Article 250. Figure 250-14

*CAUTION: Often the term ground or grounded is used when the proper term would be bond or bonded. For example, 404.9(B) specifies that "snap switches, including dimmer and similar control switches, shall be effectively grounded." Naturally, we are not expected to intentionally connect the switch to earth; we shall "bond" the metal yoke of the switch to a low-impedance path so that dangerous voltage from a ground fault can be quickly removed by opening the circuit protection device. Figure 250-15*

**AUTHOR'S COMMENT:** One of the key points in understanding the difference between grounding and bonding is that grounding provides a path to earth for lightning, whereas bonding provides the low-impedance path necessary to quickly remove dangerous voltage from metal parts by quickly opening the circuit protection device.

**Tip No. 1:** To help keep the subjects straight, the graphics in this textbook have a color-coded border:

- Dark orange border indicates that the graphic relates to grounding.
- Green border indicates that the graphic relates to bonding.
- Yellow border indicates an improper installation where neutral return current flows on the metal parts of the electrical system.
- No color border indicates important information but not specifically about grounding or bonding.

**Tip No. 2:** Where a grounding electrode symbol is displayed in the graphic, it will indicate its purpose as to whether it is a lightning electrode, a voltage stabilization electrode, or a supplemental or signal reference electrode. Figure 250-16

**AUTHOR'S COMMENT:** Grounding electrodes serve no part in clearing ground faults of premises wiring systems operating at under 600V.

**Tip No. 3:** Throughout this book, where the *NEC* references the terms "ground," "grounded," or "grounding," and the intent is "bond," "bonded" or "bonding," I will add (bond), (bonded), or (bonding) to help make the rule easier to understand. For example, snap switches, including dimmer and similar control switches, shall be effectively grounded (bonded).

# Article 250
# Grounding and Bonding

The purpose and objective of "Article 250 – Grounding" is to ensure that electrical installations are safe from electric shock and fires by limiting voltage imposed by lightning and line surges. Though not listed in the title of Article 250, yet included in the requirement, "bonding" is the intentional connection of metal parts to form a low-impedance effective ground-fault current path to remove dangerous voltage from metal parts from a ground fault.

**AUTHOR'S COMMENT:** The grounding and bonding rules covered in this book apply to solidly grounded alternating-current systems under 600V, such as 120/240V, 208Y/120V and 480Y/277V. Other system configurations, such as 3-wire corner-grounded delta systems, ungrounded systems, or high-impedance grounded neutral systems are permitted by the *National Electrical Code*, but they are typically limited to 3-phase industrial applications and not covered in this book.

## Part I. General

### 250.1 Scope

Part I contains the general requirements for grounding and bonding and the remaining parts contain specific grounding and bonding requirements such as:

(1) Systems and equipment required, permitted, or not permitted to be grounded.

(2) Which circuit conductor is required to be grounded on grounded systems.

(3) The location of grounding (bonding) connections.

(4) How to size grounding and bonding conductors.

(5) Methods of grounding and bonding.

### 250.2 Definitions

**Effective Ground-Fault Current Path.** An intentionally constructed, permanent, low-impedance conductive path designed to carry fault current from the point of a ground fault on a wiring system to the grounded (neutral) point at the electrical supply source. Figure 250-17

An effective ground-fault current path is created when all non-current-carrying electrically conductive materials of an electrical installation are bonded together and to the grounded (neutral) conductor at the electric supply. Effective bonding is accomplished through the use of equipment grounding (bonding) conductors, metallic raceways, connectors, couplings, metallic-sheathed cable with approved fittings and other approved devices recognized for this purpose [250.18].

**AUTHOR'S COMMENT:** A ground-fault current path is only effective when it is properly sized so that it will safely carry the maximum fault current likely to be imposed on it. See 250.4(A)(5) and 250.122 for additional details.

**Ground Fault (Line-to-Case Fault).** An unintentional, electrically conducting connection between an ungrounded conductor of an electrical circuit and metallic enclosures, metallic raceways, or metallic equipment. Figure 250-18

**AUTHOR'S COMMENT:** Line-to-case ground faults are not always of the low-impedance type; they might be of the high-impedance arcing type, which are difficult to clear before a fire destroys the equipment as well as the property. High impedance, in this case, occurs when improper bonding techniques have been

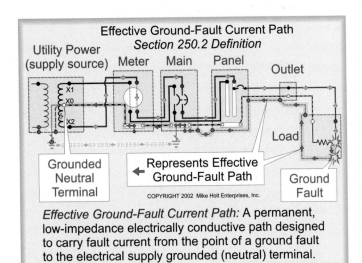

*Effective Ground-Fault Current Path:* A permanent, low-impedance electrically conductive path designed to carry fault current from the point of a ground fault to the electrical supply grounded (neutral) terminal.

**Figure 250–17**

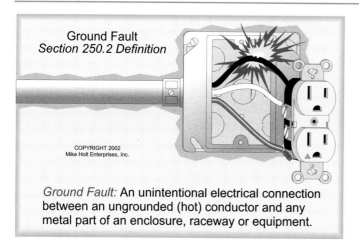

Ground Fault
*Section 250.2 Definition*

COPYRIGHT 2002
Mike Holt Enterprises, Inc.

*Ground Fault:* An unintentional electrical connection between an ungrounded (hot) conductor and any metal part of an enclosure, raceway or equipment.

**Figure 250–18**

used. This is a particular problem for 480V solidly grounded systems and that is why the *NEC* requires equipment ground-fault protection for larger installations. See 230.95. Another way of reducing this hazard is by the installation of high-impedance neutral systems. See 250.36 for the use of current-limiting fuses. This topic is beyond the scope of this book.

**Ground-Fault Current Path.** An electrically conductive path from the point of a ground fault (line-to-case fault) on a wiring system through conductors, or equipment extending to the grounded (neutral) terminal at the electrical supply source.

> FPN: The ground-fault current paths could consist of grounding and bonding conductors, metallic raceways, metallic cable sheaths, electrical equipment and other electrically conductive material, such as metallic water and gas piping, steel-framing members, stucco mesh, metal ducting, reinforcing steel, or shields of communications cables.

**AUTHOR'S COMMENT:** The difference between an "effective ground-fault current path" and "ground-fault current path" is that the effective ground-fault current path is "intentionally" made for the purpose of clearing a fault. The ground-fault current path is simply the path that ground-fault current will flow on to the power supply during a ground fault.

### 250.3 Other Code Sections

Other rules that contain additional grounding and bonding requirements listed in Table 250.3 include:

- Agricultural Building Equipotential Planes, 547.9 and 547.10. Figure 250-19
- Audio Equipment, 640.7
- Hazardous (classified) Locations, 501.16, 502.16 and 503.16

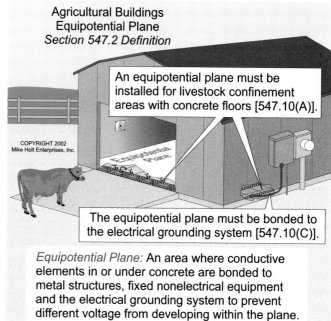

Agricultural Buildings
Equipotential Plane
*Section 547.2 Definition*

An equipotential plane must be installed for livestock confinement areas with concrete floors [547.10(A)].

COPYRIGHT 2002
Mike Holt Enterprises, Inc.

The equipotential plane must be bonded to the electrical grounding system [547.10(C)].

*Equipotential Plane:* An area where conductive elements in or under concrete are bonded to metal structures, fixed nonelectrical equipment and the electrical grounding system to prevent different voltage from developing within the plane.

**Figure 250–19**

- Panelboards, 408.20
- Receptacles, 406.3, 406.9, 517.13
- Receptacle Cover Plates, 406.5
- Swimming Pools and Spas, 680.23(F)(2), 680.24(D) and 680.25(B)
- Switches, 404.9(B) and 517.13

### 250.4 General Requirements for Grounding and Bonding

The following explains the purpose of grounding and bonding of electrical systems and equipment to ensure a safe installation.

**(A) Grounded Systems.**

**AUTHOR'S COMMENT:** The term "electrical system" as used in this subsection refers to the "power source" such as a transformer, generator or photovoltaic system, not the circuit wiring and/or the equipment.

**(1) Grounding of Electrical Systems.** Electrical power supplies such as the utility transformer shall be grounded to earth to help limit high voltage imposed on the system windings from lightning or line surges. Figure 250-20

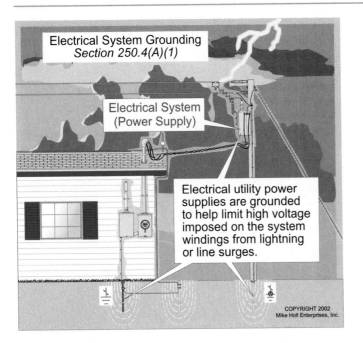

**Figure 250-20**

**AUTHOR'S COMMENT:** Grounding System for **Lightning** – The electric discharge from lightning is typically from a negative charged cloud to a positive charged earth surface, but it can be from the earth's surface to a cloud, or it can be from cloud to cloud as well as cloud to space. When the negative capacitive voltage charge of a cloud exceeds the dielectric strength of the air between the cloud and the earth, an arc will occur between the clouds and the earth in an attempt to equalize the difference of potential between the two objects. When this occurs, high voltages, often over 20,000V, drives high amperages of current (as much as 40,000A) into the earth for a fraction of a second. Figure 250-21

Typically, utility wiring outside will be struck by lightning and it's critical that these systems be grounded to the earth to assist the flow of lightning into the earth.

**Grounding System for Line Surges** – When a utility high-voltage ground fault occurs, the voltage on the other phases will rise for the duration of the fault (typically 3 to 12 cycles). This voltage surge during the utility ground fault will be transformed into an elevated surge voltage on the secondary, which can cause destruction of electrical and particularly electronic equipment in the premises. Studies have shown that the lower the resistance of the utility grounding system, the lower the voltage surge.

Electrical systems (power supplies) are grounded (actually bonded) to stabilize the system voltage during normal operation. Figure 250-22

*CAUTION: According to IEEE Std. 242 "Buff Book," if a ground fault is intermittent or allowed to continue on an ungrounded system, the system could be subjected to possible severe system overvoltage-to-ground, which can be as high as six or eight times the phase voltage. This excessive voltage can puncture conductor insulation and result in additional ground faults. System overvoltage-to-ground is caused by repetitive charging of the system capacitance or by resonance between the system capacitance and the inductances of equipment in the system.*

*In addition, IEEE Std. 142 "Green Book" states that "Field experience and theoretical studies have shown that arcing, restriking, or vibrating ground faults on ungrounded systems (actually unbonded systems) can, under certain conditions, produce surge voltages as high as six times normal. Neutral (system) grounding (actually bonding) is effective in*

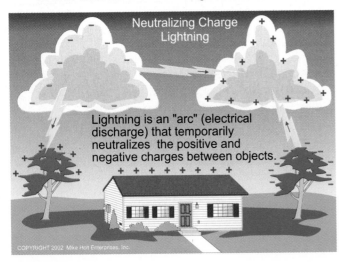

**Figure 250-21**

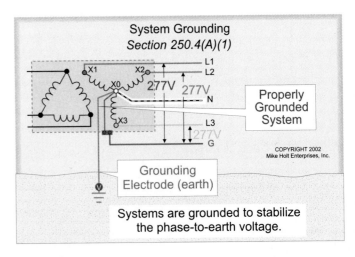

**Figure 250-22**

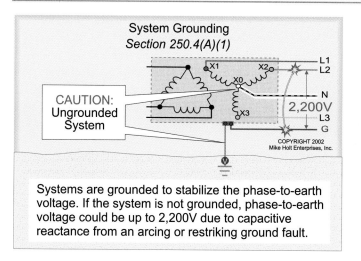

System Grounding
Section 250.4(A)(1)

CAUTION:
Ungrounded
System

2,200V

COPYRIGHT 2002
Mike Holt Enterprises, Inc.

Systems are grounded to stabilize the phase-to-earth voltage. If the system is not grounded, phase-to-earth voltage could be up to 2,200V due to capacitive reactance from an arcing or restriking ground fault.

**Figure 250-23**

reducing transient voltage buildup from such intermittent ground faults by reducing neutral displacement from ground potential and reducing destructive effectiveness of any high-frequency voltage oscillations following each arc initiation or restrike." Figure 250-23

**AUTHOR'S COMMENT:** The danger of overvoltage occurs in systems that are intended to be ungrounded as well as those systems that were supposed to be grounded but were not. Elevated voltage-to-ground is beyond the scope of this book. To obtain more information on this subject, visit http://www.mike-holt.com/Newsletters/highvolt.htm.

**(2) Grounding of Electrical Equipment.** To help limit the voltage impressed on metal parts from lightning, non-current-carrying conductive metal parts of electrical equipment in or on a building or structure shall be grounded to earth. Figure 250-24

**AUTHOR'S COMMENT:** Grounding of electrical equipment to earth is not for the purpose of clearing a ground fault.

Metal parts of electrical equipment in a building or structure are grounded to earth by electrically connecting the building or structure disconnecting means [225.31 or 230.70] with a grounding electrode conductor [250.64(A)] to an appropriate grounding electrode (earth) identified in 250.52 [250.24(A) and 250.32(B)].

DANGER: Failure to ground the metal parts of electrical equipment to earth could result in elevated voltage from lightning entering the building or structure, via metal raceways or cables, seeking a path to the earth. The high voltage on the metal parts from

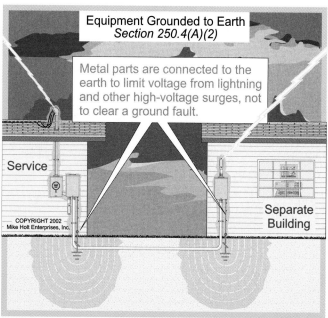

Equipment Grounded to Earth
Section 250.4(A)(2)

Metal parts are connected to the earth to limit voltage from lightning and other high-voltage surges, not to clear a ground fault.

Service

COPYRIGHT 2002
Mike Holt Enterprises, Inc.

Separate
Building

**Figure 250-24**

lightning can result in electric shock and fires, as well as the destruction of electrical equipment from lightning. Figure 250-25

Grounding of metal parts of electrical equipment also helps prevent the buildup of high-voltage static charges on metal parts. Grounding is often required in areas where the discharge (arcing) of the voltage buildup could cause failure of electronic equipment being assembled on a production line or a fire and explosion in a hazardous classified area. See 500.4 FPN 3.

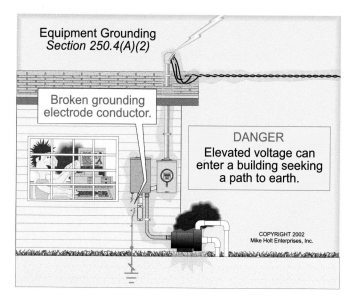

Equipment Grounding
Section 250.4(A)(2)

Broken grounding electrode conductor.

DANGER
Elevated voltage can enter a building seeking a path to earth.

COPYRIGHT 2002
Mike Holt Enterprises, Inc.

**Figure 250-25**

**AUTHOR'S COMMENT:** Grounding metal parts of electrical equipment to earth does not protect electrical or electronic equipment from lightning voltage transients (high-frequency voltage impulses) on the circuit conductors inside the building or structure. To protect electronic and electrical equipment from high-voltage transients, proper transient voltage surge protection devices should be installed in accordance with Article 280 at service equipment and Article 285 at the panelboards.

To provide proper operation of transient voltage surge protection devices, the resistance of the grounding electrode (earth) should be as low as practical. Most specifications for communications systems installations (cell towers) require the ground resistance to be 5Ω, sometimes as little as 3Ω and on some rare occasions 1Ω! To achieve and maintain a low resistive ground, special grounding configurations, design, equipment and measuring instruments must be used. This is beyond the scope of this book.

**(3) Bonding of Electrical Equipment.** To remove dangerous voltage caused by ground faults, the metal parts of electrical raceways, cables, enclosures or equipment shall be bonded together. In addition, the metal parts shall be bonded to the grounded (neutral) terminal of the electrical supply source in accordance with 250.142. Figure 250-26

**AUTHOR'S COMMENT:** An effective ground-fault current path [250.2] is created when all non-current-carrying electrically conductive materials are bonded together and to the grounded (neutral) terminal at the electric supply.

**(4) Bonding of Electrically Conductive Materials.** To remove dangerous voltage caused by ground faults, electrically conductive metal water piping, sprinkler piping, metal gas piping, and other metal piping as well as exposed structural steel members that are likely to become energized shall be bonded in accordance with 250.104. Figure 250-27

**AUTHOR'S COMMENT:** The phrase "that are likely to become energized" is subject to interpretation by the Authority Having Jurisdiction (AHJ). See 250.104 for additional details.

**(5) Effective Ground-Fault Current Path.** Electrical raceways, cables, enclosures and equipment as well as other electrically conductive material "likely to become energized" shall be installed in a manner that creates a permanent, low-impedance path that has the capacity to safely carry the maximum ground-fault current likely to be imposed on it [110.10]. The purpose of this path is to facilitate the operation of overcurrent devices if a ground fault occurs to the metal parts. Clearing ground faults is accomplished by bonding all of the metal parts of electrical equipment and conductive material likely to become energized to the power-supply grounded (neutral) terminal. Figure 250-28

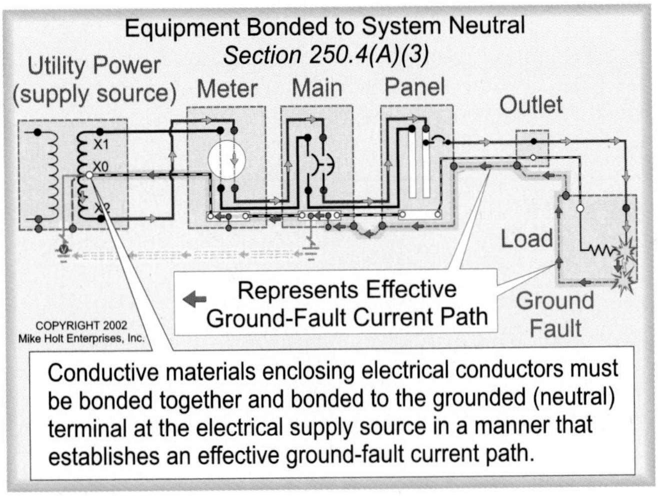

### Equipment Bonded to System Neutral
*Section 250.4(A)(3)*

Utility Power (supply source) — Meter — Main — Panel — Outlet

← Represents Effective Ground-Fault Current Path

Load

Ground Fault

COPYRIGHT 2002 Mike Holt Enterprises, Inc.

Conductive materials enclosing electrical conductors must be bonded together and bonded to the grounded (neutral) terminal at the electrical supply source in a manner that establishes an effective ground-fault current path.

**Figure 250–26**

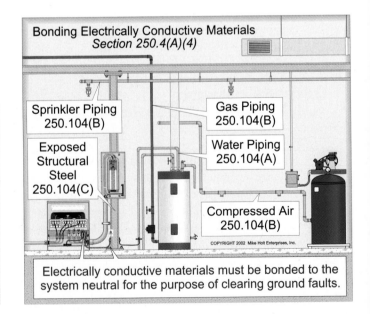

### Bonding Electrically Conductive Materials
*Section 250.4(A)(4)*

Sprinkler Piping 250.104(B)

Gas Piping 250.104(B)

Exposed Structural Steel 250.104(C)

Water Piping 250.104(A)

Compressed Air 250.104(B)

COPYRIGHT 2002 Mike Holt Enterprises, Inc.

Electrically conductive materials must be bonded to the system neutral for the purpose of clearing ground faults.

**Figure 250–27**

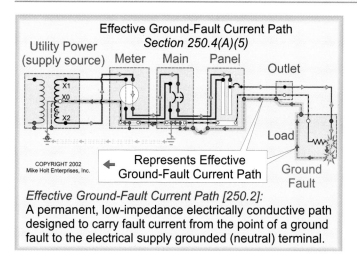

Effective Ground-Fault Current Path [250.2]:
A permanent, low-impedance electrically conductive path designed to carry fault current from the point of a ground fault to the electrical supply grounded (neutral) terminal.

**Figure 250-28**

The *NEC* does permit a ground rod at a pole [250.54] but the *Code* does not allow the earth to be used as the sole equipment grounding (bonding) conductor. An equipment grounding (bonding) conductor of a type specified in 250.118 is ALWAYS required. Figure 250-29

*CAUTION: Because the earth is a poor conductor whose resistivity does not permit sufficient fault current to flow back to the power supply [IEEE Std. 142 Section 2.2.8], a ground rod will not serve to clear a ground fault and dangerous touch voltage will remain on metal parts if an effective ground-fault current path is not provided. For more information on this topic, visit http://www.mikeholt.com/ Newsletters/GroundResistance.htm. Figure 250-30*

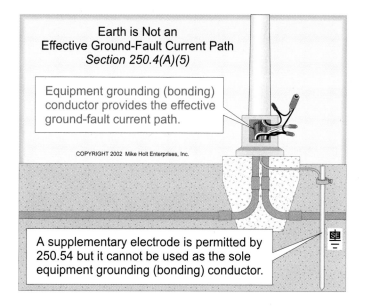

A supplementary electrode is permitted by 250.54 but it cannot be used as the sole equipment grounding (bonding) conductor.

**Figure 250-29**

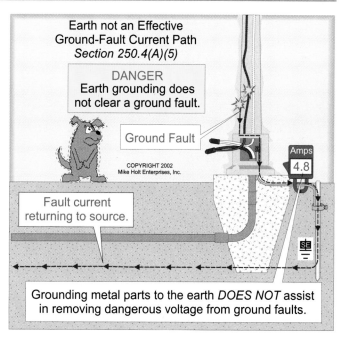

Grounding metal parts to the earth *DOES NOT* assist in removing dangerous voltage from ground faults.

**Figure 250-30**

**Question:** What is the maximum current that could flow though a ground rod if the ground rod has an impedance of $25\Omega$ and the system voltage is 120/240V?

(a) 4.8A                    (b) 24A

(c) 48A                     (d) 96A

**Answer:** (a) 4.8A

$I = E/Z$, $E = 120V$, $Z = 25\Omega$

$I = 120V/25\Omega = 4.8A$

*DANGER: Because the resistance of the earth is so great (10 to 500$\Omega$), very little current will return to the power supply via the earth if the earth is the only ground-fault return path. The result is that the circuit overcurrent protection device will not open and metal parts will remain energized at a lethal level waiting for someone to make contact with them and the earth. Therefore, a ground rod cannot be used to lower touch voltage to a safe value for metal parts that are not bonded to an effective ground-fault current path. To understand how a ground rod is useless in reducing touch voltage to a safe level, let's review the following:*

*• What is touch voltage?*

*• At what level is touch voltage hazardous?*

*• How earth surface voltage gradients operate.*

*1. Touch Voltage - The IEEE definition of touch voltage is "the potential (voltage) difference between a grounded (bonded) metallic structure and a point on the earth 3 ft from the structure."*

*2. Hazardous Level - NFPA 70E - Standard for Electrical Safety Requirements for Employee Workplaces, cautions that death and/or severe electric shock can occur whenever the touch voltage exceeds 30V.*

*3. Surface Voltage Gradients - According to IEEE Std. 142 "Green Book" [4.1.1], the resistance of the soil outward from a ground rod is equal to the sum of the series resistances of the earth shells. The shell nearest the rod has the highest resistance and each successive shell has progressively larger areas and progressively lower resistances. The following table lists the percentage of total resistance and the touch voltage based on a 120V fault. The table's percentage of resistance is based on a 10 ft ground rod having a diameter of 5/8 inches.*

*Don't worry if you don't understand the above statement, just review the table below with Figure 250-31*

| Distance from Rod | Resistance | Touch Voltage |
|---|---|---|
| 1 Foot (Shell 1) | 68% | 82V |
| 3 Feet (Shells 1 and 2) | 75% | 90V |
| 5 Feet (Shells 1, 2 and 3) | 86% | 103V |

*With the intention of providing a safer installation, many think a ground rod can be used to reduce touch voltage. However, as we can see in the above table, the voltage gradient of the earth drops off so rapidly that a person in contact with an energized object can receive a lethal electric shock one foot away from an energized object if the metal parts are not bonded to an effective ground-fault current path.*

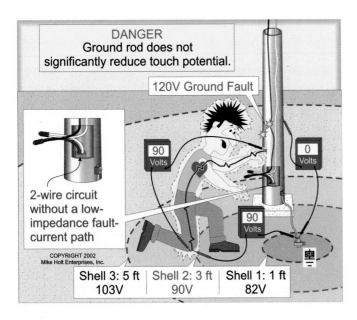

**Figure 250–31**

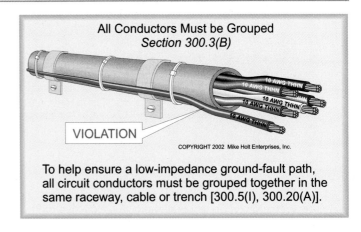

**Figure 250–32**

*Scary as it might be, the accepted grounding practice for street lighting and traffic signaling for many parts of the United States was to use the ground rod as the only ground-fault current return path. That is, the metal pole of a light fixture or traffic signal is grounded to a ground rod and an effective ground-fault current path is not provided (no equipment grounding conductor)! I'm sure there are thousands of energized metal poles, just waiting for someone to make contact with them and this is one of the reasons so many people get killed with street lighting and traffic signal poles in the United States. For a case study on this subject, visit www.mikeholt.com/Newsletters/dadecounty.htm.*

**AUTHOR'S COMMENT:** Another factor necessary to help ensure a low-impedance ground-fault path is that all circuit conductors, ungrounded, grounded and the equipment grounding (bonding) conductor shall be grouped together in the same raceway, cable or trench [300.3(B), 300.5(I), 300.20(A)]. Figure 250-32

### 250.4(A) Summary

(1) An electrical power supply shall be grounded to stabilize the system voltage.

(2) Metal parts of electrical equipment at a building or structure disconnect shall be grounded to assist lightning to earth.

(3) Electrically bonding non-current-carrying parts of the electrical wiring system to an effective ground-fault current path is required so that a ground fault can be quickly cleared by opening the circuit overcurrent protection device.

(4) Electrically bonding conductive piping and structure steel that may become energized to the effective

ground-fault current path is required so that a ground fault can be quickly cleared by opening the circuit over-current protection device.

(5) Create an effective ground-fault current path for metal parts of equipment enclosures, raceways, and equipment as well as metal piping and structural steel. The effective ground-fault current path shall be sized to withstand high fault current [110.10 and 250.122].

## 250.6. Objectionable (Neutral) Current

**(A)** **Preventing Objectionable Current.** To prevent a fire, electric shock, improper operation of circuit protection devices, as well as improper operation of sensitive equipment, the grounding of electrical systems and the bonding of equipment shall be done in a manner that prevents objectionable (neutral) current from flowing on conductive materials, electrical equipment, or on grounding and bonding paths.

## Objectionable (Neutral) Current

Objectionable current on grounding and bonding paths occur when:

1. Improper neutral-to-case bonds are made.
2. There are errors in the wiring installation.
3. Using the equipment-grounding conductor to carry neutral current.

## Improper Neutral-to-Case Bond [250.142]

**Panelboards** – Bonding of the neutral terminal to the case of a panelboard that is not part of service equipment or a separately derived system creates a parallel path, which allows objectionable neutral current to flow on the metal parts of electrical equipment as well as the grounding and bonding conductors. Figure 250-33

**Disconnects** – Where an equipment grounding (bonding) conductor is run with the feeder conductors to a separate building [250.32(B)(1)], a common and dangerous mistake is to make a neutral-to-case bond in the separate building disconnect, which allows objectionable neutral current to flow on the metal parts of electrical equipment as well as on the grounding and bonding conductors. Figure 250-34

**Separately Derived Systems** – The neutral-to-case bonding jumper for a separately derived system, such as that derived from a transformer, generator, or uninterruptible power supply (UPS) shall be installed either at the source of the separately derived system or at the first disconnect electrically downstream, but not at both locations in accordance with 250.30(A)(1).

**Transformers** – If a neutral-to-case bond is made at both the transformer and at the secondary panelboard/disconnect, then objectionable neutral current will flow on the metal parts of electrical equipment as well as the grounding and bonding conductors. Figure 250-35

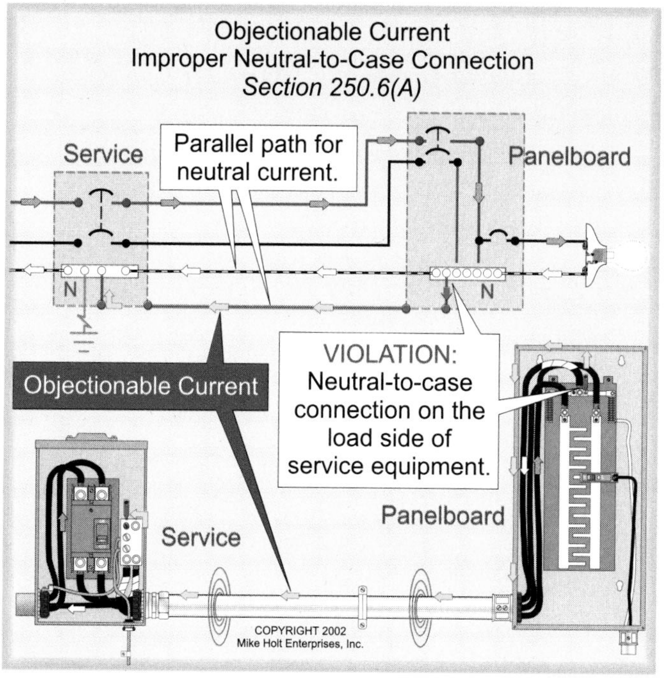

**Objectionable Current**
**Improper Neutral-to-Case Connection**
*Section 250.6(A)*

Service

Parallel path for neutral current.

Panelboard

N

N

Objectionable Current

VIOLATION:
Neutral-to-case connection on the load side of service equipment.

Service

Panelboard

COPYRIGHT 2002
Mike Holt Enterprises, Inc.

**Figure 250–33**

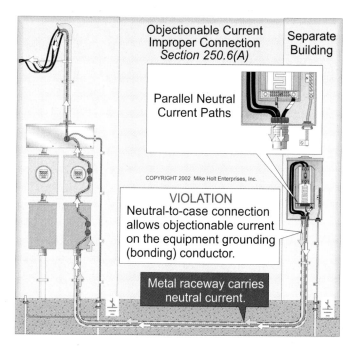

Objectionable Current
Improper Connection
*Section 250.6(A)*

Separate Building

Parallel Neutral Current Paths

COPYRIGHT 2002 Mike Holt Enterprises, Inc.

VIOLATION
Neutral-to-case connection allows objectionable current on the equipment grounding (bonding) conductor.

Metal raceway carries neutral current.

**Figure 250–34**

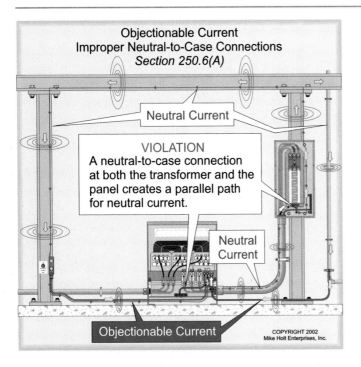

Objectionable Current
Improper Neutral-to-Case Connections
*Section 250.6(A)*

Neutral Current

VIOLATION
A neutral-to-case connection
at both the transformer and the
panel creates a parallel path
for neutral current.

Neutral
Current

Objectionable Current

COPYRIGHT 2002
Mike Holt Enterprises, Inc.

**Figure 250–35**

Generator – If the grounded (neutral) conductor in a transfer switch is not opened with the ungrounded conductors, then the grounded (neutral) from the generator will be solidly connected to the utility's service grounded (neutral) conductor. Under this condition, the generator is not a separately derived system, and a neutral-to-case bond shall not be made at the generator or at the generator disconnect [250.20(D) FPN 1]. If a neutral-to-case bond is made at the generator or generator disconnect, then objectionable neutral current will flow on the metal parts of electrical equip-

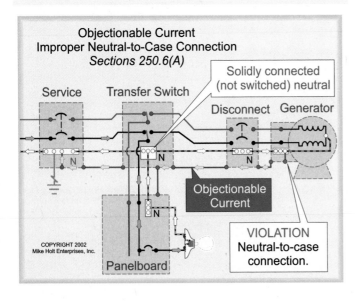

Objectionable Current
Improper Neutral-to-Case Connection
*Sections 250.6(A)*

Service   Transfer Switch

Solidly connected
(not switched) neutral

Disconnect   Generator

N        N           N

Objectionable
Current

COPYRIGHT 2002
Mike Holt Enterprises, Inc.

N

Panelboard

VIOLATION
Neutral-to-case
connection.

**Figure 250–36**

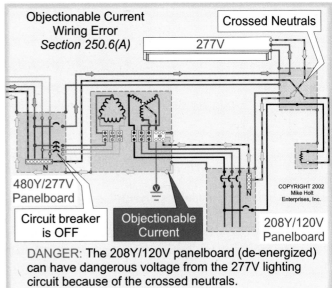

Objectionable Current
Wiring Error
*Section 250.6(A)*

Crossed Neutrals

277V

480Y/277V
Panelboard

N

Circuit breaker
is OFF

Objectionable
Current

COPYRIGHT 2002
Mike Holt
Enterprises, Inc.

N

208Y/120V
Panelboard

DANGER: The 208Y/120V panelboard (de-energized)
can have dangerous voltage from the 277V lighting
circuit because of the crossed neutrals.

**Figure 250–37**

ment as well as on the grounding and bonding conductors. Figure 250-36

### Errors in the Wiring Installation

**Mixing Neutrals.** The *NEC* does not prohibit the mixing of circuit conductors from different systems in the same raceway or enclosure [300.3(C)(1)]. As a result, mistakes can be made where the grounded (neutral) conductors from different systems are crossed (mixed). When this occurs, the grounding and bonding path will carry objectionable neutral current, even when it appears that all circuits have been de-energized. Figure 250-37

### Using Equipment Grounding (Bonding) Conductor for Neutral Current

This often happens when a 120V circuit is required at a location where a neutral conductor is not available. Example: A 240V

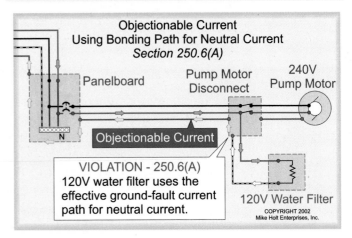

Objectionable Current
Using Bonding Path for Neutral Current
*Section 250.6(A)*

Panelboard

Pump Motor
Disconnect

240V
Pump Motor

N

Objectionable Current

VIOLATION - 250.6(A)
120V water filter uses the
effective ground-fault current
path for neutral current.

120V Water Filter

COPYRIGHT 2002
Mike Holt Enterprises, Inc.

**Figure 250–38**

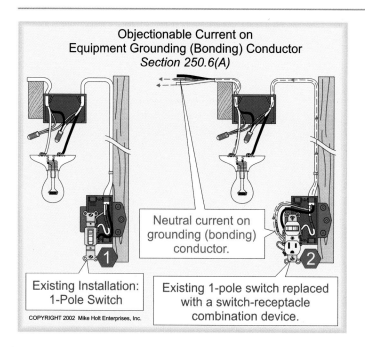

**Figure 250-39**

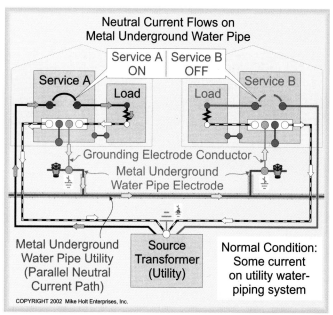

**Figure 250-40**

time clock motor is replaced with a 120V time clock motor and the equipment grounding conductor is used to feed one side of the 120V time clock. Another example is a 120V water filter wired to a 240V well-pump motor circuit and the equipment grounding conductor is used for the neutral. Figure 250-38

Using the bonding path for the neutral is also seen in ceiling fan installations where the equipment grounding (bonding) conductor is used as a neutral and the white wire is used as the switch leg for the light, or where a receptacle is added to a switch outlet that doesn't have a neutral conductor. Figure 250-39

> **AUTHOR'S COMMENT:** Neutral currents always flow on a communiity metal underground water-piping system where the water service to all of the buildings is metallic. This occurs because the underground water pipe and the service neutral conductors are in parallel with each other. Figure 250-40

## Dangers of Objectionable (neutral) Current

Objectionable neutral current can cause shock hazard, fire hazard, improper operation of sensitive electronic equipment, and improper operation of circuit protection devices.

### Shock Hazard

Objectionable current on metal parts of electrical equipment can create a condition where electric shock and even death from ventricular fibrillation can occur. Figure 250-41 shows an example where a person becomes in series with the neutral current path of a 120V circuit.

### Fire Hazard

Fire occurs when the temperature rises to a level sufficient to cause ignition of adjacent combustible material in an area that contains sufficient oxygen. In an electrical system, heat is generated whenever current flows. Improper wiring, resulting in the flow of neutral current on grounding and bonding paths can cause

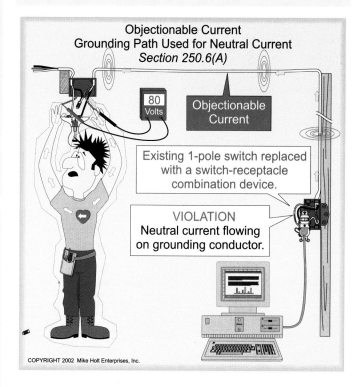

**Figure 250-41**

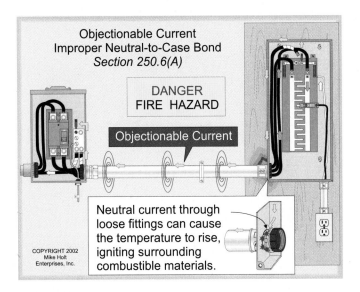

**Figure 250–42**

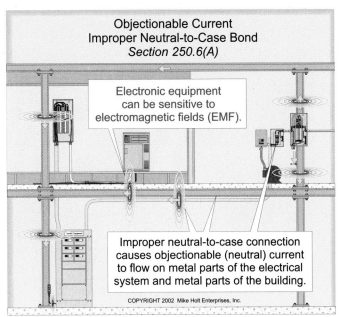

**Figure 250–43**

the temperature at loose connections to rise to a level that can cause a fire. In addition, arcing at loose connections is particularly dangerous in areas that contain easily ignitable and explosive gases, vapors, or dust. Figure 250-42

### Improper Operation of Circuit Protection Devices

Nuisance tripping of a protection device equipped with ground-fault protection can occur if neutral current returns on the equipment grounding (bonding) conductor, instead of the neutral conductor because of improper neutral-to-case bonds. A circuit breaker with ground-fault protection (480Y/277V, 3-phase system over 1,000A per 230.95) uses either the residual current method or the zero sequence method to detect a ground fault. For the zero sequence method, the ground-fault trip unit sums the currents in the three phase conductors and the neutral. When no ground fault is present, the summation of currents flowing on A+B+C+N will equal zero. Any current flow not equal to zero is considered a ground fault. The residual method is used only at the service as it measures current flowing through the main bonding jumper.

Where improper neutral-to-case bonds have been made, objectionable neutral current will flow on the equipment grounding (bonding) conductor in parallel with the grounded (neutral) conductor. Depending on the impedance of this path versus the neutral conductor path, the ground-fault protective relay may see current flow above its pickup point and cause the protective device to open the circuit.

If a ground fault occurs and there are improper neutral-to-case bonds, the protection relay might not operate because some of the ground-fault current will return on the neutral conductor bypassing the ground-fault protective device.

### Improper Operation of Sensitive Electronic Equipment

When objectionable neutral current travels on the metal parts of electrical equipment, the electromagnetic field generated from alternating-circuit conductors will not cancel. This uncanceled current flowing on metal parts of electrical equipment and conductive building parts causes elevated electromagnetic fields in the building. These low frequency electromagnetic fields can negatively impact the performance of sensitive electronic devices, particularly video monitors and medical equipment. For more information visit www.mikeholt.com/Powerquality/Powerquality.htm. Figure 250-43

**(B) Stopping Objectionable Current.** If improper neutral-to-case bonds result in an objectionable flow of current on grounding or bonding conductors, simply remove or disconnect the improper neutral-to-case bonds.

**(C) Temporary Currents Not Classified as Objectionable Currents.** Temporary ground-fault current on the effective ground-fault current path until the circuit overcurrent protection device opens removing the fault, is not classified as objectionable current. Figure 250-44

**(D) Electromagnetic Interference (Electrical Noise).** Currents that cause noise or data errors in electronic equipment are not considered objectionable currents. Figure 250-45

**AUTHOR'S COMMENT:** Some sensitive electronic equipment manufacturers require their equipment to be isolated from the equipment bonding conductor, yet they require the equipment to be grounded to an inde-

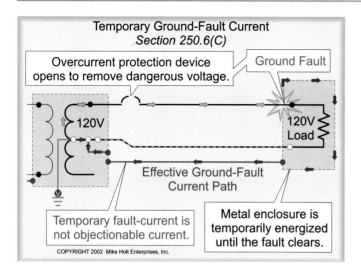

**Temporary Ground-Fault Current**
*Section 250.6(C)*

Overcurrent protection device opens to remove dangerous voltage.

Ground Fault

120V

120V Load

Effective Ground-Fault Current Path

Temporary fault-current is not objectionable current.

Metal enclosure is temporarily energized until the fault clears.

COPYRIGHT 2002 Mike Holt Enterprises, Inc.

**Figure 250–44**

pendent grounding system. This practice is very dangerous and violates the *NEC* because the earth will not provide the low-impedance path necessary to clear a ground fault [250.4(A)(5)]. See 250.54 for the proper application of a supplementary electrode and 250.96(D) and 250.146(D) for the requirements of isolated equipment grounding (bonding) conductors for sensitive electronic equipment. Figure 250-46

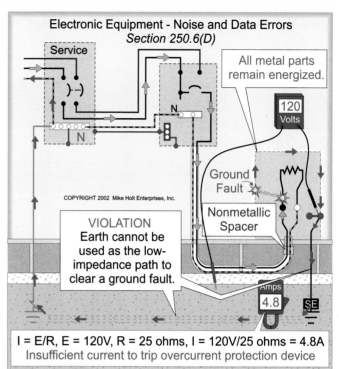

**Electronic Equipment - Noise and Data Errors**
*Section 250.6(D)*

Service

All metal parts remain energized.

N

120 Volts

Ground Fault

Nonmetallic Spacer

**VIOLATION**
Earth cannot be used as the low-impedance path to clear a ground fault.

Amps 4.8

SE

$I = E/R$, $E = 120V$, $R = 25$ ohms, $I = 120V/25$ ohms $= 4.8A$
Insufficient current to trip overcurrent protection device

**Figure 250–46**

### 250.8 Termination of Grounding and Bonding Conductors

Equipment grounding (bonding) conductors, grounding electrode conductors and bonding jumpers shall terminate by exothermic welding, listed pressure connectors of the set screw or compression type, listed clamps, or other listed fittings. Sheet-metal screws shall not be used for the termination of grounding (or bonding) conductors. Figure 250-47

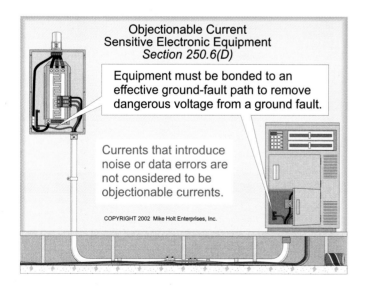

**Objectionable Current**
**Sensitive Electronic Equipment**
*Section 250.6(D)*

Equipment must be bonded to an effective ground-fault path to remove dangerous voltage from a ground fault.

Currents that introduce noise or data errors are not considered to be objectionable currents.

COPYRIGHT 2002 Mike Holt Enterprises, Inc.

**Figure 250–45**

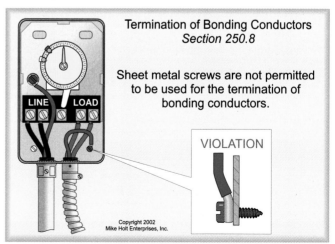

**Termination of Bonding Conductors**
*Section 250.8*

Sheet metal screws are not permitted to be used for the termination of bonding conductors.

LINE    LOAD

**VIOLATION**

Copyright 2002
Mike Holt Enterprises, Inc.

**Figure 250–47**

## 250.10 Protection of Grounding Fittings

Ground clamps and other grounding fittings shall be protected from physical damage by:

(1) Locating the grounding fitting where they are not likely to be damaged

(2) Enclosing the grounding fittings in metal, wood, or equivalent protective covering

**AUTHOR'S COMMENT:** Grounding fittings are permitted to be buried or encased in concrete if installed in accordance with 250.53(G), 250.68(A) Ex. and 250.70.

## 250.12 Clean Surface

Nonconductive coatings such as paint, lacquer and enamel shall be removed on equipment to be grounded or bonded to ensure good electrical continuity, or the termination fittings shall be designed so as to make such removal unnecessary [250.53(A) and 250.96(A)].

**AUTHOR'S COMMENT:** Some feel that "tarnish" on copper water pipe should be removed before making a grounding termination. This is a judgment call by the AHJ.

## Part II. System and Equipment Grounding

### 250.20 Alternating-Current Systems to be Grounded

System (power supply) grounding is the intentional connection of one terminal of a power supply to the earth for the purpose of stabilizing the phase-to-earth voltage during normal operation [250.4(A)(1)].

**(A) AC Circuits of Less than 50V.** Alternating-current circuits supplied from a transformer that operate at less than 50V are not required to be grounded unless:

(1) The primary is supplied from a 277V or 480V circuit.

(2) The primary is supplied from an ungrounded power supply.

**AUTHOR'S COMMENT:** Typically, circuits operating at less than 50V are not grounded because they are not supplied from a 277V or 480V system, nor are they supplied from an ungrounded system. Figure 250-48

**(B) AC Systems Over 50V.** Alternating-current systems over 50V that require a grounded (neutral) conductor shall have the grounded neutral terminal of the power supply grounded

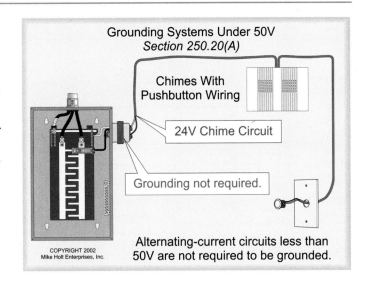

*Figure 250–48*

to earth in accordance with 250.4(A)(1) and 250.30(A)(1). Such systems include: Figure 250-49

• 120V or 120/240V single-phase systems

• 208Y/120V or 480Y/277V, 4-wire, 3-phase, wye-connected systems

• 120/240V 4-wire, 3-phase, delta-connected systems

**AUTHOR'S COMMENT:** Other power supply systems, such as a corner-grounded delta-connected system, are permitted to be grounded [250.26(4)], but this is beyond the scope of this book.

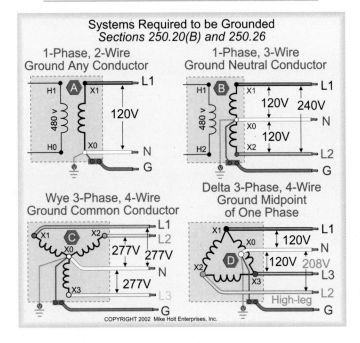

*Figure 250–49*

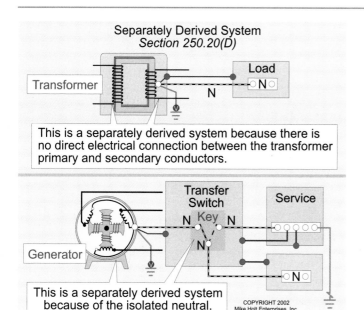

**Separately Derived System**
*Section 250.20(D)*

This is a separately derived system because there is no direct electrical connection between the transformer primary and secondary conductors.

This is a separately derived system because of the isolated neutral.

COPYRIGHT 2002
Mike Holt Enterprises, Inc.

*Figure 250–50*

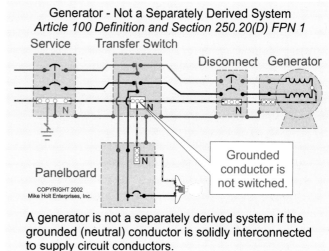

**Generator - Not a Separately Derived System**
*Article 100 Definition and Section 250.20(D) FPN 1*

Grounded conductor is not switched.

COPYRIGHT 2002
Mike Holt Enterprises, Inc.

A generator is not a separately derived system if the grounded (neutral) conductor is solidly interconnected to supply circuit conductors.

*Figure 250–51*

**(D) Separately Derived Systems.** Separately derived systems, which are required to be grounded by 250.20(A) or (B), shall be grounded (and bonded) in accordance with the requirements contained in 250.30.

**AUTHOR'S COMMENT:** According to Article 100, a separately derived system is a premises wiring system that has no direct electrical connection between the systems, including the grounded (neutral) conductor. Transformers are typically separately derived because the primary and secondary conductors are electrically isolated from each other. Generators that supply a transfer switch that opens the grounded (neutral) conductor are also separately derived. Figure 250-50

FPN 1: A generator is not a separately derived system if the grounded (neutral) conductor from the generator is solidly connected to the supply system grounded (neutral) conductor. In other words, if the transfer switch does not open the neutral conductor, then the generator will not be a separately derived system. Figure 250-51

**AUTHOR'S COMMENT:** This fine print note points out that when a generator is not a separately derived system, the grounding and bonding requirements contained in 250.30 do not apply and a neutral-to-case connection shall not be made at the generator [250.6(A) and 250.142].

FPN 2: If the generator transfer switch does not open the grounded (neutral) conductor, then the grounded (neutral) conductor will be required to carry fault current back to the generator. Under this condition, the grounded (neutral) conductor shall be sized no smaller than required for the unbalanced load by 220.22 and in addition, it shall be sized no smaller than required by 250.24(B) [445.13]. Figure 250-52

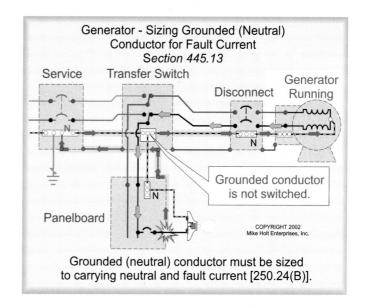

**Generator - Sizing Grounded (Neutral) Conductor for Fault Current**
*Section 445.13*

Grounded conductor is not switched.

COPYRIGHT 2002
Mike Holt Enterprises, Inc.

Grounded (neutral) conductor must be sized to carrying neutral and fault current [250.24(B)].

*Figure 250–52*

## 250.24 Grounding and Bonding at Service Equipment

The metal parts of electrical equipment shall be grounded to earth to protect persons from electric shock and to protect property from fires by limiting voltage on the metal parts from lightning [250.4(A)(2)].

**(A) Grounding.** Services supplied from a grounded utility transformer shall have the grounded (neutral) conductor grounded to any of the following grounding electrodes:

- Metal Underground Water Pipe [250.52(A)(1)]
- Effectively Grounded Metal Frame of the Building or Structure [250.52(A)(2)]
- Concrete-Encased Grounding Electrode [250.52(A)(3)]
- Ground Ring [250.52(A)(4)]

Where none of the above grounding electrodes are available, then one or more of the following grounding electrodes shall be installed:

- Ground Rod [250.52(A)(5)]
- Grounding Plate Electrodes 250.52(A)(6)
- Other Local Metal Underground Systems or Structures [250.52(A)(7)]

**AUTHOR'S COMMENT:** The grounding of the grounded (neutral) conductor to earth at service equipment is intended to help the utility limit the voltage imposed by lightning, line surges, or unintentional contact with higher-voltage lines by shunting potentially dangerous energy into the earth. In addition, grounding of the grounded (neutral) conductor to earth helps the electric utility clear high-voltage ground faults when they occur.

**(1) Accessible Location.** A grounding electrode conductor shall connect the grounded (neutral) conductor at service equipment to the grounding electrode. This connection shall be at any accessible location, from the load end of the service drop or service lateral, up to and including the service disconnecting means. Figure 250-53

**AUTHOR'S COMMENT:** Some inspectors require the grounding electrode conductor to terminate to the grounded (neutral) conductor at the meter enclosure and others require this connection at the service disconnect. However, the *Code* allows this grounding connection at either the meter enclosure or the service disconnect.

**(4) Main Bonding Jumper.** The grounding electrode conductor can terminate to the equipment grounding terminal, if the equipment grounding terminal is bonded to the service equipment enclosure [250.28].

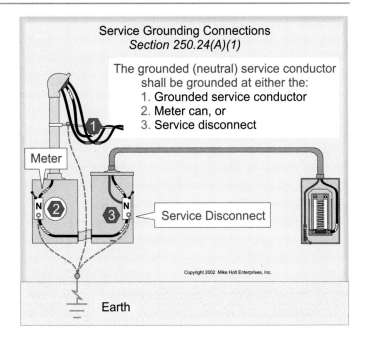

**Figure 250–53**

**(5) Load-Side Bonding Connections.** A neutral-to-case bond shall not be made on the load side of the service disconnecting means, except as permitted in 250.30(A)(1) for separately derived systems, 250.32(B)(2) for separate buildings, or 250.142(B) Ex. 2 for meter enclosures. Figure 250-54

**AUTHOR'S COMMENT:** If a neutral-to-case bond is made on the load side of service equipment, objectionable neutral current will flow on conductive metal parts of electrical equipment in violation of 250.6(A) [250.142]. Objectionable current on metal parts of electrical equipment can create a condition where

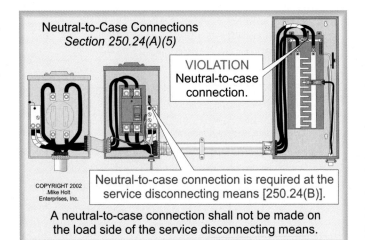

**Figure 250–54**

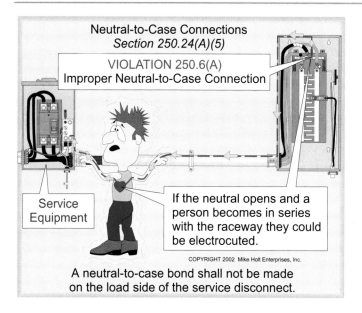

**Figure 250–55**

electric shock and even death from ventricular fibrillation can occur if a neutral-to-case connection is made. Figure 250-55

**(B)  Grounded (neutral) Conductor Brought to Each Service**. Because electric utilities are not required to install an equipment grounding (bonding) conductor, services supplied from a grounded utility transformer shall have a grounded (neutral) conductor run from the electric utility transformer

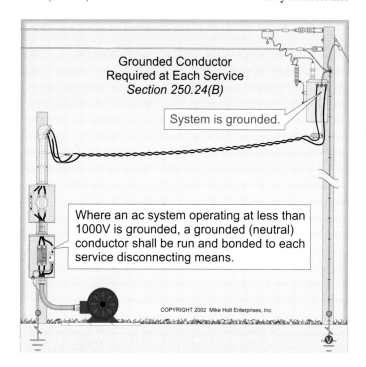

**Figure 250–56**

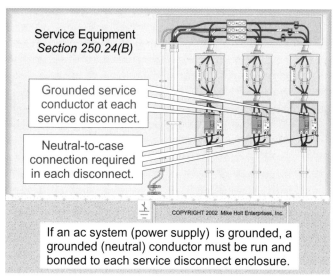

**Figure 250–57**

to each service disconnecting means. The grounded (neutral) conductor shall be bonded to the enclosure of each disconnecting means. Figures 250-56 and 250-57

**AUTHOR'S COMMENT:** It is critical that the metal parts of service equipment be bonded to the grounded (neutral) conductor (effective ground-fault current path) to ensure that dangerous voltage from a ground fault will be quickly removed [250.4(A)(3) and 250.4(A)(5)]. To accomplish this, the grounded (neutral) conductor shall be run to service equipment from the electric utility, even

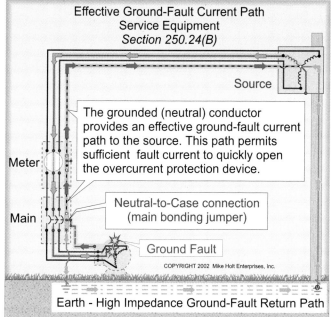

**Figure 250–58**

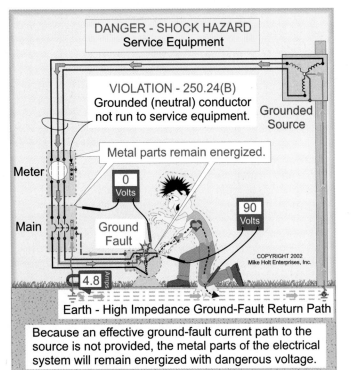

**Figure 250–59**

when there are no line-to-neutral loads being supplied! Figure 250-58

*DANGER: If the grounded (neutral) service conductor is not run between the electric utility and service equipment, there would be no low-impedance effective ground-fault current path. In the event of a ground fault, the circuit protection device will not open and metal parts will remain energized. Figure 250-59*

**(1) Minimum Size Grounded (neutral) Conductor.** Because the grounded (neutral) service conductor is required to serve as the effective ground-fault current path, it shall be sized so that it can safely carry the maximum ground-fault current likely to be imposed on it [110.10 and 250.4(A)(5)]. This is accomplished by sizing the grounded (neutral) conductor in accordance with Table 250.66, based on the total area of the largest ungrounded conductor. In addition, the grounded (neutral) conductors shall have the capacity to carry the maximum unbalanced neutral current in accordance with 220.22. Figure 250-60

**Question:** What is the minimum size grounded (neutral) service conductor required for a 400A, 3-phase, 480V service where the ungrounded service conductors are 500 kcmil and the maximum unbalanced load is 100A? Figure 250-61

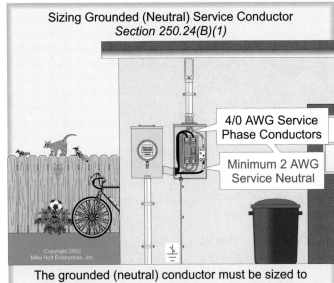

The grounded (neutral) conductor must be sized to carrying the maximum unbalanced load [220.22] and it cannot be smaller than the grounding electrode conductor size from 250.66.

**Figure 250–60**

(a) 3 AWG                     (b) 2 AWG

(c) 1 AWG                     (d) 1/0 AWG

**Answer:** (d) 1/0 AWG

Table 250.66 = 1/0 AWG. The unbalanced load requires a 3 AWG rated for 100A in accordance with Table 310.16, but 1/0 AWG is required to accommodate the maximum possible fault current [310.4]. At the service, the grounded (neutral) conductor also serves as the effective ground-fault current path to the power source.

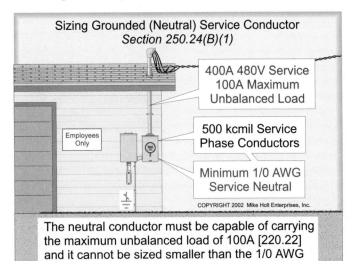

The neutral conductor must be capable of carrying the maximum unbalanced load of 100A [220.22] and it cannot be sized smaller than the 1/0 AWG grounding electrode conductor size from 250.66.

**Figure 250–61**

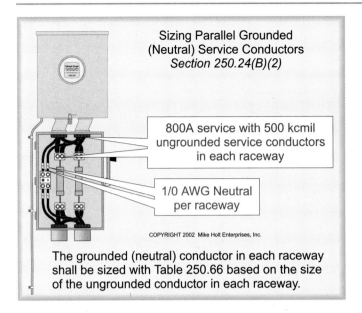

Sizing Parallel Grounded
(Neutral) Service Conductors
*Section 250.24(B)(2)*

800A service with 500 kcmil
ungrounded service conductors
in each raceway

1/0 AWG Neutral
per raceway

COPYRIGHT 2002 Mike Holt Enterprises, Inc.

The grounded (neutral) conductor in each raceway
shall be sized with Table 250.66 based on the size
of the ungrounded conductor in each raceway.

*Figure 250–62*

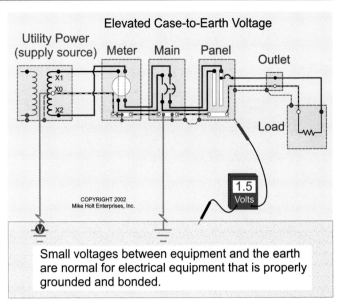

Elevated Case-to-Earth Voltage

Small voltages between equipment and the earth
are normal for electrical equipment that is properly
grounded and bonded.

*Figure 250–63*

**(2) Parallel Grounded Conductor.** Where service conductors are installed in parallel, a grounded (neutral) conductor shall be installed in each raceway, and each grounded (neutral) conductor shall be sized in accordance with Table 250.66 based on the size of the largest ungrounded conductor in the raceway.

**Question:** What is the minimum size grounded (neutral) service conductor required for an 800A, 480V, 3Ø service installed in two raceways, if the maximum unbalanced neutral load is 100A? The ungrounded service conductors in each raceway are 500 kcmil. Figure 250-62

(a) 3 AWG     (b) 2 AWG

(c) 1 AWG     (d) 1/0 AWG

**Answer:** (d) 1/0 AWG per raceway, 310.4 and Table 250.66

## Danger of Open Service Neutral

The bonding of the grounded (neutral) conductor to the service disconnect enclosure creates a condition where ground faults can be quickly cleared and the elevated voltage on the metal parts will not be much more than a few volts. Figure 250-63

**Shock Hazard.** However, if the grounded (neutral) service conductor, which serves as the effective ground-fault current path, is opened, a ground fault cannot be cleared and the metal parts of electrical equipment, as well as metal piping and structural steel, will become and remain energized providing the potential for electric shock. Figure 250-64

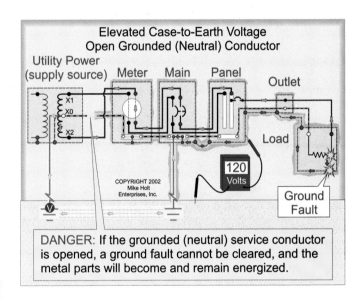

Elevated Case-to-Earth Voltage
Open Grounded (Neutral) Conductor

DANGER: If the grounded (neutral) service conductor
is opened, a ground fault cannot be cleared, and the
metal parts will become and remain energized.

*Figure 250–64*

When the service grounded (neutral) conductor is open, objectionable neutral current flows onto the metal parts of the electrical system because a neutral-to-case connection (main bonding jumper) is made at service equipment. Under this condition, dangerous voltage will be present on the metal parts providing the potential for electric shock as well as fires. This dangerous electrical shock condition is of particular concern in buildings with pools, spas and hot tubs. Figure 250-65

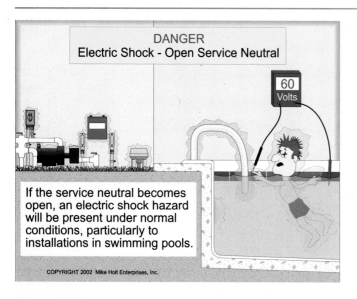

Figure 250–65

**AUTHOR'S COMMENT:** To determine the actual voltage on the metal parts from an open service grounded (neutral) conductor, you need to do some fancy math calculations with a spreadsheet to accommodate the variable conditions. Visit www.NECcode.com and go to the Free Stuff link to download a spreadsheet for this purpose.

**Fire Hazard.** If the grounded (neutral) service conductor is open, neutral current flows onto the metal parts of the electrical system. When this occurs in a wood frame construction building or structure, neutral current seeking a return path to the power supply travels into moist wood members. After many years of this current flow, the wood can be converted into charcoal (wood with no moisture) because of the neutral current flow, which can result in a fire. This condition is called pyroforic-carbonization. Figure 250-66

**AUTHOR'S COMMENT:** We can't create an acceptable graphic to demonstrate how pyroforic-carbonization causes a fire by an open service neutral. However, if you would like to order a video showing actual fires caused by pyroforic-carbonization, call 1-888-NEC-CODE.

### 250.28 Main Bonding Jumper

At service equipment, a main bonding jumper shall bond the metal service disconnect enclosure to the grounded (neutral) conductor. When equipment is listed for use as service equipment as required by 230.66, the main bonding jumper will be supplied by the equipment manufacturer [408.3(C)]. Figure 250-67

**AUTHOR'S COMMENT:** The main bonding jumper serves two very important needs. First, it establishes a connection between the equipment enclosure and the earth through the grounding electrode conductor to dissipate lightning and other high-voltage surges [250.4(A)(2)]. Secondly, it establishes a connection between the effective ground-fault current path and the service grounded (neutral) conductor to clear a ground fault. Figure 250-68

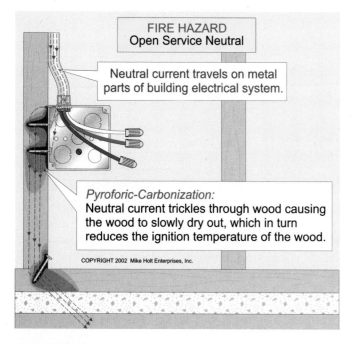

Figure 250–66

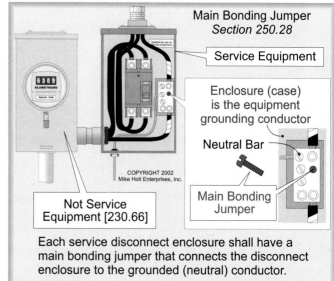

Figure 250–67

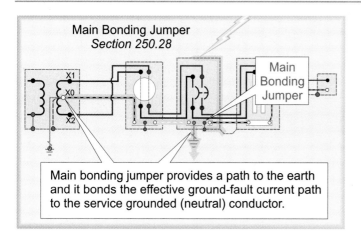

**Figure 250–68**

An effective ground-fault current path cannot be provided if a main bonding jumper is not installed in an installation where a CT enclosure is used. The result is that metal parts of the electrical installation will remain energized with dangerous voltage from a ground fault. Figure 250-69

**(A) Material.** The main bonding jumper shall be a wire, bus, or screw of copper or other corrosion-resistant material.

**(B) Construction.** Where a main bonding jumper is a screw, the screw shall be identified with a green finish that shall be visible with the screw installed.

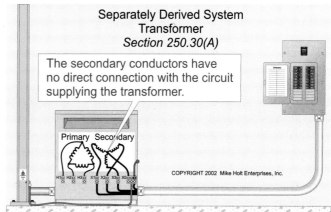

Separately Derived System [Article 100]: A wiring system whose power has no direct electrical connection to the supply conductors originating in another system.

**Figure 250–70**

### 250.30 Grounding (and Bonding) Separately Derived Systems

**AUTHOR'S COMMENT:** A separately derived system is a premises wiring system that has no direct electrical connection to conductors originating from another system. See Article 100 definition and 250.20(D). All transformers, except an autotransformer, are separately derived because the primary supply conductors do not have any direct electrical connection to the secondary conductors. Figure 250-70

A generator, a converter winding or a solar photovoltaic system can only be a separately derived system if the grounded (neutral) conductor is opened in the transfer switch [250.20(D) FPN 1]. Figure 250-71

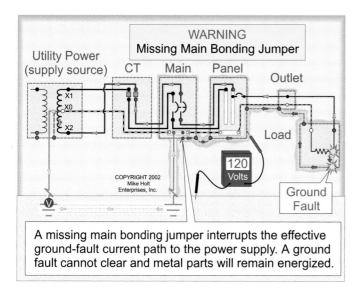

A missing main bonding jumper interrupts the effective ground-fault current path to the power supply. A ground fault cannot clear and metal parts will remain energized.

**Figure 250–69**

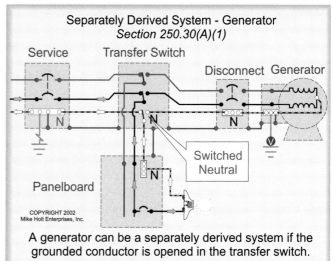

A generator can be a separately derived system if the grounded conductor is opened in the transfer switch.

**Figure 250–71**

If a generator, which is not part of a separately derived system, is bonded in accordance with 250.30(A), dangerous objectionable neutral current will flow on the bonding paths in violation of 250.6(A). Figure 250-72

All online UPS systems are separately derived even if the input and output voltages are the same. An automatic transfer switch has no impact on this determination. This is because an isolation transformer is provided as part of the module. Utilize caution when connecting these systems.

**(A) Grounded Systems.** Separately derived systems that operate at over 50V [250.20(A) and 250.112(I)] shall comply with the bonding and grounding requirements of 250.30(A)(1) through (A)(6).

**AUTHOR'S COMMENT:** Bonding the metal parts on the secondary of the separately derived system to the secondary grounded (neutral) terminal ensures that dangerous voltage from a ground fault on the secondary can be quickly removed by opening the secondary circuit's overcurrent protection device [250.2(A)(3)]. In addition, separately derived systems are grounded to stabilize the line-to-earth voltage during normal operation [250.4(A)(1)]. Figure 250-73

**(1) Bonding Jumper (Neutral-to-Case Connection).** To provide the effective ground-fault current path necessary to clear a ground fault on the secondary side of the separately derived system, the metal parts of electrical equipment shall be bonded to the grounded (neutral) terminal of the separately derived system. The bonding

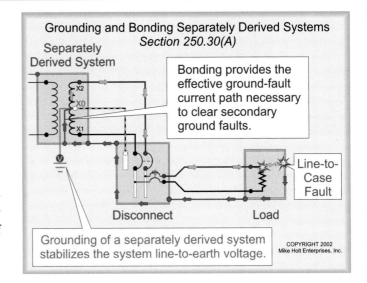

Figure 250–73

jumper used for this purpose shall be sized in accordance with Table 250.66 based on the area of the largest derived ungrounded conductor. Figure 250-74

**Question:** What size bonding jumper is required for a 45 kVA transformer if the secondary conductors are 3/0 AWG? Figure 250-75

(a) 4 AWG                 (b) 3 AWG

(c) 2 AWG                 (d) 1 AWG

**Answer:** (a) 4 AWG, Table 250.66

*DANGER: If a bonding jumper is not installed from the equipment grounding (bonding) conductor to the*

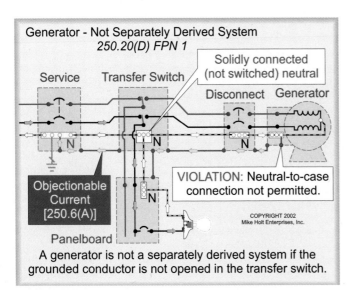

Figure 250–72

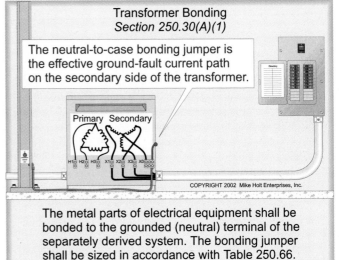

Figure 250–74

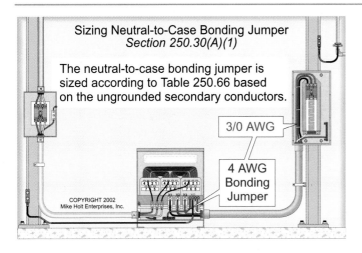

Sizing Neutral-to-Case Bonding Jumper
Section 250.30(A)(1)

The neutral-to-case bonding jumper is sized according to Table 250.66 based on the ungrounded secondary conductors.

3/0 AWG

4 AWG Bonding Jumper

COPYRIGHT 2002 Mike Holt Enterprises, Inc.

**Figure 250-75**

grounded (neutral) terminal of the separately derived system, then a ground fault cannot be cleared and the metal parts of electrical equipment, as well as metal piping and structural steel, will remain energized providing the potential for electric shock as well as fires. Figure 250-76

**AUTHOR'S COMMENT:** The neutral-to-case bonding jumper establishes the effective ground-fault current path for the equipment grounding (bonding) conductor on the secondary and the separately derived system (secondary). To protect against a primary ground fault, the primary circuit conductors shall contain an effective ground-fault current path. Figure 250-77

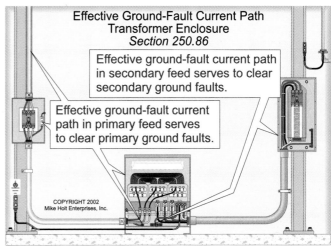

Effective Ground-Fault Current Path
Transformer Enclosure
Section 250.86

Effective ground-fault current path in secondary feed serves to clear secondary ground faults.

Effective ground-fault current path in primary feed serves to clear primary ground faults.

COPYRIGHT 2002 Mike Holt Enterprises, Inc.

**Figure 250-77**

The point of connection for the separately derived system neutral-to-case bond shall be made at the same location where the separately derived grounding electrode conductor terminates in accordance with 250.30(A)(2)(a). Figure 250-78

The neutral-to-case bond can be made at the source of a separately derived system or at the first system disconnecting means, but not at both locations. Figure 250-79

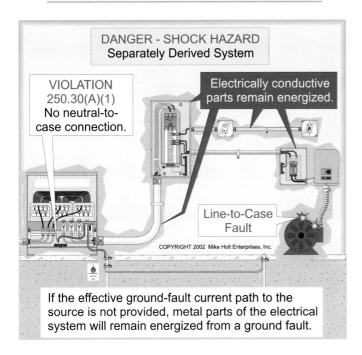

DANGER - SHOCK HAZARD
Separately Derived System

VIOLATION
250.30(A)(1)
No neutral-to-case connection.

Electrically conductive parts remain energized.

Line-to-Case Fault

COPYRIGHT 2002 Mike Holt Enterprises, Inc.

If the effective ground-fault current path to the source is not provided, metal parts of the electrical system will remain energized from a ground fault.

**Figure 250-76**

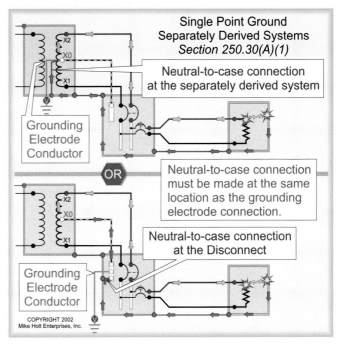

Single Point Ground
Separately Derived Systems
Section 250.30(A)(1)

Neutral-to-case connection at the separately derived system

Grounding Electrode Conductor

OR

Neutral-to-case connection must be made at the same location as the grounding electrode connection.

Neutral-to-case connection at the Disconnect

Grounding Electrode Conductor

COPYRIGHT 2002 Mike Holt Enterprises, Inc.

**Figure 250-78**

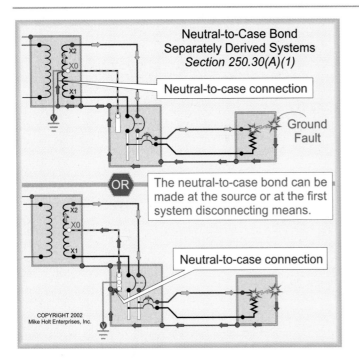

**Figure 250–79**

CAUTION: The neutral-to-case bond for a separately derived system cannot be made at more than one location, because doing so results in a parallel path(s) for neutral current. Multiple neutral current return paths to the grounded (neutral) terminal of the power supply can create dangerous objectionable current flow on grounding and bonding paths in violation of 250.6 and 250.142(A). Figure 250-80

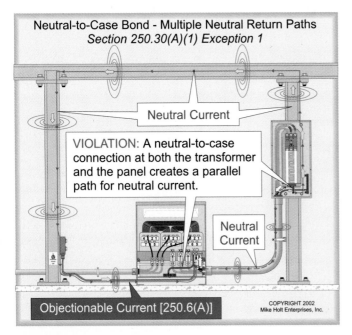

**Figure 250–80**

*Exception No. 2:* The bonding jumper for a system rated not more than 1,000 VA shall not be smaller than the derived phase conductors, and shall not be smaller than 14 AWG copper.

**(2) Grounding.** To stabilize the line-to-earth voltage during normal operation, a grounding electrode conductor shall ground the separately derived system grounded (neutral) conductor to a suitable grounding electrode.

(a) Single Separately Derived System. The grounding electrode conductor for a single separately derived system shall be sized in accordance with 250.66, based on the area of the largest separately derived ungrounded conductor. This conductor shall ground the grounded (neutral) conductor of the separately derived system to a suitable grounding electrode as specified in 250.30(A)(4). Figure 250-81

**AUTHOR'S COMMENT:** The grounding electrode conductor connection shall terminate directly to the grounded (neutral) terminal, not to the separately derived system enclosure.

To prevent objectionable current from flowing on grounding and bonding conductors, the grounding electrode conductor shall terminate at the same point on the separately derived system where the

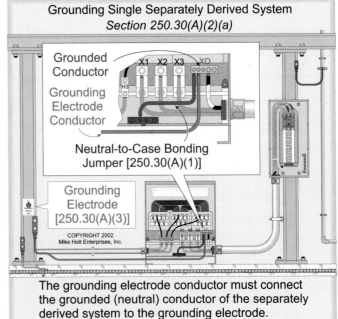

**Figure 250–81**

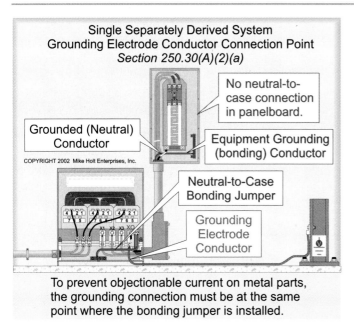

Single Separately Derived System
Grounding Electrode Conductor Connection Point
*Section 250.30(A)(2)(a)*

No neutral-to-case connection in panelboard.

Grounded (Neutral) Conductor

COPYRIGHT 2002 Mike Holt Enterprises, Inc.

Equipment Grounding (bonding) Conductor

Neutral-to-Case Bonding Jumper

Grounding Electrode Conductor

To prevent objectionable current on metal parts, the grounding connection must be at the same point where the bonding jumper is installed.

**Figure 250–82**

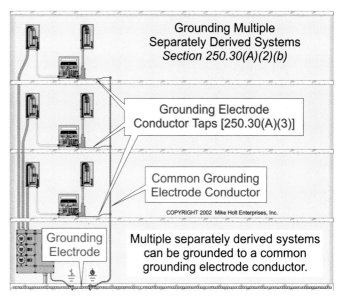

Grounding Multiple
Separately Derived Systems
*Section 250.30(A)(2)(b)*

Grounding Electrode Conductor Taps [250.30(A)(3)]

Common Grounding Electrode Conductor

COPYRIGHT 2002 Mike Holt Enterprises, Inc.

Grounding Electrode

Multiple separately derived systems can be grounded to a common grounding electrode conductor.

**Figure 250–83**

neutral-to-case bonding jumper is installed. Figure 250-82

**Exception:** A grounding electrode conductor is not required for a system rated not more than 1,000 VA. However, the system shall be bonded in accordance with 250.30(A)(1) Ex. 2.

(b)  Multiple Separately Derived Systems. Where multiple separately derived systems are grounded to a common grounding electrode conductor as provided in 250.30(A)(3), the common grounding electrode conductor shall be sized in accordance with Table 250.66 based on the total circular mil area of the separately derived ungrounded conductors from all of the separately derived systems. Figure 250-83

**(3) Grounding Electrode Taps.** A grounding electrode tap from a separately derived system to a common grounding electrode conductor shall be permitted to ground the grounded (neutral) terminal of the separately derived system to a common grounding electrode conductor.

(a)  Tap Conductor Size. Each grounding electrode tap conductor shall be sized in accordance with 250.66, based on the size of the largest separately derived ungrounded conductor of the separately derived system.

(b)  Connections. All grounding electrode tap connections shall be made at an accessible location by a listed irreversible compression connector, listed

connections to copper busbars, or by exothermic welding. Grounding electrode tap conductors shall be grounded to the common grounding electrode conductor as specified in 250.30(A)(2)(b) in such a manner that the common grounding electrode conductor is not spliced.

(c)  Installation. The common grounding electrode conductor and the grounding electrode taps to each separately derived system shall be:

•  Copper where within 18 in. of earth [250.64(A)].

•  Securely fastened to the surface on which it is carried and adequately protected if exposed to physical damage [250.64(B)].

•  Installed in one continuous length without a splice or joint, unless spliced by irreversible compression-type connectors listed for the purpose or by the exothermic welding process [250.64(C)].

•  Metal enclosures (such as a raceway) enclosing a common grounding electrode conductor shall be made electrically continuous from the point of attachment to cabinets or equipment to the grounding electrode and shall be securely fastened to the ground clamp or fitting [250.64(E)].

**(4) Grounding Electrode Conductor.** The grounding electrode conductor shall terminate to a suitable grounding electrode that is located as close as practicable and preferably in the same area as the grounding electrode conductor termination to the grounded (neutral) conductor. The grounding electrode shall be the nearest one of the following:

(1) Effectively grounded metal member of the structure.

(2) Effectively grounded metal water pipe, within 5 ft from the point of entrance into the building.

*Exception:* For industrial and commercial buildings where conditions of maintenance and supervision ensure that only qualified persons service the installation, the grounding electrode conductor can terminate on the metal water-pipe system at any point, if the entire length of the interior metal water pipe that is being used for the grounding electrode is exposed.

(3) Where none of the grounding electrodes as listed in (1) or (2) above are available, one of the following grounding electrodes shall be used:

• A concrete-encased grounding electrode encased by at least 2 in. of concrete, located within and near the bottom of a concrete foundation or footing that is in direct contact with earth, consisting of at least 20 ft of one or more bare or zinc galvanized or other electrically conductive coated steel reinforcing bars or rods of not less than $1/2$ in. in diameter, or consisting of at least 20 ft of bare copper conductor not smaller than 4 AWG [250.52(A)(3)].

• A ground ring encircling the building or structure, buried at least 30 in., consisting of at least 20 ft of bare copper conductor not smaller than 2 AWG [250.52(A)(4) and 250.53(F)].

• A ground rod having not less than 8 ft of contact with the soil [250.52(A)(5) and 250.53(G)].

• A buried ground plate electrode with not less than 2 sq ft of exposed surface area [250.52(A)(6)].

• Other metal underground systems or structures, such as piping systems and underground tanks [250.52(A)(7)].

FPN: Interior metal water piping in the area served by a separately derived system shall be bonded to the grounded (neutral) conductor at the separately derived system in accordance with the requirements contained in 250.104(A)(4).

**(5) Equipment Bonding Jumper Size.** Where an equipment bonding jumper is run with the derived phase conductors from the source of a separately derived system to the first disconnecting means, it shall be sized in accordance with Table 250.66, based on the total area of the largest separately derived ungrounded conductors.

**(6) Grounded (neutral) Conductor.** Where the neutral-to-case bond is made at the first system disconnecting means, instead of at the source of the separately derived system, the following requirements shall apply: Figure 250-84

(a) Routing and Sizing. Because the grounded (neutral) conductor is to serve as the effective ground-fault current path, the grounded (neutral) conductor shall be routed with the secondary conductors, and it shall be sized no smaller than specified in Table 250.66, based on the largest derived ungrounded conductor.

(b) Parallel Conductors. If the secondary conductors are installed in parallel, the grounded (neutral) secondary conductor in each raceway shall be sized based on the area of the largest derived ungrounded conductors in the raceway.

---

**AUTHOR'S COMMENT:** When the neutral-to-case bonding jumper is located in the first system disconnecting means, the grounding electrode conductor shall terminate at the same location to prevent objectionable current from flowing on grounding and bonding paths [250.30(A)(2)(a)].

---

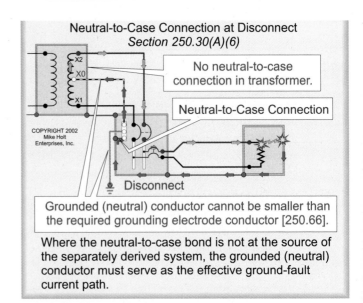

Neutral-to-Case Connection at Disconnect
*Section 250.30(A)(6)*

No neutral-to-case connection in transformer.

Neutral-to-Case Connection

COPYRIGHT 2002
Mike Holt
Enterprises, Inc.

Disconnect

Grounded (neutral) conductor cannot be smaller than the required grounding electrode conductor [250.66].

Where the neutral-to-case bond is not at the source of the separately derived system, the grounded (neutral) conductor must serve as the effective ground-fault current path.

**Figure 250–84**

## 250.32 Grounding and Bonding at Separate Buildings and Structures

**(A) Grounding.** Metal parts of electrical equipment in separate buildings or structures supplied by a feeder shall be grounded to a suitable grounding electrode. The metal parts of a building or structure shall be grounded to earth to protect persons and property by limiting voltage on the metal parts from lightning [250.4(A)(2)]. Figure 250-85

Any of the following can serve as a suitable grounding electrode [250.52(A)] for this purpose:

- Metal Underground Water Pipe [250.52(A)(1)]

- Effectively Grounded Metal Frame of the Building or Structure [250.52(A)(2)]

- Concrete-encased grounding electrode [250.52(A)(3)]

- Ground Ring [250.52(A)(4)]

Where none of the above grounding electrodes are available, then one or more of the following grounding electrodes shall be installed:

- Ground Rod [250.52(A)(5)]

- Grounding Plate Electrodes [250.52(A)(6)]

- Other Local Metal Underground Systems or Structures [250.52(A)(7)]

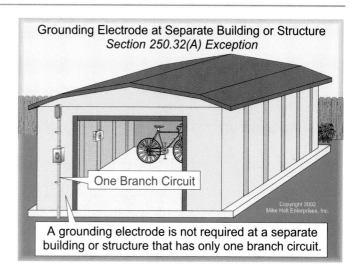

Grounding Electrode at Separate Building or Structure
Section 250.32(A) Exception

One Branch Circuit

A grounding electrode is not required at a separate building or structure that has only one branch circuit.

**Figure 250–86**

*Exception:* A grounding electrode at a separate building or structure is not required where the building or structure is served by one branch circuit containing an equipment grounding (bonding) conductor. Figure 250-86

**(B) Bonding.** To quickly clear a ground fault and remove dangerous voltage at a separate building or structure, the metal parts of electrical equipment shall be bonded to an effective ground-fault current path [250.4(A)(5)].

**AUTHOR'S COMMENT:** Keep in mind that the effective fault-current path from the separate building or structure shall be electrically continuous from the point of a ground fault to the power supply. In addition, the grounding electrode at the separate building serves no purpose in opening an overcurrent protection device in the event of a ground fault.

**(1) Equipment Grounding (bonding) Conductor.** When an equipment grounding (bonding) conductor as described in 250.118 serves as the effective ground-fault current path, it shall be installed with the feeder conductors, and it shall be bonded to the building/structure disconnect. The equipment grounding (bonding) conductor shall be sized in accordance with 250.122 based on the rating of the feeder circuit protection device. Figure 250-87

*CAUTION: To prevent dangerous objectionable neutral current from flowing on the metal parts of electrical equipment in violation of 250.6(A), the grounded (neutral) conductor at the separate building or structure shall not be bonded to the equipment grounding (bonding) conductor or to the grounding electrode system. Figures 250-88 and 250-89*

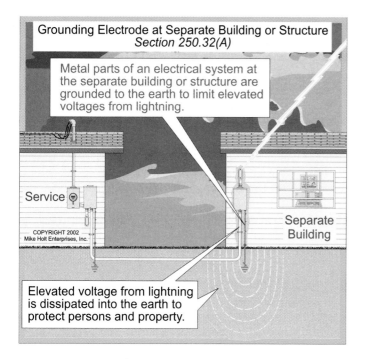

Grounding Electrode at Separate Building or Structure
Section 250.32(A)

Metal parts of an electrical system at the separate building or structure are grounded to the earth to limit elevated voltages from lightning.

Service

Separate Building

Elevated voltage from lightning is dissipated into the earth to protect persons and property.

**Figure 250–85**

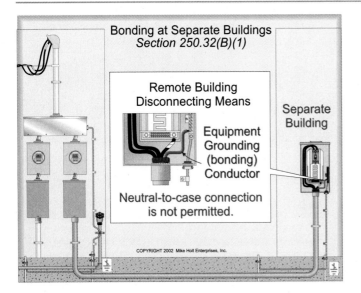

*Figure 250–87*

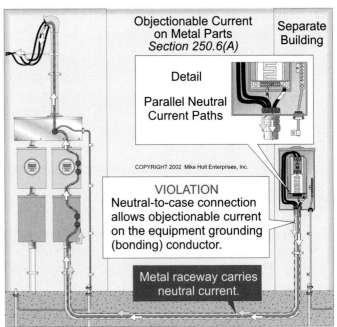

*Figure 250–89*

**(2) Grounded (neutral) Conductor.** Where an equipment grounding (bonding) conductor is not run with the feeder conductors to a building or structure disconnect, and there are no continuous metallic paths bonded to the grounding electrode system in both buildings or structures, and ground-fault protection of equipment is not installed on the service, the grounded (neutral) conductor run with the feeder conductors shall be bonded to the building or structure disconnecting means.

Because the grounded (neutral) conductor will be required to serve as the effective ground-fault current path, it shall not be smaller than the larger of: Figure 250-90

(1) To carry the maximum unbalanced neutral load in accordance with 220.22.

(2) To carry the available ground-fault current in accordance with 250.122.

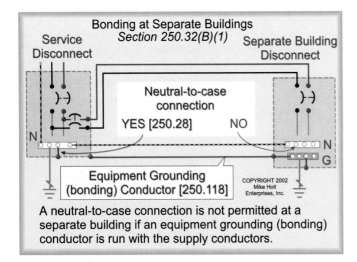

*Figure 250–88*

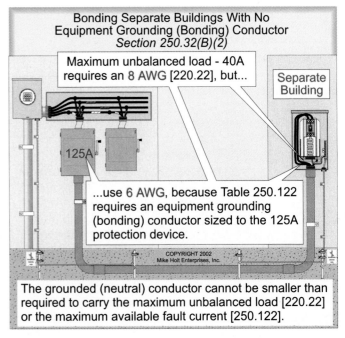

*Figure 250–90*

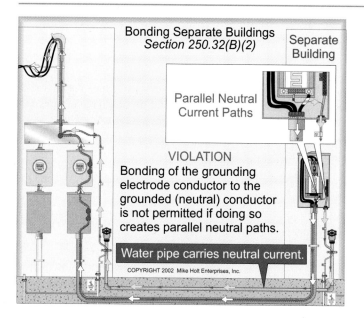

**Figure 250–91**

CAUTION: The use of the grounded (neutral) conductor as the effective ground-fault current path is permitted, but it poses potentially dangerous consequences and should only be done after careful consideration. Even if the initial installation will not result in an unacceptable parallel path for objectionable neutral current, there remains the possibility that a future installation of, say, metal piping or cables between the separate buildings or structures could reverse that situation. Figure 250-91

**AUTHOR'S COMMENT:** The preferred practice is to install an equipment grounding (bonding) conductor with the feeder conductors to the remote building or structure to serve as the effective ground-fault current path, and not bond the grounded (neutral) conductor to the building disconnect in accordance with 250.32(B)(1).

**(E)  Grounding Electrode Conductor.** The grounding electrode conductor for a separate building or structure disconnect shall be sized in accordance with 250.66, based on the size of the ungrounded feeder supply conductor. Figure 250-92

**AUTHOR'S COMMENT:** Where the grounding electrode conductor is connected to a ground rod, that portion of the conductor that is the sole connection to the ground rod is not required to be larger than 6 AWG copper [250.66(A)]. Where the grounding electrode conductor is connected to a concrete-encased grounding electrode, that portion of the conductor that is the sole connection to the concrete-encased grounding electrode is not required to be larger than 4 AWG copper [250.66(B)].

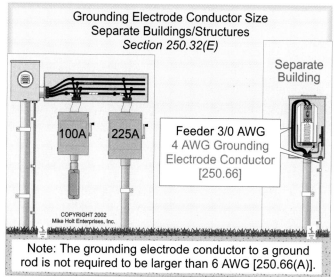

**Figure 250–92**

CAUTION: Where an equipment grounding (bonding) conductor is run with the feeder conductors to a separate building or structure, the grounding electrode conductor shall not terminate to the grounded (neutral) conductor bus. It shall terminate to the equipment grounding (bonding) terminal or bus, or to the metal disconnect enclosure by an approved device.

### 250.34 Generators – Portable and Vehicle-Mounted

**(A)  Portable Generators.** The frame of a portable generator is not required to be grounded to earth if: Figure 250-93

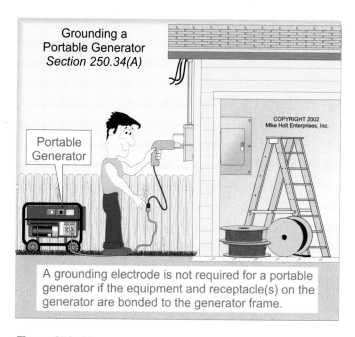

**Figure 250–93**

(1) The generator only supplies equipment or receptacles mounted on the generator, and

(2) The metal parts of generator and the equipment grounding (bonding) terminals of the receptacles are bonded to the generator frame.

**(B) Vehicle-Mounted Generators.** The frame of a vehicle-mounted generator is not required to be grounded to earth if: Figure 250-94

(1) The generator frame is bonded to the vehicle frame, and

(2) The generator only supplies equipment or receptacles mounted on the vehicle or generator, and

(3) The metal parts of the generator and the equipment grounding (bonding) terminals of the receptacles are bonded to the generator frame.

---

**AUTHOR'S COMMENT:** The bonding of the receptacle's grounding (bonding) terminal to the generator frame is done at the factory by the generator manufacturer.

---

**(C) Grounded Conductor Bonding.** If the portable generator is a separately derived system (transfer switch opens the grounded conductor), then the portable generator shall be grounded and bonded in accordance with 250.30.

---

**AUTHOR'S COMMENT:** When only a generator is used to supply a hard-wired connection to a building or structure (i.e., a construction trailer), it is a separately derived system though no transfer switch is present.

---

## Part III. Grounding Electrode System and Grounding Electrode Conductor

### Why Grounding is Important.

**Service Equipment.** The grounded (neutral) conductor at the building or structure disconnecting means shall be grounded to earth to protect persons from electric shock and to protect property from fires by limiting voltage on the metal parts from lightning [250.4(A)(1)]. Figure 250-95

**Separately Derived Systems.** The grounded (neutral) conductor from a separately derived system shall be grounded to earth to assist in stabilizing the line-to-earth voltage of the system [250.4(A)(1) and 250.30(A)(2)].

**Separate Buildings or Structures.** The metal parts of a building or structure disconnect shall be grounded to earth to protect persons from electric shock and to protect property from fires by limiting voltage on the metal parts from lightning [250.4(A)(2) and 250.32(A)].

---

**AUTHOR'S COMMENT:** Grounding to earth also assists in the reduction of radio frequency interference, static electricity and the improvement of the performance of transient voltage and lightning protection devices, thereby improving the overall reliability and performance of sensitive electronic equipment. However, for systems that operate at less than 600V, the grounding electrode system is not intended to clear a ground fault by opening the circuit's overcurrent protection device.

---

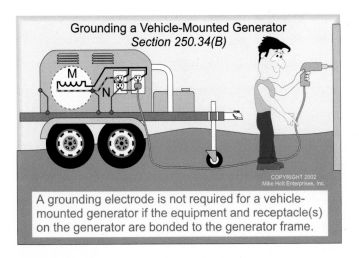

**Grounding a Vehicle-Mounted Generator**
*Section 250.34(B)*

A grounding electrode is not required for a vehicle-mounted generator if the equipment and receptacle(s) on the generator are bonded to the generator frame.

*Figure 250–94*

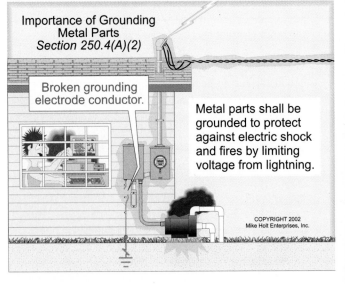

Importance of Grounding Metal Parts
*Section 250.4(A)(2)*

Broken grounding electrode conductor.

Metal parts shall be grounded to protect against electric shock and fires by limiting voltage from lightning.

*Figure 250–95*

## Soil Resistivity

The resistance of the grounding electrode system impacts the effectiveness of shunting high-voltage surges into earth. earth's ground resistance is directly impacted by the soil's resistivity, which varies throughout the world. Soil resistivity is influenced by the soil's electrolytes, which consist of moisture, minerals and dissolved salts. Because soil resistivity changes with moisture content, the resistance of any grounding system will vary with the seasons of the year. Since moisture becomes more stable at greater distances below the surface of earth, grounding systems appear to be more effective if the grounding electrode can reach the water table. In addition, having the grounding electrode below the frost line helps to ensure less deviation in the system's resistance year round.

## Measuring the Ground Resistance

The resistance of a grounding electrode can be measured by the use of a ground resistance clamp meter or a three-point fall of potential ground resistance meter.

**Ground Clamp Meter.** The ground resistance clamp meter measures the resistance of the grounding system by injecting a high-frequency signal to the utility ground (via the neutral conductor) and then measuring the strength of the return signal through earth to the grounding electrode that is being measured. Figure 250-96

**Fall of Potential Ground Resistance Meter.** The three-point fall of potential ground resistance meter determines the ground resistance by using Ohm's Law, R=E/I. This meter divides the voltage difference between the grounding

electrode to be measured and a driven potential test stake by the current flowing between the grounding electrode to be measured and a driven current test stake. The test stakes typically are made of $1/4$ in. to $3/8$ in. diameter steel rod, 14 to 24 in. long and they are driven into earth about two-thirds of their length.

The distance and alignment between the grounding electrode and the potential and current test stakes, is extremely important to the validity of the ground resistance measurements. For an 8 ft ground rod, the accepted practice is to space the current test stake 80 ft from the grounding electrode to be measured. The potential test stake is positioned (in a straight line) between the grounding electrode to be measured and the current test stake. The potential test stake should be located at approximately 62 percent of the distance that the current test stake is located from the grounding electrode. Since the current test stake is located 80 ft from the grounding electrode, the potential test stake will be about 50 ft from the electrode to be measured.

**Question:** If the voltage between the ground rod and the potential test stake is 3V and the current between the ground rod and the current test stake is 0.2A, then the ground resistance would be equal to _____. Figure 250-97

(a) 5Ω                         (b) 10Ω

(c) 15Ω                        (d) 25Ω

**Answer:** (c) 15Ω

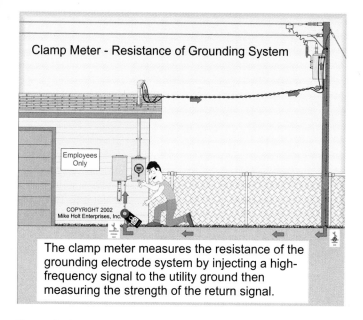

Clamp Meter - Resistance of Grounding System

Employees Only

COPYRIGHT 2002
Mike Holt Enterprises, Inc.

The clamp meter measures the resistance of the grounding electrode system by injecting a high-frequency signal to the utility ground then measuring the strength of the return signal.

**Figure 250–96**

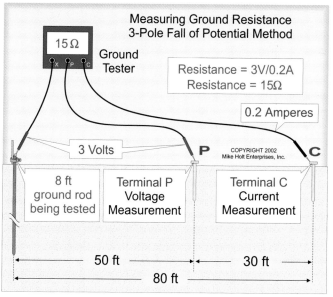

Measuring Ground Resistance
3-Pole Fall of Potential Method

15 Ω

Ground Tester

Resistance = 3V/0.2A
Resistance = 15Ω

0.2 Amperes

3 Volts

P

COPYRIGHT 2002
Mike Holt Enterprises, Inc.

C

8 ft ground rod being tested

Terminal P
Voltage Measurement

Terminal C
Current Measurement

50 ft          30 ft

80 ft

**Figure 250–97**

Resistance = E/I

E = 3V

I = 0.2A

Resistance = 3V/0.2A

Resistance = 15Ω

**AUTHOR'S COMMENT:** The three-point fall of a potential meter can only measure one electrode at a time. Two electrodes bonded together cannot be measured until they have been separated. The total resistance for two separate electrodes would be added as they would be for two resistors in parallel. For example, if the ground resistance of each electrode were 50Ω, the total resistance would be 25Ω for two grounding electrodes bonded together.

*CAUTION: The three-point fall of a potential meter can only be used on a single grounding electrode that is isolated from the electrical system. If the grounding electrode to be tested is connected to the electrical utility ground (by the neutral), then the meter will indicate that an error has occurred. To measure the ground resistance of grounding electrodes that are not isolated from the electric utility (such as at industrial facilities, commercial buildings, cell phone sites, broadcast antennas, data centers and telephone central offices), a clamp-on ground resistance tester would better serve the purpose.*

The resistance of the grounding electrode can be lowered by bonding multiple grounding electrodes that are properly spaced apart or by chemically treating earth around the grounding electrode. There are many readily available commercial products for this purpose.

## 250.50 Grounding Electrode System

If available on the premises, each item in 250.52(A)(1) through (A)(6) shall be connected together to form the grounding electrode system: Figure 250-98

- Metal Underground Water Pipe [250.52(A)(1)]

- Effectively Grounded Metal Frame of the Building or Structure [250.52(A)(2)]

- Concrete-Encased Grounding Electrode [250.52(A)(3)]

- Ground Ring [250.52(A)(4)]

Where none of the above grounding electrodes are available, then one or more of the following grounding electrodes shall be installed: Figure 250-99

- Ground Rod [250.52(A)(5)]

- Grounding Plate Electrodes 250.52(A)(6)]

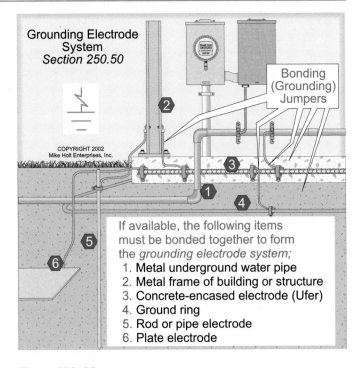

If available, the following items must be bonded together to form the *grounding electrode system;*
1. Metal underground water pipe
2. Metal frame of building or structure
3. Concrete-encased electrode (Ufer)
4. Ground ring
5. Rod or pipe electrode
6. Plate electrode

**Figure 250–98**

- Other Local Metal Underground Systems or Structures [250.52(A)(7)]

## 250.52 Grounding Electrodes

(A) **Electrodes Permitted for Grounding.**

(1) **Metal Underground Water Pipe.** A metal underground water pipe in direct contact with earth for 10 ft

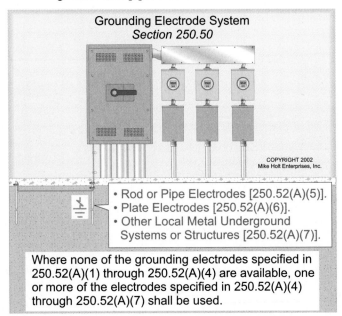

- Rod or Pipe Electrodes [250.52(A)(5)].
- Plate Electrodes [250.52(A)(6)].
- Other Local Metal Underground Systems or Structures [250.52(A)(7)].

Where none of the grounding electrodes specified in 250.52(A)(1) through 250.52(A)(4) are available, one or more of the electrodes specified in 250.52(A)(4) through 250.52(A)(7) shall be used.

**Figure 250–99**

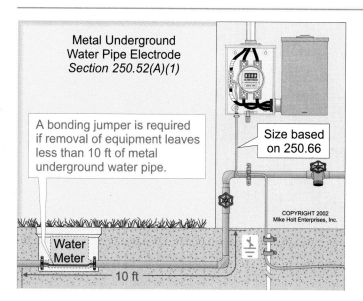

Metal Underground
Water Pipe Electrode
*Section 250.52(A)(1)*

A bonding jumper is required
if removal of equipment leaves
less than 10 ft of metal
underground water pipe.

Size based
on 250.66

COPYRIGHT 2002
Mike Holt Enterprises, Inc.

Water
Meter

10 ft

**Figure 250–100**

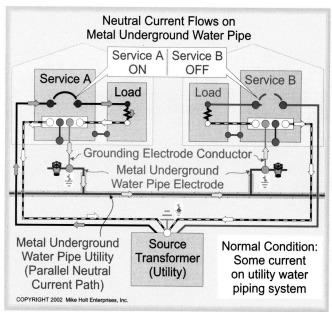

Neutral Current Flows on
Metal Underground Water Pipe

| Service A | Service B |
| ON | OFF |

Service A                                                    Service B

Load                        Load

Grounding Electrode Conductor

Metal Underground
Water Pipe Electrode

Metal Underground
Water Pipe Utility
(Parallel Neutral
Current Path)

Source
Transformer
(Utility)

Normal Condition:
Some current
on utility water
piping system

COPYRIGHT 2002  Mike Holt Enterprises, Inc.

**Figure 250–101**

or more can serve as a grounding electrode. If the 10 ft
of metal underground water pipe is interrupted, such as
with a water meter or insulating joints, it shall be made
electrically continuous with a bonding jumper sized
according to 250.66. The grounding electrode conductor shall be sized in accordance with Table 250.66.
Figure 250-100

Interior metal water piping located more than 5 ft from
the point of entrance to the building shall not be used as
a part of the grounding electrode system.

*Exception:* The grounding electrode conductor can terminate at any point on the water-pipe system for industrial and
commercial buildings where conditions of maintenance and
supervision ensure that only qualified persons service the
installation, and the entire length of the interior metal water
pipe that is being used for the grounding electrode is
exposed.

**AUTHOR'S COMMENT:** Controversy about using
the metal underground water supply piping as a grounding electrode has existed for many years. The water
supply industry feels that neutral current flowing on the
metal water-pipe system corrodes the metal. For more
information, contact the American Water Works
Association about their report *"Effects of Electrical
Grounding on Pipe Integrity and Shock Hazard,"*
Catalog No. 90702 1-800-926-7337. Figure 250-101

*CAUTION: Do not confuse the purpose of the
grounding electrode (diverting high-voltage surges
such as lightning into earth) [250.4(A)(1)] with the
bonding requirements of interior metal water-piping*

*systems (remove dangerous voltage on metal parts
from a ground fault) [250.4(A)(3) and 250.104].*

**(2) Metal Frame of the Building or Structure.** The metal
frame of the building or structure, where effectively
grounded to earth, can serve as a grounding electrode.
The grounding electrode conductor shall be sized in
accordance with Table 250.66. Figure 250-102

**AUTHOR'S COMMENT:** Exposed structural steel
shall also be bonded in accordance with 250.104(C).

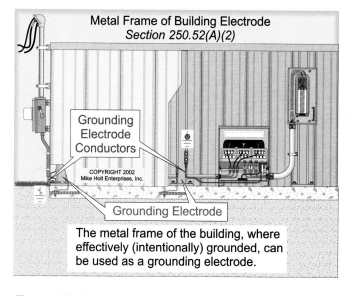

Metal Frame of Building Electrode
*Section 250.52(A)(2)*

Grounding
Electrode
Conductors

COPYRIGHT 2002
Mike Holt Enterprises, Inc.

Grounding Electrode

The metal frame of the building, where
effectively (intentionally) grounded, can
be used as a grounding electrode.

**Figure 250–102**

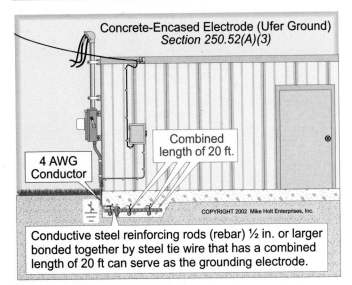

Conductive steel reinforcing rods (rebar) ½ in. or larger bonded together by steel tie wire that has a combined length of 20 ft can serve as the grounding electrode.

*Figure 250–103*

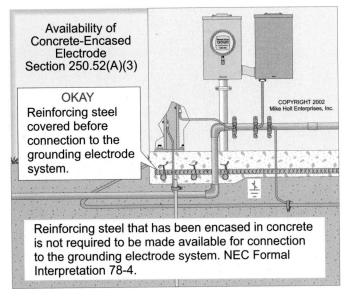

Reinforcing steel that has been encased in concrete is not required to be made available for connection to the grounding electrode system. NEC Formal Interpretation 78-4.

*Figure 250–104*

(3) **Concrete-Encased Grounding Electrode (Ufer).** Bare, galvanized or other electrically conductive reinforcing steel rods (rebar) not smaller than ½ in. diameter can be used as a grounding electrode if the total length of conductive steel rebar(s) are not less than 20 ft. The reinforcing steel rods shall be encased in at least 2 in. of concrete and located near the bottom of a foundation or footer that is in direct contact with earth. The steel rebar is not required to be one continuous length and the usual steel tie wires can be used to create the 20 ft concrete-encased grounding electrode. The grounding electrode conductor required for a concrete-encased grounding electrode is not required to be larger than 4 AWG copper [250.66(B)]. Figure 250-103

**AUTHOR'S COMMENT:** The concrete-encased grounding electrode is also called a "Ufer Ground," named after Herb Ufer, the person who determined its usefulness as a grounding electrode in the 1960s. This type of grounding electrode generally offers the lowest ground resistance for the cost. It is the grounding electrode of choice for many engineers and designers where new concrete foundations are available.

NOTICE: Where the reinforcing steel has already been covered, it is NOT required to be made available (such as chipping out the concrete to get a grounding connection). See NFPA Formal Interpretation #78-4, March of 1980. Figure 250-104

(4) **Ground Ring.** A ground ring encircling the building or structure, in direct contact with earth consisting of at least 20 ft of bare copper conductor not smaller than 2 AWG copper, can serve as a grounding electrode. The

ground ring shall be buried at a depth below earth's surface of not less than 30 in. [250.53(F)]. The grounding electrode conductor for the ground ring is not required to be larger than the conductor used for the ground ring [250.66(C)].

(5) **Ground Rod Electrodes.** Ground rod electrodes shall not be less than 8 ft in length, shall have at least 8 ft of length in contact with the soil [250.53(G)] and shall consist of the following:

(b) Rod. The ground rod shall be iron or steel at least 5/8 in. diameter, except listed stainless steel and nonferrous rods of copper, brass, or bronze shall have a diameter of at least ½ in. The grounding electrode conductor to a ground rod electrode is not required to be larger than 6 AWG copper [250.66(A)]. Figure 250-105

**AUTHOR'S COMMENT:** The diameter of a ground rod has an insignificant effect on the resistance of the grounding electrode. However, larger diameter ground rods (3/4 in. and 1 in.) are sometimes installed where mechanical strength is required or to compensate for the loss of the electrode's metal due to corrosion.

(6) **Ground Plate Electrode.** A buried iron or steel plate that has at least ¼ in. of thickness or a nonferrous metal plate at least 0.06 in. of thickness that is exposed to not less than 2 sq ft of surface area can be used as a grounding electrode. The grounding electrode conductor to a ground plate electrode is not required to be larger than 6 AWG copper [250.66(A)].

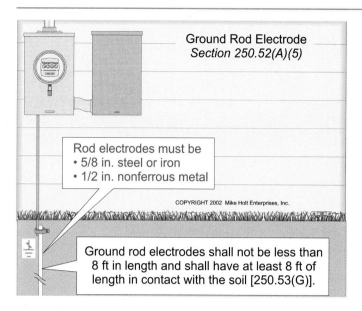

**Figure 250–105**

(7) Other Local Metal Underground Systems or Structures. Metal underground systems such as piping systems and underground tanks can be used as a grounding electrode. The grounding electrode conductor shall be sized in accordance with Table 250.66.

**(B) Electrodes Not Permitted.** None of following shall be used as a grounding electrode:

(1) Metal underground gas-piping system. Because lightning seeks a path to earth, metal underground gas-piping systems and structures shall not be used as a grounding electrode. Figure 250-106

**AUTHOR'S COMMENT:** See 250.104(B) requirements for bonding interior metal gas piping to an effective ground-fault current path.

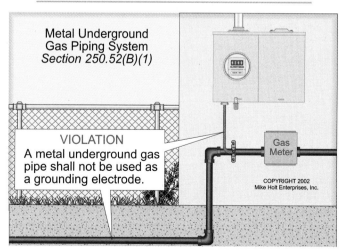

**Figure 250–106**

(2) Aluminum electrodes. Aluminum shall not be used as a grounding electrode because it corrodes much quicker than copper.

## 250.53 Installation of Grounding Electrode System.

**(A) Ground Rod Electrodes.** Where practicable, ground rod and plate electrodes shall be embedded below permanent moisture level and shall be free from nonconductive coatings such as paint or enamel [250.12].

**(C) Grounding Electrode Jumper.** The jumper used to connect grounding electrodes together to create the grounding electrode system shall be:

- Copper where within 18 in. of earth [250.64(A)].

- Securely fastened to the surface on which it is carried and adequately protected if exposed to physical damage [250.64(B)].

- Metal enclosures containing the grounding electrode conductor shall be electrically continuous from the point of attachment to cabinets or equipment to the grounding electrode and shall be securely fastened to the ground clamp or fitting [250.64(E)].

The grounding electrode jumper shall be sized in accordance with 250.66 for the grounding electrode involved and shall terminate to the grounding electrode by exothermic welding, listed lugs, listed pressure connectors, listed clamps, or other listed means. Termination fittings shall be listed for the materials of the grounding electrode and the grounding electrode conductor. When the termination of the conductor to the grounding electrode is encased in concrete or buried, the termination fittings shall be listed and identified for this purpose [250.70].

**(D) Metal Underground Water Pipe [250.52(A)(1)].**

(1) **Continuity.** The grounding path to earth or the bonding connection to interior piping [250.104] shall not be dependent on water meters or filtering devices and similar equipment. Where required, a bonding jumper shall be installed to provide the grounding or bonding path. See 250.68(B) for details. Figure 250-107

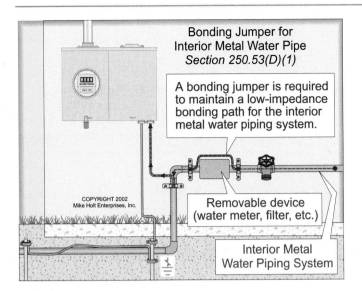

**Figure 250–107**

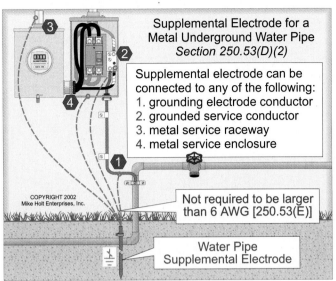

**Figure 250–108**

(2) **Supplemental Electrode Required.** The metal underground water pipe grounding electrode shall be supplemented by any of the following grounding electrode types specified in 250.52(A)(2) through (A)(7).

- Effectively Grounded Metal Frame of the Building or Structure [250.52(A)(2)]

- Concrete-Encased Grounding Electrode [250.52(A)(3)]

- Ground Ring [250.52(A)(4)]

Where none of the above grounding electrodes are available, then one or more of the following grounding electrodes shall be installed:

- Ground Rod [250.52(A)(5)]

- Grounding Plate Electrodes 250.52(A)(6)]

- Other Local Metal Underground Systems or Structures [250.52(A)(7)]

The underground water pipe supplemental grounding electrode shall terminate to the grounding electrode conductor, the grounded (neutral) service conductor, the nonflexible metal service raceway, or to any service equipment enclosure. Figure 250-108

**AUTHOR'S COMMENT:** If a building or structure uses a grounding electrode other than underground metal water pipe, such as a concrete-encased (Ufer) electrode, a supplemental grounding electrode is not required. Figure 250-109

(E) **Supplemental Grounding Electrode Jumper.** Where the supplemental grounding electrode for the underground

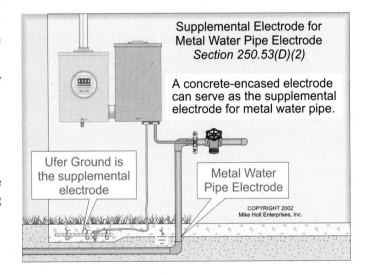

**Figure 250–109**

water pipe electrode is a ground rod, the conductor to the ground rod is not required to be larger than 6 AWG copper.

(F) **Ground Ring.** The ground ring, consisting of at least 20 ft of bare copper conductor not smaller than 2 AWG shall be buried at a depth of 30 in. [250.52(A)(4)].

(G) **Ground Rod Electrodes.** Ground rod electrodes shall be installed so that at least 8 ft of length is in contact with the soil. Where rock bottom is encountered, the ground rod shall be driven at an angle not to exceed 45 degrees from vertical. If rock bottom is encountered at an angle up to 45 degrees from vertical, the ground rod can be buried in a minimum 30 in. deep trench. Figure 250-110

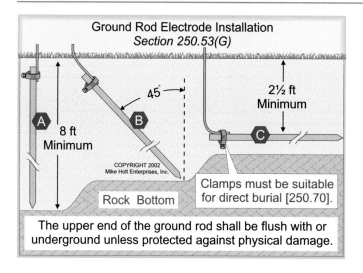

**Figure 250–110**

The upper end of the ground rod shall be flush with or underground unless the grounding electrode conductor attachment is protected against physical damage as specified in 250.10. See 250.52(A)(5).

**AUTHOR'S COMMENT:** When the grounding electrode attachment fitting is located underground, it shall be listed for direct soil burial [250.68(A) Ex. and 250.70].

**(H) Ground Plate Electrode.** A ground plate electrode of not less than 2 sq ft of surface to exterior soils shall be installed so that it is not less than 30 in. below the surface of earth [250.52(A)(6)].

## 250.54 Supplementary Electrodes

Supplementary electrodes (electrodes not required by the *NEC*) can be connected to the equipment grounding (bonding) conductor. The supplementary electrode is not required to be directly connected to the grounding electrode system nor is the conductor to the supplementary electrode required to be sized in accordance with 250.66. Figure 250-111

If the supplementary electrode is constructed from a ground rod, it is not required to comply with the 25Ω resistance requirement of 250.56. In addition, earth shall not be used as the equipment grounding (bonding) conductor [250.4(A)(5)].

*CAUTION: The "supplementary" electrode should not be confused with the "supplemental" grounding electrode for the underground metal water pipe grounding electrode required by 250.53(D)(2).*

**AUTHOR'S COMMENT:** An example of a supplementary electrode would be a ground rod installed next to a machine tool or other equipment sensitive to radio frequency (RF) interference. A supplementary electrode is not required to be bonded to the building grounding electrode system, but the equipment shall be bonded to an effective ground-fault current path so that dangerous voltage from a ground fault can be quickly removed [250.4(A)(5)]. Figure 250-112

## 250.56 Resistance of Ground Rod Electrode

When the resistance of a single ground rod is over 25Ω, one additional electrode shall be installed at least 6 ft away to augment the ground rod electrode. Figure 250-113

FPN: Spacing of grounding electrodes farther than 6 ft apart should help in lowering the total ground resistance.

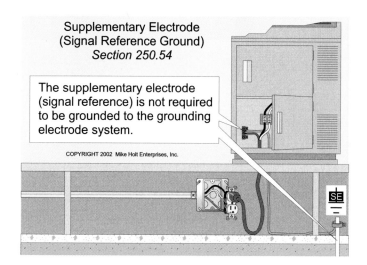

**Figure 250–111**

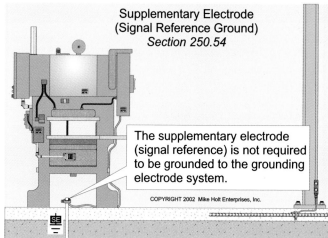

**Figure 250–112**

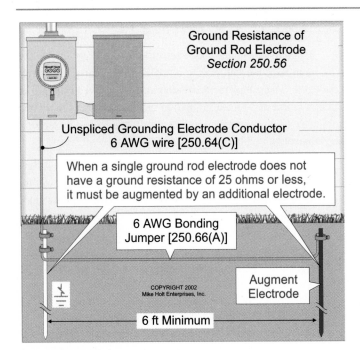

Ground Resistance of
Ground Rod Electrode
*Section 250.56*

Unspliced Grounding Electrode Conductor
6 AWG wire [250.64(C)]

When a single ground rod electrode does not
have a ground resistance of 25 ohms or less,
it must be augmented by an additional electrode.

6 AWG Bonding
Jumper [250.66(A)]

COPYRIGHT 2002
Mike Holt Enterprises, Inc.

Augment
Electrode

6 ft Minimum

**Figure 250–113**

---

**AUTHOR'S COMMENT:** The *NEC* does not require more than two ground rods to be installed, even if the total resistance of the two parallel ground rods exceeds 25Ω.

---

## 250.58 Common Grounding Electrode

Where a building or structure is supplied with multiple services or feeders as permitted by 225.30 and 230.2, a single grounding electrode shall be used for the multiple building disconnecting means as required in 250.24 and 250.32. The most practical method of meeting this requirement is to ground each of the disconnects to a common concrete-encased grounding electrode [250.52(A)(3)].

Where separate grounding electrodes are installed to meet this requirement, they shall be connected together in accordance with 250.53(C). Figure 250-114

---

**AUTHOR'S COMMENT:** Potentially dangerous neutral current can flow on the grounding electrode conductor when multiple services are grounded to the same single electrode. This occurs because neutral current from each service has multiple parallel return paths back to the electric utility power supply. This is of particular concern if the grounded (neutral) conductor from one of the services is opened. Figure 250-115

---

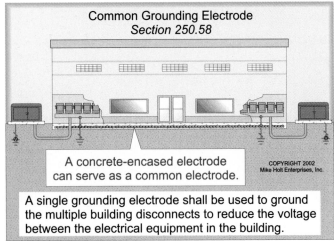

Common Grounding Electrode
*Section 250.58*

A concrete-encased electrode
can serve as a common electrode.

COPYRIGHT 2002
Mike Holt Enterprises, Inc.

A single grounding electrode shall be used to ground
the multiple building disconnects to reduce the voltage
between the electrical equipment in the building.

**Figure 250–114**

## 250.60 Lightning Protection System Grounding Electrode

A lightning protection system installed in accordance with NFPA 780 is intended to protect the building structure, not the electrical wiring or equipment within the structure. The grounding electrode for a lightning protection system shall not be used for grounding wiring systems and equipment [250.50]. Figure 250-116

---

**AUTHOR'S COMMENT:** The lightning protection system, if installed, shall be connected to the building or structure grounding electrode system [250.106]. The grounding electrode for the lightning protection system shall be kept at least 6 ft from the building grounding electrode system [250.53(B)].

---

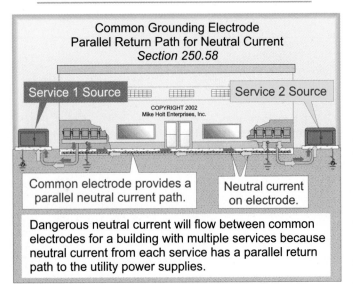

Common Grounding Electrode
Parallel Return Path for Neutral Current
*Section 250.58*

Service 1 Source                     Service 2 Source

COPYRIGHT 2002
Mike Holt Enterprises, Inc.

Common electrode provides a
parallel neutral current path.

Neutral current
on electrode.

Dangerous neutral current will flow between common
electrodes for a building with multiple services because
neutral current from each service has a parallel return
path to the utility power supplies.

**Figure 250–115**

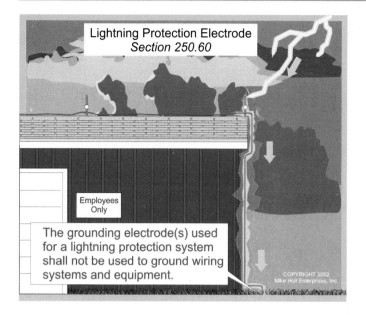

**Figure 250–116**

## 250.62 Grounding Electrode Conductor - Material

The grounding electrode conductor can be solid or stranded, insulated or bare, and it shall be copper, except aluminum can be used if installed at least 18 in. above earth in accordance with 250.64(A). Figure 250-117

> **AUTHOR'S COMMENT:** The *NEC* does not contain any identification requirement for the grounding electrode conductor, but the generally accepted practice is to identify this conductor with green marking tape.

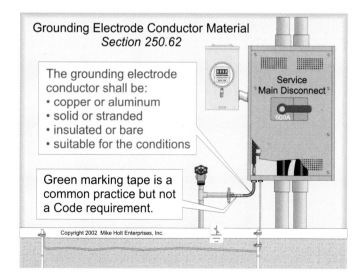

**Figure 250–117**

## 250.64 Grounding Electrode Conductor Installation

**(A) Aluminum Grounding Electrode Conductor.** Aluminum grounding electrode conductors cannot be in contact with earth, masonry, or where subject to corrosive conditions. When used outdoors, the termination of the aluminum grounding electrode conductor to the grounding electrode shall be at least 18 in. above earth.

**(B) Grounding Electrode Conductor Protection.** Grounding electrode conductors 8 AWG and smaller shall be installed in rigid metal conduit, intermediate metal conduit, rigid nonmetallic conduit or electrical metallic tubing. Grounding electrode conductors 6 AWG and larger that are not subject to physical damage shall be permitted to be run exposed along the surface if securely fastened to the construction.

> **AUTHOR'S COMMENT:** The *NEC* does not specify a burial depth for the grounding electrode conductor when it is directly buried in earth. Also, when enclosing a grounding electrode conductor in a ferromagnetic (iron/steel) raceway, the raceway shall be bonded to the grounding electrode conductor at both ends. This is to prevent an electromagnetic choke effect, which would limit the amount of current that can flow through the grounding electrode conductor [250.64(E)].

**(C) Grounding Electrode Splices and Joints.** The grounding electrode conductor shall be run in one continuous length without splice or joint, unless spliced by listed irreversible compression-type connectors or by the exothermic welding process. Figure 250-118

> **AUTHOR'S COMMENT:** The *NEC* does not require the grounding electrode conductor to be run unbroken to all grounding electrodes. See 250.64(F) for details.

**(D) Grounding Electrode Tap Conductors.** When a service consists of multiple disconnecting means as permitted in 230.40, a grounding electrode tap from each disconnect to a grounding electrode conductor is permitted.

The grounding electrode tap from each disconnect shall be sized to the largest ungrounded conductor serving that disconnect in accordance with 250.66. The grounding electrode conductor for the grounding electrode taps is also sized in accordance with 250.66, but its size is based on the service conductors feeding all the service disconnects.

The grounding electrode tap shall terminate to the grounding electrode conductor in such a manner that there will be no splices or joints in the grounding electrode conductor. In addition, the grounding electrode tap shall not take place within the service disconnect enclosure. Figure 250-119

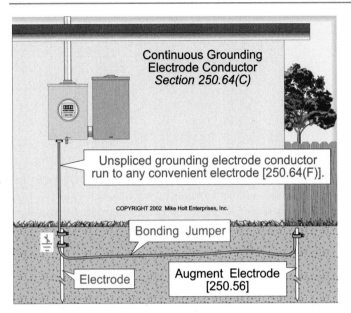

**Figure 250–118**

**(E) Raceway Protection for Grounding Electrode Conductor.** Ferromagnetic (iron/steel) raceways containing the grounding electrode conductors shall be electrically continuous from the cabinets or equipment to the grounding electrode by bonding each end of the raceway to the grounding electrode conductor. Figure 250-120

**AUTHOR'S COMMENT:** If a ferromagnetic (iron/steel) raceway containing a grounding electrode conductor is not bonded at both ends to the grounding electrode conductor, the effectiveness of the grounding electrode could be significantly reduced. This is because

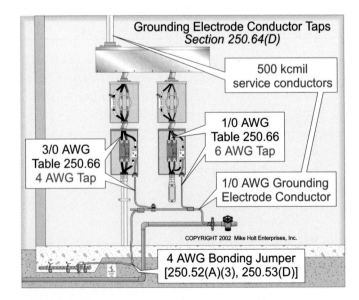

**Figure 250–119**

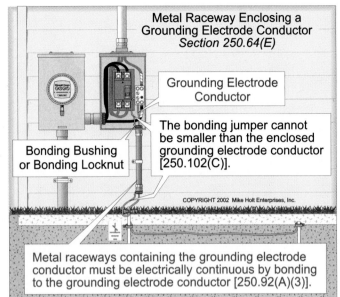

**Figure 250–120**

a single conductor carrying lightning current in a ferromagnetic (iron/steel) raceway causes the raceway to act as an inductor, which can severely limit, or choke, the current flow through the grounding electrode conductor during a lightning strike. IEEE Standard 142 "Green Book" states that the inductive choke can reduce the current by 97 percent.

**(F) To Electrode(s).** The grounding electrode conductor shall be permitted to be run to any convenient grounding electrode available in the grounding electrode system.

### 250.66 Grounding Electrode Conductor - Sizing

Except as permitted in 250.66(A) through (C), the size of the grounding electrode conductor shall not be less than given in Table 250.66.

**Table 250.66 Grounding Electrode Conductor Size**

| Equivalent Ungrounded Conductor Area* | Copper Grounding Electrode Conductor |
|---|---|
| 2 or smaller | 8 AWG |
| 1 or 1/0 | 6 AWG |
| 2/0 or 3/0 | 4 AWG |
| Over 3/0 through 350 kcmil | 2 AWG |
| Over 350 kcmil through 600 kcmil | 1/0 AWG |
| Over 600 kcmil through 1100 kcmil | 2/0 AWG |
| Over 1100 kcmil | 3/0 AWG |

*The grounding electrode conductor is based on the total area of the parallel ungrounded conductors.*

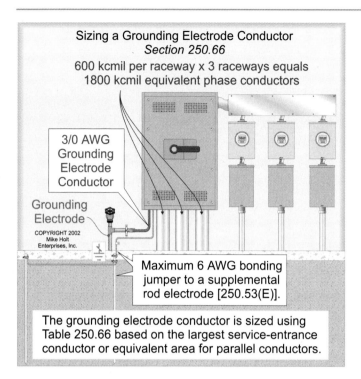

Sizing a Grounding Electrode Conductor
Section 250.66

600 kcmil per raceway x 3 raceways equals
1800 kcmil equivalent phase conductors

3/0 AWG
Grounding
Electrode
Conductor

Grounding
Electrode

COPYRIGHT 2002
Mike Holt
Enterprises, Inc.

Maximum 6 AWG bonding
jumper to a supplemental
rod electrode [250.53(E)].

The grounding electrode conductor is sized using
Table 250.66 based on the largest service-entrance
conductor or equivalent area for parallel conductors.

**Figure 250–121**

**Question:** What size grounding electrode conductor is required for a 1,200A service that is supplied with three parallel raceways each containing 600 kcmil conductors per phase? Figure 250-121

(a) 1 AWG                    (b) 1/0 AWG

(c) 2/0 AWG                  (d) 3/0 AWG

**Answer:** (d) 3/0 AWG, Table 250.66

**(A) Ground Rod [250.52(A)(5)].** Where the grounding electrode conductor is connected to a ground rod, that portion of the grounding electrode conductor that is the sole connection to the ground rod is not required to be larger than 6 AWG copper. Figure 250-122

**(B) Concrete-Encased Grounding Electrode (Ufer Ground) [250.52(A)(3)].** Where the grounding electrode conductor is connected to a concrete-encased grounding electrode, that portion of the grounding electrode conductor that is the sole connection to the concrete-encased grounding electrode is not required to be larger than 4 AWG copper. Figure 250-123

**(C) Ground Ring [250.52(A)(4)].** Where the grounding electrode conductor is connected to a ground ring, that portion of the conductor that is the sole connection to the ground ring is not required to be larger than the conductor used for the ground ring.

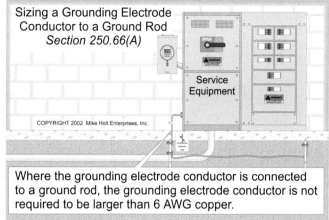

Sizing a Grounding Electrode
Conductor to a Ground Rod
Section 250.66(A)

Service
Equipment

COPYRIGHT 2002  Mike Holt Enterprises, Inc.

Where the grounding electrode conductor is connected
to a ground rod, the grounding electrode conductor is not
required to be larger than 6 AWG copper.

**Figure 250–122**

**AUTHOR'S COMMENT:** A ground ring encircling the building or structure in direct contact with earth shall consist of at least 20 ft of bare copper conductor not smaller than 2 AWG [250.52(A)(4)].

## 250.68 Grounding Electrode Conductor Termination

**(A) Attachment Fitting.** The termination of the grounding electrode conductor to the grounding electrode attachment fitting shall be accessible.

*Exception:* An encased or buried grounding electrode conductor connection to a concrete-encased, driven, or buried grounding electrode is not required to be accessible. See 250.53(G) and 250.70. Figure 250-124

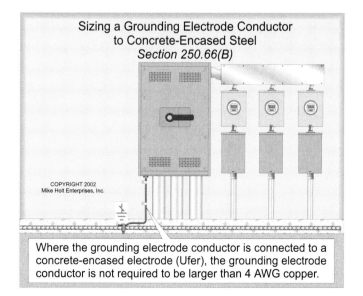

Sizing a Grounding Electrode Conductor
to Concrete-Encased Steel
Section 250.66(B)

COPYRIGHT 2002
Mike Holt Enterprises, Inc.

Where the grounding electrode conductor is connected to a
concrete-encased electrode (Ufer), the grounding electrode
conductor is not required to be larger than 4 AWG copper.

**Figure 250–123**

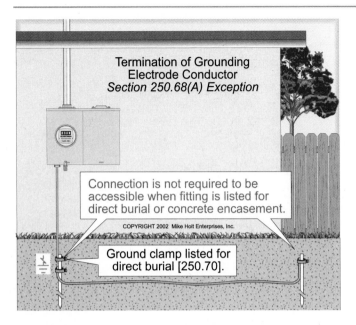

**Figure 250–124**

**(B)** **Effective Grounding Path.** The connection of a conductor to a grounding electrode shall be made in a manner that will ensure a permanent and effective path. To ensure the grounding path for the underground water pipe grounding electrode, a bonding jumper shall be provided around insulated water pipe joints and equipment likely to be disconnected for repairs or replacement. Figure 250-125

---

**AUTHOR'S COMMENT:** Continuity of the grounding connection to the grounding electrode or the bonding connection to interior piping shall not rely on water meters or filtering devices and similar equipment [250.53(D)(1)].

---

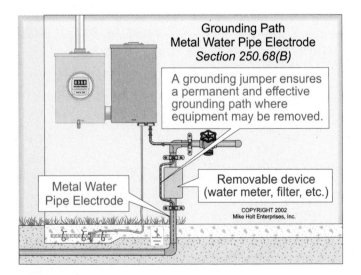

**Figure 250–125**

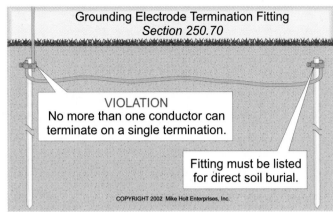

**Figure 250–126**

## 250.70 Grounding Electrode Termination Fitting

The grounding electrode conductor shall terminate to the grounding electrode by exothermic welding, listed lugs, listed pressure connectors, listed clamps, or other listed means. Termination fittings shall be listed for the materials of the grounding electrode and the conductor. Figure 250-126

When the termination of the conductor to a grounding electrode is encased in concrete or buried, the termination fittings shall be listed and identified for this purpose. No more than one conductor can terminate on a single termination, unless the termination is listed for multiple connections [110.14(A)].

---

**AUTHOR'S COMMENT:** Termination fittings listed for concrete encasement or direct burial grounding electrodes are typically made from brass or bronze, not steel or aluminum alloy.

---

## Part IV. Enclosure, Raceway and Service Cable

## 250.80 Service Enclosures

Metal raceways and enclosures containing service conductors shall be grounded (bonded) to an effective ground-fault current path.

*Exception:* Isolated metal elbows installed underground isolated from possible contact with persons by a minimum of 18 in. of earth cover are not required to be grounded (bonded). Figure 250-127

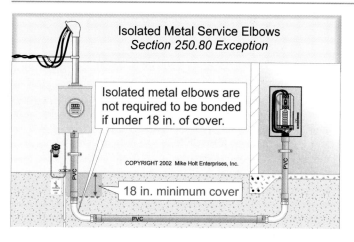

**Figure 250–127**

## 250.86 Other Enclosures

Metal raceways and enclosures containing electrical conductors shall be grounded (bonded) to an effective ground-fault current path.

*Exception No. 2:* Short sections of metal raceways used for the support or physical protection of cables are not required to be grounded (bonded). Figure 250-128

*Exception No. 3:* Isolated metal elbows installed underground isolated from possible contact with persons by a minimum 18 in. of earth cover to any part of the elbow are not required to be grounded (bonded). Figure 250-129

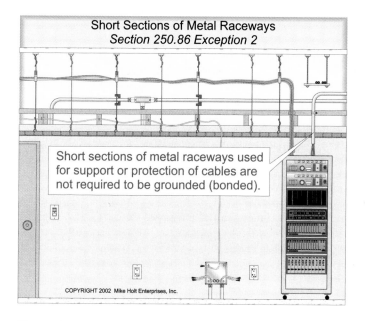

**Figure 250–128**

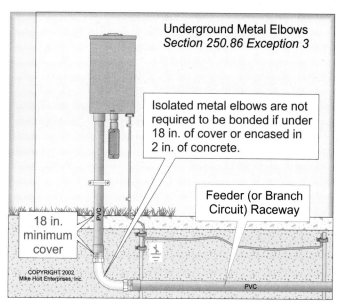

**Figure 250–129**

## Part V. Bonding

### 250.90 General Bonding Requirements

Electrical equipment and wiring and other electrically conductive material likely to become energized shall be bonded in a manner that creates a permanent, low-impedance path to facilitate the operation of overcurrent devices under fault conditions. This path shall safely carry the maximum ground-fault current likely to be imposed on it [110.10] from wherever a ground fault may occur to the power supply [250.4(A)(3) through 250.4(A)(5)]. This is accomplished by bonding the metal parts to the power supply grounded (neutral) terminal at service equipment [250.24(B)] or at a separately derived system [250.30(A)(1)] in accordance with 250.4(A)(5). Figure 250-130

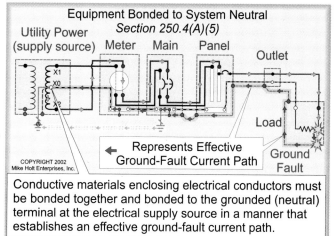

**Figure 250–130**

## 250.92 Assured Bonding – Services

**(A)  Assured Bonding of Service Enclosures and Raceways.**

A ground fault at service equipment, enclosures, or raceways containing service conductors can cause tremendous destruction if not cleared quickly. Because of this danger, the *NEC* requires the following metal parts containing service conductors to be bonded to an effective ground-fault current path in accordance with the assured bonding requirements contained in 250.92(B): Figure 250-131

(1)  Metal raceways containing service conductors.

(2)  Enclosures such as disconnects, meter enclosures, or wireways containing service conductors.

(3)  Metal raceway containing the grounding electrode conductor [250.64(E) and 250.102(C)].

**AUTHOR'S COMMENT:** Bonding is extremely important anywhere in an electrical system but there are some areas where regular bonding methods might not be sufficient. In areas such as services, 277V or 480V circuits and some hazardous (classified) locations, extra measures must be taken to ASSURE that the bonding connections will withstand the ground-fault conditions and prevent arcing from a ground fault. When I use the term "assured bonding" in this textbook, it is intended to help clarify the special care required to assure a safe installation. See 250.97 and 250.100.

Note: Figure 250-131 shows a green background around the service enclosures, metal raceways enclosing service conductors and the metal raceway enclosing the grounding electrode conductor. The green background indicates the portion of the electrical installation that the assured bonding methods apply to.

**AUTHOR'S COMMENT:** Assured bonding does not apply to raceways or enclosures containing feeder and branch-circuit conductors. Figure 250-132

**AUTHOR'S COMMENT:** The electrical system's impedance increases with increasing service conductor length. The amount of impedance at any point in the electrical wiring will determine the available fault current at that point. The service, which is the part of the premises wiring closest to the power source, has the lowest impedance of the premises wiring; therefore, the highest available fault current of the premises wiring [110.9 and 110.10]. Bonding at services shall be able to withstand these higher current values in the event of a ground fault. Figure 250-133

**(B)  Methods of Assured Bonding of Service Equipment and Raceways**

Enclosures and raceways containing service conductors shall be assured bonded to an effective ground-fault current path by one of the following methods:

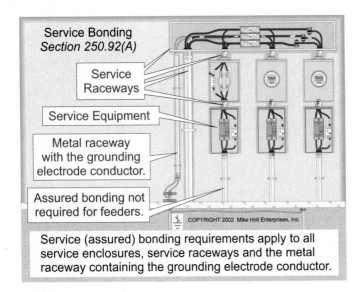

**Figure 250–131**

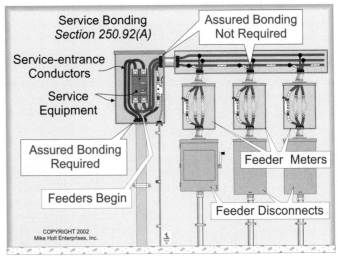

**Figure 250–132**

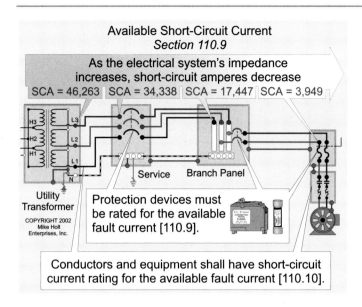

**Available Short-Circuit Current**
*Section 110.9*

As the electrical system's impedance increases, short-circuit amperes decrease

SCA = 46,263 | SCA = 34,338 | SCA = 17,447 | SCA = 3,949

Protection devices must be rated for the available fault current [110.9].

Conductors and equipment shall have short-circuit current rating for the available fault current [110.10].

*Figure 250–133*

(1) Bonding enclosures and raceways to the grounded (neutral) service conductor by exothermic welding, listed pressure connectors, listed clamps, or other listed fittings [250.8]. Figure 250-134

**AUTHOR'S COMMENT:** An equipment bonding conductor is not required within rigid nonmetallic conduit containing service conductors [250.142(A)(1) and 352.60 Ex. 2] because fault current uses the grounded (neutral) service conductor as the effective ground-fault current path to the power supply. See 250.24(B). Figure 250-135

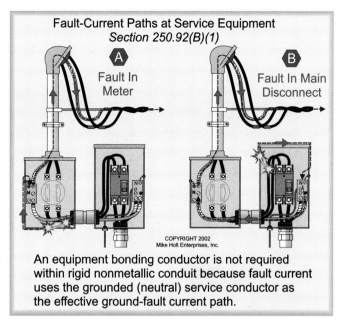

**Fault-Current Paths at Service Equipment**
*Section 250.92(B)(1)*

Ⓐ Fault In Meter

Ⓑ Fault In Main Disconnect

An equipment bonding conductor is not required within rigid nonmetallic conduit because fault current uses the grounded (neutral) service conductor as the effective ground-fault current path.

*Figure 250–135*

(2) Raceways are considered suitably bonded by utilizing threaded couplings or threaded bosses on enclosures where made up wrenchtight. Figure 250-136

(3) Raceways are considered suitably bonded by connections utilizing threadless raceway couplings and connectors where made up wrenchtight. See Figure 250-136

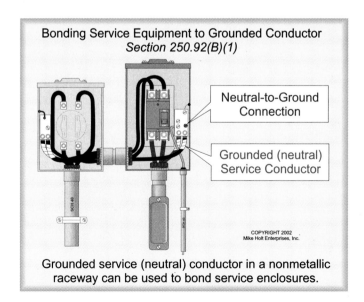

**Bonding Service Equipment to Grounded Conductor**
*Section 250.92(B)(1)*

Neutral-to-Ground Connection

Grounded (neutral) Service Conductor

Grounded service (neutral) conductor in a nonmetallic raceway can be used to bond service enclosures.

*Figure 250–134*

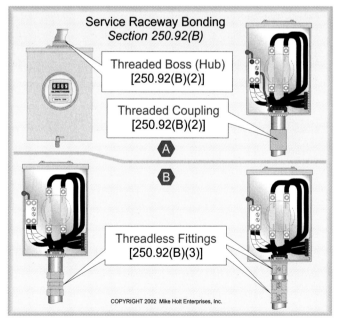

**Service Raceway Bonding**
*Section 250.92(B)*

Threaded Boss (Hub) [250.92(B)(2)]

Threaded Coupling [250.92(B)(2)]

Ⓐ

Ⓑ

Threadless Fittings [250.92(B)(3)]

*Figure 250–136*

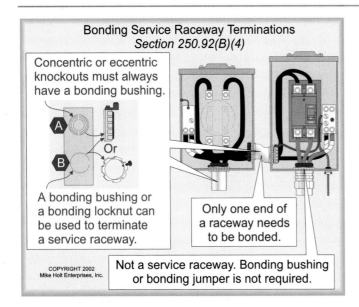

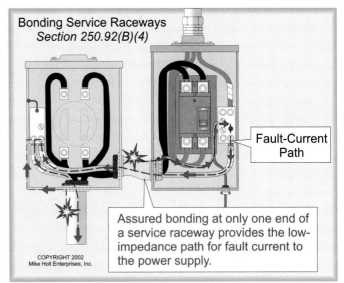

**Figure 250–137**

**Figure 250–138**

(4) When a metal service raceway terminates to an enclosure with a ringed knockout, one end of the service raceway shall be bonded with a bonding jumper sized in accordance with Table 250.66 [250.102(C)]. Figure 250-137

When a metal service raceway terminates to an enclosure without a ringed knockout, one end of the raceway shall be bonded by a bonding jumper or a bonding-type locknut because standard locknuts are not suitable for this purpose. A bonding-type locknut differs from a standard-type locknut in that it has a bonding screw with a sharp point that drives into the metal enclosure. See Figure 250-137.

**AUTHOR'S COMMENT:** Assured bonding one end of a service raceway provides the low-impedance path for fault current to return to the utility power supply. Figure 250-138

## 250.94 Grounding of Communications Systems

An accessible and external means shall be provided at service equipment or at the disconnect of separate buildings for the purpose of grounding communications systems.

The external grounding means can be one of the following: Figure 250-139

(1) An exposed, nonflexible metallic raceway,

(2) An exposed grounding electrode conductor,

(3) An approved external connection.

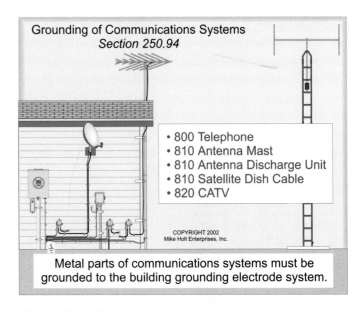

**Figure 250–139**

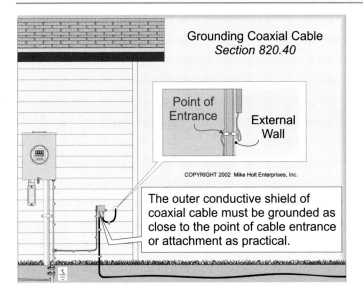

**Figure 250–140**

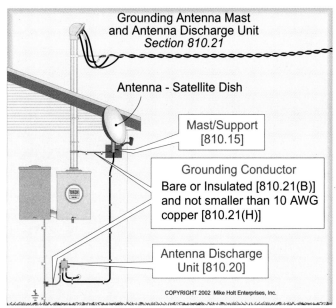

**Figure 250–142**

FPN No. 2: All communications systems that enter a building shall be grounded to the building or structure grounding electrode system in accordance with:

• CATV 820.40, Figure 250-140

• Telephone circuits, 800.40, Figure 250-141

• Antennas/satellite dishes, 810.21, Figure 250-142

In addition, any grounding electrode added for communications systems shall be bonded with a 6 AWG or larger conductor to the building or structure grounding electrode system [800.40(D), 810.21(J), 820.40(D) and 830.40(D)]. Figure 250-143

**AUTHOR'S COMMENT:** Grounding of power and communications systems to the same single point ground helps in reducing the difference in voltage potentials between the systems. This can be very important where these different systems are connected together; for example, in a computer. Figure 250-144

*WARNING: Failure to properly ground the communications systems to the building or structure grounding electrode system in accordance with the require-*

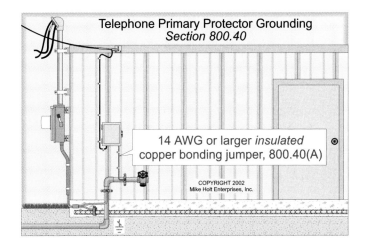

**Figure 250–141**

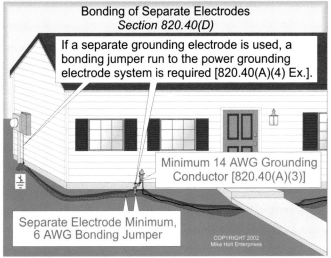

**Figure 250–143**

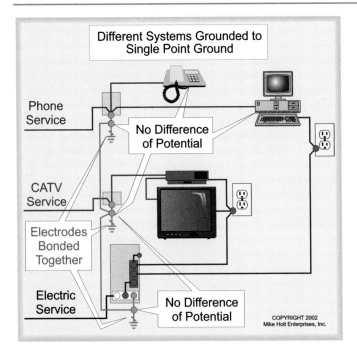

**Figure 250–144**

ments contained in Chapter 8 of the NEC can result in electric shock, the destruction of electrical components and possibly a fire from lightning. Figure 250-145

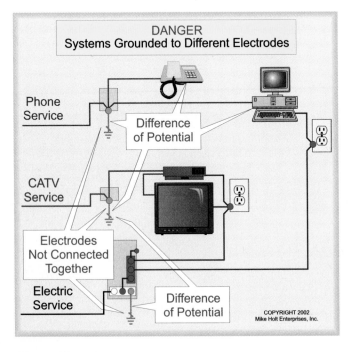

**Figure 250–145**

**AUTHOR'S COMMENT:** Other important communications grounding standards that should be considered include:

- BICSI – *Telecommunications Distribution Methods Manual (TDMM)*

- NFPA 77 – *Recommended Practices for Static Electricity*

- NFPA 780 – *Lightning Protection Systems for Buildings*

- IEEE Std. 142 "Green Book" – *Grounding of Industrial and Commercial Power Systems*

- IEEE Std. 1100 "Emerald Book" – *Powering and Grounding Sensitive Electronic Equipment*

- IEEE Std. 93-12 – *Grounding and Bonding Government Telecommunications Systems*

- TIA/EIA Std. 607 – *Commercial Building Grounding for Telecommunications Systems*

- UL Std. 497 – *Protectors for Paired Conductor Communication Circuits*

## 250.96 Bonding Other Enclosures

(A) **General Requirements.** All metal parts intended to serve as the effective ground-fault current path, including raceways, cables, equipment and enclosures shall be bonded together to ensure electrical continuity and that they have the capacity to conduct safely any fault current likely to be imposed on them in accordance with 110.10, 250.4(A)(5) and Note to Table 250.122.

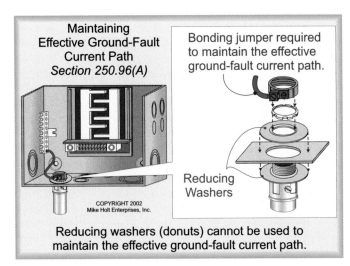

**Figure 250–146**

**AUTHOR'S COMMENT:** Reducing washers (donuts) do not provide an effective ground-fault current path [250.4(A)(5)]. Figure 250-146

Nonconductive coatings such as paint, lacquer and enamel on equipment shall be removed to ensure an effective ground-fault current path, or the termination fittings shall be designed so as to make such removal unnecessary [250.12].

**AUTHOR'S COMMENT:** The practice of driving a locknut tight with a screwdriver and pliers is considered sufficient in removing paint and other nonconductive finishes to ensure an effective ground-fault current path.

**(B) Isolated Grounding (bonding) Circuits.** Where required for the reduction of electrical noise, the metal raceway containing circuit conductors for sensitive electronic equipment can be isolated from the sensitive equipment by the use of a nonmetallic raceway fitting at the point of equipment termination. However, the metal raceway shall contain an insulated equipment grounding (bonding) conductor installed in accordance with 250.146(D) that terminates to the sensitive equipment. Figure 250-147

**AUTHOR'S COMMENT:** The insulated equipment grounding (bonding) conductor can originate at the neutral terminal (point) of the service or separately derived system and it may pass through a panelboard without a ground-to-case connection [408.20 Ex.]. In addition, the insulated equipment grounding (bonding) conductor can originate at the panelboard that supplied the circuit. Figure 250-148

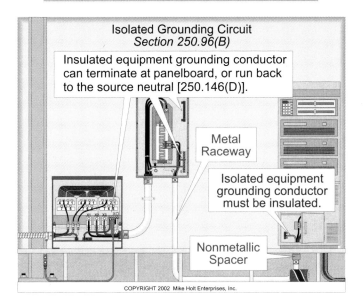

**Figure 250–147**

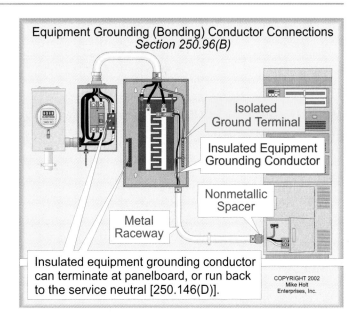

**Figure 250–148**

FPN: Use of an isolated equipment grounding (bonding) conductor does not relieve the requirement for bonding of the raceway system to an effective ground-fault current path [250.86].

**AUTHOR'S COMMENT:** This rule contains the requirements for isolated grounding (bonding) circuits, while 250.146(D) contains the requirements for isolated ground receptacles.

*DANGER: Some digital equipment manufacturers insist that their equipment be electrically isolated from the building or structure grounding and bonding parts. This is a dangerous and unwise practice, and it violates 250.4(A)(5), which prohibits earth to be used*

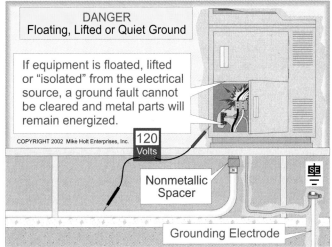

**Figure 250–149**

*as the sole effective ground-fault current path. If the metal enclosures of electrical equipment are isolated or floated as desired by some sensitive equipment manufacturers, the metal parts will remain energized from a ground fault. Figure 250-149*

**AUTHOR'S COMMENT:** For more information on how to properly ground and bond sensitive electronic equipment, go to http://www.NECcode.com and visit the power quality link.

## 250.97 Assured Bonding – Raceways and Cables Containing 277V Circuits

Metal raceways or cables containing 277V or 480V circuits at ringed knockout terminations shall be assured bonded to the enclosure with a bonding jumper sized in accordance with Table 250.122 [250.102(D)]. Figure 250-150

**AUTHOR'S COMMENT:** Bonding is extremely important anywhere in an electrical system but there are some areas where regular bonding methods might not be sufficient. In areas such as services, 277V or 480V circuits, and some hazardous (classified) locations, extra measures must be taken to ASSURE that the bonding connections will withstand the ground-fault conditions. The term "assured bonding" should help to clarify where special care is needed to assure a safe electrical installation. See 250.92 and 250.100.

Bonding jumpers for raceways and cables containing 277V or 480V circuits are required, at ringed knockout terminations, to ensure that the effective ground-fault current path has the capacity to safely conduct the maximum ground-fault current likely to be imposed on it to the power supply in accordance with 110.10 and 250.4(A)(5).

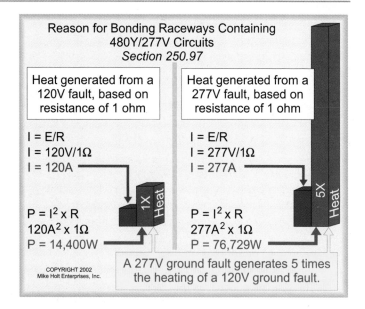

**Figure 250–151**

**AUTHOR'S COMMENT:** Ringed knockouts are not listed to withstand the heat generated by a 277V ground fault. This is because a 277V ground fault generates five times the heat as compared to a 120V ground fault (I²R). Figure 250-151

**Exception.** A bonding jumper is not required for metal raceways and cables where ringed knockouts are not encountered, or where the box is listed for this purpose. Figure 250-152

## 250.100 Assured Bonding – Hazardous (Classified) Locations

Because of the explosive conditions associated with electrical installations in hazardous (classified) locations, electrical continuity of non-current-carrying metal parts of equipment and race-

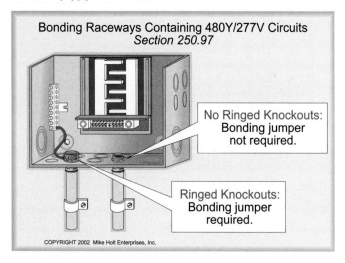

**Figure 250–150**

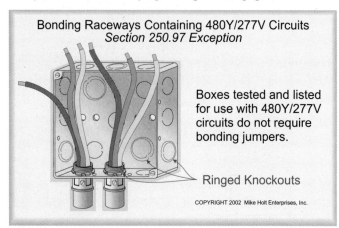

**Figure 250–152**

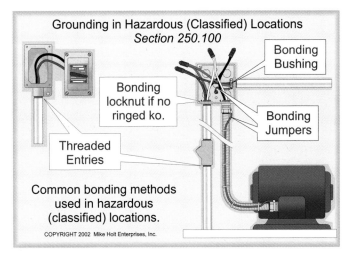

**Figure 250-153**

ways shall be ensured by threading the conduit into an enclosure, by the use of a bonding jumper at raceway terminations, or by a bonding-type locknut where ringed knockouts are not encountered. Standard locknuts are not suitable for this purpose. Figure 250-153

**AUTHOR'S COMMENT:** Hazardous (classified) location bonding requirements apply to all intervening raceways, fittings, boxes, and enclosures between Class locations and the point of grounding (and bonding) for service equipment separately derived systems. See 501.16(A) for Class I locations, 502.16 for Class II locations and 503.16 for Class III locations for specific requirements. Figure 250-154

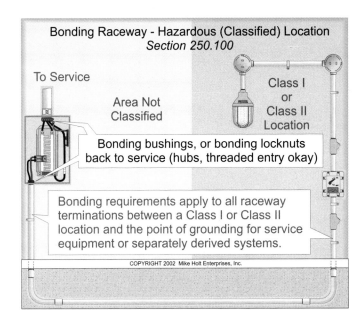

**Figure 250-154**

## 250.102 Bonding Jumper Requirements

**(A) Bonding Material.** Equipment bonding jumpers shall be of copper or other corrosive-resistant material.

**(B) Bonding Jumper Attachment.** Equipment bonding jumpers shall be connected by exothermic welding, listed pressure connectors, listed clamps, or other listed means. Sheet metal screws shall not be used for this purpose [250.8].

**(C) Service Bonding Jumper.** Bonding jumpers for service raceways shall be sized based on the size of the ungrounded service conductors within the service raceway in accordance with Table 250.66. Where service conductors are paralleled in two or more raceways or cables, the bonding jumper for each raceway or cable shall be sized based on the ungrounded service conductors in each raceway or cable.

**Question:** What size service bonding jumper is required for a metal raceway containing 600 kcmil conductors? Figure 250-155

| | |
|---|---|
| (a) 1 AWG | (b) 1/0 AWG |
| (c) 2/0 AWG | (d) 3/0 AWG |

**Answer:** (b) 1/0 AWG

The bonding jumper for a metal service raceway enclosing the grounding electrode conductor shall be the same size or larger than the required enclosed grounding electrode conductor. See 250.64(E) and 250.92(A)(3) for additional requirements.

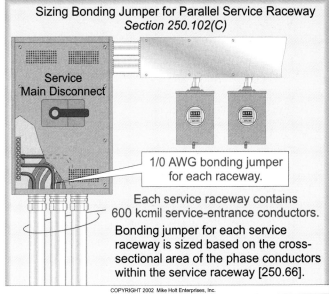

**Figure 250-155**

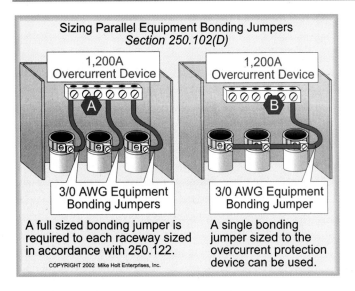

Sizing Parallel Equipment Bonding Jumpers
*Section 250.102(D)*

1,200A Overcurrent Device **A**

1,200A Overcurrent Device **B**

3/0 AWG Equipment Bonding Jumpers

3/0 AWG Equipment Bonding Jumper

A full sized bonding jumper is required to each raceway sized in accordance with 250.122.

A single bonding jumper sized to the overcurrent protection device can be used.

COPYRIGHT 2002 Mike Holt Enterprises, Inc.

**Figure 250–156**

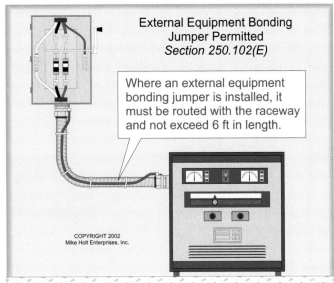

External Equipment Bonding Jumper Permitted
*Section 250.102(E)*

Where an external equipment bonding jumper is installed, it must be routed with the raceway and not exceed 6 ft in length.

COPYRIGHT 2002
Mike Holt Enterprises, Inc.

**Figure 250–157**

**AUTHOR'S COMMENT:** It would be easier and more cost effective to install the grounding electrode conductor in a nonmetallic raceway [250.64(B)] rather than a metal raceway. This is because a metal raceway is required to be bonded to the grounding electrode conductor at both ends [250.64(E)]. If the grounding electrode conductor is required to be installed in a metal raceway by the designer, a bonding locknut would be the most efficient and effective method to bond the metal service raceway to the service equipment enclosure [250.92(A)(3) and 250.92(B)].

**(D) Equipment Bonding Jumper Size.** Equipment bonding jumpers on the load side of service equipment shall be sized in accordance with 250.122 based on the size of the circuit protection device rating. Figure 250-156A

A single equipment bonding jumper can be used to bond two or more raceways or cables where the bonding jumper is sized for the largest overcurrent device protecting the circuits. See Figure 250-156B

**(E) Installation.** An equipment bonding jumper can be installed outside a raceway if its length does not exceed 6 ft and the bonding jumper is routed with the raceway. Figure 250-157

*Exception:* An equipment bonding jumper longer than 6 ft shall be permitted at outside pole locations to bond isolated sections of metal raceways or elbows in exposed risers of metal raceways The equipment bonding jumper shall be sized in accordance with Table 250.66 [250.102(C)]. Figure 250-158

**AUTHOR'S COMMENT:** Metal service elbows having at least 18 in. of cover, are not required to be bonded [250.80 Ex.]. Figure 250-159

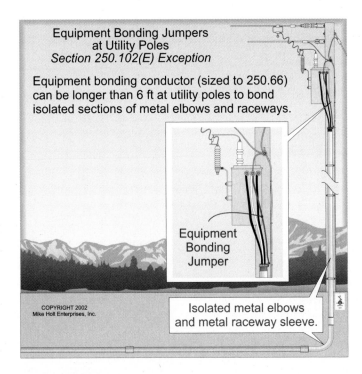

Equipment Bonding Jumpers at Utility Poles
*Section 250.102(E) Exception*

Equipment bonding conductor (sized to 250.66) can be longer than 6 ft at utility poles to bond isolated sections of metal elbows and raceways.

Equipment Bonding Jumper

COPYRIGHT 2002
Mike Holt Enterprises, Inc.

Isolated metal elbows and metal raceway sleeve.

**Figure 250–158**

**250.104 Bonding of Piping Systems and Exposed**

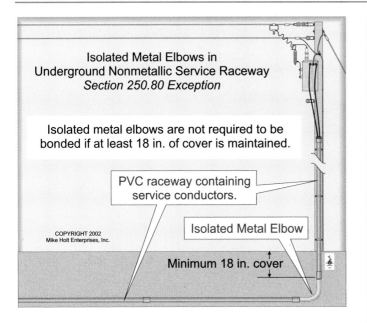

Isolated Metal Elbows in
Underground Nonmetallic Service Raceway
*Section 250.80 Exception*

Isolated metal elbows are not required to be
bonded if at least 18 in. of cover is maintained.

PVC raceway containing
service conductors.

Isolated Metal Elbow

COPYRIGHT 2002
Mike Holt Enterprises, Inc.

Minimum 18 in. cover

*Figure 250–159*

## Structural Steel.

To remove dangerous voltage from a ground fault, metal piping and exposed structural building steel shall be bonded to an effective ground-fault current path in accordance with 250.4(A)(4).

**(A) Interior Metal water-piping system.** Interior metal water-piping systems shall be bonded in accordance with (1), (2), (3), or (4) below.

**(1) Building Supplied by a Service.** The interior metal water-piping system of a building or structure shall be bonded to the service equipment enclosure, the grounded (neutral) service conductor, or the grounding electrode or grounding electrode conductor where the grounding electrode conductor is sized in accordance with Table 250.66. Figure 250-160

**Question:** What size bonding jumper is required for the interior metal water-piping system if the service conductors are 4/0 AWG? Figure 250-161

(a) 6 AWG              (b) 4 AWG

(c) 2 AWG              (d) 1/0 AWG

**Answer:** (c) 2 AWG, Table 250.66

**AUTHOR'S COMMENT:** Where hot and cold water pipes are electrically conductive together (usually at the shower or sink), only one bonding jumper is required, either to the cold or to the hot water pipe.

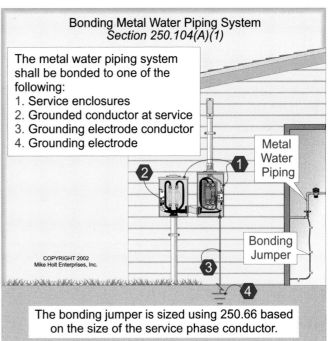

Bonding Metal Water Piping System
*Section 250.104(A)(1)*

The metal water piping system
shall be bonded to one of the
following:
1. Service enclosures
2. Grounded conductor at service
3. Grounding electrode conductor
4. Grounding electrode

Metal
Water
Piping

Bonding
Jumper

COPYRIGHT 2002
Mike Holt Enterprises, Inc.

The bonding jumper is sized using 250.66 based
on the size of the service phase conductor.

*Figure 250–160*

**(2) Multiple Occupancy Building.** When the interior metal water-piping system in each occupancy is metal-lically isolated from all other occupancies, the interior metal water-piping system for each occupancy shall be permitted to be bonded to the occupancy panelboard bonding jumper for this application and shall be sized

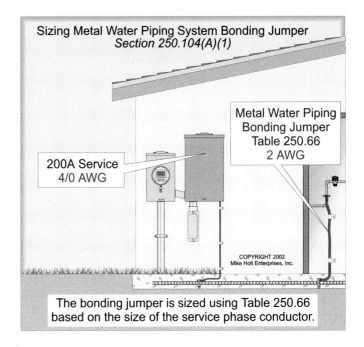

Sizing Metal Water Piping System Bonding Jumper
*Section 250.104(A)(1)*

Metal Water Piping
Bonding Jumper
Table 250.66
2 AWG

200A Service
4/0 AWG

COPYRIGHT 2002
Mike Holt Enterprises, Inc.

The bonding jumper is sized using Table 250.66
based on the size of the service phase conductor.

*Figure 250–161*

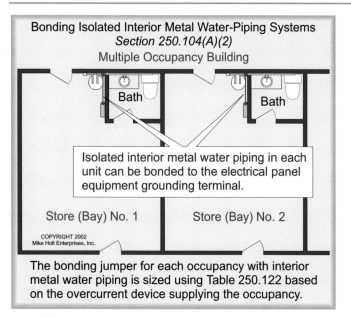

Bonding Isolated Interior Metal Water-Piping Systems
*Section 250.104(A)(2)*
Multiple Occupancy Building

Bath                    Bath

Isolated interior metal water piping in each unit can be bonded to the electrical panel equipment grounding terminal.

Store (Bay) No. 1        Store (Bay) No. 2

COPYRIGHT 2002
Mike Holt Enterprises, Inc.

The bonding jumper for each occupancy with interior metal water piping is sized using Table 250.122 based on the overcurrent device supplying the occupancy.

*Figure 250–162*

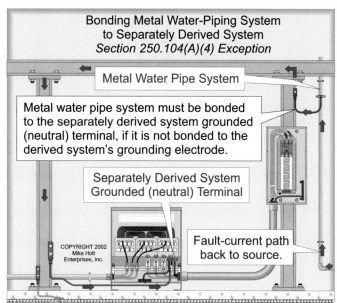

Bonding Metal Water-Piping System
to Separately Derived System
*Section 250.104(A)(4) Exception*

Metal Water Pipe System

Metal water pipe system must be bonded to the separately derived system grounded (neutral) terminal, if it is not bonded to the derived system's grounding electrode.

Separately Derived System
Grounded (neutral) Terminal

COPYRIGHT 2002
Mike Holt
Enterprises, Inc.

Fault-current path back to source.

*Figure 250–163*

in accordance with Table 250.122, based on the size of the main overcurrent device supplying the occupancy. Figure 250-162

(3) **Building or Structure Supplied by a Feeder.** The interior metal water-piping system of a building or structure that is supplied by a feeder shall be bonded to the building disconnect enclosure, the supply equipment bonding conductor, or the grounding electrode system. The bonding jumper shall be sized in accordance with Table 250.66 based on the size of the feeder circuit conductors that supply the building or structure.

(4) **Separately Derived Systems.** Interior metal water-pipe systems in the area served by a separately derived system shall be bonded to the grounded (neutral) terminal of the separately derived system at the same location where the grounding electrode conductor is connected. The bonding jumper shall be sized in accordance with Table 250.66 based on the size of the secondary circuit conductors that supply the area.

*Exception:* The interior metal water-pipe system can be bonded to the metal frame of a building that serves as the grounding electrode for the separately derived system [250.30(A)(4)(1)]. Figure 250-163

**AUTHOR'S COMMENT:** The metal frame of a building that serves as the grounding electrode for the separately derived system shall be effectively grounded to earth in accordance with 250.52(A)(1).

(B) **Other Metal-Piping Systems.** Metal-piping systems, such as gas or air piping which may become energized, shall be bonded to the service equipment enclosure, the grounded (neutral) service conductor, or the grounding electrode or grounding electrode conductor where the grounding electrode conductor is of sufficient size. The bonding jumper shall be sized in accordance with Table 250.122 using the rating of the circuit that may energize the piping. The equipment grounding (bonding) conductor for the circuit that may energize the piping can serve as the bonding means. Figure 250-164

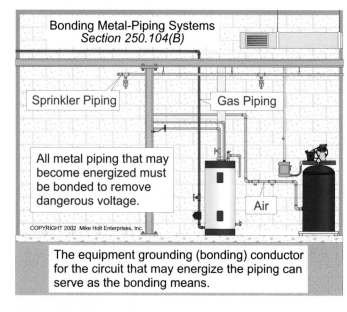

Bonding Metal-Piping Systems
*Section 250.104(B)*

Sprinkler Piping            Gas Piping

All metal piping that may become energized must be bonded to remove dangerous voltage.

Air

COPYRIGHT 2002  Mike Holt Enterprises, Inc.

The equipment grounding (bonding) conductor for the circuit that may energize the piping can serve as the bonding means.

*Figure 250–164*

**AUTHOR'S COMMENT:** The phrase "that may become energized" is subject to interpretation by the AHJ. Naturally, all metal-piping systems could become energized if the conditions are right (or, perhaps, if the conditions are wrong). I know of a case where metal gas piping for a restaurant, with no electrical appliances, became energized and killed a kitchen worker. A rat ate through the insulation of a nonmetallic-sheathed cable (naturally, the hot side), and the exposed conductor energized the metal gas-piping system.

FPN: Bonding of all metal piping and metal air ducts within the building provides an additional degree of safety, but this is not an *NEC* requirement.

**(C)   Structural Steel.** Exposed structural steel that is interconnected to form a steel building frame that is not part of the grounding electrode system shall be bonded to the service equipment enclosure, the grounded (neutral) service conductor, or the grounding electrode or grounding electrode conductor, where the grounding electrode conductor is sized in accordance with Table 250.66. Figure 250-165

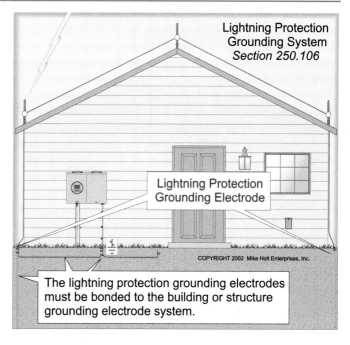

The lightning protection grounding electrodes must be bonded to the building or structure grounding electrode system.

*Figure 250–166*

**AUTHOR'S COMMENT:** This rule is not intended to require the bonding of the metal skin of a wood frame building or metal framing members (studs), but it would be a good idea.

### 250.106 Lightning Protection System

Where a lightning protection system is installed, the grounding electrode(s) for this system shall be bonded to the building or structure grounding electrode system. Figure 250-166

FPN No. 1: The grounding electrode used for a lightning protection system shall not be used as the required building or structure grounding electrode [250.60]. See NFPA 780 Standard for the Installation of Lightning Protection Systems for additional details on grounding and bonding requirements for lightning protection.

FPN No. 2: Metal raceways, enclosures, frames and other non-current-carrying metal parts of electric equipment may require bonding or spacing from the lightning protection conductors in accordance with NFPA 780. Separation from lightning protection conductors is typically 6 ft through air, or 3 ft through dense materials, such as concrete, brick, or wood. Figure 250-167

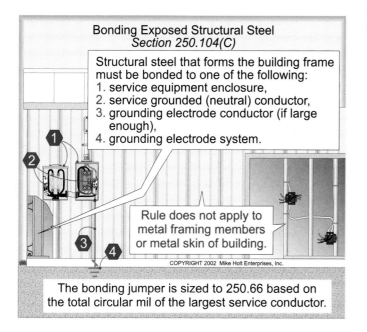

*Figure 250–165*

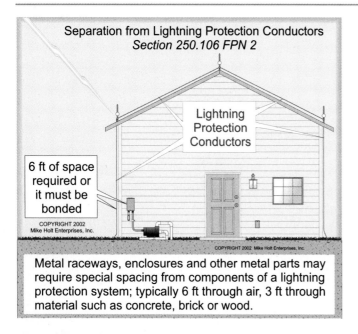

**Figure 250–167**

## Part VI. Equipment Grounding (Bonding) and Equipment Grounding (Bonding) Conductors

### 250.118 Types of Equipment Grounding (Bonding) Conductors

The equipment grounding (bonding) conductor, which serves as the effective ground-fault current path, shall be one or a combination of the following: Figure 250-168

(1) Conductor. A bare or insulated conductor sized in accordance with 250.122, installed within the raceway, cable or trench with the circuit conductors [300.3(B), 300.5(I) and 300.20(A)].

(2) Rigid Metal Conduit (RMC)

(3) Intermediate Metal Conduit (IMC)

(4) Electrical Metallic Tubing (EMT)

(5) Flexible Metal Conduit (FMC) where both the conduit and fittings are listed for grounding (bonding).

**AUTHOR'S COMMENT:** FMC is not listed for grounding (bonding).

(6) Flexible Metal Conduit (FMC) not listed for grounding (bonding) can serve as the effective ground-fault current path if [348.60]: Figure 250-169

a. The conduit terminates in fittings listed for grounding (bonding).

b. The circuit conductors are protected by overcurrent devices rated 20A or less.

c. The combined length of FMC in the same ground return path does not exceed 6 ft. Figure 250-170

d. The conduit is not installed where flexibility at terminations is required.

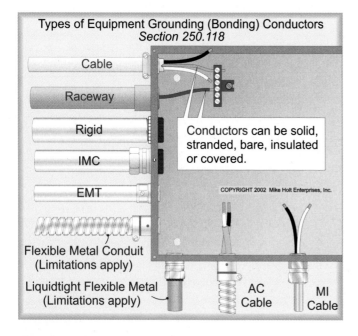

**Figure 250–168**

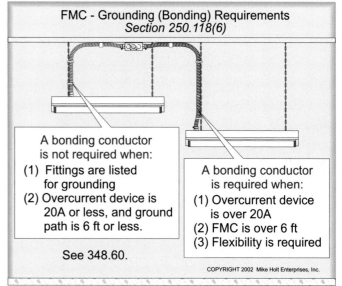

**Figure 250–169**

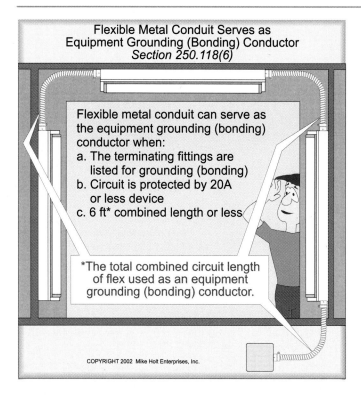

**Figure 250–170**

(7) Liquidtight Flexible Metal Conduit (LFMC) can serve as the effective ground-fault current path if [350.60]: Figure 250-171

  a. The conduit terminates in fittings listed for grounding (bonding).

  b. For trade sizes $^3/_8$ through $^1/_2$, the circuit conductors are protected by overcurrent devices rated 20A or less.

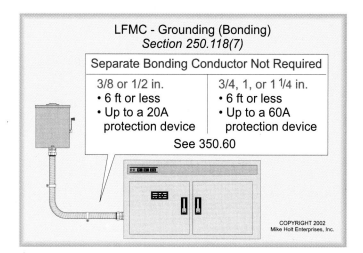

**Figure 250–171**

c. For trade sizes $^3/_4$ through $1^1/_4$, the circuit conductors are protected by overcurrent devices rated 60A or less.

d. The combined length of LFMC in the same ground return path does not exceed 6 ft.

e. The conduit is not installed where flexibility at terminations is required.

(9) Armored Cable (AC)

**AUTHOR'S COMMENT:** Interlocked AC is manufactured with an internal bonding strip of aluminum that is in direct contact with the interlocked metal armor. The combination of the bonding strip and the interlocked metal armor makes the cable suitable as an effective ground-fault current path [320.108]. Figure 250-172

(10) The copper metal sheath of Mineral Insulated Cable (MI) can serve as the effective ground-fault current path.

(11) Metal Clad Cable (MC) where listed and identified for grounding (bonding) can serve as the effective ground-fault current path if:

  a. Interlocked metal tape-type MC cable contains an equipment grounding (bonding) conductor.

**AUTHOR'S COMMENT:** The metal armor of interlocked MC Cable is not suitable as an effective ground-fault current path because it does not have an internal bonding strip like AC cable. An equipment grounding (bonding) conductor is required within this cable to serve as a suitable effective ground-fault current path. Figure 250-173

  b. The MC cable is constructed of the smooth or corrugated tube type.

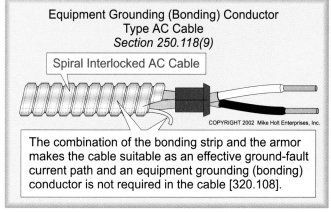

**Figure 250–172**

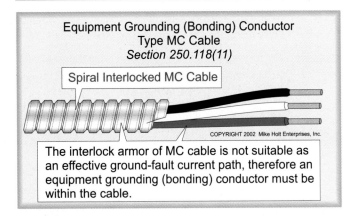

**Equipment Grounding (Bonding) Conductor**
**Type MC Cable**
*Section 250.118(11)*

Spiral Interlocked MC Cable

COPYRIGHT 2002 Mike Holt Enterprises, Inc.

The interlock armor of MC cable is not suitable as an effective ground-fault current path, therefore an equipment grounding (bonding) conductor must be within the cable.

*Figure 250–173*

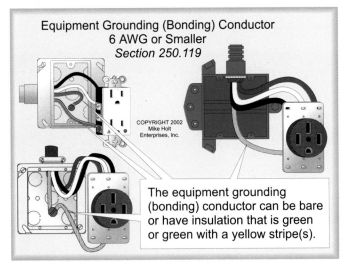

**Equipment Grounding (Bonding) Conductor**
**6 AWG or Smaller**
*Section 250.119*

COPYRIGHT 2002 Mike Holt Enterprises, Inc.

The equipment grounding (bonding) conductor can be bare or have insulation that is green or green with a yellow stripe(s).

*Figure 250–174*

**AUTHOR'S COMMENT:** The sheath of smooth or corrugated tube MC Cable is suitable as the effective ground-fault current path, therefore an internal equipment grounding (bonding) conductor is not required within the cable.

(12) Cable trays as permitted in 392.3(C) and 392.7.

(14) Other electrically continuous metal raceways listed for grounding (bonding) can serve as the effective ground-fault current path.

**AUTHOR'S COMMENT:** This would include metal wireways, surface metal raceways, multioutlet assemblies, metal underfloor raceways, etc.

### 250.119 Identification of Equipment Grounding (Bonding) Conductor

Equipment grounding (bonding) conductors can be bare or insulated. Where insulated, the equipment grounding (bonding) conductor shall have a continuous outer finish that is either green or green with one or more yellow colored stripes. Figure 250-174

**AUTHOR'S COMMENT:** Although not a *Code* requirement, isolated grounding (bonding) circuits [250.96(B)] and isolated ground receptacles [250.146(D)] frequently use a green insulated conductor for equipment bonding with a green with yellow stripe(s) insulated conductor for sensitive electronic equipment. Figure 250-175

**(A) Conductors Larger Than 6 AWG.** An insulated or covered equipment grounding (bonding) conductor larger than 6 AWG can be permanently identified by one of the following:

(1) Stripping the insulation or covering from the entire exposed length.

(2) Coloring the exposed insulation or covering green.

(3) Marking the exposed insulation or covering with green tape or green adhesive labels.

**AUTHOR'S COMMENT:** Although not a *Code* requirement, grounding electrode conductors are sometimes marked using green tape. In this textbook we have chosen to use a dark orange color to indicate grounding and green for bonding.

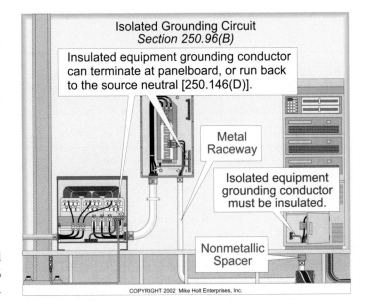

**Isolated Grounding Circuit**
*Section 250.96(B)*

Insulated equipment grounding conductor can terminate at panelboard, or run back to the source neutral [250.146(D)].

Metal Raceway

Isolated equipment grounding conductor must be insulated.

Nonmetallic Spacer

COPYRIGHT 2002 Mike Holt Enterprises, Inc.

*Figure 250–175*

## 250.120 Installation of Equipment Grounding (bonding) Conductor

**(A) Wiring Methods.** Wiring methods that serve as the equipment grounding (bonding) conductor, such as EMT, RMC, AC cable, etc, [250.118] shall be electrically continuous [300.10], and mechanically continuous [300.12]. Termination fittings shall be approved for use with the type raceway or cable [300.15(A)] and all connections, joints and fittings shall be made tight using suitable tools.

## 250.122 Sizing Equipment Grounding (Bonding) Conductor

**(A) General.** The equipment grounding (bonding) conductor shall be sized in accordance with Table 250.122, but the equipment grounding (bonding) conductor is not required to be larger than the circuit conductors [250.102(D)]. Figure 250-176

> **AUTHOR'S COMMENT:** The equipment grounding (bonding) conductor for motor branch circuits are based on the motor short-circuit ground-fault protection device rating. See comments and example following 250.122(D).

### Table 250.122

| Protection Rating | Copper Conductor |
|---|---|
| 15 | 14 AWG |
| 20 | 12 AWG |
| 30 | 10 AWG |
| 40 | 10 AWG |
| 60 | 10 AWG |
| 100 | 8 AWG |
| 200 | 6 AWG |
| 300 | 4 AWG |
| 400 | 3 AWG |
| 500 | 2 AWG |
| 600 | 1 AWG |
| 800 | 1/0 AWG |
| 1,000 | 2/0 AWG |
| 1,200 | 3/0 AWG |

Note: Where necessary to comply with 110.10 and 250.4(A)(5), the equipment grounding (bonding) conductor shall be sized larger than given in this table.

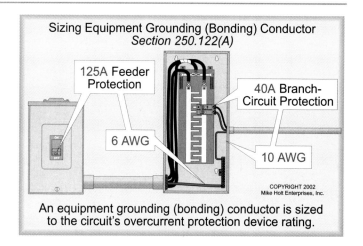

Sizing Equipment Grounding (Bonding) Conductor
Section 250.122(A)

125A Feeder Protection

40A Branch-Circuit Protection

6 AWG

10 AWG

COPYRIGHT 2002
Mike Holt Enterprises, Inc.

An equipment grounding (bonding) conductor is sized to the circuit's overcurrent protection device rating.

**Figure 250–176**

> *WARNING: Equipment grounding (bonding) conductors shall be capable of safely conducting any ground-fault current likely to be imposed on them to ensure that the overcurrent protection device will quickly clear the ground fault [(250.4(A)5]. If the equipment grounding (bonding) conductor is not sized to withstand the ground-fault currents, the conductor may burn clear before the protective device responds. For more information on this subject, http://www.NECcode.com/Newsletters/7-21-99.htm.*

> **AUTHOR'S COMMENT:** A factor that must be considered when sizing equipment grounding (bonding) conductors is the terminal contact resistance. When an equipment bonding conductor carries fault current, the contact resistance of the conductor in the terminal might be less after the fault clears and the terminal cools. To ensure proper conductor contact resistance, bonding conductors should be sized so that when they carry fault current, the temperature of the conductor and the terminal will not rise above 250°C (point at which copper softens). To ensure that equipment grounding (bonding) conductors can maintain proper contact resistance, they should be sized so that their circular mil cross-sectional area is not smaller than the available ground-fault current[1] times the following multiplier.

| Protection Type | Rating | Multiplier |
|---|---|---|
| Breakers/Class H fuses – Opens 1/2 cycle | 100A | 1.56 |
| Circuit breakers - Opens in 1 cycle | 200A | 1.70 |
| Circuit breakers - Opens in 2 cycle | 400A | 2.40 |
| Circuit breakers - Opens in 3 cycle | 1,200A | 2.94 |

[1]An Excel or Lotus spreadsheet to determine ground-fault current is available for free athttp://www.mikeholt.com/free/free.htm.

Example: A 60A circuit breaker with an available ground-fault current of 7,000A at the load should have the equipment grounding (bonding) conductor sized no smaller than: 7,000A X 1.56 = 10,920 cm. According to Chapter 9, Table 8, this would be 8 AWG. Figure 250-177

Where current-limiting fuses [240.2] provide circuit overcurrent protection, or where exothermic welding is used for conductor termination, bonding conductors are not required to be increased in size. For more information on the effects of ground-fault current and conductor sizing, visit http://www.mikeholt.com/Newsletters/7-21-99.htm

**(B)   Increased in Size.** When ungrounded circuit conductors are increased in size for any reason, the equipment grounding (bonding) conductor shall be increased in size in proportion to the increase in size of the ungrounded conductors.

**AUTHOR'S COMMENT:** Ungrounded circuit conductors could be increased in size to accommodate voltage drop, harmonic currents, fault currents, etc. For more information on sizing conductors for voltage drop, see http://www.mikeholt.com/studies/vd.htm.

**Question:** If the ungrounded circuit conductors for a 40A circuit are increased in size from 8 AWG to 6 AWG, the equipment grounding (bonding) conductor shall be increased in size from 10 AWG to _____. Figure 250-178

(a) 10 AWG                     (b) 8 AWG

(c) 6 AWG                      (d) 4 AWG

**Answer:** (b) 8 AWG

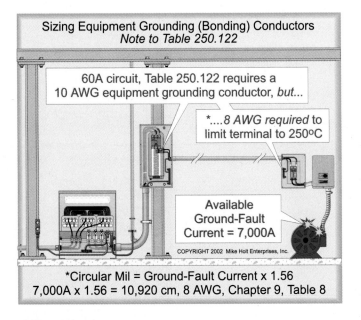

Figure 250–177

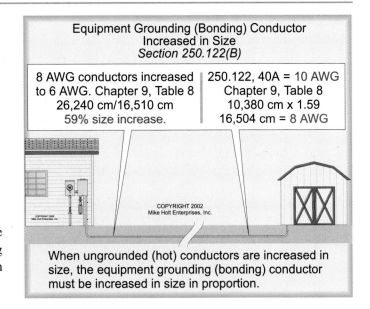

When ungrounded (hot) conductors are increased in size, the equipment grounding (bonding) conductor must be increased in size in proportion.

**Figure 250–178**

According to Chapter 9, Table 8, the circular mil area of 6 AWG is 59 percent greater than 8 AWG (26,240 cm/16,510 cm). According to Table 250.122, the equipment grounding (bonding) conductor for a 40A protection device shall be 10 AWG (10,380 cm), but it shall be increased in size by a multiplier of 1.59. This results in an 8 AWG conductor: 10,380 cm × 1.59 = 16,504 cm

**(C)   Multiple Circuits in One Raceway.** When multiple circuits are installed in a single raceway, only one equipment grounding (bonding) conductor sized in accordance with Table 250.122 based on the largest overcurrent device protecting the circuit conductors is required. Figure 250-179

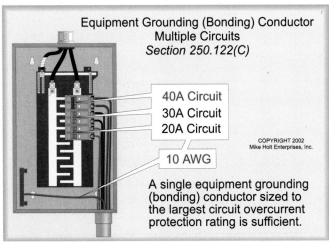

Figure 250–179

**(D) Motor Circuits.** Where the motor short-circuit ground-fault protection device consists of an instantaneous trip circuit breaker or a motor short-circuit protector, the equipment grounding (bonding) conductor size shall be permitted to be based on the rating of the motor overload protective device.

**AUTHOR'S COMMENT:** This permissive requirement only applies when instantaneous trip overcurrent devices are used for the motor short-circuit ground-fault protection device, which is very rare. Typically, inverse-ime breakers or dual-element fuses are used for motor short-circuit ground-fault protection.

**Question:** What size equipment grounding (bonding) conductor is required for a motor with a 45A protection device and 12 AWG branch-circuit conductors? Figure 250-180

(a) 14 AWG                      (b) 12 AWG

(c) 10 AWG                      (d) none of these

**Answer:** (b) 12 AWG. Table 250.122 specifies that a 10 AWG equipment grounding conductor is required based on 45A short-circuit and ground-fault protection device, but the equipment grounding (bonding) conductor is not required to be larger then the 12 AWG circuit conductors [250.122(A)].

**AUTHOR'S COMMENT:** For more details on how to size motor-circuit conductors and protection devices, see 430.22 and 430.52.

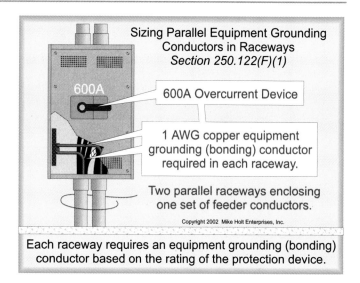

Sizing Parallel Equipment Grounding Conductors in Raceways
*Section 250.122(F)(1)*

600A

600A Overcurrent Device

1 AWG copper equipment grounding (bonding) conductor required in each raceway.

Two parallel raceways enclosing one set of feeder conductors.

Copyright 2002 Mike Holt Enterprises, Inc.

Each raceway requires an equipment grounding (bonding) conductor based on the rating of the protection device.

**Figure 250–181**

**(F) Parallel Runs.** When circuit conductors are run in parallel in accordance with 310.4, an equipment grounding (bonding) conductor shall be installed in each raceway or cable.

(1) The equipment grounding (bonding) conductor in each raceway or cable shall be sized to the ampere rating of the overcurrent device protecting the circuit conductors. Figure 250-181

**AUTHOR'S COMMENT:** The 1/0 AWG minimum parallel conductor size specified in 310.4 does not apply to equipment grounding (bonding) conductors.

### 250.126 Identification of Wiring Device Equipment Grounding (Bonding) Terminal

The terminal for the equipment grounding (bonding) conductor shall be:

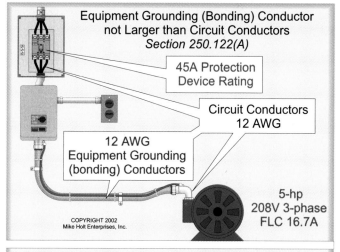

Equipment Grounding (Bonding) Conductor not Larger than Circuit Conductors
*Section 250.122(A)*

45A Protection Device Rating

Circuit Conductors 12 AWG

12 AWG Equipment Grounding (bonding) Conductors

COPYRIGHT 2002 Mike Holt Enterprises, Inc.

5-hp 208V 3-phase FLC 16.7A

Equipment (bonding) grounding conductor is sized to Table 250.122 based on the protection device rating, but it is not required to be larger than the branch-circuit conductors [250.122].

**Figure 250–180**

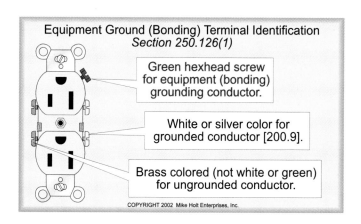

Equipment Ground (Bonding) Terminal Identification
*Section 250.126(1)*

Green hexhead screw for equipment (bonding) grounding conductor.

White or silver color for grounded conductor [200.9].

Brass colored (not white or green) for ungrounded conductor.

COPYRIGHT 2002 Mike Holt Enterprises, Inc.

**Figure 250–182**

(1) A green, hexagonal, not readily removable terminal screw head. Figure 250-182

(2) A green, hexagonal, not readily removable terminal nut.

(3) A green pressure wire connector. If the terminal for the equipment grounding (bonding) conductor is not visible, the conductor entrance hole shall be marked with the word green or ground, the letters G or GR or the grounding symbol, or otherwise identified by a distinctive green color.

## Part VII. Methods of Equipment Grounding (Bonding)

### 250.130 Equipment Grounding (Bonding) Conductor Connections

(C) **Nongrounding Receptacle Replacement or Branch-Circuit Extensions.** Where a nongrounding receptacle is replaced or when a branch circuit extension is made from an outlet box that does not contain an equipment grounding (bonding) conductor, the grounding (bonding) contacts of a grounding-type receptacle shall be bonded to the: Figure 250-183

(1) Grounding electrode system [250.50].

(2) Grounding electrode conductor.

(3) Panelboard equipment grounding (bonding) terminal.

(4) Grounded (neutral) service conductor within the service equipment enclosure.

> FPN: A nongrounding-type receptacle can be replaced with a grounding-type receptacle without a grounding (bonding) connection if the receptacle is GFCI-protected. See 406.3(D). Figure 250-184

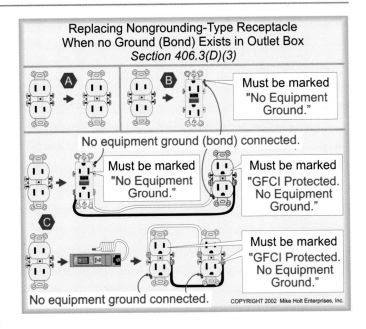

Figure 250–184

**AUTHOR'S COMMENT:** It is permitted to run that equipment grounding (bonding) conductor to any of the locations specified in 250.130(C)(1) through (4), where installing a circuit extension from an existing outlet that does not have means for bonding. Figure 250-185

### 250.132 Bonding – Short Sections of Raceway

Short sections of metal raceway or cable armor where required to be grounded (bonded) shall be bonded in accordance with 250.134.

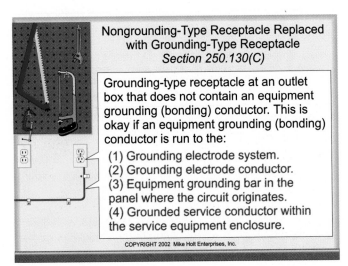

Figure 250–183

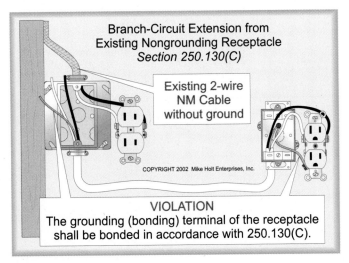

Figure 250–185

**AUTHOR'S COMMENT:** Isolated metal elbows in a nonmetallic raceway are not required to be bonded if installed in accordance with 250.80 Ex. or 250.86 Ex., 3. In addition, short sections of metal raceways used to protect or support cable are not required to be bonded [250.86 Ex. 2].

## 250.134 Bonding – Fixed Equipment Grounding

Metal parts of fixed electrical equipment, raceways and enclosures shall be grounded (bonded) in accordance with (A) or (B), except where bonded to the grounded (neutral) conductor as permitted by 250.142.

**(A) Raceways or Cables.** Metal parts of fixed electrical equipment, raceways and enclosures shall be bonded by any equipment grounding (bonding) conductor types specified in 250.118.

**(B) By Means of an Equipment Grounding (bonding) Conductor.** Metal parts of fixed electrical equipment, raceways and enclosures shall be bonded by means of an equipment grounding (bonding) conductor installed with the circuit conductors in the same raceway, cable tray, trench, cable, or cord. Figure 250-186

**AUTHOR'S COMMENT:** To ensure a low-impedance ground-fault current path and to reduce inductive heating of adjacent metal parts, all circuit conductors shall be grouped together in the same raceway, cable of trench. See 300.3(B), 300.5(I) and 300.20(A) for more details.

*Exception No. 1:* The equipment grounding (bonding) conductor shall be permitted to be run separately from the circuit conductors to comply with 250.130(C),

> FPN No. 1: The equipment bonding jumper is also permitted to be outside the raceway, when installed in accordance with 250.102(E).

## 250.138 Bonding – Cord-and-Plug-Connected Equipment.

Metal parts of cord-and-plug-connected equipment shall be grounded (bonded):

**(A) Equipment Grounding (bonding) Conductor.** Metal parts of cord-and-plug-connected equipment shall be grounded (bonded) by means of an equipment grounding (bonding) conductor run with the circuit conductors in a flexible cord that is terminated in a grounding-type attachment plug.

## 250.140 Bonding – Ranges and Clothes Dryers

In existing installations, the grounded (neutral) conductor can continue to be used to ground (bond) the frame of electric ranges, wall-mounted ovens, counter-mounted cooking units, or clothes dryers [250.142(B) Ex. 1].

**AUTHOR'S COMMENT:** This unsafe practice began in 1943 and existed until 1996.

For new installations, the frame of electric ranges, wall-mounted ovens, counter-mounted cooking units, or clothes dryers shall be bonded to an equipment grounding (bonding) conductor in accordance with 250.134 or 250.138. Figure 250-187

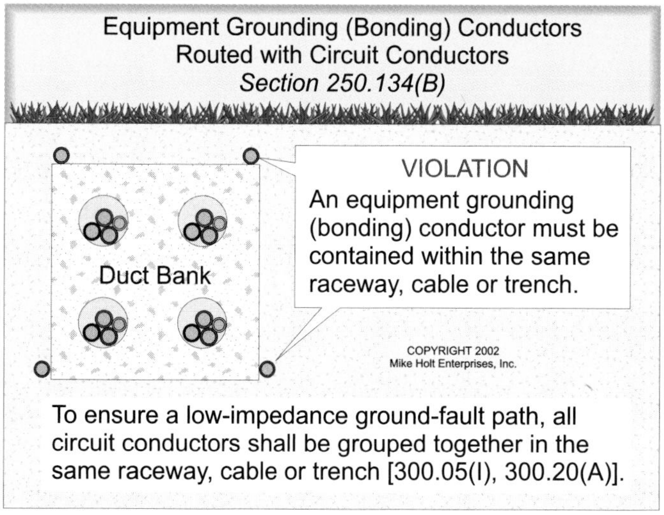

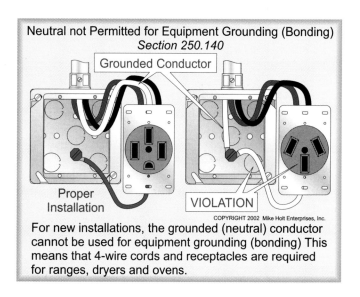

Figure 250–186

Figure 250–187

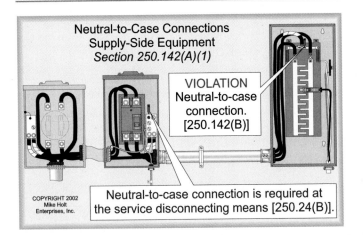

**Figure 250–188**

*CAUTION: Ranges, dryers and ovens automatically have their metal cases bonded to the grounded (neutral) conductor at the factory. This neutral-to-case bond must be removed when these appliances are installed in new construction and a 4-wire cords and receptacle shall be used.*

### 250.142 Neutral-to-Case Bond

To remove dangerous voltage from metal parts, ground faults shall be removed by bonding metal parts of wiring methods and equipment to an effective ground-fault current path.

**(A) Supply Side Equipment.**

   **(1) Service Equipment [250.24(B)].** Because a separate equipment grounding (bonding) conductor is not run from the utility to electrical services, the grounded (neutral) service conductor serves as the effective ground-fault current path. Figure 250-188

   The neutral-to-case connection at service equipment is the main bonding jumper required by 250.28 at each service enclosure.

**AUTHOR'S COMMENT:** Because the grounded (neutral) conductor is bonded to the case at service equipment, the grounded (neutral) terminal bar can be used for the termination of equipment grounding (bonding) conductors. See 408.20. Figure 250-189

   **(2) Separate Buildings and Structures [250.32(B)(2)].** If no equipment grounding (bonding) conductor is run with the feeder circuit conductors to a remote building or structure, a neutral-to-case bond is required at the building or structure disconnect if the installation complies with all of the requirements contained in 250.32(B)(2).

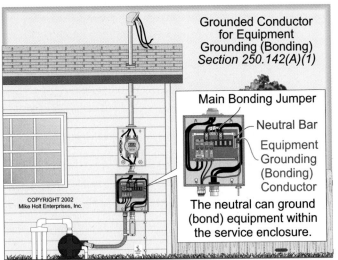

**Figure 250–189**

   **(3) Separately Derived Systems [250.30(A)(1)].** For separately derived systems, the effective ground-fault current path is established when the metal enclosure of the separately derived system is bonded to the grounded (neutral) terminal of the derived system. This neutral-to-case bond can be made at either the separately derived system or at the first system disconnect after the separately derived system, but not at both locations.

*DANGER: Failure to install the neutral-to-case bond at a separately derived system as required by 250.30(A)(1) can create a condition where dangerous touch voltage will remain on the metal parts of the electrical system from a ground fault. Figure 250-190*

**(B) Load-Side Equipment.** To prevent a fire, electric shock, elevated magnetic fields, or improper operation of sensitive equipment, equipment and circuit conductors shall be bonded in a manner that prevents objectionable neutral current from flowing on conductive materials, electrical equipment, or on grounding and bonding paths [250.6]. This is accomplished by not having a neutral-to-case bond connection on the load side of the service equipment, except as permitted by 250.142(A)(1) through (3). Figure 250-191

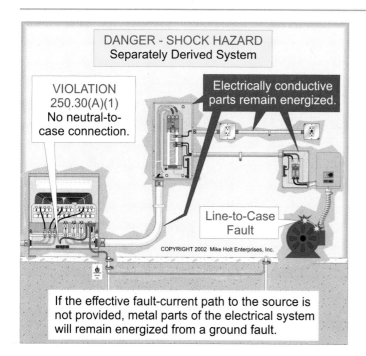

**Figure 250–190**

*Exception No. 1.* The grounded (neutral) conductor can serve as the effective ground-fault current path for existing ranges, dryers and ovens as limited in 250.140.

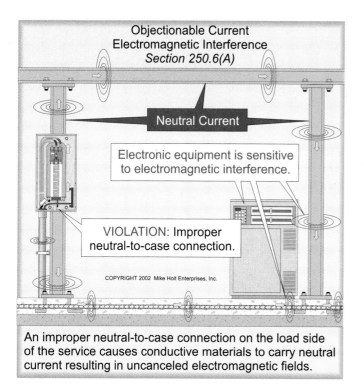

An improper neutral-to-case connection on the load side of the service causes conductive materials to carry neutral current resulting in uncanceled electromagnetic fields.

**Figure 250–191**

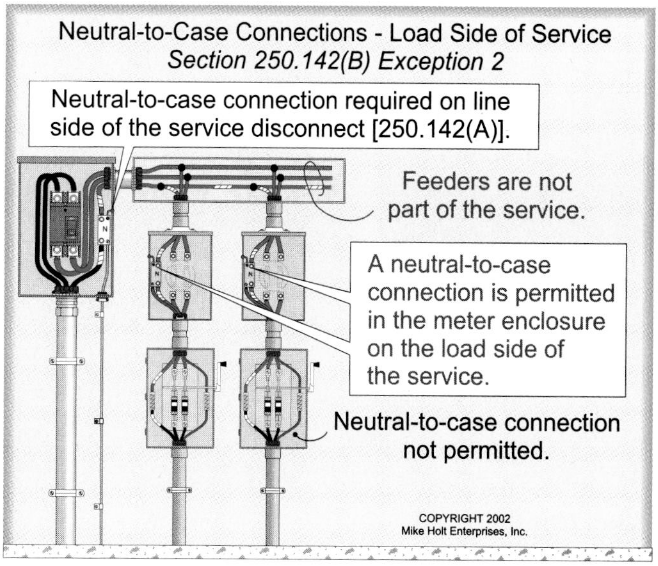

**Figure 250–192**

*Exception No. 2.* A neutral-to-case bond is permitted at meter enclosures on the load side of the service disconnect if: Figure 250-192

(a) No service ground-fault protection is installed,

(b) All meter enclosures are located near the service disconnecting means, and

(c) The size of the grounded (neutral) circuit conductor is not smaller than specified in Table 250.122.

### 250.146 Connecting Receptacle Grounding (Bonding) Terminal

Receptacles shall have their grounding (bonding) contacts bonded to an effective ground-fault current path to ensure that dangerous voltage on metal parts from a ground fault will be quickly removed [250.4(A)(5)]. This shall be accomplished by bonding the receptacle's grounding (bonding) terminal to a grounded (bonded) metal box, unless the receptacle's grounding (bonding) terminal is bonded by one of the methods provided in (A) through (D). See 406.3 for additional details. Figure 250-193

**AUTHOR'S COMMENT:** The position of the grounding (bonding) terminal of a receptacle can be up, down, or sideways. Proposals to specify the mounting position of the grounding (bonding) terminal were all rejected. For more information on this topic, visit: http://www.mikeholt.com/Newsletters/9-23-99.htm. Figure 250-194

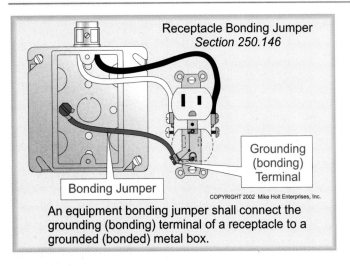

Receptacle Bonding Jumper
*Section 250.146*

Bonding Jumper

Grounding (bonding) Terminal

COPYRIGHT 2002 Mike Holt Enterprises, Inc.

An equipment bonding jumper shall connect the grounding (bonding) terminal of a receptacle to a grounded (bonded) metal box.

*Figure 250–193*

**(A) Surface-Mounted Box.** A bonding jumper is not required for a grounding-type receptacle if the outlet box is surface mounted and there is direct metal-to-metal contact between the receptacle mounting yoke and the grounded (bonded) metal box. Figure 250-195

**AUTHOR'S COMMENT:** This rule does not apply to flush or recessed boxes. Figure 250-196

Receptacles that are mounted to a cover instead of a box [406.4(C)] shall have the grounding (bonding) terminal bonded to an effective ground-fault current path. Figure 250-197

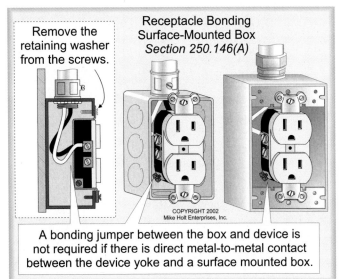

Remove the retaining washer from the screws.

Receptacle Bonding
Surface-Mounted Box
*Section 250.146(A)*

COPYRIGHT 2002
Mike Holt Enterprises, Inc.

A bonding jumper between the box and device is not required if there is direct metal-to-metal contact between the device yoke and a surface mounted box.

*Figure 250–195*

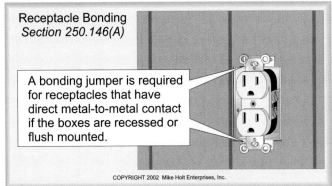

Receptacle Bonding
*Section 250.146(A)*

A bonding jumper is required for receptacles that have direct metal-to-metal contact if the boxes are recessed or flush mounted.

COPYRIGHT 2002 Mike Holt Enterprises, Inc.

*Figure 250–196*

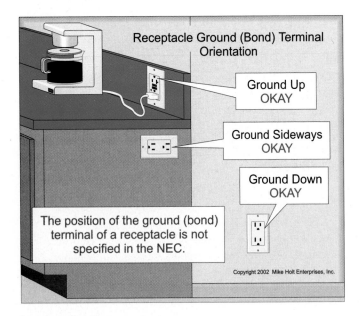

Receptacle Ground (Bond) Terminal Orientation

Ground Up
OKAY

Ground Sideways
OKAY

Ground Down
OKAY

The position of the ground (bond) terminal of a receptacle is not specified in the NEC.

Copyright 2002 Mike Holt Enterprises, Inc.

*Figure 250–194*

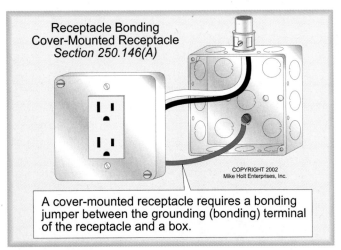

Receptacle Bonding
Cover-Mounted Receptacle
*Section 250.146(A)*

COPYRIGHT 2002
Mike Holt Enterprises, Inc.

A cover-mounted receptacle requires a bonding jumper between the grounding (bonding) terminal of the receptacle and a box.

*Figure 250–197*

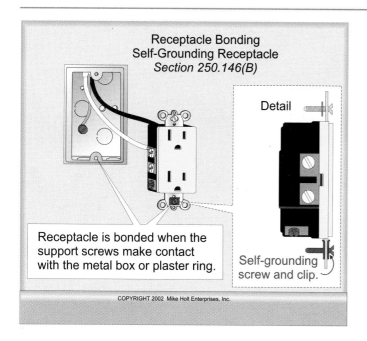

Receptacle Bonding
Self-Grounding Receptacle
Section 250.146(B)

Detail

Receptacle is bonded when the support screws make contact with the metal box or plaster ring.

Self-grounding screw and clip.

COPYRIGHT 2002 Mike Holt Enterprises, Inc.

**Figure 250–198**

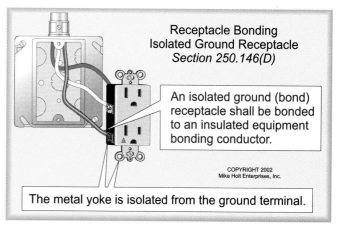

Receptacle Bonding
Isolated Ground Receptacle
Section 250.146(D)

An isolated ground (bond) receptacle shall be bonded to an insulated equipment bonding conductor.

COPYRIGHT 2002
Mike Holt Enterprises, Inc.

The metal yoke is isolated from the ground terminal.

**Figure 250–199**

**(B) Self-Grounding Receptacles.** A bonding jumper is not required for a receptacle that has its mounting yoke listed to establish the bonding path between the device yoke and a grounded (bonded) outlet box. Figure 250-198

*WARNING: Outlet boxes set back more than $1/4$ in. from the finish mounting surface violates 314.20. Replacing the self-grounding (bonding) receptacle mounting screws with longer screws to reach the box would violate 110.3(B), which requires equipment to be installed in accordance with instructions included in the listing or labeling.*

**(C) Floor Boxes.** A receptacle bonding jumper is not required for a floor box that is listed to provide the bonding continuity between a grounded (bonded) box and the receptacle mounting yoke.

**(D) Isolated Receptacles.** Receptacles intended for the reduction of electrical noise shall be identified by an orange triangle located on the face of the receptacle [406.2(D)]. The grounding (bonding) terminal for an isolated ground receptacle is insulated from its metal mounting yoke; therefore, the grounding (bonding) terminal of the receptacle shall be bonded to an insulated equipment grounding (bonding) conductor. Figure 250-199

The *NEC* does not require each isolated ground receptacle to be on its own individual branch circuit; therefore, multiple isolated ground receptacles can be connected to the same insulated equipment grounding (bonding) conductor.

**AUTHOR'S COMMENT:** Section 250.146(D) contains the requirements for isolated (insulated) ground receptacles and 250.96(B) contains the requirements for isolated grounding circuits.

*CAUTION: The insulated equipment grounding (bonding) conductor for an isolated ground receptacle can originate at the neutral terminal of the service or separately derived system. The insulated equipment grounding (bonding) conductor may pass through a panelboard [408.20 Ex.], or it could terminate at the panelboard that supplied the circuit. Figure 250-200*

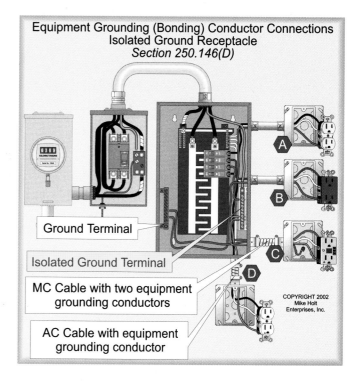

Equipment Grounding (Bonding) Conductor Connections
Isolated Ground Receptacle
Section 250.146(D)

Ground Terminal

Isolated Ground Terminal

MC Cable with two equipment grounding conductors

AC Cable with equipment grounding conductor

COPYRIGHT 2002
Mike Holt
Enterprises, Inc.

**Figure 250–200**

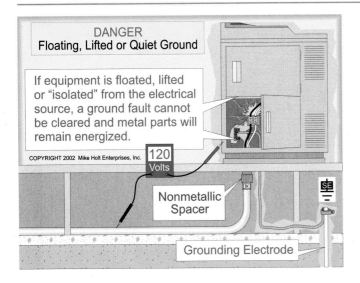

**Figure 250–201**

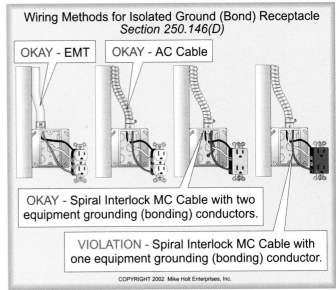

**Figure 250–202**

*DANGER: Some digital equipment manufacturers insist that their equipment be electrically isolated from the building's equipment grounding (bonding) conductor. This is a dangerous practice and it violates 250.4(A)(5), If the metal enclosure of electrical equipment is isolated or floated (as desired by some sensitive equipment manufacturers), the metal parts will remain energized from a ground fault. Figure 250-201*

*An independent supplementary electrode is permitted [250.54], but the equipment must be bonded to an effective ground-fault current path so that a ground fault can be quickly cleared [250.4(A)(5)].*

*WARNING: The outer sheath of interlocked type MC cable is not listed as an equipment grounding (bonding) conductor [250.118(11)]; therefore, this wiring method is not permitted to supply an isolated (insulated) ground receptacle unless it contains two equipment grounding (bonding) conductors. Interlocked-type AC cable containing a single insulated equipment grounding (bonding) conductor is permitted to supply isolated (insulated) ground receptacles, because the armor of type AC cable is listed as an equipment grounding (bonding) conductor [250.118(9)]. Figure 250-202*

**AUTHOR'S COMMENT:** When should an isolated grounding receptacle be installed and how should the system be designed? These questions are design issues and cannot be answered based on *NEC* requirements [90.1(C)]. IEEE Std. 1100 (emerald book) *Powering and Grounding Sensitive Electronic Equipment* states, "The results from the use of the isolated (insulated) ground method range from no observ-

able effects, the desired effects, or worse noise conditions than when standard equipment grounding (bonding) configurations are used to serve electronic load equipment [Section 8.5.3.2]. Additionally, IG circuits are not a panacea for all branch-circuit grounding concerns. They are most effective where served from a dedicated, separately derived locally grounded (and bonded) source. IG circuits will not improve grounding (and bonded) conditions when served from sources which, due to improper wiring or faulty load equipment, already have currents flowing on grounding (and bonding) conductors."

In reality, few electrical installations truly require an isolated ground system. For those systems that could benefit from an isolated ground system, engineering opinions differ as to what is a proper design. Making matters worse – of those that are properly designed, few are installed correctly and even fewer are properly maintained.

For more information on how to properly ground (and bond) sensitive electronic equipment, go to http://mikeholt.com and visit the power quality link.

FPN: Metal raceways and enclosures containing an insulated equipment grounding (bonding) conductor for an isolated ground receptacle shall be bonded to an effective ground-fault current path in accordance with 250.4(A)(3) and 250.86.

## 250.148 Continuity and Attachment of Equipment Grounding (Bonding) Conductors to Boxes.

Where circuit conductors are spliced, or terminated on equipment within a box, equipment grounding (bonding) conductors associated with those circuit conductors shall be spliced within or joined to the box with devices suitable for that purpose. Figure 250-203

Splices for equipment grounding (bonding) conductors shall be made with a listed splicing device in accordance with 110.14(B) except that insulation on the wire connector shall not be required. The grounding (bonding) connections shall be such that the disconnection or the removal of a receptacle, luminaire, or other device fed will not interfere with or interrupt the grounding (bonding) continuity. Figure 250-204

*Exception:* The equipment grounding (bonding) conductor for isolated ground receptacles [250.146(D)] is not required to be spliced with, or to, other equipment grounding (bonding) conductors, nor is it required to terminate to the grounded (bonded) box.

**AUTHOR'S COMMENT:** Wire connectors of any color can be used with equipment bonding conductor splices, but green wire connectors can only be used with equipment bonding conductors. Figure 250-205

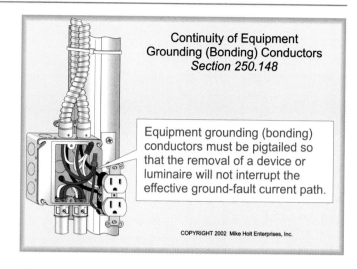

*Figure 250–204*

**(A) Metal Boxes.** When equipment grounding (bonding) conductors are installed in metal boxes, an electrical connection shall be made between the equipment grounding (bonding) conductors and a grounded (bonded) metal box by means of a screw that shall be used for no other purpose, or a listed equipment grounding (bonding) conductor device.

**AUTHOR'S COMMENT:** Equipment bonding conductors cannot terminate to a cable clamp, or a 8-32 screw that secures the plaster (mud) ring to the box. Figure 250-206

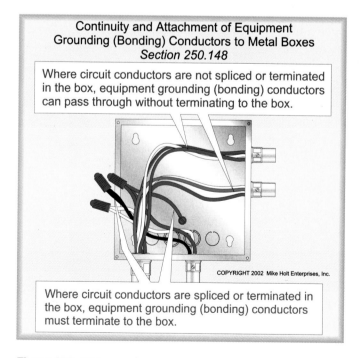

*Figure 250–203*

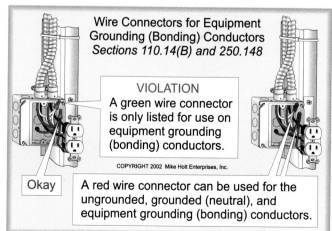

*Figure 250–205*

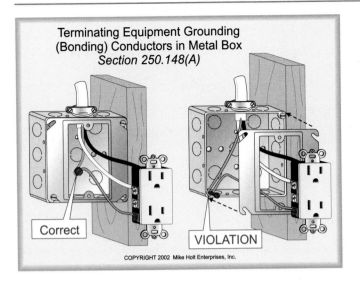

**Figure 250–206**

# Article 250

1. A ground fault is a(n) _____ electrical connection between an ungrounded (hot) conductor and metallic enclosures, metallic raceways, metallic equipment, or earth.
   (a) deliberate       (b) intentional       (c) designed       (d) unintentional

2. An effective ground-fault current path is an intentionally constructed low-impedance path designed and intended to carry fault current from the point of a line-to-case fault on a wiring system to _____.
   (a) ground       (b) earth
   (c) the electrical supply (neutral)       (d) none of these

3. An effective ground-fault current path is created when all electrically conductive materials that are likely to be energized are bonded together and to the _____.
   (a) ground       (b) earth
   (c) electrical supply (neutral)       (d) none of these

4. Electrical systems that are grounded shall be connected to earth in a manner that will _____.
   (a) limit voltages due to lightning, line surges, or unintentional contact with higher-voltage lines
   (b) stabilize the voltage-to-ground during normal operation
   (c) facilitate overcurrent protection device operation in case of ground faults
   (d) a and b

5. Non-current-carrying conductive materials enclosing electrical conductors or equipment, or forming part of such equipment, shall be connected together to the _____ in a manner that establishes an effective ground-fault current path.
   (a) ground       (b) earth
   (c) electrical supply (neutral)       (d) none of these

6. Electrical equipment and wiring, and other electrically conductive material likely to become energized, shall be installed in a manner that creates _____ likely to be imposed on it from any point on the wiring system where a ground fault may occur to the electrical supply source.
   (a) a permanent path
   (b) a low-impedance path
   (c) the capability of safely carrying the ground-fault current
   (d) all of these

7. The earth can be used as the sole equipment grounding conductor.
   (a) True       (b) False

8. Grounding electrode conductor fittings shall be protected from physical damage by being enclosed in _____.
   (a) metal       (b) wood
   (c) the equivalent of a or b       (d) none of these

9. _____ on equipment to be grounded shall be removed from contact surfaces to ensure good electrical continuity.
   (a) Paint       (b) Lacquer       (c) Enamel       (d) any of these

10. When grounding service-supplied alternating-current systems, the grounding electrode conductor shall be connected (bonded) to the grounded service conductor (neutral) at the _____.
    (a) load end of the service drop       (b) meter equipment
    (c) service disconnect       (d) any of these

11. The grounding electrode conductor at the service is permitted to terminate on an equipment grounding terminal bar if a (main) bonding jumper is installed between the grounded conductor bus and the equipment grounding terminal.
    (a) True       (b) False

12. Where an ac system operating at less than 1,000V is grounded at any point, the _____ conductors shall be run to each service disconnecting means and shall be bonded to each disconnect enclosure.
(a) ungrounded      (b) grounded      (c) grounding      (d) none of these

13. Where the service-entrance phase conductors are installed in parallel, the size of the grounded conductor in each raceway shall be based on the size of the ungrounded service-entrance conductor in the raceway but not smaller than _____.
(a) 6 AWG      (b) 1 AWG      (c) 1/0 AWG      (d) none of these

14. The grounding electrode conductor for a single separately derived system must connect the grounded (neutral) conductor of the derived system to the grounding electrode.
(a) True      (b) False

15. The grounding electrode for a separately-derived system shall be as near as practicable to, and preferably in, the same area as the grounding electrode conductor connection to the system. The grounding electrode shall be the nearest one of the following:
(a) An effectively grounded metal member of the building structure.
(b) An effectively grounded metal water pipe, but only if it's within 5 ft from the point of entrance into the building.
(c) Any metal structure that is effectively grounded.
(d) a or b

16. A grounding electrode is required if a building or structure is supplied by a feeder, or by more than one branch circuit.
(a) True      (b) False

17. The size of the grounding electrode conductor for a building or structure supplied by a feeder cannot be smaller than that identified in _____ based on the largest ungrounded supply conductor.
(a) Section 250.66      (b) Section 250.122      (c) Table 250.66      (d) not specified

18. Because of the increasing use of nonmetallic repairs to the interior metal water pipes, interior metal water piping located more than _____ from the point of entrance to the building shall not be used as a part of the grounding electrode system or as a conductor to interconnect electrodes that are part of the grounding electrode system.
(a) 2 ft      (b) 4 ft      (c) 5 ft      (d) 6 ft

19. The metal frame of a building that is effectively grounded is considered part of the grounding electrode system.
(a) True      (b) False

20. Grounding electrodes that consist of driven rods require a minimum of _____ of contact with the soil.
(a) 10 ft      (b) 8 ft      (c) 6 ft      (d) 12 ft

21. A metal underground water pipe shall be supplemented by an additional electrode of a type specified in 250.52(A)(2) through (A)(7). Where the supplemental electrode is a rod, pipe or plate electrode, that portion of the bonding jumper that is the sole connection to the supplemental grounding electrode shall not be required to be larger than _____ copper wire.
(a) 8 AWG      (b) 6 AWG      (c) 4 AWG      (d) 1 AWG

22. The upper end of the rod electrode shall be _____ ground level unless the aboveground end and the grounding electrode conductor attachment are protected against physical damage.
(a) above      (b) flush with      (c) below      (d) b or c

23. Supplementary electrodes for electrical equipment are not required to be bonded to the grounding electrode system. The bonding jumper to the supplemental electrode can be any size. The 25Ω resistance requirement of 250.56 does not apply.
(a) True      (b) False

24. Where the resistance-to-ground of a single rod electrode exceeds 25Ω, _____.
(a) other means besides made electrodes must be used in order to provide grounding
(b) at least one additional electrode must be added
(c) no additional electrodes are required
(d) the electrode can be omitted

25.  When multiple ground rods are used for a grounding electrode, they shall be separated not less than _____ apart.
     (a) 6 ft              (b) 8 ft              (c) 20 ft              (d) 12 ft

26.  Metal enclosures for grounding electrode conductors shall be electrically continuous from the point of attachment to cabinets or equipment to the grounding electrode.
     (a) True              (b) False

27.  A service that contains 12 AWG service-entrance conductors, as permitted by 230.23(B) Ex., shall require a grounding electrode conductor sized no larger than _____.
     (a) 6              (b) 4              (c) 8              (d) 10

28.  What size copper grounding electrode conductor is required for a service that has three sets of 500 kcmil copper conductors per phase?
     (a) 1 AWG              (b) 1/0 AWG              (c) 2/0 AWG              (d) 3/0 AWG

29.  In an ac system, the size of the grounding electrode conductor to a concrete-encased electrode shall not be required to be larger than _____ copper wire.
     (a) 4 AWG              (b) 6 AWG              (c) 8 AWG              (d) 10 AWG

30.  The connection of the grounding electrode conductor to a buried grounding electrode (driven ground rod) shall be made with a listed terminal device that is accessible.
     (a) True              (b) False

31.  A metal elbow that is installed in an underground installation of rigid nonmetallic conduit and is isolated from possible contact by a minimum cover _____ to any part of the elbow shall not be required to be grounded.
     (a) of 6 in.              (b) of 12 in.              (c) of 18 in.              (d) according to Table 300.5

32.  Service raceways threaded into metal service equipment such as bosses (hubs) are considered to be effectively _____ to the service metal enclosure.
     (a) attached              (b) bonded              (c) grounded              (d) none of these

33.  Metal raceways, cable trays, cable armor, cable sheath, enclosures, frames, fittings, and other metal non-current-carrying parts that serve as the grounding conductor must be _____ together to ensure electrical continuity and have the capacity to conduct safely any fault current likely to be imposed.
     (a) grounded              (b) effectively bonded              (c) soldered or welded              (d) any of these

34.  For circuits over 250 volts-to-ground (480Y/277V), electrical continuity can be maintained between a box or enclosure where no knockouts are encountered and a metal conduit by _____.
     (a) threadless fittings for cables with metal sheath
     (b) double locknuts on threaded conduit (one inside and one outside the box or enclosure)
     (c) fittings that have shoulders that seat firmly against the box locknut on the inside.
     (d) all of these

35.  What is the minimum size copper bonding jumper for a service raceway containing 4/0 AWG THHN aluminum conductors?
     (a) 6 AWG aluminum              (b) 3 AWG copper              (c) 4 AWG aluminum              (d) 4 AWG copper

36.  What is the minimum size copper equipment bonding jumper required for equipment connected to a 40A circuit?
     (a) 12 AWG              (b) 14 AWG              (c) 8 AWG              (d) 10 AWG

37.  The equipment bonding jumper can be installed on the outside of a raceway providing the length of the run is not more than _____ and the bonding jumper is routed with the raceway.
     (a) 12 in.              (b) 24 in.              (c) 36 in.              (d) 72 in.

38. The metal water-piping systems must be bonded to the _____.
    (a) grounded conductor at the service
    (b) service equipment enclosure
    (c) equipment grounding bar or bus at any panelboard within the building
    (d) a and b

39. Metal gas piping can be considered bonded by the circuit's equipment grounding conductor of the circuit that may energize the piping.
    (a) True                  (b) False

40. The lightning protection system grounding electrode _____ be bonded to the building grounding electrode system.
    (a) shall            (b) shall not           (c) can            (d) none of these

41. Flexible metal conduit that is not listed for grounding can be used for grounding if the length in any ground return path does not exceed 6 ft and the circuit conductors contained in the conduit are protected by overcurrent devices rated at _____ or less.
    (a) 15A              (b) 20A                 (c) 30A           (d) 60A

42. The equipment grounding conductor shall be identified by _____.
    (a) a continuous outer-green finish
    (b) being bare
    (c) a continuous outer-green finish with one or more yellow stripes
    (d) any of these

43. What size equipment grounding conductor is required for a nonmetallic raceway that contains the following three circuits? Circuit 1 - 12 AWG protected by a 20A device; Circuit 2 - 10 AWG protected by a 30A device; Circuit 3 - 8 AWG protected by a 40A device.
    (a) 10 AWG           (b) 6 AWG               (c) 8 AWG         (d) 12 AWG

44. Where conductors are run in parallel in multiple raceways or cables, the equipment grounding conductor, where used, shall be run in parallel in each raceway or cable.
    (a) True                  (b) False

45. A grounding-type receptacle can replace a nongrounding type receptacle at an outlet box that does not contain an equipment grounding conductor if the equipment grounding conductor is connected to a _____.
    (a) grounding electrode system as described in 250.50
    (b) grounding electrode conductor
    (c) equipment grounding terminal bar within the enclosure where the branch circuits for the receptacle or branch circuit originates
    (d) any of these

46. A grounded circuit conductor shall not be used for grounding non-current-carrying metal parts of equipment on the load side of _____.
    (a) the service disconnecting means
    (b) separately-derived system disconnecting means
    (c) overcurrent protection devices for separately-derived system not having a main disconnecting means
    (d) all of these

47. An equipment bonding jumper shall be used to connect the grounding terminal of a grounding-type receptacle to a grounded box. Where the box is surface mounted, direct metal-to-metal contact between the device yoke and the box shall be permitted to ground the receptacle to the box.
    (a) True                  (b) False

48. Contact devices or yokes designed and listed for the purpose shall be permitted in conjunction with the supporting screws to establish the grounding circuit between the device yoke and flush-type boxes.
    (a) True                    (b) False

49. Where circuit conductors are spliced within a box, or terminated on equipment within or supported by a box, any separate equipment grounding conductors associated with those circuit conductors shall be spliced or joined within the box or to the box with devices suitable for the use.
    (a) True                    (b) False

50. When equipment grounding conductor(s) are installed in a metal box, an electrical connection is required between the equipment grounding conductor and the metal box enclosure by means of a _____.
    (a) grounding screw                    (b) soldered connection
    (c) listed grounding device            (d) a or c

# Article 280
# Surge Arresters

## Part I. General

### 280.1 Scope

This Article covers the installation and connection requirements for surge arresters that are permanently installed on premises wiring systems.

### 280.2. Definition

A surge arrester is a protective device for limiting surge voltages by discharging or bypassing surge current, and it also prevents continued flow of follow current while remaining capable of repeating these functions.

### 280.3 Number Required

Where used, a surge arrester shall be connected to each ungrounded conductor of the system and a single surge arrester shall be permitted to protect a number of circuits.

### 280.4 Surge Arrester Selection

**(A)** **Circuits of Less Than 1000 Volts.** The rating of the surge arrester shall be equal to or greater than the maximum phase-to-ground voltage at the point of connection.

Surge arresters installed on circuits of less than 1000 volts shall be listed to a ANSI/IEEE testing standard.

**AUTHOR'S COMMENT:** A surge arrester is listed to be installed on the line side of service equipment and it is intended to afford protection against surge-related damage to secondary distribution wiring systems and/or to equipment connected thereto from lightning.

FPN No. 2: See the manufacturer's application rules for the selection of an arrester to be used at a particular location.

## Part II. Installation

### 280.11 Location

Surge arresters shall be permitted to be located indoors or outdoors.

### 280.12 Routing of Connections

The conductors for the surge arresters shall not be longer than necessary, and unnecessary bends should be avoided.

## Part III. Connecting Surge Arresters

### 280.21 Installed at Services of Less Than 1,000V

The conductors for a surge arrester installed on the line side of service equipment shall not be smaller than 14 AWG copper, and the arrester grounding conductor shall be connected to one of the following:

(1)  Grounded (neutral) service conductor.

(2)  Grounding electrode conductor.

(3)  Grounding electrode for the service.

(4)  Equipment grounding terminal in the service equipment.

### 280.22 Installed on the Load Side of Services of Less Than 1000 Volts

The conductors for a surge arrester installed on the load side of service equipment shall not be smaller than 14 AWG copper. A surge arrester shall be permitted to be connected between any two conductors — ungrounded conductor(s), grounded conductor and grounding conductor.

### 280.25 Grounding

Grounding conductors for surge arresters shall not be run in metal enclosures unless bonded to both ends of such enclosure.

## Article 280

1. A surge arrester is a protective device for limiting surge voltages by _____ or bypassing surge current.
   (a) decreasing      (b) discharging      (c) limiting      (d) derating

2. Line and ground-connecting conductors for a surge arrester shall not be smaller than _____ .
   (a) 14 AWG      (b) 12 AWG      (c) 10 AWG      (d) 8 AWG

3. The conductor between a lightning surge arrester and the line and the grounding connection shall not be smaller than (installations operating at 1 kV or more) _____ copper.
   (a) 4 AWG      (b) 6 AWG      (c) 8 AWG      (d) 2 AWG

# Article 285
# Transient Voltage Surge Suppressors (TVSSs)

**AUTHOR'S COMMENT:** A surge arrester and a TVSS both operate in the same way: They shunt high voltage surges to ground [280.2 and 285.2]. A surge arrester is "designed to IEEE/ANSI standards" to be installed on the line side of service equipment (without protection). A TVSS device is "listed" for installation on the load side of service equipment, where short-circuit and ground-fault protection is provided.

## I. General

### 285.1 Scope

This Article covers the installation and connection requirements for TVSSs that are permanently installed on premises wiring systems. Figure 285-1

**AUTHOR'S COMMENT:** The scope of Article 285 applies to devices that are listed as TVSS devices. It does not apply to devices that incorporate a TVSS device, such as a cord-and-plug connected TVSS unit, a receptacle, or an appliance that has integral TVSS protection. For more information about TVSS devices, visit www.mikeholt.com/Powerquality/Powerquality.htm

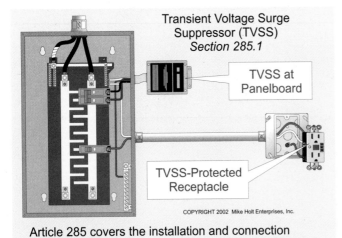

Transient Voltage Surge
Suppressor (TVSS)
*Section 285.1*

TVSS at
Panelboard

TVSS-Protected
Receptacle

COPYRIGHT 2002 Mike Holt Enterprises, Inc.

Article 285 covers the installation and connection requirements for TVSSs that are permanently installed. It does not apply to TVSS-protected receptacles.

*Figure 285-1*

### 285.2 Definition

**Transient Voltage Surge Suppressor (TVSS).** A protective device for limiting transient voltages by diverting or limiting surge current; it also prevents continued flow of follow current while remaining capable of repeating these functions.

### 285.3 Uses Not Permitted.

A TVSS shall not be used in the following:

(1) Circuits exceeding 600 volts

(2) Ungrounded electrical systems as permitted in 250.21

(3) Where the rating of the TVSS is less than the maximum continuous phase-to-ground power frequency voltage available at the point of application

FPN: For further information on TVSSs, see NEMA LS 1-1992, Standard for Low Voltage Surge Suppression Devices. The selection of a properly rated TVSS is based on criteria such as maximum continuous operating voltage and the magnitude and duration of overvoltages at the suppressor location.

### 285.4 Number Required.

Where used, a surge arrester shall be connected to each ungrounded conductor of the system.

### 285.5 Listing

A TVSS shall be a listed device.

**AUTHOR'S COMMENT:** The UL standard covers transient voltage surge suppressors intended to limit the maximum amplitude of transient voltage surges on power lines to specified values. They are not intended to function as lightning arresters.

The effect of the adequacy of the voltage suppression level to protect connected equipment from voltage surges has not been evaluated.

## 285.6 Short Circuit Current Rating.

TVSS devices shall be marked with their short circuit current rating, and they shall not be installed where the available fault current is in excess of that rating. This marking requirement shall not apply to receptacles.

*WARNING: TVSS devices of the series type are susceptible to high fault currents if located near service equipment, and a hazard would be present if the device rating is less than the available fault current.*

## Part II. Installation

### 285.11 Location.

TVSSs shall be permitted to be located indoors or outdoors.

### 285.12 Routing of Connections

The conductors for the TVSS shall not be any longer than necessary and unnecessary bends should be avoided.

## Part III. Connecting Transient Voltage Surge Suppressors

### 285.21 Connection

Where a TVSS is installed, it shall be connected as follows:

**(A) Location.**

  **(1) Service Supplied Building or Structure.** A TVSS can be connected anywhere on the premises wiring system, except on the line side of the service disconnect overcurrent device. Figure 285-2

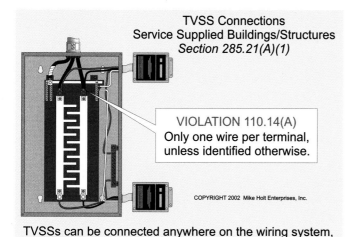

**TVSS Connections**
**Service Supplied Buildings/Structures**
*Section 285.21(A)(1)*

VIOLATION 110.14(A)
Only one wire per terminal,
unless identified otherwise.

COPYRIGHT 2002 Mike Holt Enterprises, Inc.

TVSSs can be connected anywhere on the wiring system, except on the line side of service overcurrent device.

*Figure 285-2*

**AUTHOR'S COMMENT:** Care shall be taken to ensure that no more than one conductor terminates to a terminal, unless the terminal is identified otherwise, in accordance with the requirements of 110.14(A).

*Exception:* A TVSS dual rated as a surge arrester [Article 280] can be connected to the line side of the service overcurrent device.

**AUTHOR'S COMMENT:** TVSSs are listed to be located only on the load side of service equipment. TVSS devices cannot be installed on the line side of the building or structure overcurrent device because of the concern that they might be exposed to lightning-induced surges.

  **(2) Feeder Supplied Building or Structure.** A TVSS can be connected anywhere on the premises wiring system, but not on the line side of the building or structure disconnect overcurrent device.

**AUTHOR'S COMMENT:** TVSS devices cannot be installed on the line side of the building or structure overcurrent device because there is concern that they might be exposed to lightning-induced surges

  **(3) Separately Derived System.** A TVSS can be connected anywhere on the premises wiring of the separately derived system, but not on the line side of the separately derived system overcurrent device.

**(B) Conductor Size.** Line and ground connecting conductors shall not be smaller than 14 AWG copper.

**(C) Connection Between Conductors.** A TVSS shall be permitted to be connected between any two conductors — ungrounded conductor(s), grounded conductor and grounding conductor.

### 285.25 Grounding

Grounding conductors for surge arresters shall not be run in metal enclosures unless bonded to both ends of such enclosure.

# Article 285

1. The scope of Article 285 applies to devices such as cord-and-plug-connected TVSS units or receptacles or appliances that have integral TVSS protection.
   (a) True            (b) False

2. A TVSS device must be listed.
   (a) True            (b) False

3. A TVSS must be marked with its short-circuit current rating and, cannot be installed where the available fault current is in excess of that rating.
   (a) True            (b) False

4. The conductors for the TVSS cannot be any longer than _____, and unnecessary bends should be avoided.
   (a) 6 in.      (b) 12 in.      (c) 18 in.      (d) none of these

5. A TVSS can be connected anywhere on the premises wiring system.
   (a) True            (b) False

# Chapter 3 – Wiring Methods and Materials

Chapter 2 was a bit of an uphill climb, because many rules had a kind of abstract or elusive quality to them. Chapter 3, on the other hand, involves wiring methods, such as conductors, cables, boxes, raceways, and fittings. The actual type of wiring method used depends on several factors, such as Code requirements, environment, need, and cost.

## Article 300. Wiring Methods

Article 300 contains the general requirements for all wiring methods included in the *NEC*. However, this article does not apply to signal and communications systems as covered in Chapters 7 and 8.

## Article 310. Conductors for General Wiring

This Article contains the general requirements for conductors, such as insulation markings, ampacity ratings, and conductor use. Article 310 does not apply to conductors that are part of cable assemblies, flexible cords, fixture wires, or conductors that are an integral part of equipment [90.6, 300.1(B)].

## Article 314. Outlet, Device, Pull and Junction Boxes, Conduit Bodies, Fittings and Manholes

Article 314 contains installation requirements for outlet boxes, pull and junction boxes, as well as conduit bodies and manholes.

## Article 320. Armored Cable (Type AC)

Armored cable is an assembly of insulated conductors, 14 AWG through 1 AWG, that are individually wrapped within waxed paper. The conductors are contained within a flexible spiral metal (steel or aluminum) sheath that interlocks at the edges. Armored cable has an outside appearance like flexible metal conduit. Many electricians call this metal cable BX®.

## Article 330. Metal-Clad Cable (Type MC)

Metal-clad cable encloses one or more insulated conductors in a metal sheath of either corrugated or smooth copper or aluminum tubing, or spiral interlocked steel or aluminum. The physical characteristics of MC cable make it a versatile wiring method that can be used in almost any location and for almost any application. The most common type of MC cable is the interlocking type, which has an appearance similar to armored cable or flexible metal conduit.

## Article 334. Nonmetallic-Sheathed Cable (Type NM)

Nonmetallic-sheathed cable is a wiring type enclosing two or three insulated conductors, 14 AWG through 2 AWG, within a nonmetallic outer jacket. Because this cable is nonmetallic, it contains a separate equipment grounding conductor. Nonmetallic-sheathed cable is a common wiring method used for residential and commercial branch circuits. Most electricians call this wiring Romex®.

## Article 336. Power and Control Tray Cable (Type TC)

TC cable is a factory assembly of two or more insulated conductors under a nonmetallic sheath, for installation in cable trays, in raceways, or where supported by a messenger wire.

## Article 338. Service-Entrance Cable (Type SE and USE)

Service-entrance cable can be a single conductor or multi-conductor assembly within an overall nonmetallic covering. This cable is used primarily for services not over 600V, but can be used for feeders and branch circuits.

## Article 340. Underground Feeder Cable (Type UF)

UF cable is a moisture, fungus, and corrosion-resistant cable suitable for direct burial in the earth and comes in sizes 14 AWG through 4/0 AWG [340.104]. The covering of multiconductor UF cable is molded plastic that encapsulates the insulated conductors.

## Article 342. Intermediate Metal Conduit (Type IMC)

IMC is a circular metal raceway with an outside diameter the same as rigid metal conduit (RMC). The wall thickness of IMC is thinner than RMC, so it has a greater interior cross-sectional area. IMC is lighter than RMC, and it can be used in all of the same locations as RMC. IMC also uses a different steel alloy, which results in it being more rigid than RMC, even though the walls are thinner.

## Article 344. Rigid Metal Conduit (Type RMC)

RMC is a circular metal raceway with an outside diameter the same as IMC. The wall thickness of RMC is greater than IMC, so it has a smaller interior cross-sectional area. RMC is heavier than IMC, and it can be used in any location.

## Article 348. Flexible Metal Conduit (Type FMC)

FMC is a raceway of circular cross section made of helically- wound, interlocked metal strip of either steel or aluminum. It is commonly called "Greenfield" or simply "Flex."

## Article 350. Liquidtight Flexible Metal Conduit (Type LFMC)

LFMC is a listed raceway of circular cross section having an outer liquidtight, nonmetallic, sunlight-resistant jacket over an inner flexible metal core with associated couplings, connectors, and fittings and approved for the installation of electric conductors. LFMC is commonly called Sealtite® or simply "liquidtight." LFMC is of similar construction to FMC, but has an outer liquidtight thermoplastic covering.

## Article 352. Rigid Nonmetallic Conduit (Type RNC)

RNC conduit is a listed nonmetallic raceway of circular cross section with integral or associated couplings, approved for the installation of electrical conductors. Typically this is PVC.

## Article 354. Nonmetallic Underground Conduit with Conductors (Type NUCC)

Nonmetallic underground conduit with conductors is a factory assembly of conductors or cables inside a nonmetallic, smooth wall conduit with a circular cross section. The nonmetallic conduit is manufactured from a material that is resistant to moisture and corrosive agents. It shall also be capable of being supplied on reels without damage or distortion and shall be of sufficient strength to withstand abuse, such as impact or crushing, in handling and during installation without damage to conduit or conductors.

## Article 356. Liquidtight Flexible Nonmetallic Conduit (Type LFNC)

LFNC is a listed raceway of circular cross section having an outer liquidtight, nonmetallic, sunlight-resistant jacket over an inner flexible core with associated couplings, connectors, and fittings and approved for the installation of electric conductors.

Type LFNC-A (orange color). A smooth seamless inner core and cover bonded together and having one or more reinforcement layers between the core and covers.

Type LFNC-B (gray color). A smooth inner surface with integral reinforcement within the conduit wall.

Type LFNC-C (black color). A corrugated internal and external surface without integral reinforcement within the conduit wall.

## Article 358. Electrical Metallic Tubing (Type EMT)

EMT is a listed metallic tubing of circular cross section raceway approved for the installation of electrical conductors. Compared to RMC and IMC, EMT is relatively easy to bend, cut, and ream and because it is not threaded; all connectors and couplings are of the threadless type.

## Article 362. Electrical Nonmetallic Tubing (Type ENT)

Electrical nonmetallic tubing is a pliable, corrugated, circular raceway made of polyvinyl chloride (PVC). It's often called "smurf pipe," because when it originally came out at the height of popularity of the children's characters the Smurfs, it was available only in blue.

## Article 376. Metal Wireways

A metal wireway is a sheet metal trough with hinged or removable covers for housing and protecting electric wires and cable in which conductors are placed after the wireway has been installed.

## Article 378. Nonmetallic Wireways

A nonmetallic wireway is a flame-retardant trough with hinged or removable covers for housing and protecting electric wires and cable and in which conductors are placed after the wireway has been installed as a complete system.

## Article 380. Multioutlet Assembly

A multioutlet assembly is a surface, flush, or freestanding raceway designed to hold conductors and receptacles, assembled in the field or at the factory.

## Article 384. Strut-Type Channel Raceway

Strut-type channel raceway is a metallic raceway intended to be mounted to the surface or suspended, with associated accessories, in which conductors are placed after the raceway has been installed as a complete system.

## Article 386. Surface Metal Raceways

A surface metal raceway is a metallic raceway intended to be mounted to the surface with associated accessories in which conductors are placed after the raceway has been installed as a complete system.

## Article 388. Surface Nonmetallic Raceways

A surface nonmetallic raceway is a nonmetallic raceway intended to be mounted to the surface with associated accessories in which conductors are placed after the raceway has been installed as a complete system.

## Article 392. Cable Trays

A cable tray system is a unit or assembly of units or sections and associated fittings forming a structural system used to securely fasten or support cables and raceways. A cable tray is not a raceway.

# Article 300
# Wiring Methods

## Part I. General Requirements

### 300.1 Scope

**(A) Wiring Installations.** Article 300 contains the general requirements for power and lighting wiring methods identified as suitable for use by the *NEC*.

**AUTHOR'S COMMENT:** The requirements of Article 300 do not apply to the wiring methods for signaling and communications systems, unless a specific reference is made to a section in Article 300. See:

- CATV, 820.3
- Class 2 and 3 Circuits, 725.3
- Communications (telephone), 800.52(A)(2)
- Fiber Optics, 770.3
- Fire Alarm Circuits, 760.3

**(B) Integral Parts of Equipment.** The requirements contained in Article 300 do not apply to the internal parts of electric equipment. See 90.7. Figure 300-1

**(C) Metric Designators and Trade Sizes.** Metric designators and trade sizes for conduit, tubing, and associated fittings and accessories are designated in the following table:

| Trade Size | Metric |
|---|---|
| 3/8 | 12 |
| 1/2 | 16 |
| 3/4 | 21 |
| 1 | 27 |
| 1 1/4 | 35 |
| 1 1/2 | 41 |
| 2 | 53 |
| 2 1/2 | 63 |
| 3 | 78 |
| 3 1/2 | 91 |
| 4 | 103 |
| 5 | 129 |
| 6 | 155 |

**AUTHOR'S COMMENT:** Since compliance with either the metric or the trade size measurement system constitutes compliance with this Code, this book places the trade size unit first with the metric in parentheses, or it uses the trade size unit only [90.9(D)].

### 300.3 Conductors

**(A) Conductors.** Individual conductors shall be installed within a raceway, cable, or enclosure.

**(B) Circuit Conductors Grouped Together.** All conductors of a circuit shall be installed in the same raceway, cable, trench, cord, or cable tray, except as permitted by the following: Figure 300-2

**AUTHOR'S COMMENT:** The reason all circuit conductors should be in the same raceway or cable is to minimize induction heating of metallic raceways and enclosures and to reduce impedance should a ground fault develop.

**Examination of Equipment**
*Section 90.7*

Listed factory-installed internal wiring, or the construction of equipment, need not be inspected at the time of installation, except to detect alterations or damage.

*Figure 300-1*

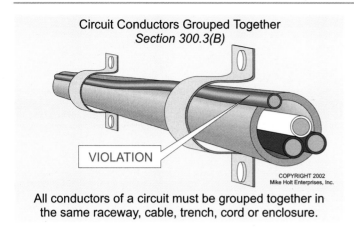

Circuit Conductors Grouped Together
*Section 300.3(B)*

VIOLATION

COPYRIGHT 2002
Mike Holt Enterprises, Inc.

All conductors of a circuit must be grouped together in the same raceway, cable, trench, cord or enclosure.

**Figure 300-2**

**(1) Paralleled Installations.** Conductors can run in parallel in accordance with 310.4. Each parallel set shall have all circuit conductors within the same raceway, auxiliary gutter, cable tray, trench, or cable. Figure 300-3

*Exception:* Conductors installed in nonmetallic raceways run underground shall be permitted to be arranged as isolated phase installations in accordance with 300.5(I).

**(2) Grounding Conductors.** Equipment grounding and bonding jumpers can be located outside a flexible raceway, if the equipment or bonding jumper is installed in accordance with 250.102(E). Figure 300-4

**(3) Nonferrous Wiring Methods.** Circuit conductors can be run in different raceways (Phase A in raceway 1, Phase B in raceway 2, etc.) if the raceway is nonmetallic or nonmagnetic and the installation complies with the requirements 300.20(B) to reduce or eliminate inductive heating. See 300.5(I) Ex. 2.

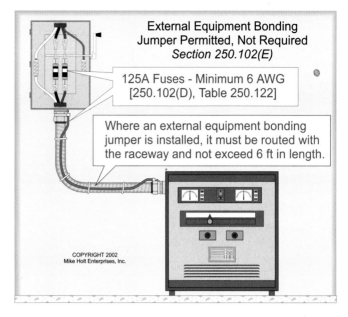

External Equipment Bonding
Jumper Permitted, Not Required
*Section 250.102(E)*

125A Fuses - Minimum 6 AWG
[250.102(D), Table 250.122]

Where an external equipment bonding jumper is installed, it must be routed with the raceway and not exceed 6 ft in length.

COPYRIGHT 2002
Mike Holt Enterprises, Inc.

**Figure 300-4**

**(C) Conductors of Different Systems.**

(1) Power conductors of different systems can occupy the same raceway, cable, or enclosure if all conductors have an insulation voltage rating not less than the maximum circuit voltage. Figure 300-5

**AUTHOR'S COMMENT:** Control, signal and communications wiring must be separated from power and lighting circuits so they are not accidentally energized by the higher voltage conductors. The following Code sections prohibit the mixing of signaling and communications conductors with power conductors:

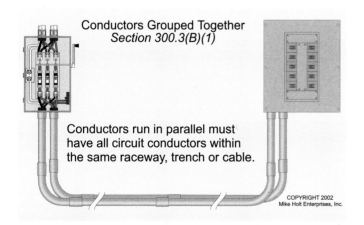

Conductors Grouped Together
*Section 300.3(B)(1)*

Conductors run in parallel must have all circuit conductors within the same raceway, trench or cable.

COPYRIGHT 2002
Mike Holt Enterprises, Inc.

**Figure 300-3**

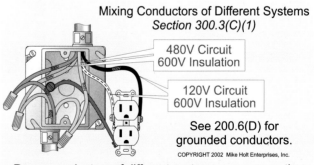

Mixing Conductors of Different Systems
*Section 300.3(C)(1)*

480V Circuit
600V Insulation

120V Circuit
600V Insulation

See 200.6(D) for grounded conductors.

COPYRIGHT 2002  Mike Holt Enterprises, Inc.

Power conductors of different systems can occupy the same raceway, cable, or enclosure if all conductors have an insulation voltage rating not less than the maximum circuit voltage.

**Figure 300-5**

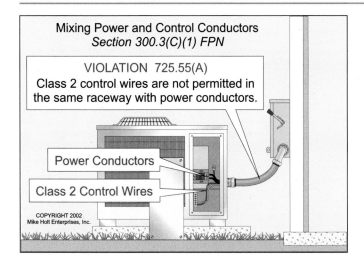

**Figure 300-6**

- Class 1, Class 2 and Class 3 Control, 725.55(A). Figure 300-6
- Fire Alarm, 760.55(A)
- Instrument Tray Cable, 727.5
- Sound Systems, 640.9(C)

Exceptions to the above Code sections allow power conductors to terminate to listed signaling equipment, if the power conductors maintain a minimum of $1/4$ in. separation from the low voltage and limited-energy conductors.

## 300.4 Protection Against Physical Damage

Conductors and equipment shall be protected when subject to physical damage. See 110.27(B).

**(A) Cables and Raceways Through Wood Members.** When the following cables or raceways are installed through wood members, they shall comply with the requirements of (1) and (2). Figure 300-7

- Armored Cable, Article 320
- Electrical Nonmetallic Tubing, Article 362
- Flexible Metal Conduit, Article 348
- Liquidtight Flexible Conduit, Article 350
- Metal-Clad Cable, Article 330
- Nonmetallic-Sheathed Cable, Article 334
- Service-Entrance Cable, Article 338

**(1) Drilling Holes in Wood Members.** When drilling holes through wood framing members for the above cables or raceways, the edge of the holes shall be at least $1^1/4$ in. from the edge of the wood member. If the edge of the hole is less than $1^1/4$ in. from the edge, a $1/16$

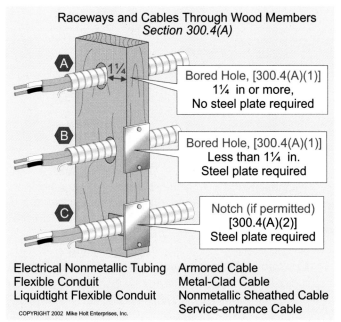

**Figure 300-7**

in. thick steel plate of sufficient length and width shall be installed to protect the wiring method from screws and nails.

**(2) Notching Wood Members.** Where notching of wood framing members for cables and raceways is permitted by the building code, cables and raceways laid in these wood notches shall have a $1/16$ in. thick steel plate of sufficient length and width to protect the wiring method from screws and nails.

*CAUTION: When drilling or notching wood members, be sure to check with the building code inspector to ensure that you don't damage or weaken the structure.*

**(B) NM Cable and ENT Through Metal Framing Members.**

**(1) Nonmetallic-Sheathed Cable (NM).** Where NM cables pass through factory or field openings in metal members, the cable shall be protected by listed bushings or listed grommets that cover metal edges. The protection fitting shall be securely fastened in the opening before installation of the cable. Figure 300-8

**AUTHOR'S COMMENT:** Control, signal and communications cables are not required to maintain any separation from power cables or raceways. See:

- CATV, 820.52(A)(2) Ex. 1
- Communications, 800.52(A)(2) Ex. 1
- Control/Signaling, 725.55(J)
- Fire Alarm, 760.55(G)

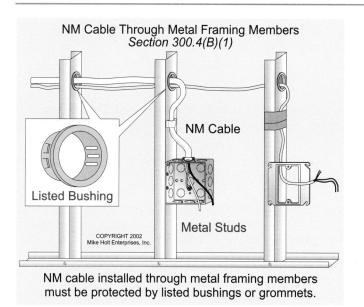

NM Cable Through Metal Framing Members
Section 300.4(B)(1)

NM Cable

Listed Bushing

Metal Studs

COPYRIGHT 2002
Mike Holt Enterprises, Inc.

NM cable installed through metal framing members
must be protected by listed bushings or grommets.

**Figure 300-8**

- Intrinsically Safe, 504.30(A)(2) Ex.
- Network Broadband, 830.58(A)(2) Ex. 1
- Radio and Television, 810.18(B) Ex. 1

**(2) NM Cable and ENT.** Where nails or screws are likely to penetrate NM cable or ENT, a steel sleeve, steel plate, or steel clip not less than $1/16$ in. in thickness shall be installed to protect the cable or tubing.

**(C) Behind Suspended Ceilings.** Wiring methods, such as cables or raceways installed behind panels designed to allow access, shall be supported in accordance with their applicable article. Figure 300-9

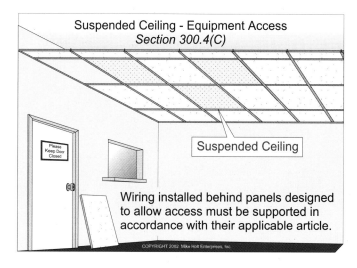

Suspended Ceiling - Equipment Access
Section 300.4(C)

Please
Keep Door
Closed

Suspended Ceiling

Wiring installed behind panels designed
to allow access must be supported in
accordance with their applicable article.

COPYRIGHT 2002 Mike Holt Enterprises, Inc.

**Figure 300-9**

**AUTHOR'S COMMENT:** This rule does not apply to control, signal and communications cables, but similar requirements are contained as follows:

CATV, 820.5 and 820.6

Communications Systems, 800.5 and 800.6

Control and Signaling, 725.5 and 725.6

Fiber Optical Cable, 770.7 and 770.8

Fire Alarm, 760.5 and 760.6

Network Broadband, 830.6 and 830.7

Sound Systems, 640.5 and 640.6

**(D) Cables and Raceways Parallel to Framing Members.** Cables or raceways run parallel to framing members shall be protected where likely to be penetrated by nails or screws, by installing the wiring method so it is not less than $1^1/_4$ in. from the nearest edge of the framing member or is protected by a $^1/_{16}$ in. thick steel plate. For example, when running NM cable on the interior of masonry walls, keep the cable at least $1^1/_4$ in. from the furring strips. Figure 300-10

**AUTHOR'S COMMENT:** This rule does not apply to control, signal and communications cables, but similar requirements are contained as follows:

- CATV, 820.6
- Communications Systems, 800.6
- Control and Signaling, 725.6
- Fiber Optical Cable, 770.8

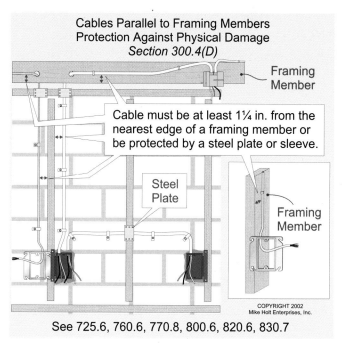

Cables Parallel to Framing Members
Protection Against Physical Damage
Section 300.4(D)

Framing
Member

Cable must be at least 1¼ in. from the
nearest edge of a framing member or
be protected by a steel plate or sleeve.

Steel
Plate

Framing
Member

COPYRIGHT 2002
Mike Holt Enterprises, Inc.

See 725.6, 760.6, 770.8, 800.6, 820.6, 830.7

**Figure 300-10**

- Fire Alarm, 760.6
- Network Broadband, 830.7
- Sound Systems, 640.6

*Exception No. 1:* Protection is not required for rigid metal conduit, intermediate metal conduit, rigid nonmetallic conduit, or electrical metallic tubing.

**(E) Cables and Raceways Installed in Groove.** Cables and raceways installed in a groove shall be protected by a $1/16$ in. thick steel plate or sleeve, or by $1\frac{1}{4}$ in. of free space. An example would be if NM cable is installed in a groove cut into the styrofoam-type insulation building block structure and then covered with wallboard.

*Exception:* Protection is not required if the cable is installed in rigid metal conduit, intermediate metal conduit, rigid nonmetallic conduit or electrical nonmetallic tubing.

**(F) Insulating Bushings.** Conductors 4 AWG and larger entering an enclosure shall be protected from abrasion during and after installation by a fitting that provides a smooth, rounded insulating surface, such as an insulating bushing. Figure 300-11

**AUTHOR'S COMMENT:** Rigid nonmetallic conduit male-adapter termination fittings, at times, are considered to provide the required smooth rounded insulating surface, but not always. Check with the AHJ.

*Exception:* Insulating bushings are not required where a raceway terminates in a threaded raceway entry that provides a smooth, rounded, or flared surface for the conductors. Examples would be a meter hub fitting or a Meyers hub-type fitting.

## 300.5 Underground Installations

**(A) Minimum Burial Depths.** When cables and raceways are run underground, they shall have a minimum "cover" in accordance with Table 300.5.

"Cover" is defined as the distance from the top of the underground cable raceway to the surface of finish grade [Table 300.5 Note 1]. Figure 300-12

### Table 300.5 Minimum Cover Requirements In Inches
#### Figure 300-13

| Location Circuit | Buried Cables | Metal Raceway | Nonmetallic Raceway | 120V 20A GFC |
|---|---|---|---|---|
| Under Building | 0 | 0 | 0 | 0 |
| Dwelling Unit | 18 | 18 | 18 | 12 |
| Under Roadway | 24 | 24 | 24 | 24 |
| Other Locations | 24 | 6 | 18 | 12 |

Bushing Requirements
Raceway Terminations
*Section 300.4(F)*

Bushing Not Required

| Threaded IMC Threaded Rigid | ANY Wire Size | |
| IMC - Rigid | 4 AWG and LARGER | IMC - 342.46 RMC - 344.46 |
| IMC - Rigid | 6 AWG and SMALLER | |
| EMT | 4 AWG and LARGER | EMT |
| EMT | 6 AWG and SMALLER | |
| PVC - ENT | 4 AWG and LARGER | PVC - 352.46 ENT - 362.46 |
| PVC - ENT | 6 AWG and SMALLER | |

COPYRIGHT 2002
Mike Holt Enterprises, Inc.

**Figure 300-11**

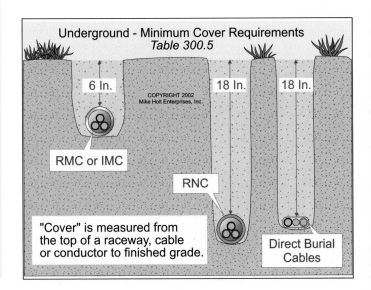

Underground - Minimum Cover Requirements
*Table 300.5*

6 In.    18 In.    18 In.

RMC or IMC

RNC

"Cover" is measured from the top of a raceway, cable or conductor to finished grade.

Direct Burial Cables

COPYRIGHT 2002
Mike Holt Enterprises, Inc.

**Figure 300-12**

### Underground - Cover Requirements Summary

| | UF or USE Cables or Conductors | RMC or IMC | RNC (PVC) not encased in concrete | Residential 15- & 20A GFCI Branch-Circuits | Landscape Lighting 30V or less |
|---|---|---|---|---|---|
| Applications NOT listed below | 24 | 6 | 18 | 12 | 6 |
| TRENCH not less than 2 in. of concrete | 18 | 6 | 12 | 6 | 6 |
| Under a BUILDING | Raceway Only | No Depth | No Depth | Raceway Only | Raceway Only |
| STREET Driveway Parking Lot | 24 | 24 | 24 | 24 | 24 |
| DRIVEWAYS One - Two Family | 18 | 18 | 18 | 12 | 18 |
| SOLID ROCK With not less than 2 in. of concrete | Raceway Only | | | Raceway Only | Raceway Only |

COPYRIGHT 2002 Mike Holt Enterprises, Inc.   Not all the applications listed on Table 300.5 are shown in this diagram.

**Figure 300-13**

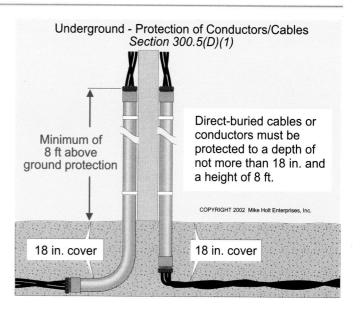

### Underground - Protection of Conductors/Cables
*Section 300.5(D)(1)*

Minimum of 8 ft above ground protection

Direct-buried cables or conductors must be protected to a depth of not more than 18 in. and a height of 8 ft.

COPYRIGHT 2002 Mike Holt Enterprises, Inc.

18 in. cover                18 in. cover

**Figure 300-15**

---

**AUTHOR'S COMMENT:** The cover requirements of this section do not apply to the following signaling and communications wiring. Figure 300-14

- CATV, 90.3
- Class 2 and 3 Circuits, 725.3
- Communications, 90.3
- Fiber Optics, 770.3
- Fire Alarm Circuits, 760.3

**(C) Cables Under Buildings.** Cables run under a building shall be in a raceway that extends past the outside walls of the building.

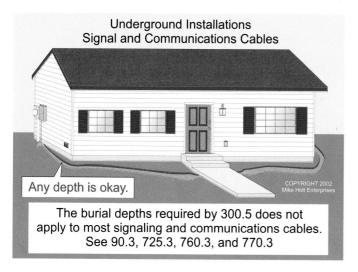

### Underground Installations
### Signal and Communications Cables

Any depth is okay.

COPYRIGHT 2002 Mike Holt Enterprises

The burial depths required by 300.5 does not apply to most signaling and communications cables. See 90.3, 725.3, 760.3, and 770.3

**Figure 300-14**

**(D) Protecting Underground Cables and Conductors From Damage.**

**(1) Emerging From Grade.** Direct-buried cables or conductors that emerge from the ground shall be protected against physical damage by installing them in an enclosure or raceway. Protection is not required to extend a depth of more than 18 in. below grade. The protection above ground shall extend to a height of at least 8 ft. Figure 300-15

**(3) Service Conductors.** Service conductors, not under the exclusive control of the electric utility, buried 18 in. or more below grade shall have their location identified by a warning ribbon placed in the trench at least 1 ft above the underground installation. Figure 300-16

**AUTHOR'S COMMENT:** It will be next to impossible to comply with this requirement when service conductors are installed via directional boring equipment.

**(4) Physical Protection.** Where direct-buried cables or conductors are subject to physical damage, they shall be installed in rigid metal conduit, intermediate metal conduit, or Schedule 80 rigid nonmetallic conduit.

**(E) Splices and Taps Underground.** Direct-buried conductors or cables can be spliced or tapped underground without a splice box [300.15(G)], if the splice or tap is made in accordance with the requirements of 110.14(B).

**(F) Backfill.** Backfill material for underground wiring shall not damage the underground cable raceway or contribute to the corrosion of the metal raceway.

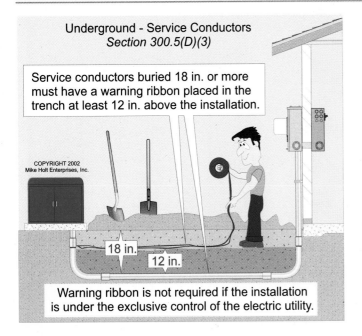

Underground - Service Conductors
*Section 300.5(D)(3)*

Service conductors buried 18 in. or more must have a warning ribbon placed in the trench at least 12 in. above the installation.

COPYRIGHT 2002
Mike Holt Enterprises, Inc.

18 in.

12 in.

Warning ribbon is not required if the installation is under the exclusive control of the electric utility.

**Figure 300-16**

---

**AUTHOR'S COMMENT:** Large rocks, chunks of concrete, steel rods, mesh, and other sharp-edged objects shall not be used for backfill material, because they can damage the underground conductors, cables, or raceways.

---

**(G) Raceway Seals.** Where moisture could enter a raceway and contact energized live parts, seals shall be installed at one or both ends of the raceway.

---

**AUTHOR'S COMMENT:** This is a common problem for equipment located downhill from the supply source or in underground equipment rooms. See 230.8 for service raceway seals and 300.7(A) for different temperature area seals.

---

FPN: Hazardous explosive gases or vapors require the sealing of underground conduits or raceways entering the building in accordance with 501.5 and 501.11.

---

**AUTHOR'S COMMENT:** It is not the intent of this FPN to imply that seal-offs of the types required in hazardous (classified) locations be installed in unclassified locations, except as required in Chapter 5. This also does not imply that the sealing material provide a watertight seal, but that it only prevents moisture from entering.

---

**(H) Bushing.** Raceways that terminate underground shall have a bushing or fitting at the end of the raceway to protect emerging cables or conductors.

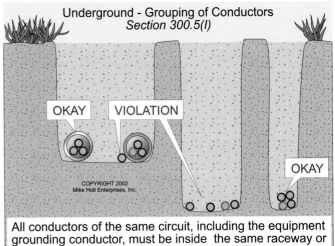

Underground - Grouping of Conductors
*Section 300.5(I)*

OKAY      VIOLATION

OKAY

COPYRIGHT 2002
Mike Holt Enterprises, Inc.

All conductors of the same circuit, including the equipment grounding conductor, must be inside the same raceway or in close proximity to each other. See 300.3(B).

**Figure 300-17**

**(I) Conductors Grouped Together.** All conductors of a circuit shall be installed in the same raceway, cable, trench, cord, or cable tray to minimize induction heating of metallic raceways and enclosures, and to maintain a low impedance ground-fault current path. See 250.102(E), 300.3(B), and 392.8(D). Figure 300-17

*Exception No. 1:* Conductors can be installed in parallel in accordance with 310.4. See 300.3(B)(1) Ex. No. 1.

*Exception No. 2:* Individual sets of parallel circuit conductors (isolated phase installations) can be installed in underground nonmetallic raceways, if inductive heating at the raceway terminations can be reduced by complying with 300.20. Figure 300-18

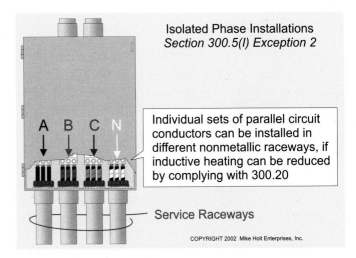

Isolated Phase Installations
*Section 300.5(I) Exception 2*

A  B  C  N

Individual sets of parallel circuit conductors can be installed in different nonmetallic raceways, if inductive heating can be reduced by complying with 300.20

Service Raceways

COPYRIGHT 2002  Mike Holt Enterprises, Inc.

**Figure 300-18**

**AUTHOR'S COMMENT:** Installing phase and neutral wires in different nonmetallic raceways makes it easier to terminate larger parallel sets of conductors, but it will result in higher levels of electromagnetic fields (EMF), which can cause computer monitors to jitter in a distracting manner.

**(J) Ground Movement.** Direct-buried conductors, cables, or raceways, subject to movement by settlement or frost, shall be arranged to prevent damage to conductors or equipment connected to the wiring.

**(K) Directional Boring.** Cables or raceways installed using directional boring equipment shall be approved by the AHJ for this purpose.

**AUTHOR'S COMMENT:** Directional boring technology uses a directional drill (meaning it is steered continuously) from point "A" to point "B." When the drill head comes out of the earth at point "B," it is replaced with a back-reamer and the duct or conduit being installed is attached to it. The size of the boring rig (hp, torque, and pull-back power) comes into play along with the type of soil in determining the type of raceways required. For telecom work, multiple poly innerducts are pulled in at one time. If a major crossing is encountered, such as an expressway, railroad or river, an outer duct may be installed to create a permanent sleeve for the innerducts.

"Innerduct" and "outerduct" are terms usually associated with fiber-optic installations while unitduct has current-carrying conductors factory installed. All of these come in various sizes. Galvanized rigid steel conduit and Schedule 40 and 80 PVC are installed extensively with directional boring installations.

## 300.6 Protection Against Corrosion

Metal raceways, cable trays, cablebus, auxiliary gutters, cable armor, boxes, cable sheathing, cabinets, elbows, couplings, fittings, supports, and support hardware shall be of materials suitable for the environment in which they are to be installed. Figure 300-19

**(A) General.** Steel raceways, enclosures, cables, fittings and support hardware shall be protected against corrosion inside and outside by a coating of approved corrosion-resistant material such as zinc, cadmium, or enamel. Where corrosion protection is necessary (underground and wet locations) and the conduit is threaded in the field, the threads shall be coated with an approved electrically conductive, corrosion-resistant compound (cold zinc).

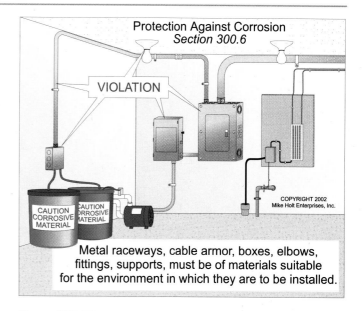

Protection Against Corrosion
*Section 300.6*

VIOLATION

CAUTION CORROSIVE MATERIAL

CAUTION CORROSIVE MATERIAL

COPYRIGHT 2002
Mike Holt Enterprises, Inc.

Metal raceways, cable armor, boxes, elbows, fittings, supports, must be of materials suitable for the environment in which they are to be installed.

**Figure 300-19**

**AUTHOR'S COMMENT:** For more information, visit www.mikeholt.com/Newsletters/300.6A.pdf

**(B) In Concrete or Direct Contact with the Earth.** Metal raceways, boxes, fittings, supports, and support hardware can be installed in concrete or in direct contact with the earth, or other areas subject to severe corrosive influences, if the metal parts are judged suitable for the condition, or where provided with corrosion protection approved for the condition.

**(C) Indoor Wet Locations.** At least $1/4$ in. of air space shall be provided between equipment and the wet surface of an indoor installation. See 312.2(A) for similar requirements for outdoor wet installations for cabinets, cutout boxes and meter cans.

*Exception:* The $1/4$ in. air space does not apply to nonmetallic equipment, raceways, and cables.

## 300.7 Raceways Exposed to Different Temperatures

**(A) Sealing.** Where portions of a cable, raceway or sleeve are known to be subjected to different temperatures and condensation is known to be a problem, the raceway or sleeve shall be filled with a material approved by the AHJ, to prevent the circulation of warm air to a colder section of the raceway or sleeve. An explosionproof seal is not required for this purpose. Figure 300-20

**(B) Expansion Fittings.** Raceways shall be provided with expansion fittings where necessary to compensate for thermal expansion and contraction. Figure 300-21

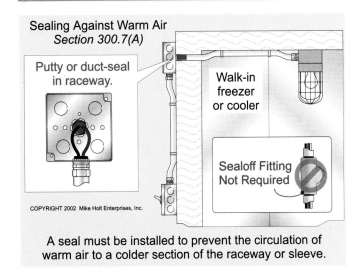

Sealing Against Warm Air
*Section 300.7(A)*

Putty or duct-seal in raceway.

Walk-in freezer or cooler

Sealoff Fitting Not Required

COPYRIGHT 2002 Mike Holt Enterprises, Inc.

A seal must be installed to prevent the circulation of warm air to a colder section of the raceway or sleeve.

**Figure 300-20**

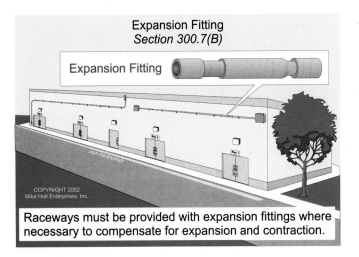

Expansion Fitting
*Section 300.7(B)*

Expansion Fitting

COPYRIGHT 2002
Mike Holt Enterprises, Inc.

Raceways must be provided with expansion fittings where necessary to compensate for expansion and contraction.

**Figure 300-21**

FPN: Table 352.44(A) provides the expansion characteristics for PVC rigid nonmetallic conduit. The expansion characteristics for metal raceways (EMT IMC and RMC) are determined by multiplying the values from Table 352.44(A) by a multiplier of 0.20.

**AUTHOR'S COMMENT:** Where an expansion fitting is used with a metal raceway, a bonding jumper is required to maintain the equipment grounding path. See 250.98.

## 300.8 Not Permitted in Raceways

Raceways are designed for the exclusive use of electrical conductors and cables, and shall not contain nonelectrical components, such as lines for steam, water, air, gas, drainage, etc. Figure 300-22

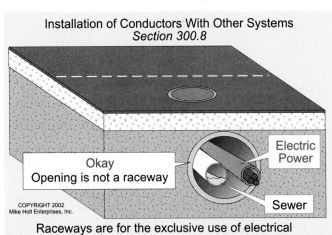

Installation of Conductors With Other Systems
*Section 300.8*

Okay
Opening is not a raceway

Electric Power

Sewer

COPYRIGHT 2002
Mike Holt Enterprises, Inc.

Raceways are for the exclusive use of electrical conductors and cables, and must not contain lines for steam, water, air, gas, drainage, etc.

**Figure 300-22**

## 300.10 Electrical Continuity

All metal raceways, cable, boxes, fittings, cabinets, and enclosures for conductors shall be mechanically and metallically joined together to form a continuous low-impedance ground-fault current path. The ground-fault current path shall have adequate capacity to carry any fault likely to be imposed on it. See 110.10, 250.4(A)(3). and 250.22 Figure 300-23

***Exception No. 1:*** Short lengths of metal raceways used for the support or protection of cables are not required to be electrically continuous, nor are they required to be bonded. See 250.86 Ex. 2, and 300.12 Ex. Figure 300-24

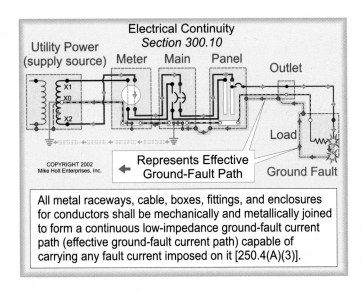

Electrical Continuity
*Section 300.10*

Utility Power (supply source)    Meter    Main    Panel    Outlet

X1
X0
X2

Load

COPYRIGHT 2002
Mike Holt Enterprises, Inc.

← Represents Effective Ground-Fault Path

Ground Fault

All metal raceways, cable, boxes, fittings, and enclosures for conductors shall be mechanically and metallically joined to form a continuous low-impedance ground-fault current path (effective ground-fault current path) capable of carrying any fault current imposed on it [250.4(A)(3)].

**Figure 300-23**

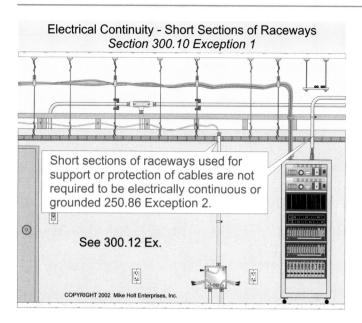

**Electrical Continuity - Short Sections of Raceways**
*Section 300.10 Exception 1*

Short sections of raceways used for support or protection of cables are not required to be electrically continuous or grounded 250.86 Exception 2.

See 300.12 Ex.

COPYRIGHT 2002 Mike Holt Enterprises, Inc.

**Figure 300-24**

## 300.11 Securing and Supporting

**(A) Secured in Place.** Raceways, cable assemblies, boxes, cabinets and fittings shall be securely fastened in place. The ceiling support wires or celing grid shall not be used to support raceways and cables.

**(1) Fire-Rated Assembly.** Electrical wiring within the cavity of a fire-rated floor-ceiling or roof-ceiling assembly can be supported to independent support wires that are secured at both ends. The independent support wires shall be distinguishable from the suspended ceiling support wires by color, tagging, or other effective means. Figure 300-25

**(2) Nonfire-Rated Assembly.** Electrical wiring located within the cavity of a non–fire-rated floor-ceiling or roof-ceiling assembly can be supported to independent support wires that are secured at both ends. Support wires within nonfire-rated assemblies are not required to be distinguishable from the suspended ceiling framing support wires.

**AUTHOR'S COMMENT:** Outlet boxes [314.23(D)] and luminaires can be supported to the suspended ceiling if securely fastened to the ceiling framing member [410.16(C)].

**(B) Raceways Used for Support.** Raceways shall not be used for the support of other raceways, cables, or other equipment, except as permitted by the following: Figure 300-26

**AUTHOR'S COMMENT:** This rule does not apply to signal and communications cables, but similar requirements are contained as follows:

- CATV, 820.52(D)
- Communications, 800.52(E)
- Fire Alarm, 760.57
- Network Broadband, 830.58(D)

**(2) Class 2 and 3 Circuits.** Class 2 and 3 cables can be supported to the raceway that supplies power to the equipment controlled by the circuit.

**AUTHOR'S COMMENT:** Since Class 2 and Class 3 cables cannot be installed in the same raceway with the power conductors [725.55(A)], the next best thing is to attach them to the raceway. See 725.58. Figure 300-27

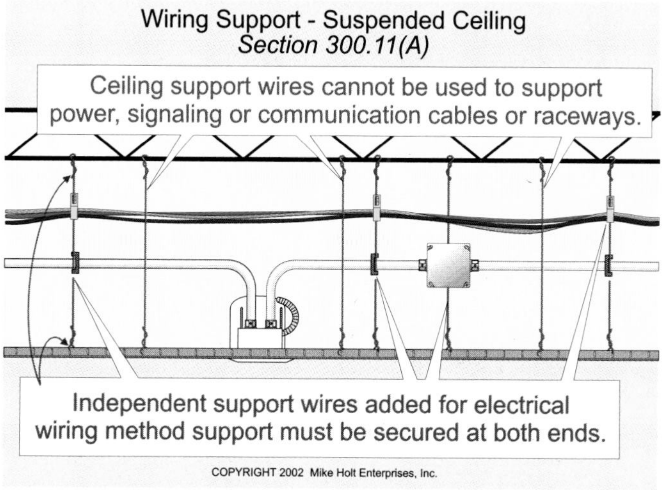

**Wiring Support - Suspended Ceiling**
*Section 300.11(A)*

Ceiling support wires cannot be used to support power, signaling or communication cables or raceways.

Independent support wires added for electrical wiring method support must be secured at both ends.

COPYRIGHT 2002 Mike Holt Enterprises, Inc.

**Figure 300-25**

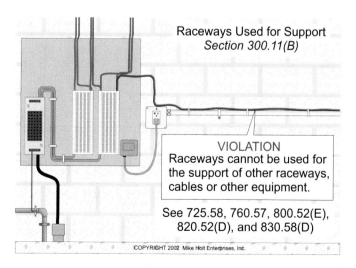

**Raceways Used for Support**
*Section 300.11(B)*

VIOLATION
Raceways cannot be used for the support of other raceways, cables or other equipment.

See 725.58, 760.57, 800.52(E), 820.52(D), and 830.58(D)

COPYRIGHT 2002 Mike Holt Enterprises, Inc.

**Figure 300-26**

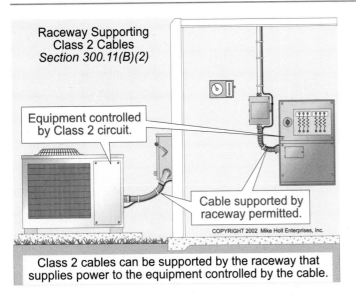

Raceway Supporting
Class 2 Cables
*Section 300.11(B)(2)*

Equipment controlled
by Class 2 circuit.

Cable supported by
raceway permitted.

COPYRIGHT 2002 Mike Holt Enterprises, Inc.

Class 2 cables can be supported by the raceway that
supplies power to the equipment controlled by the cable.

**Figure 300-27**

   (3) **Boxes Supported by Conduits.** Raceways can be used as a means of support for threaded boxes and conduit bodies [314.23(E) and (F)] and luminaires [410.16(F)].

(C) **Cables Not Used as Means of Support.** AC, NM, or MC cable shall not be used to support other cables, raceways, or nonelectrical equipment.

## 300.12 Mechanical Continuity

   Raceways and cable sheaths shall be mechanically continuous between terminations. See 300.10 and 314.17(B) and (C). Figure 300-28

   *Exception:* Short sections of raceways used to provide support or protection of cable from physical damage are not required to be mechanically continuous. See 250.86 Ex. 2, and 300.10 Ex. 1. Figure 300-29

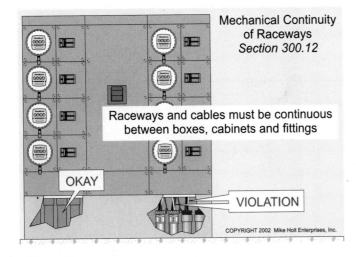

Mechanical Continuity
of Raceways
*Section 300.12*

Raceways and cables must be continuous
between boxes, cabinets and fittings

OKAY

VIOLATION

COPYRIGHT 2002 Mike Holt Enterprises, Inc.

**Figure 300-28**

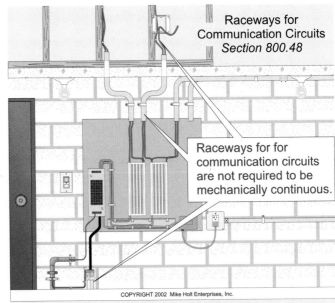

Raceways for
Communication Circuits
*Section 800.48*

Raceways for for
communication circuits
are not required to be
mechanically continuous.

COPYRIGHT 2002 Mike Holt Enterprises, Inc.

Chapter 3 requirements do not apply unless
specifically referenced in Article 800 [800.52(A)(2)].
Where raceways are used for communication circuits,
mechanical continuity [300.12] is not required.

**Figure 300-29**

## 300.13 Splices and Pigtails

(A) **Splices.** Conductors in raceways shall be continuous between all points of the system. Splices are not permitted in the raceway, except as permitted by 376.56, 378.56, 384.56, 386.56, or 388.56. See 300.15. Figure 300-30

(B) **Pigtail Neutrals.** In multiwire branch circuits, the removal of a wiring device, such as a receptacle, shall not cause an interruption of continuity for the grounded (neutral) conductor. Therefore, the grounded (neutral) conductors shall be spliced together, and a pigtail shall be provided for device terminations. Figure 300-31

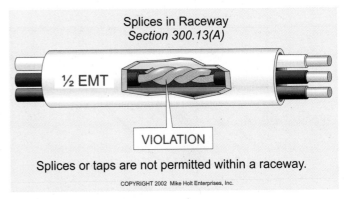

Splices in Raceway
*Section 300.13(A)*

½ EMT

VIOLATION

Splices or taps are not permitted within a raceway.

COPYRIGHT 2002 Mike Holt Enterprises, Inc.

**Figure 300-30**

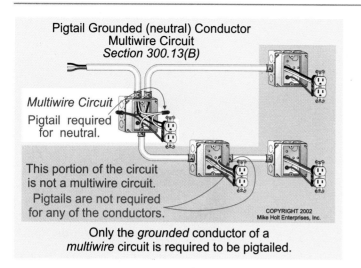

Pigtail Grounded (neutral) Conductor
Multiwire Circuit
*Section 300.13(B)*

*Multiwire Circuit*
Pigtail required for neutral.

This portion of the circuit is not a multiwire circuit.
Pigtails are not required for any of the conductors.

COPYRIGHT 2002
Mike Holt Enterprises, Inc.

Only the *grounded* conductor of a *multiwire* circuit is required to be pigtailed.

**Figure 300-31**

**AUTHOR'S COMMENT:** The opening of the ungrounded or grounded (neutral) conductor of a 2-wire circuit during the replacement of a device does not cause a safety hazard, so pigtailing of these conductors is not required [110.14(B)].

*CAUTION: If the continuity of the grounded (neutral) conductor of a multiwire circuit is interrupted (open), there could be a fire and/or destruction to electrical equipment resulting from over or undervoltage.*

*Example: A 3-wire, 120/240V circuit supplies a 1,200W, 120V hair dryer and a 600W, 120V television. If the grounded (neutral) conductor is interrupted, it will cause the 120V television to operate at 160V and consume 1,067W of power (instead of 600W) for only a few seconds before it burns up. Figure 300-32*

Step 1. *Determine the resistance of each appliance,*
$R = E^2/P$.
*Hair Dryer = 120V$^2$/1,200W, R = 12 ohms*
*Television = 120V$^2$/600W, R = 24 ohms*

Step 2. *Determine the current of the circuit, I = E/R.*
*I = 240V/36 ohms (12 ohms + 24 ohms),*
*= 6.7A*

Step 3. *Determine the operating voltage for each appliance, E = I × R.*
*Hair Dryer = 6.7A × 12 ohms, E = 80V*
*Television = 6.7A × 24 ohms, E = 160V*

*WARNING: Failure to terminate the ungrounded (hot) conductors to separate phases could cause the grounded (neutral) conductor to become overloaded from excessive neutral current, and the insulation could be damaged or destroyed. Conductor over-heating is known to decrease insulating material*

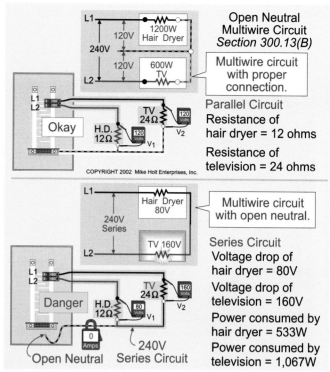

Open Neutral
Multiwire Circuit
*Section 300.13(B)*

Multiwire circuit with proper connection.

**Parallel Circuit**
Resistance of hair dryer = 12 ohms
Resistance of television = 24 ohms

Multiwire circuit with open neutral.

**Series Circuit**
Voltage drop of hair dryer = 80V
Voltage drop of television = 160V
Power consumed by hair dryer = 533W
Power consumed by television = 1,067W

COPYRIGHT 2002 Mike Holt Enterprises, Inc.

**Figure 300-32**

*service life, potentially resulting in a fire from arcing faults in hidden locations. We do not know just how long conductor insulation will last, but heat does decrease its life span. Figure 300-33*

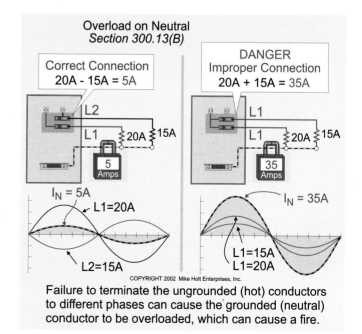

Overload on Neutral
*Section 300.13(B)*

Correct Connection
20A - 15A = 5A

DANGER
Improper Connection
20A + 15A = 35A

$I_N$ = 5A
L1=20A
L2=15A

$I_N$ = 35A
L1=15A
L1=20A

COPYRIGHT 2002 Mike Holt Enterprises, Inc.

Failure to terminate the ungrounded (hot) conductors to different phases can cause the grounded (neutral) conductor to be overloaded, which can cause a fire.

**Figure 300-33**

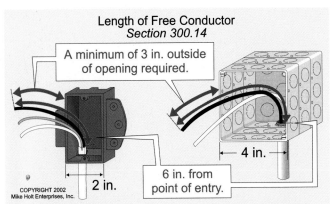

Figure 300-34

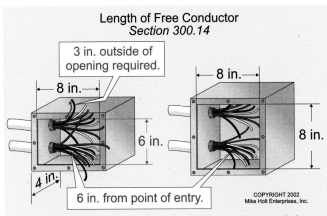

Figure 300-35

## 300.14 Length of Free Conductors

At least 6 in. of free conductor, measured from the point in the box where the conductors enter the enclosure, shall be left at each outlet, junction, and switch point for splices or terminations of luminaires or devices.

Boxes that have openings less than 8 in. shall have at least 6 in. of free conductor, measured from the point where the conductors enter the box, and not less than 3 in. of free conductor outside the box opening. Figure 300-34

> **AUTHOR'S COMMENT:** There's no NEC limit on the number of extension rings permitted on an outlet box, however there shall be at least 3 in. of free conductors outside the opening of the finel extension ring, but check with the AHJ.

**Example.** A 6 × 8 × 4 in. box would require approximately 7 in. of free conductor (4 in. within the box and 3 in. outside the box). Figure 300-35

Boxes with openings of 8 in. and larger shall have at least 6 in. of free conductor, measured from the point where the conductors enter the box.

**Example.** An 8 × 8 × 4 in. box requires 6 in. of free conductor, measured from the point where the conductors enter the box. The 3 in. outside the box rule does not apply to enclosures that have openings of 8 in. and lerger.

*Exception:* This rule does not apply to conductors that pass through a box without splice or termination.

## 300.15 Boxes or Conduit Bodies

A box shall be installed at each splice or termination point, except as permitted for: Figure 300-36

- Cabinet or Cutout Boxes, 312.8
- Conduit Bodies, 314.16(C), Figure 300-37
- Luminaires, 410.31
- Surface Raceways, 352.7
- Wireways, 376.56

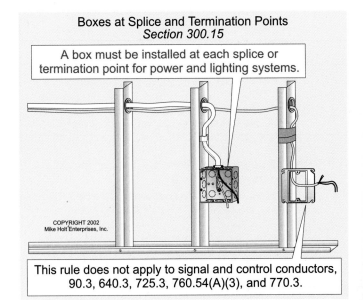

Figure 300-36

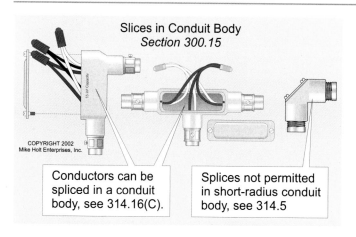

**Slices in Conduit Body**
*Section 300.15*

Conductors can be spliced in a conduit body, see 314.16(C).

Splices not permitted in short-radius conduit body, see 314.5

**Figure 300-37**

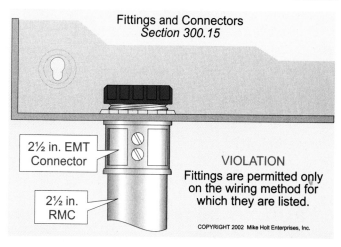

**Fittings and Connectors**
*Section 300.15*

2½ in. EMT Connector

2½ in. RMC

**VIOLATION**
Fittings are permitted only on the wiring method for which they are listed.

COPYRIGHT 2002 Mike Holt Enterprises, Inc.

**Figure 300-39**

**AUTHOR'S COMMENT:** Boxes are not required for signal and communications cables Figure 300-38

- CATV, 90.3
- Class 2 and 3 Control and Signaling, 725.3
- Communications, 90.3
- Fiber Optical Cable, 770.3
- Sound Systems, 640.3(A)

**Fittings and Connectors.** Fittings shall only be used with the specific wiring methods for which they are designed and listed. For example, NM cable connectors shall not be used with AC cable and EMT fittings shall not be used with RMC or IMC, unless listed for the purpose. Figure 300-39

**AUTHOR'S COMMENT:** Rigid nonmetallic conduit couplings and connectors (RNC) can be used with electrical nonmetallic tubing (ENT) if the proper glue is used in accordance with manufacturer's instructions [110.3(B)]. See 362.48.

**(C) Raceways for Support or Protection.** When a raceway is used for the support or protection of cables, a fitting to reduce the potential for abrasion shall be placed at the location the cables enter the raceway. See 250.86 Ex. 2, 300.5(D), 300.10 Ex. 1, and 300.12 Ex. for more details.

**(G) Underground Splices.** A box or conduit body is not required where a splice is made underground, if the conductors are spliced with a splicing device that is listed for direct burial. See 110.14(B) and 300.5(E).

**(I) Enclosures.** A box or conduit body is not required where a splice is made in a cabinet or in cutout boxes containing switches or overcurrent protection devices, if the splices or taps do not fill the wiring space at any cross section to more than 75 percent, and the wiring at any cross section does not exceed 40 percent. See 312.8 and 404.3. Figure 300-40

### 300.17 Raceway Sizing

Raceways shall be large enough to permit the installation and removal of conductors without damaging the conductor's insulation.

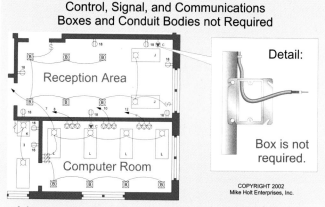

**Control, Signal, and Communications Boxes and Conduit Bodies not Required**

Reception Area

Computer Room

Detail:

Box is not required.

COPYRIGHT 2002
Mike Holt Enterprises, Inc.

A box or conduit body is not required at each conductor splice, connection point, junction point or pull point for most control signal and communications circuits.

**Figure 300-38**

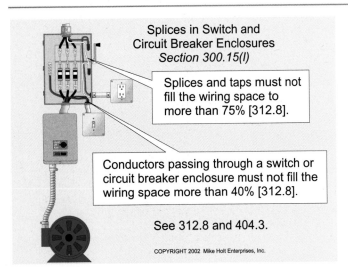

Figure 300-40

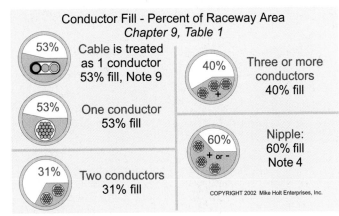

Figure 300-42

**Same Size Conductors.** When all conductors in a raceway are the same size and insulation, the number of conductors permitted can be determined by locating the answer in Annex C.

**Question:** How many 12 AWG THHN conductors can be installed in ¾ in. Electrical Metallic Tubing? Figure 300-41

(a) 12                    (b) 13

(c) 14                    (d) 16

**Answer:** (d) 16 conductors, Annex C, Table C1

**Different Size Conductors.** When the conductors in a raceway are of different sizes, the raceway shall limit conductor fill to the following percentages in accordance with Table 1 of Chapter 9. These percentages are based on conditions where the length of the conductor and number of raceway bends are within reasonable limits. Figure 300-42

### Table 1, Chapter 9

| Number | Percent Fill |
| --- | --- |
| 1 Conductor | 53% |
| 2 Conductors | 31% |
| 3 or more | 40% |

**AUTHOR'S COMMENT:** Where a raceway has a maximum length of 24 in., it can be conductor filled to 60 percent of its total cross-sectional area [Chapter 9, Table 1, Note 4]. Figure 300-43

The first step in raceway sizing is determining the total area of conductors (Chapter 9, Table 5 for insulated conductors and Chapter 9, Table 8 for bare stranded or solid conductors). Figure 300-44

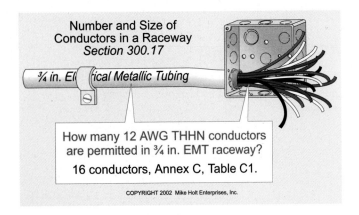

Figure 300-41

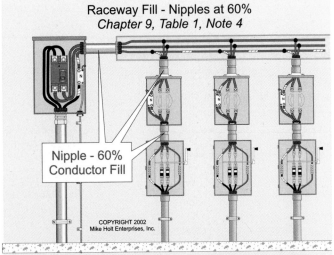

A raceway that is 24 in. or less in length can be filled with conductors up to 60% of its cross-sectional area.

Figure 300-43

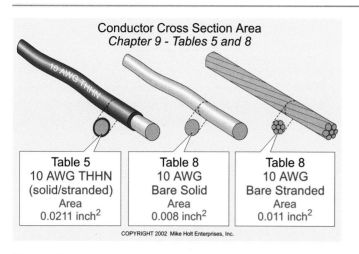

## Conductor Cross Section Area
### Chapter 9 - Tables 5 and 8

| Table 5<br>10 AWG THHN<br>(solid/stranded)<br>Area<br>0.0211 inch$^2$ | Table 8<br>10 AWG<br>Bare Solid<br>Area<br>0.008 inch$^2$ | Table 8<br>10 AWG<br>Bare Stranded<br>Area<br>0.011 inch$^2$ |

COPYRIGHT 2002 Mike Holt Enterprises, Inc.

**Figure 300-44**

The second step is to select the raceway from Chapter 9, Table 4, in accordance with the percent fill listed in Chapter 9, Table 1. Figure 300-45

**Question:** What size rigid nonmetallic Schedule 40 conduit is required for the following THHN conductors: 3 – 500 kcmils, 1 – 250 kcmil, and 1 – 3 AWG? Figure 300-46

(a) 2 in.          (b) 3 in.

(c) 4 in.          (d) none of these

**Answer:** (b) 3 in.

Step 1. Determine the total area of conductors [Chapter 9, Table 5]

| 500 kcmil THHN | $0.7073 \times 3 =$ | 2.1219 in$^2$ |
| 250 kcmil THHN | $0.3970 \times 1 =$ | 0.3970 in$^2$ |
| 3 AWG THHN | $0.0973 \times 1 =$ | 0.0973 in$^2$ |
| Total Area | | 2.6162 in$^2$ |

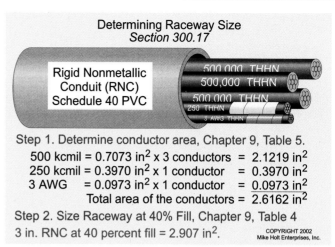

## Determining Raceway Size
### Section 300.17

Rigid Nonmetallic Conduit (RNC) Schedule 40 PVC

500,000 THHN
500,000 THHN
500,000 THHN
250 THHN
3 AWG THHN

Step 1. Determine conductor area, Chapter 9, Table 5.

| 500 kcmil = 0.7073 in$^2$ x 3 conductors | = 2.1219 in$^2$ |
| 250 kcmil = 0.3970 in$^2$ x 1 conductor | = 0.3970 in$^2$ |
| 3 AWG    = 0.0973 in$^2$ x 1 conductor | = 0.0973 in$^2$ |
| Total area of the conductors | = 2.6162 in$^2$ |

Step 2. Size Raceway at 40% Fill, Chapter 9, Table 4
3 in. RNC at 40 percent fill = 2.907 in$^2$.

COPYRIGHT 2002 Mike Holt Enterprises, Inc.

**Figure 300-46**

Step 2. Select raceway at 40 percent fill [Chapter 9, Table 4].

3 in. PVC at 40 percent fill of 2.907 in$^2$.

## 300.18 Inserting Conductors in Raceways

**(A) Complete Runs.** To protect conductor insulation from abrasion during installation, raceways shall be mechanically completed between the pulling points before conductors are installed. See 300.10 and 300.12. Figure 300-47

## 300.19 Supporting Conductors in Vertical Raceways

**(A) Spacing Intervals.** If the vertical rise of a raceway exceeds the values of Table 300.19(A), the conductors shall be supported at the top or as close to the top as practical. Figure 300-48

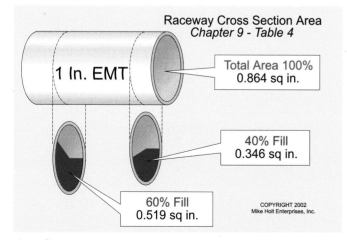

## Raceway Cross Section Area
### Chapter 9 - Table 4

1 In. EMT

Total Area 100%
0.864 sq in.

40% Fill
0.346 sq in.

60% Fill
0.519 sq in.

COPYRIGHT 2002 Mike Holt Enterprises, Inc.

**Figure 300-45**

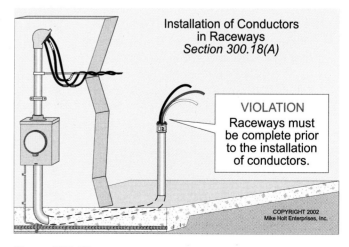

## Installation of Conductors in Raceways
### Section 300.18(A)

VIOLATION
Raceways must be complete prior to the installation of conductors.

COPYRIGHT 2002 Mike Holt Enterprises, Inc.

**Figure 300-47**

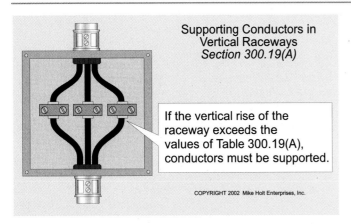

**Figure 300-48**

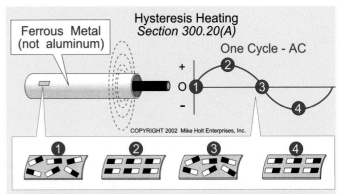

Steel and iron (ferrous) molecules align to the polarity of the magnetic field and when the field reverses, the molecules reverse their polarity. This back-and-forth alignment of the molecules heats up ferrous metals parts.

**Figure 300-49**

**AUTHOR'S COMMENT:** The weight of long vertical runs of conductors can cause the conductors to actually drop out of the raceway if they are not properly secured. There have been many cases where conductors in a vertical raceway were released from the pulling basket (at the top) without having them secured, and the conductors fell down and out of the raceway, injuring those feeding the conductors at the base.

## 300.20 Induced Currents in Metal Parts

**(A) Conductors Grouped Together.** To minimize induction heating of ferrous (iron conduit) metallic raceways and enclosures and to maintain a low-impedance ground-fault path, all conductors of a circuit shall be installed in the same raceway, cable, trench, cord, or cable tray. See 250.102(E), 300.3(B), 300.5(I), and 392.8(D).

**AUTHOR'S COMMENT:** When alternating current (ac) flows through a conductor, a pulsating or varying magnetic field is created around the conductor. This magnetic field is constantly expanding and contracting with the frequency of the ac. In the United States, the frequency is 60 cycles per second. Since it's the nature of ac to reverse polarity (120 times per second), the magnetic field surrounding the conductor reverses its direction 120 times per second. This expanding and collapsing magnetic field induces eddy currents in the ferrous metal parts surrounding the conductors, causing the metal parts to heat up from hysteresis.

Hysteresis heating affects only ferrous metals that have magnetic properties, such as steel and iron, but not aluminum. Simply put, the molecules of steel and iron align to the polarity of the magnetic field and when the magnetic field reverses, the molecules reverse their polarity as well. This back-and-forth alignment of the molecules

in ferrous metals heats up the metal. The greater the current flow, the greater the heat rise. Figure 300-49

When conductors are grouped together, the magnetic fields of the different conductors tend to cancel each other out, resulting in a reduced magnetic field around the conductors. The lower magnetic field reduces induced currents, which reduces hysteresis heating of the surrounding metal enclosure.

*WARNING: There has been much discussion in the press on the effects of electromagnetic fields on humans. According to the Institute of Electrical and Electronic Engineers (IEEE), there is insufficient information at this time to define an unsafe electromagnetic field level.*

**(B) Single Conductors.** When separate conductors are installed in a nonmetallic raceway as permitted in 300.5(I) Ex. 2, the inductive heating of the metal enclosure can be minimized by the use of aluminum locknuts and by cutting a slot between the individual holes through which the conductors pass. Figure 300-50

FPN: Because aluminum is a nonmagnetic metal, there will be no heating due to hysteresis.

**AUTHOR'S COMMENT:** Aluminum conduit, locknuts, and enclosures will carry eddy currents, but because aluminum is nonferrous, it does not heat up [300.20(B) FPN].

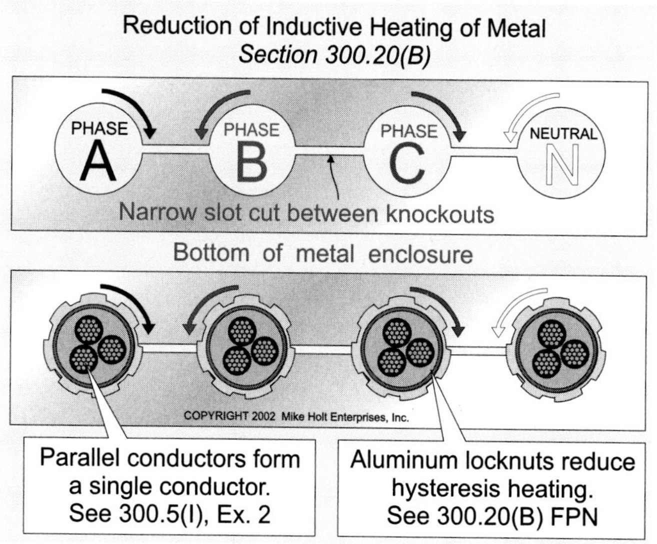

## Reduction of Inductive Heating of Metal
### Section 300.20(B)

PHASE A → PHASE B → PHASE C → NEUTRAL N

Narrow slot cut between knockouts

Bottom of metal enclosure

COPYRIGHT 2002 Mike Holt Enterprises, Inc.

Parallel conductors form a single conductor. See 300.5(I), Ex. 2

Aluminum locknuts reduce hysteresis heating. See 300.20(B) FPN

**Figure 300-50**

## 300.21 Spread of Fire or Products of Combustion

Electrical circuits and equipment shall be installed in a way that the possible spread of fire or products of combustion will not be substantially increased. This means that openings in fire-rated walls, floors, and ceilings for electrical equipment shall be sealed with an approved fire-stop. The fire-stop material shall maintain the fire resistance of the fire-rated assembly in accordance with the manufacturer's instructions.

> **AUTHOR'S COMMENT:** Fire-stop material is listed for the specific types of wiring methods and construction structures.

> FPN: Directories of electrical construction materials published by qualified testing laboratories contain listing installation restrictions necessary to maintain the fire-resistive rating of assemblies. Outlet boxes shall have a horizontal separation of not less than 24 in. when installed in a fire-rated assembly, unless an outlet box is listed for closer spacing or protected by fire-resistant "Putty Pads" in accordance with manufacturer's instructions.

> **AUTHOR'S COMMENT:** This rule applies to control, signal and communications cables. Figure 300-51
> - CATV, 820.3(B)
> - Communications, 800.52(B)
> - Control and Signaling, 725.3(A)
> - Fiber Optical Cable, 770.3(A)
> - Fire Alarm, 760.3(A)
> - Network Broadband, 830.3(A) and 830.58(B)
> - Sound Systems, 640.3(A)

## Spread of Fire or Products of Combustion
### Section 300.21

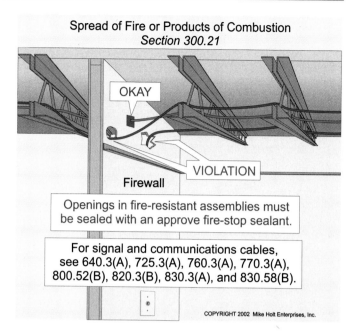

OKAY

VIOLATION

Firewall

Openings in fire-resistant assemblies must be sealed with an approve fire-stop sealant.

For signal and communications cables, see 640.3(A), 725.3(A), 760.3(A), 770.3(A), 800.52(B), 820.3(B), 830.3(A), and 830.58(B).

COPYRIGHT 2002 Mike Holt Enterprises, Inc.

**Figure 300-51**

## 300.22 Ducts, Plenums, and Air-Handling Spaces

**(A)  Ducts Used for Dust, Loose Stock, or Vapor.** Manufactured ducts that transport dust, loose stock, or flammable vapors, shall not have any wiring method installed within them. Figure 300-52

**(C)  Space Used for Environmental Air.** This section applies to space used for environmental air-handling purposes other than ducts and plenums as specified in 300.22(A) and (B). It does not apply to habitable rooms or areas of buildings, the prime purpose of which is not air handling.

## Ducts Used for Dust and Loose Stock
### Section 300.22(A)

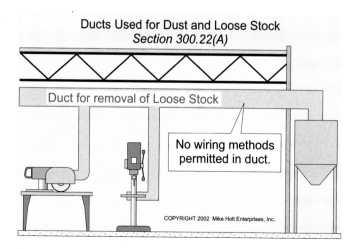

Duct for removal of Loose Stock

No wiring methods permitted in duct.

COPYRIGHT 2002 Mike Holt Enterprises, Inc.

**Figure 300-52**

**AUTHOR'S COMMENT:** The space above a suspended ceiling, or below a raised floor that is not part of an information technology equipment room, is an example of the type of space to which this section applies.

**(1) Wiring Methods Permitted.** EMT, RMC, IMC, AC, MC cable without a nonmetallic cover, and FMC can be installed in an environmental air space.

**AUTHOR'S COMMENT:** Nonmetallic raceways and cables are not permitted in an environmental air space because they give off deadly toxic fumes when burned or super heated. However, control, signal and communications cables are permitted if they are plenum rated or they shall be installed in a raceway that is suitable for this space. Figure 300-53

- CATV, 820.51(A)
- Communications, 800.53(A)
- Control and Signaling, 725.61(A)
- Fire Alarm, 760.61(A)
- Network Broadband, 830.54(B) and 830.55(B)
- Optical Fiber Cables, 770.53(A)
- Sound System Cables, 640.9(C) and 725.61(A)

**AUTHOR'S COMMENT:** A space not used for environmental air-handling purposes has no restrictions on wiring methods, and nonplenum cables can be used. Figure 300-54

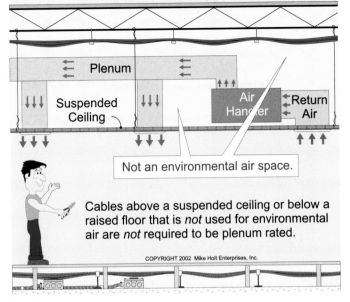

Not an environmental air space.

Cables above a suspended ceiling or below a raised floor that is *not* used for environmental air are *not* required to be plenum rated.

COPYRIGHT 2002 Mike Holt Enterprises, Inc.

**Figure 300-54**

**(D) Information Technology Equipment Room.** Wiring within a raised floor in an information technology room shall be in accordance with the requirements of 645.5(D). Figure 300-55

**AUTHOR'S COMMENT:** Control, signal and communications cables are not required to be plenum rated when installed beneath a raised floor for information technology equipment. See 645.5(D)(5)(c).

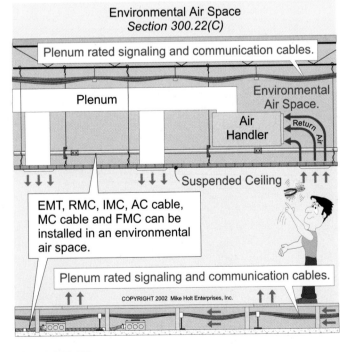

**Figure 300-53**

Signal and communications cables under a raised floor are not required to be plenum rated [645.5(D)(5)(c)], because ventilation is restricted to that room/space [645.5(D)(3)].

**Figure 300-55**

## 300.23 Panels Designed to Allow Access.

Wiring, cables and equipment installed behind panels shall be located so the panels can be removed to give access to electrical equipment. Figure 300-56

**AUTHOR'S COMMENT:** Access to equipment shall not be prohibited by an accumulation of cables that prevent the removal of suspended-ceiling panels. Control, signal, and communications cables shall be located so that the suspended-ceiling panels can be moved to provide access to electrical equipment. See:

- CATV, 820.5
- Communications, 800.5
- Class 2 and 3, 725.5
- Fiber Optical Cable, 770.7
- Fire Alarm, 760.5
- Network Broadband, 830.6
- Sound Systems, 640.5

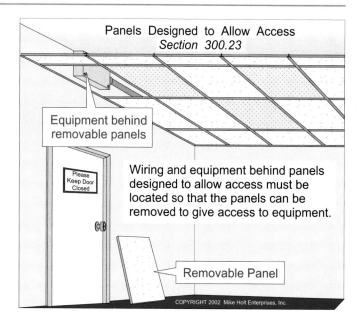

Panels Designed to Allow Access
Section 300.23

Equipment behind removable panels

Please Keep Door Closed

Wiring and equipment behind panels designed to allow access must be located so that the panels can be removed to give access to equipment.

Removable Panel

COPYRIGHT 2002 Mike Holt Enterprises, Inc.

**Figure 300-56**

# Article 300

1.  All conductors of a circuit, including the grounded and equipment grounding conductors, must be contained within the same _____.
    (a) raceway          (b) cable          (c) trench          (d) all of these

2.  Where NM cables pass through cut or drilled slots or holes in metal members, the cable needs to be protected by _____ securely covering all metal edges fastened in the opening prior to installation of the cable.
    (a) listed bushings     (b) listed grommets     (c) plates     (d) a or b

3.  Where nails or screws are likely to penetrate nonmetallic-sheathed cable or electrical nonmetallic tubing installed through metal framing members, a steel sleeve, steel plate, or steel clip not less than _____ in thickness shall be used to protect the cable or tubing.
    (a) 1/16 in.          (b) 1/8 in.          (c) 1/2 in.          (d) none of these

4.  What is the minimum cover requirement in inches for UF cable installed outdoors and underground that is supplying power to a 120V, 30A circuit?
    (a) 6 in.          (b) 12 in.          (c) 18 in.          (d) 24 in.

5.  Rigid metal conduit that is directly buried outdoors must have at least _____ of cover.
    (a) 6 in.          (b) 12 in.          (c) 18 in.          (d) 24 in.

6.  When installing raceways underground in rigid nonmetallic conduit and other approved raceways, there must be a minimum of _____ of cover.
    (a) 6 in.          (b) 12 in.          (c) 18 in.          (d) 22 in.

7.  What is the minimum cover requirement in inches for UF cable supplying power to a 120V, 15A GFCI-protected circuit outdoors under a driveway of a one-family dwelling?
    (a) 12 in.          (b) 24 in.          (c) 16 in.          (d) 6 in.

8.  UF cable used with a 24V landscape lighting system is permitted to have a minimum cover of _____
    (a) 6 in.          (b) 12 in.          (c) 18 in.          (d) 24 in.

9.  Direct-buried conductors or cables shall be permitted to be spliced or tapped without the use of splice boxes.
    (a) True          (b) False

10. Where corrosion protection is necessary and the conduit is threaded in the field, the threads must be coated with a(n) _____ electrically conductive, corrosion-resistance compound.
    (a) marked          (b) listed          (c) labeled          (d) approved

11. The provisions required for mounting conduits on indoor walls or in rooms that must be hosed down frequently is _____ between the mounting surface and the electrical equipment.
    (a) a permanent 1/4 in. airspace          (b) separated by insulated bushings
    (c) separated by noncombustible tubing     (d) none of these

12. Where portions of a cable raceway or sleeve are known to be subjected to different temperatures and where condensation is known to be a problem, as in cold storage areas of buildings or where passing from the interior to the exterior of a building, the _____ shall be filled with an approved material to prevent the circulation of warm air to a colder section of the raceway or sleeve.
    (a) raceways          (b) sleeve          (c) a or b          (d) none of these

13. Raceways shall be provided with expansion fittings where necessary to compensate for thermal expansion and contraction.
    (a) True          (b) False

14. All metal raceways, cable armor, boxes, fittings, cabinets, and other metal enclosures for conductors must be _____ joined together to form a continuous electrical conductor.
    (a) electrically          (b) permanently          (c) metallically          (d) none of these

15. In multiwire circuits, the continuity of the _____ conductor shall not be dependent upon the device connections.
    (a) ungrounded          (b) grounded          (c) grounding          (d) a and b

16. When the opening to an outlet, junction, or switch point is less than 8 inches in any dimension, each conductor shall be long enough to extend at least _____ outside the opening of the enclosure.
    (a) 0 in.          (b) 3 in.          (c) 6 in.          (d) 12 in.

17. Fittings and connectors shall be used only with the specific wiring methods for which they are designed and listed.
    (a) True          (b) False

18. The number of conductors permitted in a raceway shall be limited to _____.
    (a) permit heat to dissipate
    (b) prevent damage to insulation during installation
    (c) prevent damage to insulation during removal of conductors
    (d) all of these

19. Raceways shall be _____ between pulling points prior to the installation of conductors.
    (a) mechanically completed          (b) tested for ground faults
    (c) a minimum of 80 percent completed          (d) none of these

20. Conductors in metal raceways and enclosures shall be so arranged as to avoid heating the surrounding metal by alternating-current induction. To accomplish this, the _____ conductor(s) shall be grouped together.
    (a) phase          (b) neutral          (c) ungrounded          (d) all of these

21. _____ is a nonferrous, nonmagnetic metal that has no heating due to inductive hysteresis heating.
    (a) Steel          (b) Iron          (c) Aluminum          (d) all of these

22. No wiring of any type shall be installed in ducts used to transport _____.
    (a) dust          (b) flammable vapors          (c) loose stock          (d) all of these

23. Wiring methods permitted in the drop ceiling area used for environmental air include _____.
    (a) electrical metallic tubing
    (b) flexible metal conduit of any length
    (c) rigid metal conduit without an overall nonmetallic covering
    (d) all of these

24. The air-handling area beneath raised floors for data-processing systems is not a plenum and is not required to comply with the requirements of 300.22 but shall be permitted in accordance with Article 645.
    (a) True          (b) False

25. Wiring methods and equipment installed behind panels designed to permit access (such as drop ceilings) shall be so arranged and secured to permit the removal of panels to give access to electrical equipment.
    (a) True          (b) False

# Article 310
# Conductors For General Wiring

## 310.1 Scope

Article 310 contains the general requirements for conductors, such as insulation markings, ampacity ratings, and their use. This article does not apply to conductors that are an integral part of cable assemblies, cords, or equipment. See 90.6 and 300.1(B).

## 310.2 Conductors

**(A) Insulation.** All conductors shall be insulated and they shall be installed as part of a recognized wiring method listed in Chapter 3. See 110.8 and 300.3(A).

*Exception:* Where permitted to be bare for equipment grounding or bonding in accordance with Article 250. See 250.64 and 250.119.

**AUTHOR'S COMMENT:** Equipment grounding conductors shall be insulated for patient-care receptacles, switches and equipment [517.13(B)], and for wet-niche pool lights and pool equipment [680.23(F)(2) and 680.25(B)(2)]. Figure 310-1

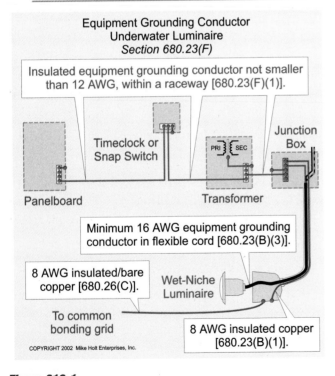

Equipment Grounding Conductor
Underwater Luminaire
*Section 680.23(F)*

Insulated equipment grounding conductor not smaller than 12 AWG, within a raceway [680.23(F)(1)].

Timeclock or Snap Switch

PRI    SEC

Junction Box

Panelboard

Transformer

Minimum 16 AWG equipment grounding conductor in flexible cord [680.23(B)(3)].

8 AWG insulated/bare copper [680.26(C)].

Wet-Niche Luminaire

To common bonding grid

8 AWG insulated copper [680.23(B)(1)].

COPYRIGHT 2002 Mike Holt Enterprises, Inc.

**Figure 310-1**

Stranded Conductors
*Section 310.3*

Listed Liquidtight

Stranded Wire

8 AWG THHN

Conductor sizes 8 AWG and larger must be stranded when installed in a raceway.

COPYRIGHT 2002 Mike Holt Enterprises, Inc.

**Figure 310-2**

## 310.3 Stranded Conductors

Conductors 8 AWG and larger shall be stranded when installed in a raceway. Figure 310-2

**AUTHOR'S COMMENT:** Bare solid conductors are often used for the grounding electrode conductor [250.62] and for pool bonding [680.26(C)].

## 310.4 Conductors in Parallel

Ungrounded and grounded (neutral) conductors sized 1/0 AWG and larger are permitted to be connected in parallel; that is, electrically joined together at both ends, forming a single electrical conductor.

When conductors are run in parallel, the current shall be evenly distributed between the individual parallel conductors. This is accomplished by ensuring that each of the conductors within a parallel set is identical to each other in:

(1) length,

(2) material,

(3) size,

(4) insulation material, and

(5) termination method.

In addition, raceways or cables containing parallel conductors shall have the same physical characteristics.

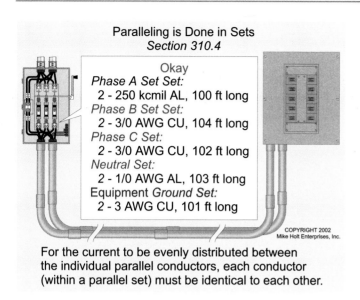

Paralleling is Done in Sets
*Section 310.4*

**Okay**
*Phase A Set Set:*
2 - 250 kcmil AL, 100 ft long
*Phase B Set Set:*
2 - 3/0 AWG CU, 104 ft long
*Phase C Set:*
2 - 3/0 AWG CU, 102 ft long
*Neutral Set:*
2 - 1/0 AWG AL, 103 ft long
Equipment *Ground Set:*
2 - 3 AWG CU, 101 ft long

COPYRIGHT 2002
Mike Holt Enterprises, Inc.

For the current to be evenly distributed between the individual parallel conductors, each conductor (within a parallel set) must be identical to each other.

**Figure 310-3**

**AUTHOR'S COMMENT:** If one set of parallel conductors are run in a metallic raceway and the other conductors are run in a nonmetallic raceway, the conductors in the metallic raceway will have an increased opposition to current flow (impedance) as compared to the conductors in the nonmetallic raceway. This will result in an unbalanced distribution of the currents between the parallel conductors.

**Paralleling is done in sets.** Conductors of one phase are not required to have the same physical characteristics as those of another phase to achieve balance. For example, a 400A feeder that has a neutral load of 240A can be in parallel as follows: Figure 310-3

Phase A, 2 – 250 kcmil THHN aluminum, 100 ft

Phase B, 2 – 3/0 AWG THHN copper, 104 ft

Phase C, 2 – 3/0 AWG THHN copper, 102 ft

Neutral, 2 – 1/0 AWG THHN aluminum, 103 ft

Equipment Ground, 2 – 3 AWG copper, 101 ft

**Equipment Grounding Conductors.** The equipment grounding conductors for circuits in parallel shall be identical to each other in length, material, size, insulation, and termination. In addition, each raceway, where required, shall have an equipment grounding conductor sized in accordance with 250.122. The minimum 1/0 AWG rule of 310.4 does not apply to equipment grounding conductors. Figure 310-4

**Ampacity Adjustment.** When more than three current-carrying conductors are run in the same raceway longer than 24 in., the ampacity adjustment factors of Table 310.15(B)(2) shall be applied. See 310.10, and 310.15 for details and examples.

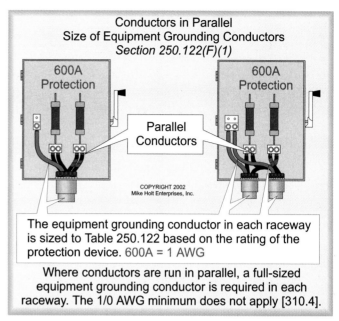

Conductors in Parallel
Size of Equipment Grounding Conductors
*Section 250.122(F)(1)*

600A Protection          600A Protection

Parallel Conductors

COPYRIGHT 2002
Mike Holt Enterprises, Inc.

The equipment grounding conductor in each raceway is sized to Table 250.122 based on the rating of the protection device. 600A = 1 AWG

Where conductors are run in parallel, a full-sized equipment grounding conductor is required in each raceway. The 1/0 AWG minimum does not apply [310.4].

**Figure 310-4**

### 310.5 Minimum Size Conductors

The smallest conductor permitted for branch circuits for residential, commercial, and industrial locations is 14 AWG copper [Table 310.5].

**AUTHOR'S COMMENT:** There is a misconception that 12 AWG copper is the smallest conductor that can be used for commercial or industrial facilities.

### 310.8 Locations

**(D) Locations Exposed to Direct Sunlight.** Insulated conductors and cables exposed to the direct rays of the sun shall be listed for sunlight resistance or listed and marked "sunlight resistant." Figure 310-5

### 310.9 Corrosive Conditions.

Conductor insulation shall be suitable for any substance that may have a detrimental effect on the conductor's insulation, such as oil, grease, vapor, gases, fumes, liquids, or other substances. See 110.11.

### 310.10 Insulation Temperature Limitation

Conductors shall not be used or installed in any way where the operating temperature exceeds that designated for the type of insulated conductor involved.

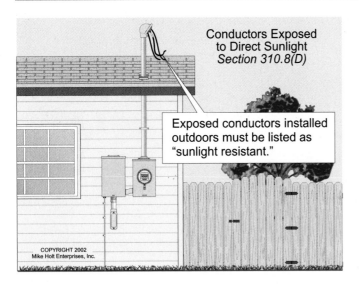

**Figure 310-5**

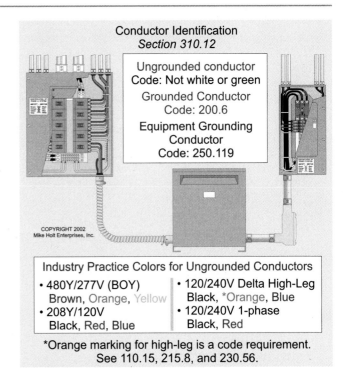

**Figure 310-6**

FPN: The temperature rating of a conductor (see Table 310.13) is the maximum temperature a conductor can withstand over a prolonged time without serious degradation.

## 310.12 Conductor Identification

(A) **Grounded Conductor.** The grounded (neutral) conductor shall be identified in accordance with 200.6.

(B) **Equipment Grounding Conductor.** The equipment grounding conductor shall be identified in accordance with 250.119.

(C) **Ungrounded Conductors.** Ungrounded conductors shall be clearly distinguishable from grounded and equipment grounding conductors.

**AUTHOR'S COMMENT:** The *NEC* does not require color coding of ungrounded conductors, except for the high-leg conductor when a grounded (neutral) conductor is present [110.15, 215.8 and 230.56]. However, electricians often use the following color system for power and lighting conductor identification: Figure 310-6

- 120/240, 1Ø; black, red, white
- 120/240, 3Ø; black, orange, blue, white
- 208Y/120; black, red, blue, white
- 480Y/277; brown, orange, yellow, gray

## 310.13 Conductor Construction

Table 310.13 contains information on insulation rating, such as operating temperature, applications, sizes, and outer cover.

**AUTHOR'S COMMENT:** The following explains the lettering on conductor insulation: Figure 310-7
Suitable for 90°C in wet locations

- F    Fixture wires (solid or 7 strand) [Table 402.3]
- FF   Flexible fixture wire (19 strands) [Table 402.3]
- No H   60°C insulation rating
- H    75°C insulation rating
- HH   90°C insulation rating
- N    Nylon outer cover
- T    Thermoplastic insulation
- W   Wet or damp locations
- X    Cross-linked polyethylene

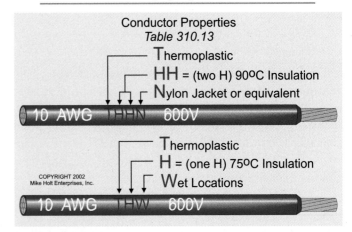

**Figure 310-7**

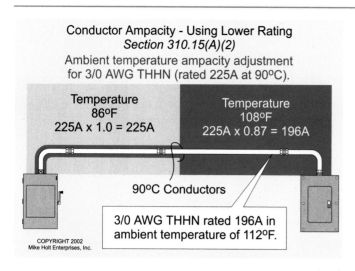

Conductor Ampacity - Using Lower Rating
Section 310.15(A)(2)
Ambient temperature ampacity adjustment
for 3/0 AWG THHN (rated 225A at 90°C).

Temperature
86°F
225A x 1.0 = 225A

Temperature
108°F
225A x 0.87 = 196A

90°C Conductors

3/0 AWG THHN rated 196A in
ambient temperature of 112°F.

COPYRIGHT 2002
Mike Holt Enterprises, Inc.

**Figure 310-8**

### 310.15 Conductor Ampacity

**(A) General Requirements**

**(1) Tables for Engineering Supervision.** The ampacity of a conductor can be determined by using the table in this article or under engineering supervision.

**(2) Conductor Ampacity – Lower Rating.** When two or more ampacities apply to a single conductor length, the lower ampacity value shall be used for the entire circuit. For example, this would occur when a conductor runs through different ambient temperatures. See 310.15(B). Figure 310-8

*Exception:* When different ampacities apply to a conductor length, the higher ampacity can be used for the entire circuit

if the reduced ampacity length is not in excess of 10 ft and its length does not exceed 10 percent of the length of the higher ampacity. Figure 310-9

**(B) Table Ampacity.** The allowable conductor ampacities listed in Table 310.16 are based on conditions where the ambient temperature is not over 86°F and no more than three current-carrying conductors are bundled together. Figure 310-10

**Question:** What is the adjusted ampacity of 3/0 AWG THHN conductors if the ambient temperature is 86°F?

(a) 175A                    (b) 200A

(c) 225A                    (d) 250A

**Answer:** (c) 225A.

**AUTHOR'S COMMENT:** When conductors are installed in an ambient temperature other than 78–86°F, the conductors' ampacities, as listed in Table 310.16, shall be adjusted in accordance with the following multipliers at the bottom of Table 310.16 for 90°C conductors such as THHN and XHHW.

| Temperature °F | Temperature °C | Multiplier |
|---|---|---|
| 70–77°F | 21-25°C | 1.04 |
| 78–86°F | 26-30°C | 1.00 |
| 87–95°F | 31-35°C | 0.96 |
| 96–104°F | 36-40°C | 0.91 |
| 105–113°F | 41-45°C | 0.87 |
| 114–122°F | 46-50°C | 0.82 |
| 123–131°F | 51-55°C | 0.76 |
| 132–140°F | 56-60°C | 0.71 |
| 141–158°F | 61-70°C | 0.58 |
| 159–176°F | 71-80°C | 0.41 |

Conductor Ampacity - Higher Rating
Section 310.15(A)(2) Exception

Higher Ampacity
12 AWG = 30A

Lower Ampacity
12 AWG = 15A
(30A x 0.5)

COPYRIGHT 2002 Mike Holt Enterprises, Inc.

The higher ampacity can be used if the length of the low ampacity is not more than 10 ft, and it is not longer than 10 percent of the higher ampacity length.

**Figure 310-9**

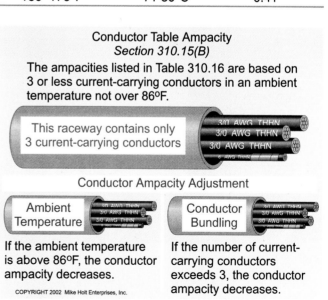

Conductor Table Ampacity
Section 310.15(B)

The ampacities listed in Table 310.16 are based on 3 or less current-carrying conductors in an ambient temperature not over 86°F.

This raceway contains only 3 current-carrying conductors

3/0 AWG THHN
3/0 AWG THHN
3/0 AWG THHN

Conductor Ampacity Adjustment

Ambient Temperature

Conductor Bundling

COPYRIGHT 2002 Mike Holt Enterprises, Inc.

If the ambient temperature is above 86°F, the conductor ampacity decreases.

If the number of current-carrying conductors exceeds 3, the conductor ampacity decreases.

**Figure 310-10**

**Question:** What is the adjusted ampacity of 3/0 AWG THHN conductors if the ambient temperature is 108°F?

(a) 173A                    (b) 180A

(c) 213A                    (d) 241A

**Answer:** (b) 180A. 225A × 0.80 = 180A

.87 = 198

**AUTHOR'S COMMENT:** When adjusting conductor ampacity, the ampacity of the conductor is based on 90°C rating of the conductor [110.14(C)].

**(2) Ampacity Adjustment**

**(a) Conductor Bundle.** Where the number of current-carrying conductors in a raceway or cable exceeds three, or where single conductors or multiconductor cables are stacked or bundled longer than 24 in., the allowable ampacity of each conductor, as listed in Table 310.16, shall be adjusted according to the following multiplying factors: Figure 310-11

| Current–Carrying | Multiplier |
| --- | --- |
| 1–3 Conductors | 1.00 |
| 4–6 Conductors | 0.80 |
| 7–9 Conductors | 0.70 |
| 10–20 Conductors | 0.50 |

**Question:** What is the adjusted ampacity of 3/0 AWG THHN conductors, if the raceway contains a total of four current-carrying conductors?

**Answer:** 225A × 0.80 = 180A

**AUTHOR'S COMMENT:** When adjusting conductor ampacity for THHN insulation, the ampacity of the conductor is based on the 90°C ampere rating of the conductor as listed in Table 310.16 [110.14(C)].

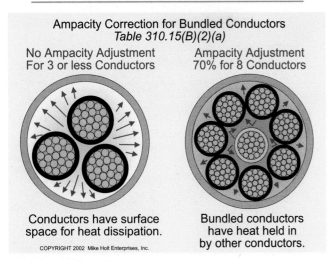

Figure 310-11

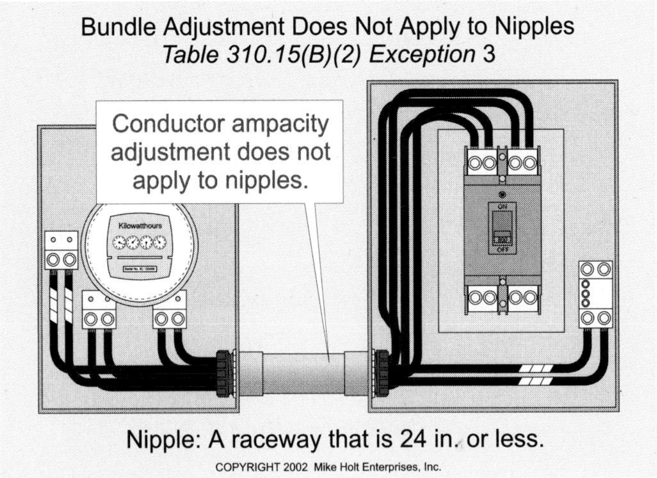

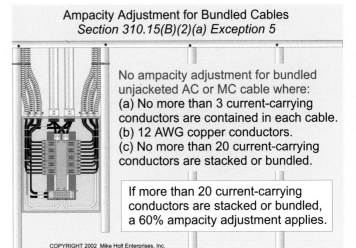

Figure 310-12

*Exception No. 3:* The ampacity adjustment factors of Table 310.15(B)(2)(a) do not apply to conductors installed in raceways having a length not exceeding 24 in. Figure 310-12

*Exception No. 5A:* The ampacity adjustment factors of Table 310.15(B)(2)(a) do not apply to AC or MC cable under the following conditions: Figure 310-13

(a) Each cable has not more than three current-carrying conductors.

(b) The conductors are 12 AWG copper.

(c) No more than 20 current-carrying conductors (ten 2-wire cables or six 3-wire cables) are bundled or stacked.

*Exception No. 5B:* When more than 20 current-carrying conductors (eleven or more 2-wire or seven or more 3-wire

Figure 310-13

cables) are bundled or stacked for more than 24 in., an ampacity adjustment factor of 60 percent shall be applied.

**Temperature and Bundling Adjustments.** Where there are more than three current-carrying conductors and the ambient temperature is not between 78–86°F, the ampacity listed in Table 310.16 shall be adjusted for both conditions.

**Question:** What is the adjusted ampacity of 3/0 AWG THHN conductors at an ambient temperature of 108°F if the raceway contains four current-carrying conductors?

(a) 157A          (b) 176A

(c) 199A          (d) 214A

**Answer:** (a) 157A. 225A × 0.87 × 0.80 = 157A

Temperature Adjustment [Table 310.16] = 0.87

Bundle Adjustment [310.15(B)(2)(a)] = 0.80

---

**AUTHOR'S COMMENT:** When adjusting conductor ampacity, the ampacity of the conductor is based on the 90°C rating of the conductor [110.14(C)].

---

(4) **Neutral Conductor.** All ungrounded conductors are considered current-carrying, and some grounded (neutral) conductors are considered current-carrying.

 (a) **Balanced Circuits.** The grounded (neutral) conductor of a 3-wire, 1Ø, 120/240V or 4-wire, 3Ø, 208Y/120V or 480Y/277V system is not considered a current-carrying conductor. Figure 310-14

 (b) **Wye 3-Wire Circuits.** The grounded (neutral) conductor of a 3-wire circuit from a 4-wire, 3Ø, 208Y/120V or 480Y/277V system is considered a current-carrying conductor.

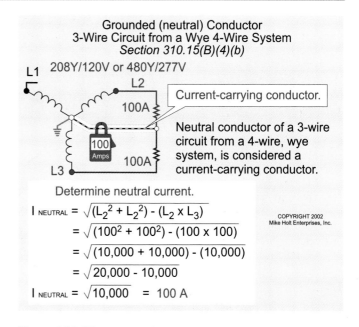

Grounded (neutral) Conductor
3-Wire Circuit from a Wye 4-Wire System
*Section 310.15(B)(4)(b)*

208Y/120V or 480Y/277V

Current-carrying conductor.

Neutral conductor of a 3-wire circuit from a 4-wire, wye system, is considered a current-carrying conductor.

Determine neutral current.

$$I_{NEUTRAL} = \sqrt{(L_2^2 + L_2^2) - (L_2 \times L_3)}$$
$$= \sqrt{(100^2 + 100^2) - (100 \times 100)}$$
$$= \sqrt{(10,000 + 10,000) - (10,000)}$$
$$= \sqrt{20,000 - 10,000}$$
$$I_{NEUTRAL} = \sqrt{10,000} = 100\ A$$

COPYRIGHT 2002
Mike Holt Enterprises, Inc.

*Figure 310-15*

---

**AUTHOR'S COMMENT:** When a 3-wire circuit is supplied from a 4-wire, 3Ø, wye-connected system, the grounded (neutral) conductor carries approximately the same current as the line-to-neutral current from the ungrounded (hot) conductors. Figure 310-15

---

 (c) **Wye 4-Wire Circuits Supplying Nonlinear Loads.** The neutral conductor of a balanced 4-wire, 3Ø, circuit that is at least 50 percent loaded with nonlinear loads is considered a current-carrying conductor.

---

**AUTHOR'S COMMENT:** Nonlinear loads supplied by 3Ø, 4-wire wye-connected power-supply systems such as 208Y/120V or 480Y/277V can produce unwanted and potentially hazardous odd triplen harmonic currents that can add on to the neutral conductor. To prevent a fire, or equipment damage from excessive harmonic neutral current, the designer should consider (1) increasing the size of the neutral conductor, or (2) installing a separate neutral for each phase. For more information, visit http://www.mikeholt.com/studies/harmonic.htm and see 210.4(A) FPN, 220.22 FPN 2, and 450.3 FPN 2. Figure 310-16

---

(5) **Grounding Conductors.** Grounding and bonding conductors are not considered current-carrying.

(6) **Dwelling Unit Feeder/Service Conductors.** One-family, two-family, or multifamily dwelling units can have their ungrounded (hot) conductors sized for 3-wire, 1Ø, 120/240V service-entrance and feeders that serve

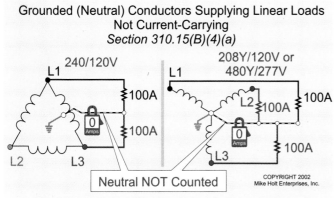

Grounded (Neutral) Conductors Supplying Linear Loads
Not Current-Carrying
*Section 310.15(B)(4)(a)*

240/120V

208Y/120V or 480Y/277V

Neutral NOT Counted

COPYRIGHT 2002
Mike Holt Enterprises, Inc.

Neutral conductor that carries only the unbalanced current is not considered a current-carrying conductor.

*Figure 310-14*

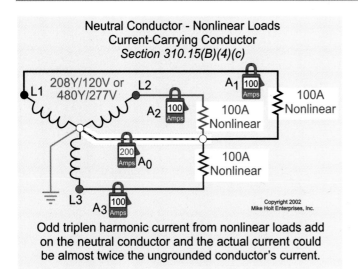

Neutral Conductor - Nonlinear Loads
Current-Carrying Conductor
*Section 310.15(B)(4)(c)*

Odd triplen harmonic current from nonlinear loads add on the neutral conductor and the actual current could be almost twice the ungrounded conductor's current.

**Figure 310-16**

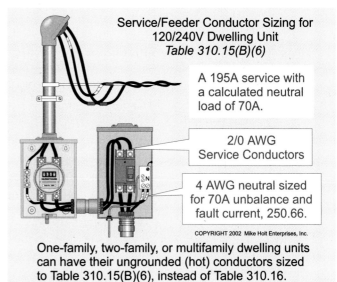

Service/Feeder Conductor Sizing for
120/240V Dwelling Unit
*Table 310.15(B)(6)*

A 195A service with a calculated neutral load of 70A.

2/0 AWG Service Conductors

4 AWG neutral sized for 70A unbalance and fault current, 250.66.

COPYRIGHT 2002 Mike Holt Enterprises, Inc.

One-family, two-family, or multifamily dwelling units can have their ungrounded (hot) conductors sized to Table 310.15(B)(6), instead of Table 310.16.

**Figure 310-17**

as the main power feeder to a dwelling unit in accordance with Table 310.15(B)(6), instead of Table 310.16.

| Amperes | Copper | Aluminum |
|---|---|---|
| 100 | 4 AWG | 2 AWG |
| 110 | 3 AWG | 1 AWG |
| 125 | 2 AWG | 1/0 AWG |
| 150 | 1 AWG | 2/0 AWG |
| 175 | 1/0 AWG | 3/0 AWG |
| 200 | 2/0 AWG | 4/0 AWG |
| 225 | 3/0 AWG | 250 kcmil |
| 250 | 4/0 AWG | 300 kcmil |
| 300 | 250 kcmil | 350 kcmil |
| 350 | 350 kcmil | 500 kcmil |
| 400 | 400 kcmil | 600 kcm |

*WARNING: Table 310.15(B)(6) does not apply to 3-wire, 1Ø, 208Y/120V systems, because the grounded (neutral) conductor in these systems carries neutral current even when the load on the phases is balanced [310.15(B))4)(6)].*

**Grounded (neutral) Conductor Sizing.** The grounded (neutral) conductor for dwelling units can be smaller than the ungrounded conductors, but in no case can it be smaller than required to carry the maximum unbalanced load as determined in 220.22. In addition, the grounded (neutral) conductor shall not be smaller than that required by 250.24(B).

**Question:** What size service/feeder conductors would be required if the total adjusted demand load for a dwelling unit equals 195A and the maximum unbalanced neutral load is 70A? Figure 310-17

(a) 1/0 AWG and 6 AWG        (b) 2/0 AWG and 4 AWG

(c) 3/0 AWG and 2 AWG        (d) 4/0 AWG and 1 AWG

**Answer:** (b) 2/0 AWG and 4 AWG

Feeder: 2/0 AWG rated 200A [Table 310.15(B)(6)]

Grounded (neutral) Conductor: 4 AWG, rated 85A at 75°C [110.14(C)(1) and Table 310.16]. 250.24(B) requires the gounded (neutral) conductor to be sized no smaller than 4 AWG based on on Table 250.66.

## Article 310

1. In general, the minimum size phase, neutral or grounded conductor permitted for use in parallel installations is _____ AWG.
   (a) 10              (b) 1              (c) 1/0              (d) 4

2. When conductors are run in parallel, the currents should be evenly divided between the individual parallel conductors so that each conductor is evenly heated. This is accomplished by ensuring that each of the conductors within a parallel set has the same_____ and all conductors terminate in the same manner.
   (a) length          (b) material       (c) cross-sectional area    (d) all of these

3. Insulated conductors and cables exposed to the direct rays of the sun must be _____.
   (a) listed                                (b) listed and marked sunlight resistant
   (c) listed for sunlight resistance        (d) b or c

4. The _____ rating of a conductor is the maximum temperature, at any location along its length, that the conductor can withstand over a prolonged period of time without serious degradation.
   (a) ambient         (b) temperature    (c) maximum withstand    (d) short-circuit

5. Which conductor has a temperature rating of 90°C?
   (a) RH              (b) RHW            (c) THHN             (d) TW

6. Lettering on conductor insulation indicates intended condition of use. THWN is rated _____.
   (a) 75°C            (b) for wet locations    (c) a and b      (d) not enough information

7. The ampacity of a conductor can be different along the length of the conductor. The higher ampacity is permitted to be used for the lower ampacity if the lower ampacity is no more than _____ ft or no more than _____ percent of the length of the higher ampacity.
   (a) 10, 20          (b) 20, 10         (c) 10, 10           (d) 15, 15

8. Where six current-carrying conductors are run in the same conduit or cable, the ampacity of each conductor shall be adjusted to a factor of _____ percent of its value.
   (a) 90              (b) 60             (c) 40               (d) 80

9. Conductor derating factors shall not apply to conductors in nipples having a length not exceeding _____
   (a) 12 in.          (b) 24 in.         (c) 36 in.           (d) 48 in.

10. In designing circuits, the current-carrying capacity of conductors should be corrected for heat at room temperatures above _____.
    (a) 30°F           (b) 86°F           (c) 94°F             (d) 75°F

# Article 312
## Cabinets, Cutout Boxes, and Meter Socket Enclosures

### 312.1 Scope

Article 312 covers the installation and construction specifications for cabinets, cutout boxes, and meter socket enclosures. Figure 312-1

**AUTHOR'S COMMENT:** See Article 100 Definitions for cabinets and cutout boxes.

### 312.2 Damp, Wet, or Hazardous (Classified) Locations

**(A) Damp and Wet Locations.** Enclosures in damp or wet locations shall prevent moisture or water from entering or accumulating within the enclosure, and they shall be weatherproof. When the enclosure is surface-mounted in a wet location, the enclosure shall be mounted with at least a $1/4$ in. air space between it and the mounting surface. See 300.6(C).

*Exception:* The $1/4$ in. air space does not apply to nonmetallic equipment, raceways, and cables.

**(B) Hazardous (classified) Locations.** Cabinets, cutout boxes, and meter socket enclosures installed in hazardous (classified) locations shall comply with the requirements of 501.4, 502.4, and 503.3(A)(1).

### 312.3 Installed in Walls

Enclosures installed in walls of concrete, tile, or other noncombustible material shall be installed so that the front edge of the enclosure is set back no more than $1/4$ in. from the finished surface. In walls constructed of wood or other combustible material, enclosures shall be flush with the finished surface or project outward.

### 312.5 Enclosures

**(A) Unused Openings.** Unused openings for conductors shall be closed with a fitting to afford protection substantially equivalent to that of the enclosures. See 110.12(A). Figure 312-2

**AUTHOR'S COMMENT:** In addition, a panel filler is required to guard against accidental contact with live parts of electrical equipment [110.27(A)].

**(C) Cable Termination.** Where a cable is used, it shall be properly secured to the enclosure with fittings that are designed and listed for the cable. See 300.12. Figure 312-3

**AUTHOR'S COMMENT:** Two NM cables are permitted to terminate into a single cable clamp that is listed for this purpose.

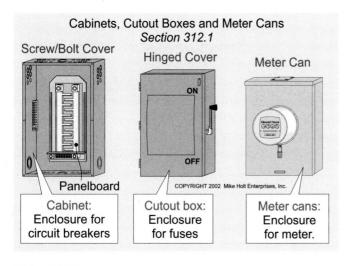

*Figure 312-1*

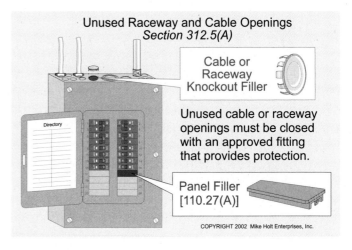

*Figure 312-2*

## Cable Terminations
### Section 312.5(C)

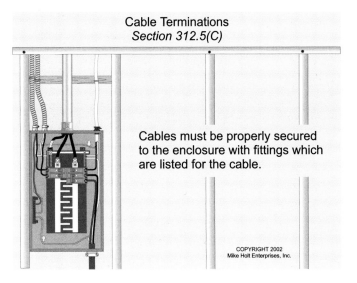

Cables must be properly secured to the enclosure with fittings which are listed for the cable.

COPYRIGHT 2002
Mike Holt Enterprises, Inc.

**Figure 312-3**

*Exception:* Cables with nonmetallic sheaths are not required to be secured to the enclosure if the cables enter the top of a surface-mounted enclosure through a nonflexible raceway not less than 18 in. or more than 10 ft long, if all of the following conditions are met: Figure 312-4

(a) Each cable is fastened within 1 ft from the raceway.

(b) The raceway does not penetrate a structural ceiling.

(c) Fittings are provided on the raceway to protect the cables from abrasion.

(d) The raceway is sealed.

(e) Each cable sheath extends at least $1/4$ in. into the panelboard.

(f) The raceway is properly secured.

## Sleeving NM Cable into Panel Cabinet
### Section 312.5(C) Exception

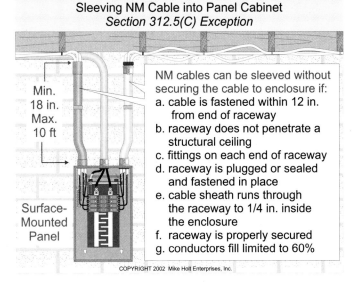

Min. 18 in.
Max. 10 ft

NM cables can be sleeved without securing the cable to enclosure if:
a. cable is fastened within 12 in. from end of raceway
b. raceway does not penetrate a structural ceiling
c. fittings on each end of raceway
d. raceway is plugged or sealed and fastened in place
e. cable sheath runs through the raceway to 1/4 in. inside the enclosure
f. raceway is properly secured
g. conductors fill limited to 60%

Surface-Mounted Panel

COPYRIGHT 2002 Mike Holt Enterprises, Inc.

**Figure 312-4**

## Cabinet and Cutout Box as Raceway
### Section 312.8

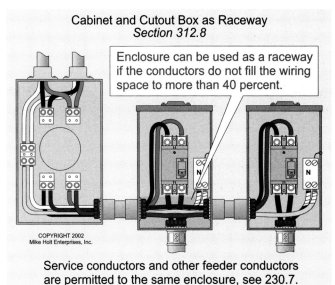

Enclosure can be used as a raceway if the conductors do not fill the wiring space to more than 40 percent.

COPYRIGHT 2002
Mike Holt Enterprises, Inc.

Service conductors and other feeder conductors are permitted to the same enclosure, see 230.7.

**Figure 312-5**

(g) Conductor fill is limited to 60 percent of the raceway cross-sectional area.

## 312.8 Used for Raceway and Splices.

Cabinets, cutout boxes, and meter socket enclosures can be used as a raceway for conductors feeding through if the conductors do not fill the wiring space at any cross section to more than 40 percent. Figure 312-5

**AUTHOR'S COMMENT:** Service conductors and other conductors are permitted to be in the same enclosure. See 230.7.

Splices and taps are permitted in cabinets, cutout boxes, or meter socket enclosures if the splice or tap does not fill the wiring space at any cross section to more than 75 percent. Figure 312-6

## Cabinet and Cutout Box
## Splices and Taps
### Section 312.8

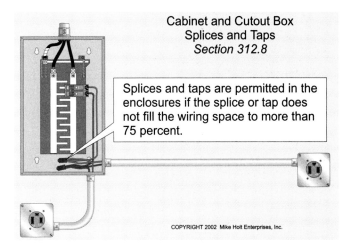

Splices and taps are permitted in the enclosures if the splice or tap does not fill the wiring space to more than 75 percent.

COPYRIGHT 2002 Mike Holt Enterprises, Inc.

**Figure 312-6**

# Article 312

1. Cabinets or cutout boxes installed in wet locations shall be _____.
   (a) waterproof      (b) raintight      (c) weatherproof      (d) watertight

2. In walls constructed of wood or other _____ material, electrical cabinets shall be flush with the finished surface or project therefrom.
   (a) nonconductive      (b) porous      (c) fibrous      (d) combustible

3. Cables entering a cutout box _____.
   (a) shall be secured independently to the cutout box
   (b) can be sleeved through a chase
   (c) shall have a maximum of two cables per connector
   (d) all of these

4. A switch enclosure (cabinet) shall not be used as a junction box, except where adequate space is provided, so that the conductors do not fill the wiring space at any cross section to more than 40 percent of the cross-sectional area of the space, and so that _____ do not fill the wiring space at any cross section to more than 75 percent of the cross-sectional area of the space.
   (a) splices      (b) taps      (c) conductors      (d) all of these

5. For a steel cabinet or cutout box, the metal shall not be less than _____ uncoated.
   (a) 0.530 in.      (b) 0.035 in.      (c) 0.053 in.      (d) 1.350 in.

# Article 314
## Outlet, Device, Pull and Junction Boxes, Conduit Bodies, and Fittings

## Part I. Scope and General

### 314.1 Scope

Article 314 contains the installation requirements for outlet boxes, conduit bodies, pull and junction boxes, and manholes. Figure 314-1

### 314.3 Nonmetallic Boxes

Nonmetallic boxes can only be used with nonmetallic cables and raceways.

**Exception No. 1:** Metal raceways and metal cables can be used with nonmetallic boxes if an internal bonding means is provided in the box between all metal entries.

### 314.5 Short-Radius Conduit Bodies

Short-radius conduit bodies such as capped elbows, handy ells, and service-entrance elbows shall not contain any splices or taps. Figure 314-2

---

**AUTHOR'S COMMENT:** Splices and taps can be made in standard conduit bodies. See 314.16(C) for specific requirements.

---

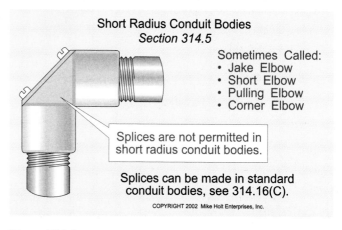

**Figure 314-2**

## Part II. Installation

### 314.15 Damp, Wet, or Hazardous (Classified) Locations

(A) **Damp and Wet Locations.** Boxes and conduit bodies in damp or wet locations shall prevent moisture or water from entering or accumulating within the enclosure, and they shall be weatherproof. Boxes, conduit bodies, and fittings installed in wet locations shall be listed for use in wet locations. Figure 314-3

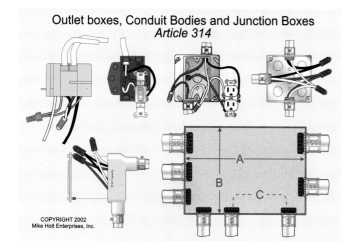

**Figure 314-1**

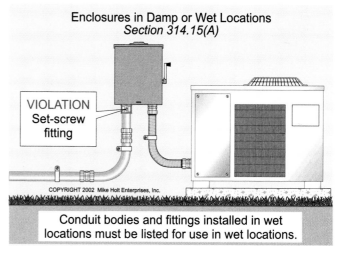

**Figure 314-3**

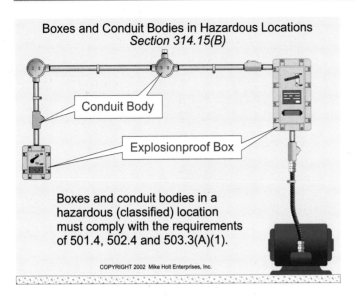

**Figure 314-4**

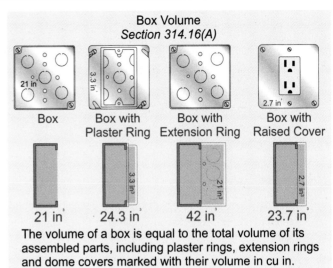

**Figure 314-5**

---

**AUTHOR'S COMMENT:** Some AHJs permit set-screw connectors to be used outdoors if the fitting is located on the bottom of the enclosure so water will not enter.

---

**(B)** **Hazardous (Classified) Locations.** Boxes and conduit bodies installed in hazardous (classified) locations shall comply with the requirements of 501.4, 502.4, and 503.3(A)(1). Figure 314-4

### 314.16 Number of 6 AWG and Smaller Conductors in Boxes and Conduit Bodies

**Boxes.** Boxes and conduit bodies containing 6 AWG and smaller conductors shall be properly sized to provide sufficient free space for all conductors, devices and fittings. In no case shall the volume of the box, as calculated in 314.16(A), be less than the fill calculation as calculated in 314.16(B).

**Conduit Bodies.** Conduit bodies shall be as calculated in accordance with 314.16(C).

---

**AUTHOR'S COMMENT:** The rules for sizing of boxes and conduit bodies containing conductors 4 AWG and larger are in accordance with 314.28.

---

**(A)** **Box Volume Calculations.** The volume of a box includes the total volume of its assembled parts, including plaster rings, extension rings and domed covers that are marked with their volume in cubic inches (cu in.) or are made from boxes the dimensions of which are listed in Table 314.16(A). Figure 314-5

**(B)** **Box Fill Calculations.** The total conductor volume is determined by adding the volumes of (1) through (5).

Wire connectors (wirenuts®), cable connectors, and raceway fittings such as locknuts or bushing are not counted [314.16(B)]. Figure 314-6

**(1)** **Conductors.** Each conductor running through a box without splice and each conductor that terminates in the box is counted as one conductor. Conductors that originate and terminate within the box, such as pigtails, are not counted at all. Figure 314-7

**Exception:** Equipment grounding conductor(s) and not more than four 16 AWG and smaller fixture wires can be omitted from box fill calculations if they enter the box from a domed luminaire or similar canopy, such as a ceiling paddle fan canopy. Figure 314-8

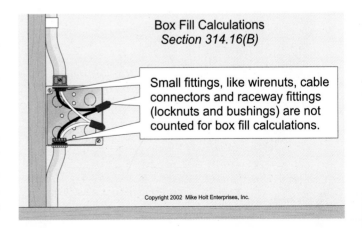

**Figure 314-6**

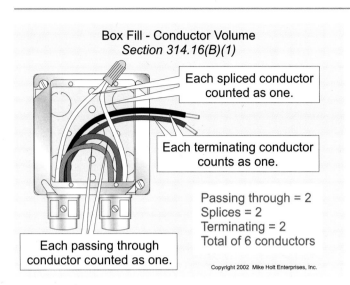

**Box Fill - Conductor Volume**
*Section 314.16(B)(1)*

Each spliced conductor counted as one.

Each terminating conductor counts as one.

Each passing through conductor counted as one.

Passing through = 2
Splices = 2
Terminating = 2
Total of 6 conductors

Copyright 2002 Mike Holt Enterprises, Inc.

*Figure 314-7*

(2) **Cable Clamps.** All internal cable clamps count as a single conductor volume, based on the largest conductor entering the box. No allowance is required for a cable connector that has its clamping mechanism outside the box.

(3) **Luminaire Stud and Hickey.** Each luminaire stud or luminaire hickey counts as a single conductor volume, based on the largest conductor entering the box. Figure 314-9

(4) **Device Yoke.** Each device yoke (regardless of the ampere rating of the device) counts as two conductors, based on the largest conductor that terminates on the device. Figure 314-10

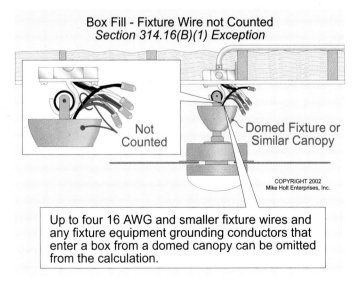

**Box Fill - Fixture Wire not Counted**
*Section 314.16(B)(1) Exception*

Not Counted

Domed Fixture or Similar Canopy

COPYRIGHT 2002
Mike Holt Enterprises, Inc.

Up to four 16 AWG and smaller fixture wires and any fixture equipment grounding conductors that enter a box from a domed canopy can be omitted from the calculation.

*Figure 314-8*

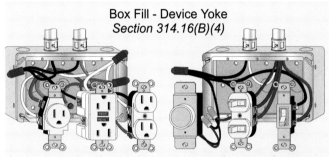

**Box Fill - Luminaire Stud and Hickey**
*Section 314.16(B)(3)*

Luminarie Stud
1 Conductor

Each luminaire stud or hickey counts as 1 conductor, based on largest conductor in the box.

COPYRIGHT 2002
Mike Holt Enterprises, Inc.

Luminaire Hickey
1 Conductor

3/8 Inch Mounting
Stem (not counted)

*Figure 314-9*

**Box Fill - Device Yoke**
*Section 314.16(B)(4)*

Each device yoke counts as two conductors, based on the largest conductor terminating on the device.

*Figure 314-10*

(5) **Grounding Conductor.** All equipment grounding or bonding conductors count as one conductor, based on the largest grounding conductor entering the box. An additional equipment grounding conductor for an isolated ground circuit counts as one conductor, based on the largest equipment grounding conductor. Figure 314-11

**AUTHOR'S COMMENT:** Table 314.16(B) contains the volume allowance required per conductor as follows:

| Conductor AWG | Volume cu in. |
|---|---|
| 18 | 1.50 |
| 16 | 1.75 |
| 14 | 2.00 |
| 12 | 2.25 |
| 10 | 2.50 |
| 8 | 3.00 |
| 6 | 5.00 |

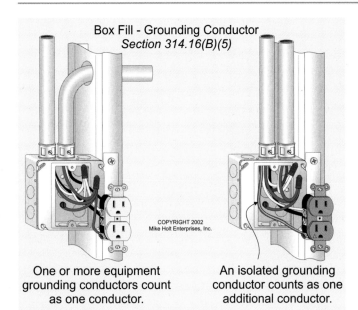

**Box Fill - Grounding Conductor**
*Section 314.16(B)(5)*

COPYRIGHT 2002
Mike Holt Enterprises, Inc.

One or more equipment grounding conductors count as one conductor.

An isolated grounding conductor counts as one additional conductor.

**Figure 314-11**

**Question:** How many 14 AWG THHN conductors can be pulled through a $4 \times 4 \times 2\frac{1}{8}$ in. deep box having a plaster ring with a marking of 3.6 cu in.? The box contains two receptacles, five 12 AWG conductors, and two 12 AWG equipment grounding conductors. Figure 314-12

**Answer:** 5 - 14 AWG THHN conductors

Step 1. Volume of the box assembly [314.16(A)].

Box 30.3 cu in. + 3.6 cu in. plaster ring = 33.9 cu in.

**AUTHOR'S COMMENT:** A $4 \times 4 \times 2\frac{1}{8}$ in. box would have a gross volume of 34 cu. in, but because the box is within this space, the actual volume is 30.3 cu in, as listed in Table 314.16(A).

Step 2. Determine the volume of the devices and conductors in the box.

| 2 receptacles | 4 – 12 AWG |
|---|---|
| 5 12 AWG THHN | 5 – 12 AWG |
| 2 12 AWG Grounds | 1 – 12 AWG |
| Total | 10 – 12 AWG |

$10 \times 2.25$ cu in. = 22.50 cu in.

Step 3. Determine the remaining volume permitted for the 14 AWG conductors.

33.9 cu in – 22.50 cu in. = 11.40 cu in.

Step 4. Determine the number of 14 AWG conductors permitted in the remaining volume.

11.40 cu in./2.00 cu in. = 5 conductors

**(C)   Conduit Bodies.**

**(2)   Splices.** Conductors 6 AWG and smaller can be spliced in a conduit body if the conductor volume limitations of 314.16(B) are complied with.

**Question:** How many 12 AWG conductors can be spliced in a 15 cu in. conduit body? Figure 314-13

(a) 4            (b) 6

(c) 8            (d) 10

**Answer:** (b) 6 conductors (15 cu in./2.25)

## 314.17 Conductors Entering Boxes or Conduit Bodies

**(B)   Metal Boxes or Conduit Bodies.** Raceways and cables shall be mechanically fastened to metal boxes or conduit bodies by fittings designed for the wiring method. See 300.12, and 300.15.

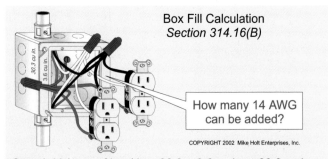

**Box Fill Calculation**
*Section 314.16(B)*

How many 14 AWG can be added?

COPYRIGHT 2002 Mike Holt Enterprises, Inc.

Step 1. Volume of box/ring: 30.3 + 3.6 cu in. = 33.9 cu in.
Step 2. Volume of existing conductors/devices = 22.5 cu in.
Step 3. Space remaining: 33.9 - 22.5 = 11.4 cu in.
Step 4. Number of 14 AWG added: 11.4/2.0 cu in. = 5

**Figure 314-12**

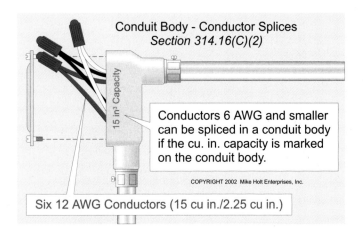

**Conduit Body - Conductor Splices**
*Section 314.16(C)(2)*

15 in³ Capacity

Conductors 6 AWG and smaller can be spliced in a conduit body if the cu. in. capacity is marked on the conduit body.

COPYRIGHT 2002  Mike Holt Enterprises, Inc.

Six 12 AWG Conductors (15 cu in./2.25 cu in.)

**Figure 314-13**

**(C) Nonmetallic Boxes.** Raceways and cables shall be securely fastened to nonmetallic boxes or conduit bodies by fittings designed for the wiring method, and NM cable shall extended at least $1/4$ in. into the nonmetallic box [300.12].

**AUTHOR'S COMMENT:** Two NM cables are permitted to terminate into a single cable clamp that is listed for this purpose.

*Exception:* NM cable terminating to a single-gang $2^1/4$ in. × 4 in. device box is not required to be secured to the box, if the cable is securely fastened within 8 in. of the box. Figure 314-14

## 314.20 Boxes Installed in Walls or Ceilings

Boxes installed in walls of concrete, tile, or other noncombustible material shall be installed so the front edge of the enclosure is set back no more than $1/4$ in. from the finished surface. In walls constructed of wood or other combustible material, boxes shall be flush with the finished surface or project therefrom.

**AUTHOR'S COMMENT:** Plaster rings are available in a variety of depths to meet the above requirements.

## 314.21 Gaps Around Boxes

To protect against electric shock and spread of fire, gaps around recessed boxes in plaster, drywall, or plasterboard surfaces shall be repaired so there will be no gaps or open spaces greater than $1/8$ in. at the edge of the box.

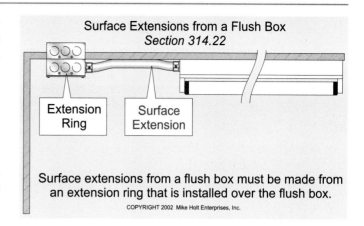

Surface Extensions from a Flush Box
*Section 314.22*

Extension Ring

Surface Extension

Surface extensions from a flush box must be made from an extension ring that is installed over the flush box.

COPYRIGHT 2002 Mike Holt Enterprises, Inc.

*Figure 314-15*

## 314.22 Surface Extensions

Surface extensions shall be made from an extension ring mounted over a flush-mounted box. Figure 314-15

*Exception:* A surface extension can be made from the cover of a flush-mounted box if the cover is designed so it is unlikely to fall off if the mounting screws become loose. The wiring method shall be flexible to permit the removal of the cover and provide access to the box interior, and grounding (bonding) continuity shall be independent of the connection between the box and the cover. Figure 314-16

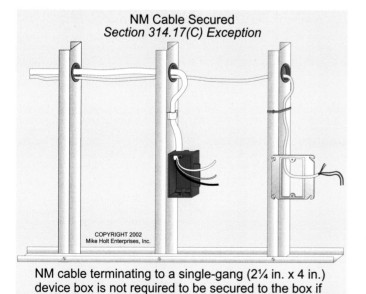

NM Cable Secured
Section 314.17(C) Exception

COPYRIGHT 2002
Mike Holt Enterprises, Inc.

NM cable terminating to a single-gang (2¼ in. x 4 in.) device box is not required to be secured to the box if the cable is securely fastened within 8 in. of the box.

*Figure 314-14*

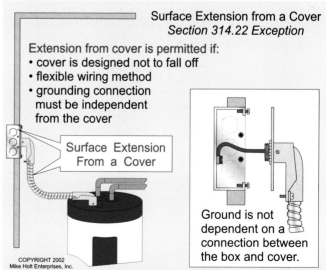

Surface Extension from a Cover
*Section 314.22 Exception*

Extension from cover is permitted if:
• cover is designed not to fall off
• flexible wiring method
• grounding connection must be independent from the cover

Surface Extension From a Cover

Ground is not dependent on a connection between the box and cover.

COPYRIGHT 2002
Mike Holt Enterprises, Inc.

*Figure 314-16*

## 314.23 Supports of Boxes and Conduit Bodies

Boxes shall be securely supported by any one of the following methods:

**(A) Surface.** Boxes can be fastened to any surface that provides adequate support.

**(B) Structural Mounting.** A box can be supported from a structural member of a building or from grade by a metal, plastic, or wood brace.

    **(1) Nails and Screws.** Nails or screws can be used to fasten boxes.

    **(2) Braces.** Metal braces no less than 0.020 in. thick and wood braces not less than a nominal 1 in. × 2 in. can be used to support a box.

**(C) Finished Surface Support.** Boxes can be secured to a finished surface (drywall or plaster walls or ceilings) by clamps, anchors, or fittings identified for the application. Figure 314-17

**(D) Suspended-Ceiling Support.** Outlet boxes can be supported to the structural or supporting elements of a suspended ceiling if securely fastened in one of the following ways:

    **(1) Ceiling Framing Members.** An outlet box can be secured to suspended-ceiling framing members by bolts, screws, rivets, clips, or other means identified for the suspended-ceiling framing member(s). Figure 314-18

**AUTHOR'S COMMENT:** Luminaires can be supported to ceiling framing members as well. See 410.16(C).

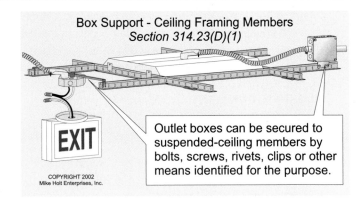

Box Support - Ceiling Framing Members
*Section 314.23(D)(1)*

Outlet boxes can be secured to suspended-ceiling members by bolts, screws, rivets, clips or other means identified for the purpose.

COPYRIGHT 2002
Mike Holt Enterprises, Inc.

**Figure 314-18**

    **(2) Independent Support Wires.** Outlet boxes can be secured, with fittings identified for the purpose, to independent support wires that are taut and secured at both ends. Figure 314-19

**AUTHOR'S COMMENT:** See 300.11(A) on the use of independent support wires for raceway and cable.

**(E) Raceway Support - Boxes and Conduit Bodies without Devices or Luminaires.** Boxes without devices or luminaires can be supported by IMC or RMC threaded wrenchtight on two different sides of the box, if each raceway is supported within 36 in. of the box. Figure 314-20

*Exception:* Conduit bodies are considered supported with two IMC, RMC, RNC, or EMT raceways.

**AUTHOR'S COMMENT:** This exception allows a conduit body to be supported by two IMC, RMC, RNC, or EMT raceways. Figure 314-21

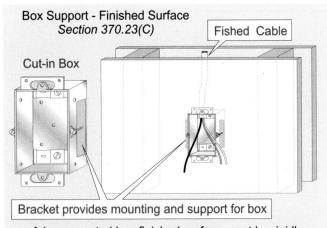

Box Support - Finished Surface
*Section 370.23(C)*

Fished Cable

Cut-in Box

Bracket provides mounting and support for box

A box mounted in a finished surface must be rigidly secured by clamps or fittings identified for the purpose.

COPYRIGHT 2002 Mike Holt Enterprises, Inc.

**Figure 314-17**

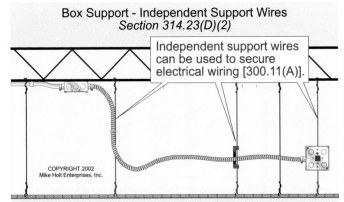

Box Support - Independent Support Wires
*Section 314.23(D)(2)*

Independent support wires can be used to secure electrical wiring [300.11(A)].

COPYRIGHT 2002
Mike Holt Enterprises, Inc.

Outlet boxes can be secured, with fittings identified for the purpose, to independent support wires that are taut and secured at both ends.

**Figure 314-19**

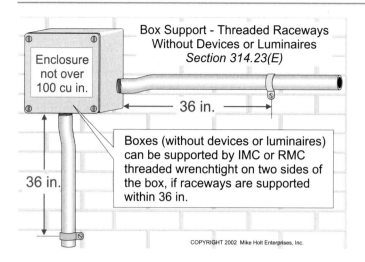

**Figure 314-20**

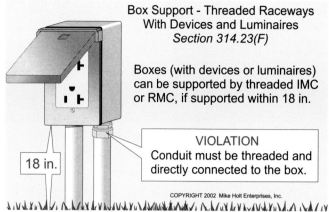

**Figure 314-22**

**(F) Raceway Support - Boxes and Conduit Bodies with Devices or Luminaires.** Boxes containing devices or luminaires are considered supported when two IMC or RMC conduits are threaded wrenchtight, if each raceway is supported within 18 in. of the box. Figure 314-22

**(G) Concrete or Masonry.** Boxes embedded in concrete or masonry are considered supported.

**(H) Pendant Cords.**

   (1) Boxes can be supported from a cord that is connected to fittings so tension is not transmitted to joints or terminals. See 400.10. Figure 314-23

## 314.25 Covers and Canopies

When the installation is complete, each outlet box shall be provided with a cover or faceplate, unless covered by a fixture canopy, lampholder, or similar device. See 410.12. Figure 314-24

**(A) Nonmetallic or Metallic.** Nonmetallic or metallic covers or plates can be used on any box, but metallic cover plates can only be used on metal boxes. See 314.28(C).

**AUTHOR'S COMMENT:** Switch cover plates shall be grounded (bonded) in accordance with 404.9(B), and receptacle cover plates shall be grounded (bonded) in accordance with 406.5(A).

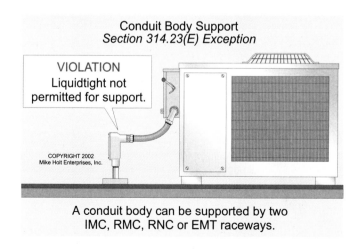

A conduit body can be supported by two IMC, RMC, RNC or EMT raceways.

**Figure 314-21**

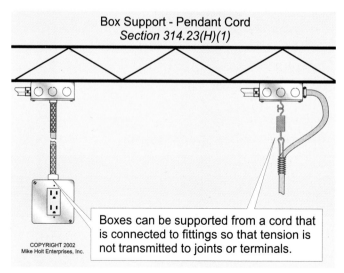

**Figure 314-23**

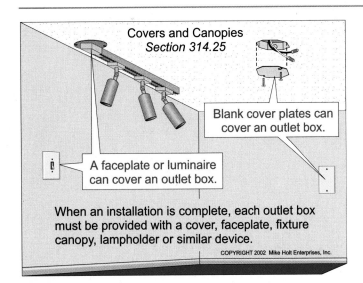

Figure 314-24

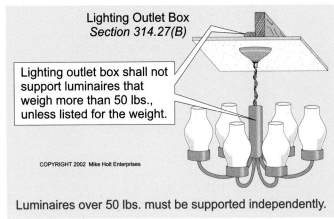

Figure 314-26

## 314.27 Outlet Boxes

**(A)** **Luminaire Support.** Lighting outlet boxes for luminaires shall be designed for the purpose and shall be installed at every outlet used for luminaires. Device outlet boxes (6-32 screws) are only to be used for the support of switches or receptacles, not luminaires or lampholders.

*Exception:* A wall-mounted luminaire weighing not more than 6 lbs. can be supported to a device box, or to a plaster ring that is secured to a box, if the luminaire is secured to the device box with two 6-32 screws. Figure 314-25

**(B)** **Maximum Luminaire Weight.** Lighting outlet boxes (8-32 and larger screws) can be used to support luminaires that weigh up to 50 lbs. Luminaires that weigh more than 50 lbs.

shall be supported independently of the lighting outlet box, unless the lighting outlet box is listed for the weight of the luminaire to be supported. Figure 314-26

**(C)** **Floor Boxes.** Floor boxes shall be specifically listed for the purpose. Figure 314-27

**(D)** **Ceiling Paddle Fan.** Where a fan box (10-24 and larger screws) is used as the sole support for a ceiling paddle fan, the fan box shall be listed for this purpose and the maximum weight of the fan. Ceiling paddle fans exceeding 35 lbs. shall be supported independently of the fan outlet box, unless the box is listed to support the weight of the fan [422.18(B)]. Figure 314-28

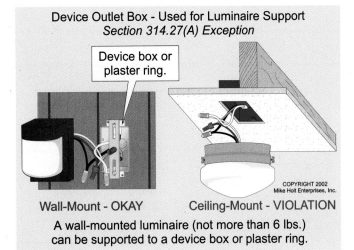

Figure 314-25

Figure 314-27

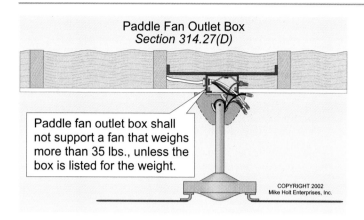

Paddle Fan Outlet Box
Section 314.27(D)

Paddle fan outlet box shall not support a fan that weighs more than 35 lbs., unless the box is listed for the weight.

COPYRIGHT 2002
Mike Holt Enterprises, Inc.

**Figure 314-28**

## 314.28 Boxes and Conduit Bodies for Conductors 4 AWG and Larger

Boxes and conduit bodies containing conductors 4 AWG and larger shall be sized so the conductor insulation will not be damaged.

**AUTHOR'S COMMENT:** Where conductors 4 AWG or larger enter a box, fitting, or other enclosure, a fitting that provides a smooth, rounded insulating surface, such as a bushing or adapter, shall be provided to protect the conduit from abrasion during and after installation. See 300.4(F).

**(A)** **Minimum Size.** For raceways containing conductors 4 AWG or larger, the minimum dimensions of boxes and conduit bodies shall comply with the following:

**(1)** **Straight Pulls.** The minimum distance from where the conductors enter to the opposite wall shall not be less than eight times the trade size of the largest raceway. Figure 314-29

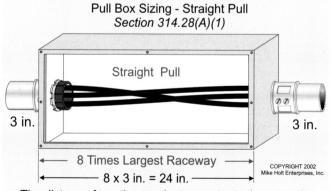

Pull Box Sizing - Straight Pull
Section 314.28(A)(1)

Straight Pull

3 in.                                 3 in.

8 Times Largest Raceway
8 x 3 in. = 24 in.

COPYRIGHT 2002
Mike Holt Enterprises, Inc.

The distance from the conductors entry to the opposite wall must not be less than 8X the largest raceway.

**Figure 314-29**

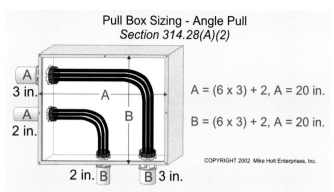

Pull Box Sizing - Angle Pull
Section 314.28(A)(2)

A
3 in.

A
2 in.

2 in. B          B 3 in.

A = (6 x 3) + 2, A = 20 in.

B = (6 x 3) + 2, A = 20 in.

COPYRIGHT 2002 Mike Holt Enterprises, Inc.

For angle pulls, the distance (measured from conductor wall entry to the opposite wall) must not be less than 6X the largest raceway, plus the sum of the diameters of the remaining raceways on the same wall and row.

**Figure 314-30**

**(2)** **Angle and U Pulls.**

**Angle Pulls.** The distance from the raceway entry to the opposite wall shall not be less than six times the trade diameter of the largest raceway, plus the sum of the diameters of the remaining raceways on the same wall and row. Figure 314-30

**U Pulls.** When a conductor enters and leaves from the same wall, the distance from where the raceways enter to the opposite wall shall not be less than six times the trade diameter of the largest raceway, plus the sum of the diameters of the remaining raceways on the same wall and row. Figure 314-31

**Rows.** Where there are multiple rows of raceway entries, each row is calculated individually, and the row that has the largest distance shall be used.

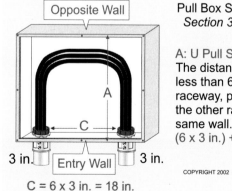

Opposite Wall

A

C

3 in.          Entry Wall          3 in.

C = 6 x 3 in. = 18 in.

Pull Box Sizing - U Pull
Section 314.28(A)(2)

A: U Pull Sizing:
The distance must not be less than 6X the largest raceway, plus the sum of the other raceways on the same wall.
(6 x 3 in.) + 3 in. = 21 in.

COPYRIGHT 2002 Mike Holt Enterprises, Inc.

The distance between raceways enclosing the same conductor is not less than 6X the largest raceway.

**Figure 314-31**

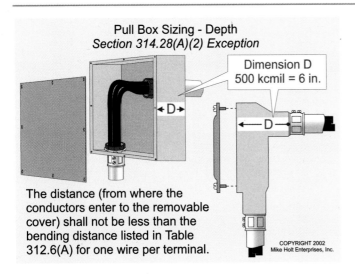

**Pull Box Sizing - Depth**
*Section 314.28(A)(2) Exception*

Dimension D
500 kcmil = 6 in.

The distance (from where the conductors enter to the removable cover) shall not be less than the bending distance listed in Table 312.6(A) for one wire per terminal.

COPYRIGHT 2002
Mike Holt Enterprises, Inc.

**Figure 314-32**

*Exception:* When conductors enter an enclosure with a removable cover such as a conduit body or wireway, the distance from where the conductors enter to the removable cover shall not be less than the bending distance as listed in Table 312.6(A) for one wire per terminal. Figure 314-32

**Distance between Raceways.** The distance between raceways enclosing the same conductor shall not be less than six times the trade diameter of the largest raceway, measured from raceway edge-to-edge.

**(3) Smaller Dimensions.** Boxes or conduit bodies of dimensions less than those required in (A)(1) and (A)(2) can be used if the enclosure is permanently marked with the maximum number and maximum size of conductors permitted.

**(C) Covers.** All pull boxes, junction boxes, and conduit bodies shall have a cover that is suitable for the conditions of use. Nonmetallic covers or plates can be used on any box, but metallic cover plates can only be used on metal boxes. See 314.25(A).

## 314.29 Wiring to be Accessible

Conduit bodies and boxes shall be installed so that the wiring can be rendered accessible without removing any part of the building, sidewalks, paving, or earth.

# Article 314

1.  Nonmetallic boxes are permitted for use with _____.
    (a) flexible nonmetallic conduit          (b) liquidtight nonmetallic conduit
    (c) nonmetallic cables and raceways       (d) all of these

2.  When counting the number of conductors in a box, a conductor running through the box is counted as _____ conductor(s).
    (a) one          (b) two          (c) zero          (d) none of these

3.  When determining the number of conductors in a box, and one or more factory or field-supplied internal cable clamps are present in the box, a double volume allowance for the clamps, in accordance with Table 314.16(B), shall be made based on the largest conductor present in the box.
    (a) True          (b) False

4.  What is the total volume in cubic inches for box fill calculation for two internal cable clamps, six 12 AWG THHN conductors, and one single-pole switch?
    (a) 12.00          (b) 13.50          (c) 14.50          (d) 20.25

5.  When a box has three equipment grounding conductors in it that have originated outside the box, the three grounding conductors are counted as _____ conductor(s) when determining the number of conductors in a box for box fill calculations.
    (a) 3          (b) 6          (c) 1          (d) 0

6.  In combustible walls or ceilings, the front edge of an outlet box or fitting may be set back _____ from the finished surface.
    (a) $3/8$ in.          (b) $1/8$ in.          (c) $1/2$ in.          (d) $1/4$ in.

7.  Only a _____ wiring method can be used for a surface extension from a cover, and the wiring method must include an equipment grounding conductor.
    (a) solid          (b) flexible          (c) rigid          (d) cord

8.  Outlet boxes can be secured to suspended-ceiling framing members by mechanical means such as _____, or other means identified for the suspended-ceiling framing member(s).
    (a) bolts          (b) screws          (c) rivets          (d) all of these

9.  Outlet boxes can be secured to independent support wires, which are taut and secured at both ends if the box is supported to the independent support wires, with fittings and methods identified for the purpose.
    (a) True          (b) False

10. Enclosures that are not over _____ cubic inches in size, having threaded entries, and not containing a device(s) or supporting a luminaire(s) or other equipment, shall be considered to be adequately supported where two or more conduits are threaded wrenchtight into the enclosure.
    (a) 50          (b) 75          (c) 100          (d) 125

11. Enclosures not over 100 cu in. that have threaded entries that support luminaires or contain devices shall be considered adequately supported where two or more conduits are threaded wrenchtight into the enclosure where each conduit is supported within _____ of the enclosure.
    (a) 12 in.          (b) 18 in.          (c) 24 in.          (d) 30 in.

12. Boxes can be supported from a multiconductor cord or cable, provided the conductors are protected from _____.
    (a) strain          (b) temperature          (c) sunlight          (d) abrasion

13. Luminaires shall be supported independently of the outlet box where the weight exceeds _____
    (a) 60 lbs.          (b) 50 lbs.          (c) 40 lbs.          (d) 30 lbs.

14. Where a box is used as the sole support of a ceiling-suspended (paddle) fan, the box shall be listed for the application and for the weight of the fan to be supported.
    (a) True          (b) False

15. When sizing a pull box in a straight run, which contains conductors of 4 AWG or larger, the length of the box shall not be less than _____ for systems not over 600V.
    (a) 8 times the diameter of the largest raceway
    (b) 6 times the diameter of the largest raceway
    (c) 48 times the outside diameter of the largest shielded conductor
    (d) 36 times the largest conductor

# Article 320
# Armored Cable (Type AC)

## Part I. General

### 320.1 Scope

This article covers the use, installation, and construction specifications of armored cable, Type AC.

### 320.2 Definition

**Armored Cable (Type AC).** A fabricated assembly of conductors in a flexible metal sheath with an internal bonding strip in intimate contact with the armor for its entire length. See 320.100. Figure 320-1

> **AUTHOR'S COMMENT:** AC cable is an assembly of insulated conductors, 14 AWG through 1 AWG, that are individually wrapped within waxed paper. The conductors are contained within a flexible metal (steel or aluminum) sheath that interlocks at the edges, giving AC cable an outside appearance like flexible metal conduit. Many electricians call this metal cable BX®. The advantages of any flexible cable system are that there is no limit to the number of bends between terminations, and the cable can be quickly installed.

## Part II. Installation

### 320.10 Uses Permitted

AC cable can be installed only where not subject to physical damage in the following locations:

(1) Exposed and concealed,

(2) Cable trays,

(3) Dry locations,

(4) Embedded in plaster or brick, except in damp or wet locations, and

(5) Air voids where not exposed to excessive moisture or dampness.

> **AUTHOR'S COMMENT:** AC cable can be installed in environmental air spaces [300.22(C)].

### 320.12 Uses Not Permitted

AC cable shall not be installed in the following locations:

(1) In theaters and similar locations, except as permitted by 518.4(A),

(2) In motion picture studios,

(3) In any hazardous location, except as permitted by 501.4(B), Ex.; 502.4(B), Ex. 1; and 504.20,

(4) Where exposed to corrosive fumes or vapors,

(5) In storage battery rooms,

(6) In hoistways or on elevators, except as permitted in 620.21, and

(7) In commercial garages where prohibited in Article 511.

### 320.15 Exposed Work

Exposed AC cable shall closely follow the surface of the building finish or running boards. AC cable can run on the bottom of floor or ceiling joists, if secured at every joist and not subject to physical damage. Figure 320-2

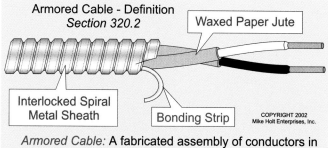

Armored Cable - Definition
*Section 320.2*

Waxed Paper Jute

Interlocked Spiral Metal Sheath

Bonding Strip

COPYRIGHT 2002
Mike Holt Enterprises, Inc.

*Armored Cable:* A fabricated assembly of conductors in a flexible metal sheath with an internal bonding strip in intimate contact with the armor for its entire length.

**Figure 320-1**

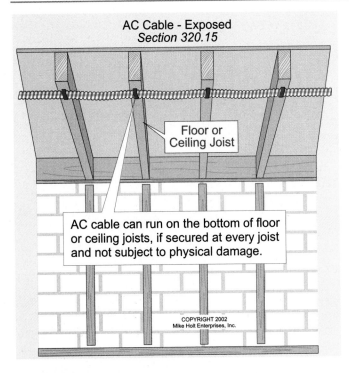

**Figure 320-2**

### 320.17 Through or Parallel to Framing Members

AC cable installed through or parallel to framing members shall be protected against physical damage from penetration by screws or nails by maintaining a $1^{1}/_{4}$ in. separation or by installing a suitable metal plate. See 300.4(A). Figure 320-3

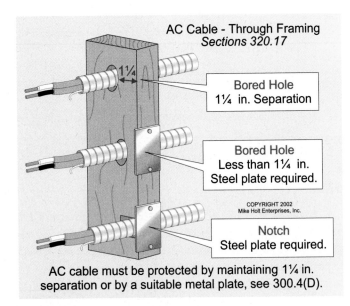

**Figure 320-3**

### 320.23 In Accessible Attics or Roof Spaces

(A) **On the Surface of Floor Joists, Rafters, or Studs.** Substantial guards shall protect the cable where run across the top of floor joists, or within 7 ft of floor or floor joists across the face of rafters or studding, or in attics and roof spaces that are accessible. Where this space is not accessible by permanent stairs or ladders, protection is required only within 6 ft of the nearest edge of the scuttle hole or attic entrance.

(B) **Along the Side of Floor Joists.** When AC cable is run on the side of rafters, studs, or floor joists, no protection is required if the cable is installed and supported so the nearest outside surface of the cable or raceway is not less than $1^{1}/_{4}$ in. from the nearest edge of the framing member where nails or screws are likely to penetrate. See 300.4(D).

### 320.24 Bends

AC shall not be bent in a manner that will damage the cable. This is accomplished by limiting the bending radius of the inner edge of the cable to not less than five times the internal diameter of the cable.

### 320.30 Secured and Supported

AC cable shall be secured by staples, cable ties, straps, hangers, or fittings designed and installed so the cable is not damaged, at intervals not exceeding $4^{1}/_{2}$ ft and within 1 ft of the cable termination. Figure 320-4

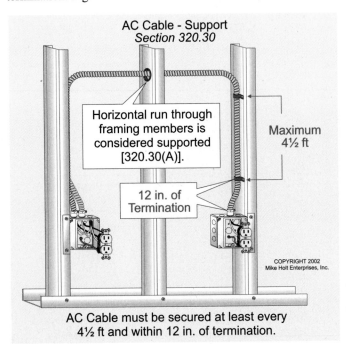

**Figure 320-4**

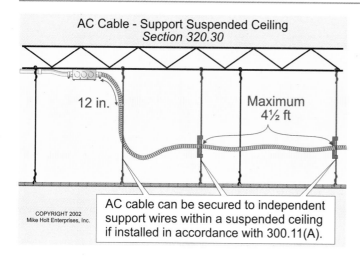

AC Cable - Support Suspended Ceiling
*Section 320.30*

12 in.

Maximum 4½ ft

COPYRIGHT 2002
Mike Holt Enterprises, Inc.

AC cable can be secured to independent support wires within a suspended ceiling if installed in accordance with 300.11(A).

**Figure 320-5**

> **AUTHOR'S COMMENT:** AC cable can be secured to independent support wires within a suspended ceiling, in accordance with 300.11(A). Figure 320-5

**(A) Horizontal Runs.** AC cable installed horizontally in bored or punched holes in wood or metal framing members, or notches in wooden members, is considered supported, but the cable shall be secured within 1 ft of termination.

**(B) Unsupported.** AC cable can be unsupported where the cable is:

(1) Fished between concealed access points in finished buildings or structures, and support is impracticable; or

(2) Not more than 2 ft in length at terminals where flexibility is necessary; or

(3) Not more than 6 ft from the last point of support within an accessible ceiling for the connection of luminaires or equipment. Figure 320-6

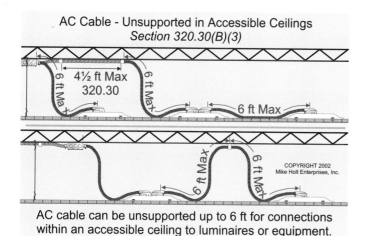

AC Cable - Unsupported in Accessible Ceilings
*Section 320.30(B)(3)*

4½ ft Max
320.30

6 ft Max

6 ft Max

6 ft Max

6 ft Max

6 ft Max

6 ft Max

COPYRIGHT 2002
Mike Holt Enterprises, Inc.

AC cable can be unsupported up to 6 ft for connections within an accessible ceiling to luminaires or equipment.

**Figure 320-6**

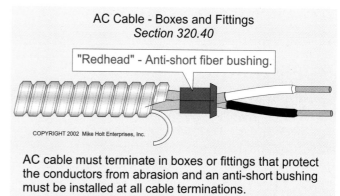

AC Cable - Boxes and Fittings
*Section 320.40*

"Redhead" - Anti-short fiber bushing.

COPYRIGHT 2002  Mike Holt Enterprises, Inc.

AC cable must terminate in boxes or fittings that protect the conductors from abrasion and an anti-short bushing must be installed at all cable terminations.

**Figure 320-7**

### 320.40 Boxes and Fittings

AC cable shall terminate in boxes or fittings specifically listed for AC cable that protect the conductors from abrasion [300.15]. An approved insulating anti-short bushing, sometimes called a "redhead," shall be installed at all AC cable terminations. The termination fitting shall permit the visual inspection of the anti-short bushing once the cable has been installed. Figure 320-7

> **AUTHOR'S COMMENT:** The aluminum-bonding strip within the cable serves no electrical purpose once outside the cable, but many electricians use it to secure the anti-short bushing to the cable.

### 320.80 Conductor Ampacities

Conductor ampacity is determined in accordance with 310.15, based on the 90°C insulation rating of the conductors and the terminal ratings of the fittings. See 110.14(C).

**(A) Thermal Insulation.** AC cable installed in thermal insulation shall have the conductors rated at 90°C, but the ampacity shall be that of 60°C conductors.

## Part III. Construction Specifications

### 320.100 Construction

AC cable shall have an armor of flexible metal tape with an internal aluminum-bonding strip in intimate contact with the armor for its entire length.

> **AUTHOR'S COMMENT:** When cutting AC cable with a hacksaw, be sure to cut only one spiral of the cable and be careful not to nick the conductors; this is done by cutting the cable at an angle. Breaking the cable spiral (bending the cable very sharply), then cutting the cable with a pair of dikes is not a good practice. The

best method to separate the cable is to use a tool specially designed for the purpose.

## 320.108 Grounding

The combination of the armor and 18 AWG aluminum-bonding strip [320.100] provides an adequate path for equipment grounding (bonding) as required by 250.4(A)(5). The effective fault-current path shall be maintained by the use of fittings that are specifically listed for AC cable [320.40]. See 300.12, 300.15, and 300.100. Figure 320-8

**AUTHOR'S COMMENT:** The internal aluminum-bonding strip is not an equipment grounding conductor, but it serves to reduce the inductive reactance of the armored spirals, to ensure that a short circuit will be cleared. Once the bonding strip exits the cable, it can be cut off because it no longer serves any purpose.

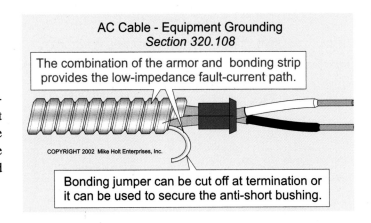

AC Cable - Equipment Grounding
*Section 320.108*

The combination of the armor and bonding strip provides the low-impedance fault-current path.

COPYRIGHT 2002 Mike Holt Enterprises, Inc.

Bonding jumper can be cut off at termination or it can be used to secure the anti-short bushing.

*Figure 320-8*

# Article 320

1.  The use of AC cable is permitted in _____ installations.
    (a) wet           (b) cable tray           (c) exposed           (d) b and c

2.  Armored cable is limited or not permitted _____.
    (a) in commercial garages           (b) where subject to physical damage
    (c) in motion picture studios           (d) all of these

3.  Exposed runs of AC cable shall closely follow the surface of the building finish or of running boards. Exposed runs shall also be permitted to be installed on the underside of joists where supported at each joist and located so as not to be subject to physical damage.
    (a) True           (b) False

4.  AC cable installed through or parallel to framing members must be protected against physical damage from penetration by screws or nails.
    (a) True           (b) False

5.  Where run across the top of floor joists, or within 7 ft of floor or floor joists across the face of rafters or studding, or in attics and roof spaces that are accessible, AC cable shall be protected by substantial guard strips that are _____.
    (a) at least as high as the cable           (b) constructed of metal
    (c) made for the cable           (d) none of these

6.  When AC cable is run across the top of a floor joist in an attic without permanent ladders or stairs, substantial guard strips within _____ of the scuttle hole shall protect the cable.
    (a) 7 ft           (b) 6 ft           (c) 5 ft           (d) 3 ft

7.  When armored cable is run on the side of rafters, studs or floor joists in an accessible attic, protection is required for the cable with running boards.
    (a) True           (b) False

8.  The radius of the curve of the inner edge of any bend shall not be less than _____ for AC cable.
    (a) five times the largest conductor within the cable
    (b) three times the diameter of the cable
    (c) five times the diameter of the cable
    (d) six times the outside diameter of the conductors

9.  AC cable shall be supported at intervals not exceeding $4^1/_2$ ft and the cable shall be secured within _____ of cable termination.
    (a) 4 in.           (b) 8 in.           (c) 9 in.           (d) 12 in.

10. Armored cable used for the connection of recessed luminaires or equipment within an accessible ceiling does not need to be secured for lengths up to _____.
    (a) 2 ft           (b) 3 ft           (c) 4 ft           (d) 6 ft

# Article 330
# Metal-Clad Cable (Type MC)

## Part I. General

### 330.1 Scope

This article covers the use, installation, and construction specifications of metal-clad cable, Type MC.

### 330.2 Definition

**Metal-Clad Cable (Type MC).** A factory assembly of one or more insulated circuit conductors with or without optical fiber members enclosed in an armor of interlocking metal tape, or a smooth or corrugated metallic sheath. Figure 330-1

> **AUTHOR'S COMMENT:** MC cable encloses one or more insulated conductors in a metal sheath of either corrugated or smooth copper or aluminum tubing, or spiral-interlocked steel or aluminum. Because the outer sheath of interlock MC cable is not listed as an equipment grounding conductor, it contains an insulated equipment grounding conductor [330.108]. The most common type of MC cable is the interlocking type, which has an appearance similar to armored cable or flexible metal conduit.

### 330.10 Uses Permitted

**(A) General Uses.** Where not subject to physical damage, metal-clad (MC) cable can be used:

(1) In branch circuits, feeders, and services.

(2) In power, lighting, control, and signal circuits.

(3) Indoors or outdoors.

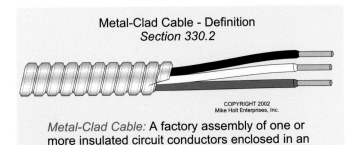

Metal-Clad Cable - Definition
Section 330.2

COPYRIGHT 2002
Mike Holt Enterprises, Inc.

*Metal-Clad Cable:* A factory assembly of one or more insulated circuit conductors enclosed in an armor of interlocking metal tape, or a smooth or

**Figure 330-1**

(4) Exposed or concealed.

(5) Directly buried (if identified for the purpose).

(6) In a cable tray.

(7) In a raceway.

(8) In open runs.

(9) As aerial cable on a messenger.

(10) In hazardous (classified) locations as permitted in 501.4(B), 502.4(B), and 503.3.

(11) Embedded in plaster or brick.

(12) In wet locations, where the cable is impervious to moisture and the conductors are listed for use in a wet location.

> **AUTHOR'S COMMENT:** MC cable can be installed in environmental air spaces [300.22(C)].

### 330.12 Uses Not Permitted

Type MC cable shall not be installed where exposed to the following destructive corrosive conditions, unless the metallic sheath is suitable for the conditions or is protected by material suitable for the conditions:

(1) Direct burial in the earth.

(2) In concrete.

(3) Where exposed to cinder fills, strong chlorides, caustic alkalis, vapors of chlorine, or hydrochloric acid.

## Part II. Installation

### 330.17 Through or Parallel to Framing Members

MC cable installed through or parallel to framing members shall be protected against physical damage from penetration by screws or nails by a $1^1/_4$ in. separation or by a suitable metal plate. See 300.4.

### 330.23 In Accessible Attics or Roof Spaces

**(A) On the Surface of Floor Joists, Rafters, or Studs.** Substantial guards shall protect the cable where run across the top of floor joists, or within 7 ft of floor or floor joists across the face of rafters or studding, or in attics and roof spaces that are accessible. Where this space is not accessible by permanent stairs or ladders, protection is required only within 6 ft of the nearest edge of the scuttle hole or attic entrance.

**(B) Along the Side of Floor Joists.** When MC cable is run on the side of rafters, studs, or floor joists, no protection is required if the cable is installed and supported so the nearest outside surface of the cable or raceway is not less than $1^{1}/_{4}$ in. from the nearest edge of the framing member where nails or screws are likely to penetrate. See 300.4(D).

## 330.24 Bends

All bends shall be made so the cable will not be damaged, and the radius of the curve of the inner edge of any bend shall not be less than:

**(A) Smooth-Sheath Cables.** Smooth-sheath cables cannot be bent in a manner that the bending radius of the inner edge of the cable is less than:

(1) Ten times the external diameter of the metallic sheath for cable not more than $^{3}/_{4}$ in. in external diameter;

(2) Twelve times the external diameter of the metallic sheath for cable more than $^{3}/_{4}$ in. but not more than $1^{1}/_{2}$ in. in external diameter; and

(3) Fifteen times the external diameter of the metallic sheath for cable more than $1^{1}/_{2}$ in. in external diameter.

**(B) Interlocked- or Corrugated-Sheath.** Interlocked- or corrugated-sheath cables shall not be bent where the bending radius of the inner edge of the cable is less than seven times the external diameter of the cable.

## 330.30 Securing and Supporting

MC cable shall be supported and secured at intervals not exceeding 6 ft. Figure 330-2

**(A) Horizontal Runs.** MC cable installed horizontally in bored or punched holes in wood or metal framing members, or notches in wooden members, is considered supported, but the cable shall be secured within 1 ft of termination. Figure 330-3

**(B) Unsupported.** MC cable can be unsupported where the cable is:

(1) **Fished.** Fished between concealed access points in finished buildings or structures and support is impracticable.

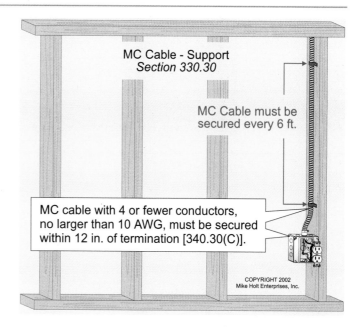

**Figure 330-2**

(2) **Accessible Ceiling.** Not more than 6 ft of unsupported MC cable is permitted from the last point of support within an accessible ceiling for the connection of luminaires or equipment. Figure 330-4

**(C) Termination.** MC cable with conductors 10 AWG and smaller shall be securely fastened within 1 ft of termination to prevent stressing the terminal fittings. See Figures 330-2 and 330-3.

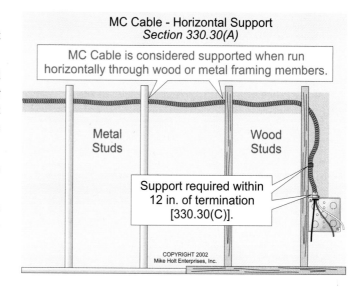

**Figure 330-3**

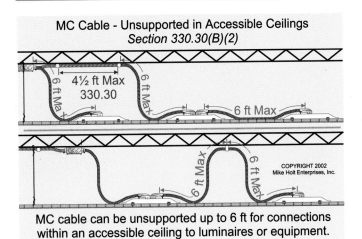

MC Cable - Unsupported in Accessible Ceilings
*Section 330.30(B)(2)*

4½ ft Max
330.30

6 ft Max

6 ft Max

COPYRIGHT 2002
Mike Holt Enterprises, Inc.

MC cable can be unsupported up to 6 ft for connections within an accessible ceiling to luminaires or equipment.

**Figure 330-4**

### 330.40 Fittings

Fittings used with MC cable shall be listed and identified for use with it [300.15]. Figure 330-5

> **AUTHOR'S COMMENT:** The NEC does not require anti-short bushings (red heads) for termination of MC cable; if supplied with the cable, it would be a good practice to use them at terminations.

### 320.80 Conductor Ampacities

Conductor ampacity is determined in accordance with 310.15, based on the 90°C insulation rating of the conductors and the terminal ratings of the fittings. See 110.14(C).

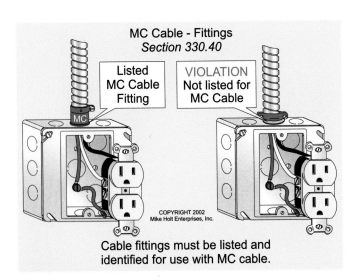

MC Cable - Fittings
*Section 330.40*

Listed MC Cable Fitting

VIOLATION Not listed for MC Cable

COPYRIGHT 2002
Mike Holt Enterprises, Inc.

Cable fittings must be listed and identified for use with MC cable.

**Figure 330-5**

**(A) Thermal Insulation.** MC cable installed in thermal insulation shall have the conductors rated at 90°C, but the ampacity shall be that of 60°C conductors in accordance with Table 310.16.

## Part III. Construction Specifications

### 330.108 Grounding

Interlocked MC cable is not listed as an equipment grounding conductor; therefore, it shall have an equipment grounding conductor installed within the metal armor wrap. See 250.118(11). Figure 330-6

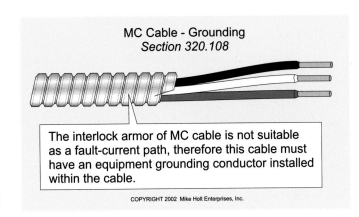

MC Cable - Grounding
*Section 320.108*

The interlock armor of MC cable is not suitable as a fault-current path, therefore this cable must have an equipment grounding conductor installed within the cable.

COPYRIGHT 2002 Mike Holt Enterprises, Inc.

**Figure 330-6**

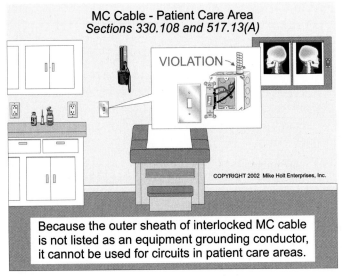

MC Cable - Patient Care Area
*Sections 330.108 and 517.13(A)*

VIOLATION

VIOLATION

COPYRIGHT 2002 Mike Holt Enterprises, Inc.

Because the outer sheath of interlocked MC cable is not listed as an equipment grounding conductor, it cannot be used for circuits in patient care areas.

**Figure 330-7**

**AUTHOR'S COMMENT:** Because the outer sheath of interlocked MC cable is not listed as an equipment grounding conductor [250-118(11)], it cannot be used for circuits in patient care areas [517.13(A)]. Figure 330-7. MC cable of the interlocked type containing two equipment grounding conductors can be used for isolated circuits as permitted by 250.146(D). Figure 330-8

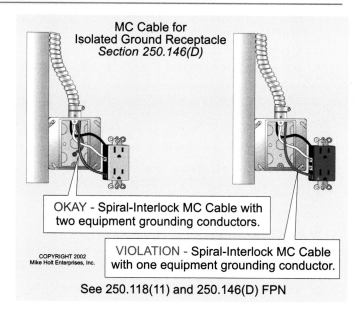

MC Cable for
Isolated Ground Receptacle
*Section 250.146(D)*

OKAY - Spiral-Interlock MC Cable with two equipment grounding conductors.

VIOLATION - Spiral-Interlock MC Cable with one equipment grounding conductor.

COPYRIGHT 2002
Mike Holt Enterprises, Inc.

See 250.118(11) and 250.146(D) FPN

*Figure 330-8*

## Article 330

1. MC cable shall not be used where exposed to the following destructive corrosive conditions, unless the metallic sheath is suitable for the conditions or is protected by material suitable for the conditions:
   (a) Direct burial in the earth.                    (b) In concrete.
   (c) In cinder fill.                                (d) all of these

2. MC cable installed through, or parallel to, framing members must be protected against physical damage from penetration by screws or nails by $1^1/_4$ in. separation or protected by a suitable metal plate.
   (a) True                    (b) False

3. Smooth-sheath MC cable that has an external diameter of not greater than 1 in. shall have a bending radius of not more than _____ times the cable external diameter.
   (a) 5             (b) 10             (c) 12             (d) 13

4. MC cable shall be supported and secured at intervals not exceeding _____.
   (a) 3 ft             (b) 6 ft             (c) 4 ft             (d) 2 ft

5. MC cable can be unsupported where it is:
   (a) Fished between concealed access points in finished buildings or structures and support is impracticable.
   (b) Not more than 2 ft in length at terminals where flexibility is necessary.
   (c) Not more than 6 ft from the last point of support within an accessible ceiling for the connection of luminaires.
   (d) a or c

# Article 334
## Nonmetallic-Sheathed Cable
## (Types NM and NMC)

## Part I. General

### 334.1 Scope

This article covers the use, installation, and construction specifications of nonmetallic-sheathed cable, Type NM.

### 334.2 Definition

NM cable is a wiring method enclosing two or three insulated conductors, 14 AWG through 2 AWG, within a nonmetallic outer cover. Because this cable is nonmetallic, it contains a separate equipment grounding conductor. NM cable is a common wiring method used for residential and commercial branch circuits. It is called Romex® by most electricians.

### 334.6 Listed

Types NM shall be listed. See 334.116.

## Part II. Installation

### 334.10 Uses Permitted

NM Cable can be used in:

(1) One- and two-family dwellings of any height. Figure 334-1

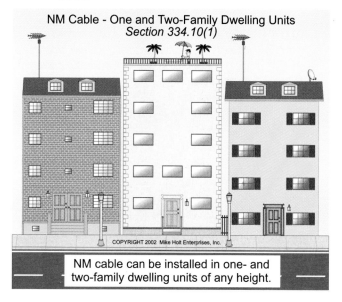

**NM cable can be installed in one- and two-family dwelling units of any height.**

**Figure 334-1**

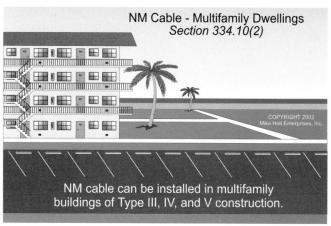

**NM cable can be installed in multifamily buildings of Type III, IV, and V construction.**

**Figure 334-2**

(2) Multifamily dwellings of Type III, IV, and V construction. Figure 334-2

(3) In other structures of Type III, IV, and V construction, concealed within walls, floors, or ceilings that provide a thermal barrier of material that has at least a 15-minute finish rating, as identified in listings of fire-rated assemblies. Figure 334-3

FPN No. 1: Building construction types are defined in NFPA 220 – Standard on Types of Building Construction or the applicable building code or both.

FPN No. 2: See Annex F for determination of building types.

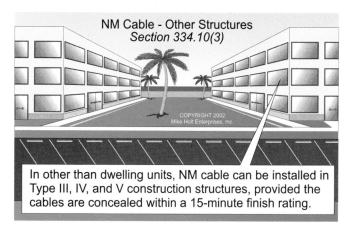

**In other than dwelling units, NM cable can be installed in Type III, IV, and V construction structures, provided the cables are concealed within a 15-minute finish rating.**

**Figure 334-3**

**(A) Type NM.** Where NM cable is permitted, it can be:

(1) Exposed or concealed in normally dry locations, except as prohibited in 334.10(3) and 334.12(1).

(2) Fished in air voids in masonry block or tile walls where such walls are not exposed or subject to excessive moisture or dampness.

**(B) NMC Cable.** The outer cover of NMC cable is similar to NM cable, except that it is corrosion resistant, as shown by the letter "C." This cable can be installed:

(1) Exposed or concealed in dry, moist, damp, and corrosive locations.

(2) Inside or outside masonry or tile walls.

(3) In chases of concrete, masonry, or adobe when protected from nails and screws with a metal plate.

## 334.12 Uses Not Permitted

**(A) Types NM, and NMC.** Types NM, and NMC cables shall not be used:

(1) Suspended ceilings, except for one- and two-family and multifamily dwellings. Figure 334-4

(2) Services [230.43].

(3) Hazardous (classified) locations of commercial garages as defined in 511.3.

(4) Places of assembly [518.4] and theaters [520.5].

(5) Motion picture studios [Article 530].

(6) Storage battery rooms [Article 480].

(7) Hoistways [Article 620].

(8) Embedded in concrete.

(9) Hazardous (classified) locations, except where permitted in the following:
   a. 501.4(B), Ex.
   b. 502.4(B), Ex. No. 1
   c. 504.20

(10) Type NM cable shall not be installed:
   a. Where exposed to corrosive fumes or vapors.
   b. Where embedded in masonry, concrete, adobe, fill, or plaster.
   c. In a shallow chase in masonry, concrete, or adobe and covered with plaster, adobe, or similar finish.
   d. Where exposed or subject to excessive moisture or dampness.

> **AUTHOR'S COMMENT:** NM cable is not permitted as a wiring method for swimming pool equipment [680.23(F)(2)], patient care areas [517.13], or environmental air space [300.22(C)]

## 334.15 Exposed

**(A) Surface of the Building.** Exposed NM cable shall closely follow the surface of the building.

**(B) Protected from Physical Damage.** Nonmetallic-sheathed cable shall be protected from physical damage by RMC, IMC, Schedule 80 RNC [352.10(F)], EMT, guard strips, or other means. When installed in a raceway, the cable shall be protected from abrasion by a fitting installed on the end of the raceway. See 300.15(C).

> **AUTHOR'S COMMENT:** Where NM cable is installed in a metal raceway for protection, the raceway is not required to be grounded (bonded), because the cable contains an equipment grounding conductor. See 250.86 Ex 2, and 300.12 Ex.

**(C) Unfinished Basements.** NM cables sized 6/2, 8/3 and larger can be mounted directly to the bottom of the joist without protection. Smaller cables shall be run through bored holes in the joist or protected by running boards.

## 334.17 Through or Parallel to Framing Members

NM cable installed through or parallel to framing members shall be protected against physical damage from penetration by screws or nails, by a $1\frac{1}{4}$ in. separation or by a suitable metal plate. See 300.4(A). Figure 334-5

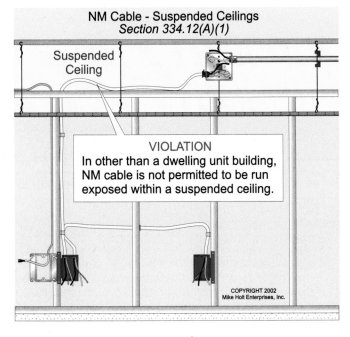

NM Cable - Suspended Ceilings
Section 334.12(A)(1)

Suspended Ceiling

**VIOLATION**
In other than a dwelling unit building, NM cable is not permitted to be run exposed within a suspended ceiling.

COPYRIGHT 2002
Mike Holt Enterprises, Inc.

**Figure 334-4**

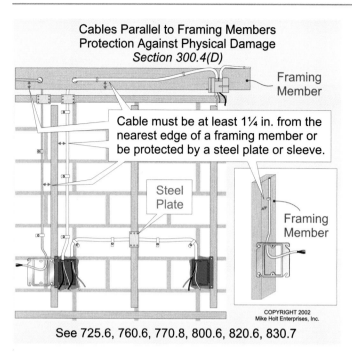

Cables Parallel to Framing Members
Protection Against Physical Damage
Section 300.4(D)

Framing Member

Cable must be at least 1¼ in. from the nearest edge of a framing member or be protected by a steel plate or sleeve.

Steel Plate

Framing Member

COPYRIGHT 2002
Mike Holt Enterprises, Inc.

See 725.6, 760.6, 770.8, 800.6, 820.6, 830.7

**Figure 334-5**

Where NM cable passes through holes in metal studs, a listed bushing or listed grommet is required [300.4(B)(1)] to be in place before the cable is installed. Figure 334-6

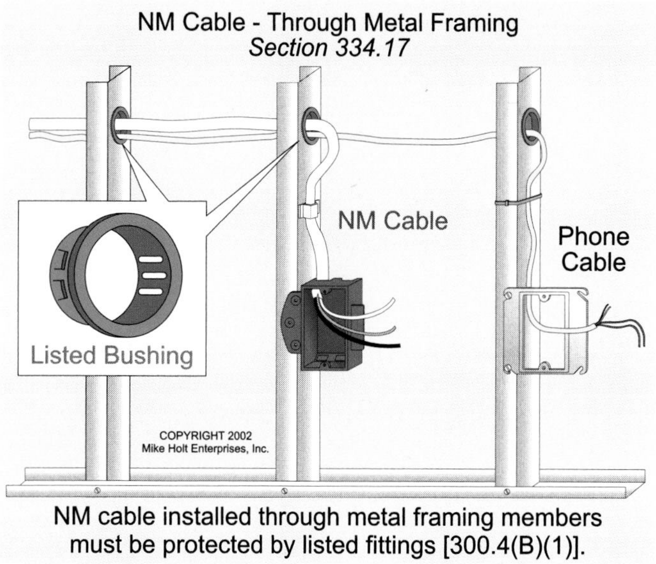

NM Cable - Through Metal Framing
Section 334.17

NM Cable

Listed Bushing

Phone Cable

COPYRIGHT 2002
Mike Holt Enterprises, Inc.

NM cable installed through metal framing members must be protected by listed fittings [300.4(B)(1)].

**Figure 334-6**

## 334.23. Attics and Roof Spaces

**(A) On the Surface of Floor Joists, Rafters, or Studs.** Substantial guards shall protect the cable where run across the top of floor joists, or within 7 ft of floor or floor joists across the face of rafters or studding, or in attics and roof spaces that are accessible. Where this space is not accessible by permanent stairs or ladders, protection is required only within 6 ft of the nearest edge of the scuttle hole or attic entrance.

**(B) Along the Side of Floor Joists.** When NM cable is run on the side of rafters, studs, or floor joists, no protection is required if the cable is installed and supported so the nearest outside surface of the cable or raceway is not less than $1^1/_4$ in. from the nearest edge of the framing member where nails or screws are likely to penetrate. See 300.4(D).

## 334.24 Bends

When the cable is bent, it shall not be damaged. The radius of the curve of the inner edge of any bend shall not be less than five times the diameter of the cable. Figure 334-7

## 334.30 Secured or Supported

Approved staples or straps shall secure NM cable or cable ties in such a manner that the cable will not be damaged. The cable shall be secured within 1 ft of every box, cabinet, enclosure, or termination fitting, except as permitted by 314.17(C) Ex. or 312.5(C) Ex., and at intervals not exceeding $4^1/_2$ ft. The cable is not to be stapled on edge. Figure 334-8

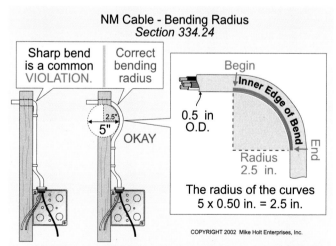

NM Cable - Bending Radius
Section 334.24

Sharp bend is a common VIOLATION.

Correct bending radius

OKAY

Begin
Inner Edge of Bend
End

0.5 in O.D.

Radius 2.5 in.

The radius of the curves
5 x 0.50 in. = 2.5 in.

COPYRIGHT 2002 Mike Holt Enterprises, Inc.

The radius of the curve of the inner edge of any bend must not be less than 5 times the diameter of the cable.

**Figure 334-7**

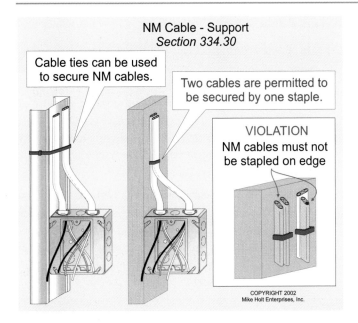

**Figure 334-8**

**(A)** **Horizontal Runs.** NM cable installed horizontally in bored or punched holes in wood or metal framing members, or notches in wooden members, is considered supported, but the cable shall be secured within 1 ft of termination. Figure 334-9

**(B)** **Unsupported.** NM cable can be unsupported where it is:

    **(1)** **Fished.** Fished between concealed access points in finished buildings or structures and support is impracticable.

    **(2)** **Accessible Ceiling.** Not more than $4^1/_2$ ft of unsupported cable is permitted from the last point of support within an accessible ceiling for the connection of luminaires or equipment.

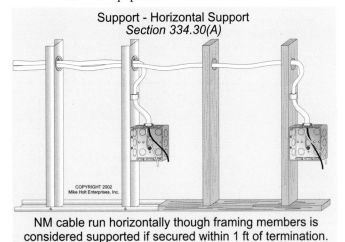

NM cable run horizontally though framing members is considered supported if secured within 1 ft of termination.

**Figure 334-9**

## 334.80 Ampacity

Conductors within NM cable shall be rated 90°C, but the ampacity of the conductors shall be based on the 60°C temperature ratings as listed in Table 310.16. The 90°C ampacity rating can be used for ampacity adjustment purposes, provided the final adjusted ampacity does not exceed the value of a 60°C rated conductor.

**AUTHOR'S COMMENT:** For more information on sizing conductors, see 110.14(C), 310.10 and 310.15 as well as http://www.mikeholt.com/studies/conductor.htm.

**Question:** What size NM cable with THHN conductors is required to supply a 10 kW, 240V, 1Ø fixed space heater that has a 3A blower motor? Terminal rating of 75°C Figure 334-10

(a) 4                      (b) 4

(c) 6                      (d) 8

**Answer:** (b) 4 AWG, According to 424.3(B), the conductors and overcurrent protection device for electric space-heating equipment shall be sized no less than 125 percent of the total heating load.

Step 1. Determine Total Load in Amperes

    10,000W/240V + 3A = 44.67A

Step 2. Size Conductor and Protection

    44.67A × 1.25 = 56A, 4 AWG THHN, rated 70A at 60°C, with 60A protection device

**AUTHOR'S COMMENT:** AC, MC, and SE cables do not have a 60°C ampacity limitation; therefore, 6/2 AC, MC, or SE cable (rated 65A at 75°C) could be used. I know this does not make sense, but it's the *Code.*

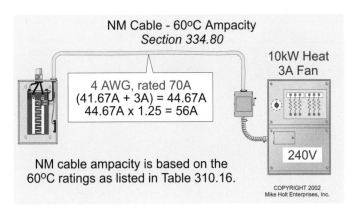

NM cable ampacity is based on the 60°C ratings as listed in Table 310.16.

**Figure 334-10**

## Article 334

1. NM cable can be installed in multifamily dwellings of Types III, IV, and V construction except as prohibited in 334.
   (a) True (b) False

2. NM cable can be installed as open runs in dropped or suspended ceilings in other than one- and two-family and multifamily dwellings.
   (a) True (b) False

3. NM cable must closely follow the surface of the building finish or running boards when run exposed.
   (a) True (b) False

4. Where NMC cable is run at angles with joists in unfinished basements, it shall be permissible to secure cables not smaller than _____ conductors directly to the lower edges of the joist.
   (a) two, 6 AWG (b) three, 8 AWG (c) three, 10 AWG (d) a or b

5. NM cable installed through or parallel to framing members must be protected against physical damage from penetration by screws or nails. Grommets or bushings for the protection of NM cables must be _____ for the purpose, and they must remain in place.
   (a) marked (b) approved (c) identified (d) listed

6. NM cables runnung horizontally through framing are considered supported and secured where such support does not exceed $4^1/_2$ ft intervals and the NM cable is securely fastened in place by an approved means within 12 in. of each box, cabinet, conduit body, or other NM cable termination.
   (a) True (b) False

7. Two conductor NM cables cannot be stapled on edge.
   (a) True (b) False

8. The ampacity of NM cable shall be that of 60°C conductors, as listed in 310.15. However, the 90°C rating can be used for ampacity derating purposes, provided the final derated ampacity does not exceed that of a _____ rated conductor.
   (a) 120°C (b) 60°C
   (c) 90°C (d) none of these

9. Insulation rating of ungrounded conductors in NM cable must be _____.
   (a) 60°C (b) 75°C (c) 90°C (d) any of these

10. The difference in the construction specifications between NM cable and NMC cable is that NMC cable is _____.
    (a) corrosion resistant (b) flame retardant (c) fungus-resistant (d) a and c

# Article 336
# Power and Control Tray Cable (Type TC)

## Part I. General

### 336.1 Scope

This article covers the use, installation, and construction specifications of power and control tray cable, Type TC.

### 336.2. Definition

TC cable is a factory assembly of two or more insulated conductors under a nonmetallic sheath, for installation in cable trays, in raceways, or where supported by a messenger wire.

## Part II. Installation

### 336.10. Uses Permitted

(1)  For power, lighting, control, and signal circuits.

(2)  In cable trays, or in raceways, or where supported in outdoor locations by a messenger wire.

(3)  In cable trays in hazardous (classified) locations as permitted in Articles 392, 501, 502, 504, and 505 in industrial establishments where the conditions of maintenance and supervision ensure that only qualified persons will service the installation.

(4)  For Class 1 circuits as permitted in Article 725.

(5)  For nonpower limited fire alarm circuits if conductors comply with 760.27.

(6)  In industrial establishments, where the conditions of maintenance and supervision ensure that only qualified persons will service the installation, and where the cable is not subject to physical damage. The cable shall be supported and secured at intervals not exceeding 6 ft. Equipment grounding for the utilization equipment shall be provided by an equipment grounding conductor within the cable.

(7)  In wet locations, where the cable is impervious to moisture and the conductors are listed for use in a wet location.

### 336.12. Uses Not Permitted

(1)  Where exposed to physical damage.

(2)  As open cable on brackets or cleats.

(3)  Exposed to direct rays of the sun, unless identified as sunlight-resistant.

(4)  Directly buried, unless identified for such use.

# *Article 336*

1.  TC cable shall be permitted to be used _____.
    (a) for power and lighting circuits
    (b) in cable trays in hazardous locations
    (c) in Class 1 control circuits
    (d) all of these

2.  TC tray cable shall not be installed _____.
    (a) where it will be exposed to physical damage
    (b) as open cable on brackets or cleats
    (c) direct buried, unless identified for such use
    (d) all of these

# Article 338
## Service-Entrance Cables
## (Types SE and USE)

## Part I. General

### 338.1 Scope

This article covers the use, installation, and construction specifications of service-entrance cable, Type SE.

### 338.2 Definition

**Service-Entrance Cable.** Service-entrance cable can be a single conductor or multiconductor assembly with or without an overall covering. This cable is used primarily for services not over 600V, but can also be used for feeders and branch circuits. Figure 338-1

**Type SE.** SE cable has a flame-retardant, moisture-resistant covering and is permitted only in aboveground installations. This cable can be used for branch circuits or feeders when installed according to 338.10.

> **AUTHOR'S COMMENT:** SER cable is simply SE cable with an insulated neutral, resulting in three insulated conductors with an uninsulated ground. SER cable is round, where 2-wire SE cable is flat.

**Type USE.** USE cable is identified for underground use. Its covering is moisture resistant but not flame retardant.

## Part II. Installation

### 338.10 Uses Permitted

**(A) Services.** Type SE and USE cable used for service-entrance conductors shall be installed in accordance with the requirements of Article 230, particularly 230.50 and 230.51.

Type USE cable used for service-laterals can emerge above ground where installed in accordance with 300.5.

**(B) Branch Circuits or Feeders.** Type SE cable can be used for interior wiring as permitted in the following. Figure 338-2

> **AUTHOR'S COMMENT:** USE cable cannot be used for interior wiring because it does not have a flame-retardant insulation rating.

**(1) Insulated Grounded (neutral) Conductor.** Service-entrance cable with an insulated grounded (neutral) conductor, such as Type SER, can be used for branch circuits and feeders, but the uninsulated conductor within the cable can be used only for equipment grounding (bonding) purposes [250.118].

**(2) Uninsulated Grounded Conductor.** Service-entrance cable with an uninsulated grounded (neutral) conductor, such as SE and USE, can be used only for branch

Service-Entrance Cable
*Section 338.2 Definition*

SE cable is typically used for services, but it can be used for interior wiring as permitted in 338.10(B).

**Figure 338-1**

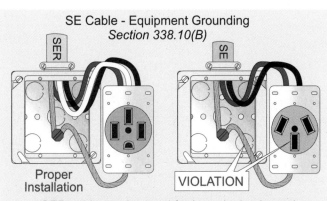

SE Cable - Equipment Grounding
*Section 338.10(B)*

SER cable can be used for branch circuits and feeders, but the uninsulated conductor can only be used for equipment grounding.

**Figure 338-2**

circuits or feeders where the uninsulated conductor is used for equipment grounding.

**(4) Installation Methods for Branch Circuits and Feeders.**

**(a) Interior Installations.** SE cable used for interior wiring shall be installed in accordance with the requirements of NM cable, Parts I and II of Article 334, excluding 334.80.

**AUTHOR'S COMMENT:** This means that when SE cable is used for branch circuits and feeders, it must be sized based on the terminal rating of the equipment in accordance with 110.14. The 60°C ampacity limitations of NM cable [334.80] do not apply to SE or SER cable. Figure 338-3

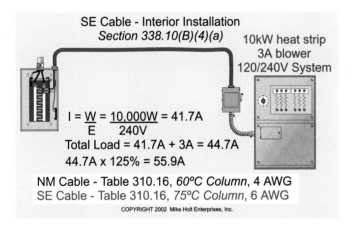

**Figure 338-3**

**(b) Exterior Installations.** SE cable installed outdoors for branch circuits and feeders shall be installed in accordance with the requirements of Article 225 – Outside Wiring. When installed underground, the cable shall comply with the underground requirements in Article 340 – Underground Feeder Cable (Type UF), and where USE cable emerges above ground, it shall be protected in accordance with 300.5(D).

**AUTHOR'S COMMENT:** SE cable is not permitted in ducts, plenums, or other spaces used for environmental air [300.22], for swimming pool wiring [680.23(F)(2)], or for wiring in patient care areas [517.13(A)].

### 338.24 Bends

Bends in cable shall be made so the protective coverings of the cable will not be damaged, and the radius of the curve of the inner edge is not less than five times the diameter of the cable.

## Article 338

1. _____ is a type of multiconductor cable permitted for use as an underground service-entrance cable.
   (a) SE          (b) NMC          (c) UF          (d) USE

2. USE or SE cable must have a minimum of _____ conductors (including the uninsulated one) in order for one of the conductors to be uninsulated.
   (a) one          (b) two          (c) three          (d) four

# Article 340
# Underground Feeder and Branch-Circuit Cable (Type UF)

## 340.1 Scope

This article covers the use, installation, and construction specifications of underground feeder and branch-circuit cable, Type UF.

## 340.2 Definition

UF cable is a moisture, fungus, and corrosion-resistant cable system suitable for direct burial in the earth. It comes in sizes 12 AWG through 4/0 AWG [340.104]. The covering of multiconductor UF cable is molded plastic that encapsulates the insulated conductors.

> **AUTHOR'S COMMENT:** Because the covering of UF cable encapsulates the insulated conductors, it's difficult to strip off the outer jacket to gain access to the conductors, but this encapsulated cover provides excellent corrosion protection. Be careful you don't damage the conductor insulation or cut yourself when you remove the encapsulated outer cover.

## 340.10 Uses Permitted

(1) Underground, in accordance with 300.5.

(2) As a single conductor in the same trench or raceway with the other circuit conductors.

(3) As interior or exterior wiring in wet, dry, or corrosive locations.

(4) As NM cable when installed in accordance with Article 334.

(5) For solar photovoltaic systems in accordance with 690.31.

(6) As single-conductor cables for nonheating leads for heating cables as provided in 424.43.

(7) Supported by cable trays.

## 340.12 Uses Not Permitted

(1) Services [230.43].

(2) Commercial garages [511.3].

(3) Theaters [520.5].

(4) Motion picture studios [530.11].

(5) Storage battery rooms [Article 480].

(6) Hoistways [Article 620].

(7) Hazardous (classified) locations.

(8) Embedded in concrete.

(9) Exposed to direct sunlight unless identified.

(10) Where subject to physical damage. Figure 340-1

(11) Overhead messenger supported wiring.

> **AUTHOR'S COMMENT:** UF cable is not permitted in ducts, plenums, or other spaces used for environmental air [300.22], swimming pool wiring [680.23(F)(2)], or wiring in patient care areas [517.13].

## 340.24 Bends

Bends in cables shall be made so the protective coverings of the cable will not be damaged, and the radius of the curve of the inner edge is not less than five times the diameter of the cable.

## 340.80 Ampacity

The ampacity of conductors in UF cable shall be based on 60°C ratings as listed in Table 310.16.

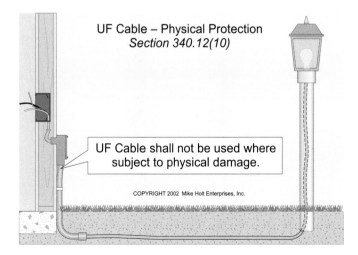

UF Cable – Physical Protection
*Section 340.12(10)*

UF Cable shall not be used where subject to physical damage.

COPYRIGHT 2002 Mike Holt Enterprises, Inc.

*Figure 340-1*

# Article 340

1.  UF cable shall not be used where subjected to physical damage. When this cable is subject to physical damage, it shall be protected by a suitable method such as a raceway.
    (a) True                  (b) False

2.  UF cable shall not be used _____.
    (a) in any hazardous (classified) location
    (b) embedded in poured cement, concrete, or aggregate
    (c) where exposed to direct rays of the sun, unless identified as sunlight-resistant
    (d) all of these

3.  The maximum size underground UF cable is _____.
    (a) 14           (b) 10           (c) 1/0           (d) 4/0

4.  The overall covering of UF cable shall be _____.
    (a) flame retardant                          (b) moisture, fungus, and corrosion-resistant
    (c) suitable for direct burial in the earth       (d) all of these

# Article 342
# Intermediate Metal Conduit (IMC)

## Part I. General

### 342.1 Scope

This article covers the use, installation, and construction specifications of Intermediate Metal Conduit (IMC) and associated fittings.

### 342.2 Definition

IMC is a listed steel raceway of circular cross section that can be threaded with integral or associated couplings, approved for the installation of electrical conductors and used with listed fittings to provide electrical continuity.

> **AUTHOR'S COMMENT:** The wall thickness of IMC is thinner than RMC, so it has a greater interior cross-sectional area and can hold more conductors. In addition, it's lighter and it can be used in all of the same locations as rigid conduit. The type of steel from which IMC is manufactured, the process by which it is made, and the corrosion protection applied are equal to or superior to that of rigid conduit. Impact, crush, and hydrostatic tests on both IMC and RMC show IMC to be equal to or better than rigid conduit.

### 342.6 Listing Requirements

IMC, factory elbows, couplings, and associated fittings shall be listed for the purpose.

## Part II. Installation

### 342.10 Uses Permitted

(A) **All Atmospheric Conditions and Occupancies.** IMC can be installed in all atmospheric conditions and occupancies.

(B) **Corrosion Environments.** IMC, elbows, couplings, and fittings can be installed in concrete, in direct contact with the earth, or in areas subject to severe corrosive influences where protected by corrosion protection and judged suitable for the condition. See 300.6

(C) **Wet Locations.** All support fittings, such as screws, straps, etc. installed in a wet location shall be made of corrosion-resistant material or be protected by corrosion-resistant coatings. See 300.6.

> *CAUTION: Supplementary coatings for corrosion protection have not been investigated by a product testing and listing agency, and these coatings are known to cause cancer in laboratory animals. I know of a case where an electrician was taken to the hospital for lead poisoning, after using a supplemental coating product in a poorly ventilated area. As with all products, be sure to read and follow all product instructions, particularly when volatile organic compounds may be in the material.*

### 342.14 Dissimilar Metals

Where practical, contact with dissimilar metals should be avoided to prevent the deterioration of the metal because of galvanic action (corrosion). However, aluminum fittings and enclosures can be used with steel IMC.

### 342.20 Size

(A) **Minimum Size.** $1/2$ in.

(B) **Maximum Size.** 4 in.

### 342.22 Conductor Fill

Raceways shall be large enough to permit the installation and removal of conductors without damaging the conductors' insulation. When all conductors in a raceway are the same size and insulation, the number of conductors permitted can be found in Annex C for the raceway type.

> **Question:** How many 10 AWG THHN conductors can be installed in 1 in. IMC?
>
> **Answer:** 18 conductors, Annex C, Table C4

> **AUTHOR'S COMMENT:** See 300.17 for additional examples on how to size raceways when conductors are not all the same size.

### 342.24 Bends

Raceway bends shall not be made in any manner that would damage the raceway or significantly change its internal diameter (no kinks). This is accomplished by complying with the bending radius requirements in Table 344.24.

**AUTHOR'S COMMENT:** This is usually not a problem because benders are made to comply with this Table, however, when using a hickey bender (short-radius bender), be careful not to over-bend the raceway.

## 342.26 Number of Bends (360°)

To reduce the stress and friction on the conductors' insulation, the maximum number of bends (including offsets) between pull points cannot exceed 360°. Figure 342-1

**AUTHOR'S COMMENT:** There is no maximum distance between pull boxes because this is a design issue, not a safety issue.

## 342.28 Reaming

When the raceway is cut in the field, reaming is required to remove the burrs and rough edges.

**AUTHOR'S COMMENT:** It's considered an accepted practice to ream small raceways with a screwdriver or the backside of Channelocks®. However, when the raceway is cut with a three-wheel pipe cutter, a reaming tool is required to remove the sharp edge of the indented raceway. In addition, where conduit is threaded in the field, the threads shall be coated with an approved electrically conductive, corrosion-resistant compound (cold zinc). See 300.6(A).

## 342.30 Secured and Supported

IMC shall be installed as a complete system [300.10, 300.12 and 300.18(A)], and it shall be securely fastened in place and supported in accordance with the following:

**(A) Securely Fastened.** IMC shall be securely fastened within 3 ft of every box, cabinet, or termination fitting. Figure 342-2

When structural members do not permit the raceway to be secured within 3 ft of a box or termination fitting, the raceway shall be secured within 5 ft of termination. Figure 342-3

**AUTHOR'S COMMENT:** Support is not required within 3 ft of a coupling.

**(B) Supports**

(1) IMC shall be supported at intervals not exceeding 10 ft.

(2) Straight horizontal runs made with threaded couplings can be supported in accordance with the distances listed in Table 344.30(B)(2).

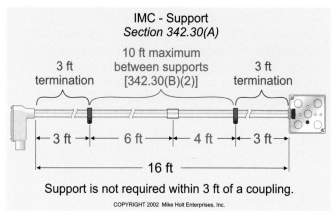

**Figure 342-2**

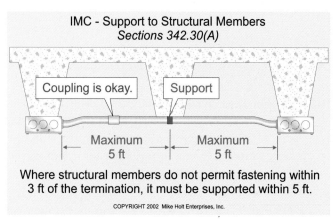

**Figure 342-3**

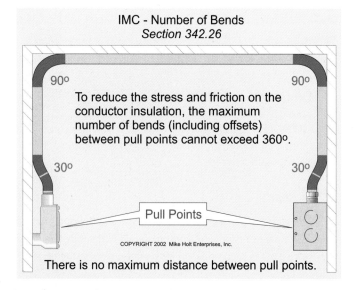

**Figure 342-1**

**Table 344.30(B)(2)**

| Conduit Size | Max. Distance Between Rigid Metal Supports |
|---|---|
| $^1/_2$ - $^3/_4$ | 10 ft |
| 1 | 12 ft |
| $1^1/_4$ - $1^1/_2$ | 14 ft |
| 2 - $2^1/_2$ | 16 ft |
| 3 and larger | 20 ft |

**(3) Vertical Risers.** Exposed vertical risers for fixed equipment can be supported at intervals not exceeding 20 ft, if the conduit is made up with threaded couplings, firmly supported at the top and bottom of the riser, and no other means of support is available. Figure 342-4

**(4) Horizontal Runs.** Conduits installed horizontally in bored or punched holes in wood or metal framing members, or notches in wooden members, are considered supported, but the raceway shall be secured within 3 ft of termination.

## 342.42 Couplings and Connectors

**(A) Installation.** Threadless couplings and connectors shall be made up wrenchtight to maintain an effective low-impedance ground-fault current path to safely conduct fault current. See 250.4(A)(5), 250.96(A), and 300.10. When IMC is installed in concrete, the fittings shall be concrete tight, and when installed in a wet location, the fittings shall be the raintight type. Threadless couplings and connectors shall not be used on threaded conduit ends unless listed for the purpose.

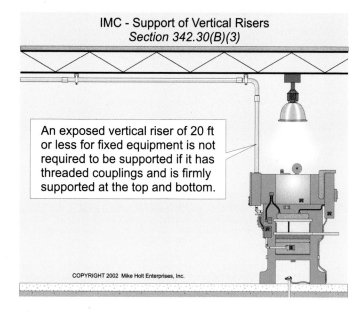

**Figure 342-4**

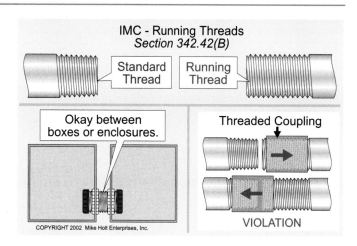

**Figure 342-5**

**AUTHOR'S COMMENT:** Some locknuts have been found to burn clear before the fault was cleared because loose threads increase the impedance of the fault-current path.

**(B) Running Threads.** Running threads are not permitted for the connection of couplings, but they can be used at other locations. Figure 342-5

## 342.46 Bushings.

To protect conductors from abrasion, a metal or plastic bushing shall be installed on conduit termination threads, unless the design of the box, fitting, or enclosure is such as to afford equivalent protection. See 300.4(F) for bushing requirements for raceway terminations. Figure 342-6

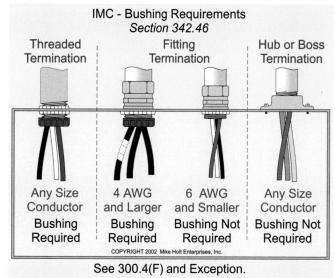

**Figure 342-6**

# Article 342

1.  When practical, contact of dissimilar metals shall be avoided anywhere in a raceway system to prevent _____.
    (a) corrosion      (b) galvanic action      (c) shorts      (d) none of these

2.  One-inch IMC raceway containing three or more conductors can be conductor filled to _____ percent.
    (a) 53      (b) 31      (c) 40      (d) 60

3.  The cross-sectional area of 1 in. IMC is approximately _____
    (a) 1.22 sq in.      (b) 0.62 sq in.      (c) 0.96 sq in.      (d) 2.13 sq in.

4.  A run of IMC shall not contain more than the equivalent of _____ quarter bend(s) including the offsets located immediately at the outlet or fitting.
    (a) 1      (b) 2      (c) 3      (d) 4

5.  IMC shall be firmly fastened within _____ of each outlet box, junction box, device box, fitting, cabinet, or other conduit termination.
    (a) 12 in.      (b) 18 in.      (c) 2 ft      (d) 3 ft

6.  One-inch IMC must be supported every _____.
    (a) 8 ft      (b) 10 ft      (c) 12 ft      (d) 14 ft

7.  For industrial machinery, straight exposed vertical risers of IMC with threaded couplings are permitted, if supported at the top and bottom no more than _____ apart.
    (a) 10 ft      (b) 12 ft      (c) 15 ft      (d) 20 ft

8.  Threadless couplings approved for use with IMC in wet locations shall be of the _____ type.
    (a) rainproof      (b) raintight      (c) moistureproof      (d) concrete-tight

9.  Threadless couplings and connectors must not be used on threaded IMC ends unless the fittings are listed for the purpose.
    (a) True      (b) False

10. Running threads of IMC shall not be used on conduit for connection at couplings.
    (a) True      (b) False

# Article 344
# Rigid Metal Conduit (RMC)

## Part I. General

### 344.1 Scope

This article covers the use, installation, and construction specifications of Rigid Metal Conduit (RMC) and associated fittings.

### 344.2 Definition

RMC is a listed metal raceway of circular cross section with integral or associated couplings, approved for the installation of electrical conductors and used with listed fittings to provide electrical continuity.

> **AUTHOR'S COMMENT:** When the mechanical and physical characteristics of RMC are desired and an aggressive environment is anticipated to be present, a type of PVC-coated raceway system is commonly used. This type of raceway is used in the petrochemical industry; the common trade name is "Plasti-bond®." The benefits of the improved corrosion protection can be achieved only when the system is properly installed. All joints shall be sealed in accordance with the manufacturer's instructions, and the coating shall not be damaged with tools such as benders, pliers and pipe wrenches. Couplings are available that have an extended skirt that shall be properly sealed after installation.

### 344.6 Listing Requirements

RMC, factory elbows, couplings, and associated fittings shall be listed for the purpose.

## Part II. Installation

### 344.10 Uses Permitted

**(A)** **All Atmospheric Conditions and Occupancies.** RMC can be installed in all atmospheric conditions and occupancies.

**(B)** **Corrosion Environments.** RMC, elbows, couplings, and fittings can be installed in concrete, in direct contact with the earth, or in areas subject to severe corrosive influences where protected by corrosion protection and judged suitable for the condition. See 300.6.

**(C)** **Wet Locations.** All support fittings, such as screws, straps, etc., installed in a wet location shall be made of corrosion-resistant material or protected by corrosion-resistant coatings. See 300.6.

> *CAUTION: Supplementary coatings for corrosion protection have not been investigated by a product testing and listing agency, and these coatings are known to cause cancer in laboratory animals.*

### 344.14 Dissimilar Metals

Where practical, contact with dissimilar metals should be avoided to prevent the deterioration of the metal because of galvanic action (corrosion). However, aluminum fittings and enclosures can be used with steel RMC.

### 344.20 Size

**(A)** **Minimum Size.** $1/2$ in.
**(B)** **Maximum Size.** 6 in.

### 344.22 Conductor Fill

Raceways shall be large enough to permit the installation and removal of conductors without damaging the conductors' insulation. When all conductors in a raceway are the same size and insulation, the number of conductors permitted can be found in Annex C for the raceway type.

> **Question:** How many 8 AWG THHN conductors can be installed in $1^1/_2$ in. RMC?
>
> **Answer:** 22 conductors, Annex C, Table C8

> **AUTHOR'S COMMENT:** See 300.17 for additional examples on how to size raceways when conductors are not all the same size.

### 344.24 Bends

Raceway bends shall not be made in any manner that would damage the raceway or significantly change its internal diameter (no kinks). This is accomplished by complying with the bending radius requirements in Table 344.24.

**AUTHOR'S COMMENT:** This is usually not a problem because benders are made to comply with this table; however, when using a hickey bender (short-radius bender), be careful not to over-bend the raceway.

## 344.26 Number of Bends (360°)

To reduce the stress and friction on the conductors' insulation, the maximum number of bends (including offsets) between pull points cannot exceed 360°. Figure 344-1

**AUTHOR'S COMMENT:** There is no maximum distance between pull boxes.

## 344.28 Reaming

When the raceway is cut in the field, reaming is required to remove the burrs and rough edges.

**AUTHOR'S COMMENT:** It's considered an accepted practice to ream small raceways with a screwdriver or the backside of Channelocks®. However, when the raceway is cut with a three-wheel pipe cutter, a reaming tool is required to remove the sharp edge of the indented raceway. In addition, where conduit is threaded in the field, the threads shall be coated with an approved electrically conductive, corrosion-resistant compound (cold zinc) see 300.6(A).

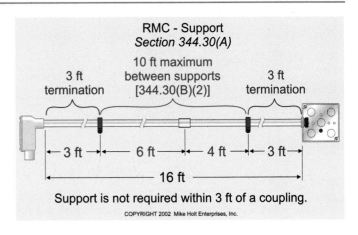

Support is not required within 3 ft of a coupling.

COPYRIGHT 2002 Mike Holt Enterprises, Inc.

**Figure 344-2**

## 344.30 Secured and Supported

RMC shall be installed as a complete system [300.10, 300.12, and 300.18(A)], and it shall be securely fastened in place and supported in accordance with the following:

**(A) Securely Fastened.** RMC shall be securely fastened within 3 ft of every box, cabinet, or termination fitting. Figure 344-2

When structural members do not permit the raceway to be secured within 3 ft of a box or termination fitting, the raceway shall be secured within 5 ft of termination. Figure 344-3

**AUTHOR'S COMMENT:** Support is not required within 3 ft of a coupling.

**(B) Supports**

(1) RMC shall be supported at intervals not exceeding 10 ft.

(2) Straight horizontal runs made with threaded couplings can be supported in accordance with the distances listed in Table 344.30(B)(2).

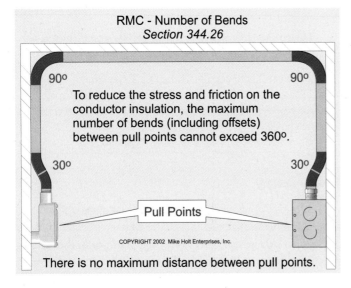

There is no maximum distance between pull points.

**Figure 344-1**

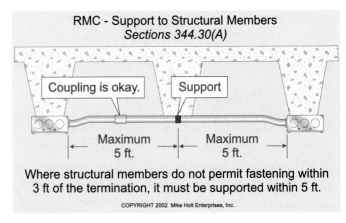

Where structural members do not permit fastening within 3 ft of the termination, it must be supported within 5 ft.

COPYRIGHT 2002 Mike Holt Enterprises, Inc.

**Figure 344-3**

**Table 344.30(B)(2)**

| Conduit Size | Max. Distance Between Rigid Metal Supports |
|---|---|
| 1/2 - 3/4 | 10 ft |
| 1 | 12 ft |
| 1 1/4 - 1 1/2 | 14 ft |
| 2 - 2 1/2 | 16 ft |
| 3 and larger | 20 ft |

**(3) Vertical Risers.** Exposed vertical risers for fixed equipment can be supported at intervals not exceeding 20 ft, if the conduit is made up with threaded couplings, firmly supported at the top and bottom of the riser, and no other means of support is available. Figure 344-4

**(4) Horizontal Runs.** Conduits installed horizontally in bored or punched holes in wood or metal framing members, or notches in wooden members, are considered supported, but the raceway shall be secured within 3 ft of termination.

## 344.42 Couplings and Connectors

**(A) Installation.** Threadless couplings and connectors shall be made up wrenchtight to maintain an effective low-impedance ground-fault current path to safely conduct fault current. See 250.4(A)(5), 250.96(A), and 300.10. When RMC is installed in concrete, the fittings shall be concrete tight, and when installed in a wet location, the fittings shall be the raintight type. Threadless couplings and connectors shall not be used on threaded conduit ends unless listed for the purpose.

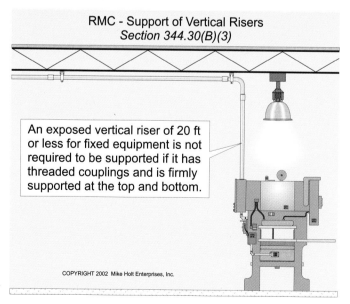

RMC - Support of Vertical Risers
Section 344.30(B)(3)

An exposed vertical riser of 20 ft or less for fixed equipment is not required to be supported if it has threaded couplings and is firmly supported at the top and bottom.

COPYRIGHT 2002 Mike Holt Enterprises, Inc.

**Figure 344-4**

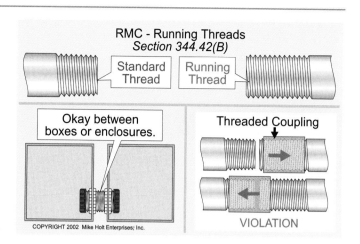

RMC - Running Threads
Section 344.42(B)

Standard Thread    Running Thread

Okay between boxes or enclosures.

Threaded Coupling

VIOLATION

COPYRIGHT 2002 Mike Holt Enterprises; Inc.

**Figure 344-5**

**AUTHOR'S COMMENT:** Some locknuts have been found to burn clear before the fault was cleared, because loose threads increase the impedance of the fault current path.

**(B) Running Threads.** Running threads are not permitted for the connection of couplings, but they can be used at other locations. Figure 344-5

## 344.46 Bushings.

To protect conductors from abrasion, a metal or plastic bushing shall be installed on conduit termination threads, unless the design of the box, fitting, or enclosure is such as to afford equivalent protection. See 300.4(F) for bushing requirements for raceway terminations without threads. Figure 344-6

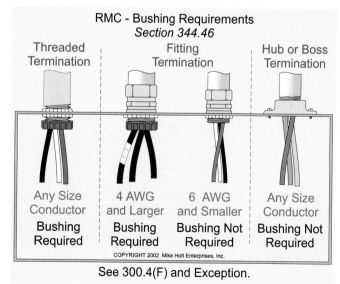

RMC - Bushing Requirements
Section 344.46

Threaded Termination — Any Size Conductor — Bushing Required

Fitting Termination — 4 AWG and Larger — Bushing Required

6 AWG and Smaller — Bushing Not Required

Hub or Boss Termination — Any Size Conductor — Bushing Not Required

COPYRIGHT 2002 Mike Holt Enterprises, Inc.

See 300.4(F) and Exception.

**Figure 344-6**

## Article 344

1.  RMC can be installed in or under cinder fill that is subjected to permanent moisture when protected on all sides by a layer of noncinder concrete not less than _____ thick.
    (a) 2 in.                (b) 4 in.                (c) 6 in.                (d) 18 in.

2.  Materials such as straps, bolts, etc., associated with the installation of RMC in a wet location are required to be _____.
    (a) weatherproof         (b) weathertight         (c) corrosion resistant    (d) none of these

3.  Aluminum fittings and enclosures shall be permitted to be used with _____ conduit.
    (a) steel rigid metal                             (b) aluminum rigid metal
    (c) rigid nonmetallic                             (d) a and b

4.  The minimum radius for a bend of 1 in. rigid conduit is _____, when using a one-shot bender.
    (a) $10^1/_2$ in.        (b) $11^1/_2$ in.        (c) $5^3/_4$ in.        (d) $9^1/_2$ in.

5.  The minimum radius of a field bend on $1^1/_4$ in. RMC is _____
    (a) 7 in.                (b) 8 in.                (c) 14 in.               (d) 10 in.

6.  Two-inch RMC shall be supported every _____.
    (a) 10 ft                (b) 12 ft                (c) 14 ft                (d) 15 ft

7.  Straight runs of 1 in. RMC using threaded couplings may be secured at intervals not exceeding _____.
    (a) 5 ft                 (b) 10 ft                (c) 12 ft                (d) 14 ft

8.  Exposed vertical risers for industrial machinery or fixed equipment can be supported at intervals not exceeding _____, if the conduit is made up with threaded couplings, firmly supported at the top and bottom of the riser, and no other means of support is available.
    (a) 6 ft                 (b) 10 ft                (c) 20 ft                (d) none of these

9.  When threadless couplings and connectors used in the installation of RMC are buried in masonry or concrete, they shall be of the _____ type.
    (a) raintight                                     (b) wet and damp location
    (c) nonabsorbent                                 (d) concrete-tight

10. Threadless couplings and connectors used with RMC and installed in wet locations shall be of the _____ type.
    (a) raintight                                     (b) wet and damp location
    (c) nonabsorbent                                 (d) weatherproof

# Article 348
# Flexible Metal Conduit (FMC)

## Part I. General

### 348.1 Scope

This article covers the use, installation, and construction specifications for Flexible Metal Conduit (FMC) and associated fittings.

### 348.2 Definition

FMC is a raceway of circular cross section made of a helically-wound, formed, interlocked metal strip of either steel or aluminum. It is commonly called "Greenfield" or simply "flex."

### 344.6 Listing Requirements

FMC and associated fittings shall be listed for the purpose.

## Part II. Installation

### 348.10 Uses Permitted

FMC can be installed exposed or concealed.

### 348.12 Uses Not Permitted

(1) In wet locations.

(2) In hoistways, other than as permitted in 620.21(A)(1).

(3) In storage battery rooms.

(4) In hazardous (classified) locations, except as permitted by 501.4(B).

(5) Exposed to material having a deteriorating effect on the installed conductors

(6) Underground or embedded in poured concrete.

(7) Where subject to physical damage.

### 348.20 Size

(A) **Minimum.** FMC in sizes less than $^1/_2$ in. shall not be used, except: Figure 348-1

(1) For enclosing the leads of motors,

(2) In lengths not in excess of 6 ft:

a. For utilizing equipment,

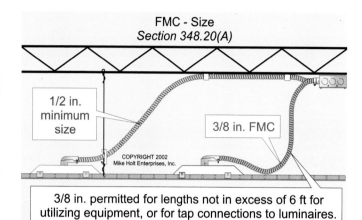

FMC - Size
Section 348.20(A)

1/2 in. minimum size

3/8 in. FMC

COPYRIGHT 2002
Mike Holt Enterprises, Inc.

3/8 in. permitted for lengths not in excess of 6 ft for utilizing equipment, or for tap connections to luminaires.

**Figure 348-1**

b. As part of a listed assembly, or

c. For tap connections to luminaires as permitted by 410.67(C).

(3) In manufactured wiring systems, 604.6(A).

(4) In hoistways, 620.21(A)(1).

(5) As part of a listed assembly to connect wired luminaire sections, 410.77(C).

(B) **Maximum.** FMC larger than 4 in. shall not be used.

### 348.22 Number of Conductors

(A) **Raceway $^1/_2$ in. and Larger.** Raceways shall be large enough to permit the installation and removal of conductors without damaging the conductors' insulation. When all conductors in a raceway are the same size and insulation, the number of conductors permitted can be found in Annex C for the raceway type.

**Question:** How many 6 AWG THHN conductors can be installed in 1 in. FMC?

**Answer:** 6 conductors, Annex C, Table C3

**AUTHOR'S COMMENT:** See 300.17 for additional examples on how to size raceways when conductors are not all the same size.

**(B) Raceway $^3/_8$ in.** The number and size of conductors in a $^3/_8$ in. FMC shall be in accordance with Table 348.22.

### 348.24 Bends

Bends shall be made so the conduit will not be damaged and its internal diameter will not be effectively reduced. The radius of the curve of the inner edge of any field bend shall not be less than shown in Table 344.22.

### 348.26 Number of Bends (360°)

Bends between pull points such as boxes and conduit bodies cannot exceed 360°, and the raceway shall be installed so the conductors can be pulled or removed without damaging the conductor insulation.

### 348.28 Trimming

The cut ends of FMC shall be trimmed to remove the rough edges, but this is not necessary where fittings that thread into the convolutions are used.

### 348.30 Secured and Supports

FMC shall be installed as a complete system [300.10, 300.12, and 300.18(A)], and it shall be securely fastened in place and supported in accordance with the following:

**(A) Securely Fastened.** FMC shall be securely fastened by an approved means within 1 ft of termination and at intervals not exceeding $4^1/_2$ ft. Figure 348-2

*Exception No. 1:* Support is not required where FMC is fished.

*Exception No. 2:* Securing the raceway is not required for lengths up to 3 ft from the last support to the termination, where flexibility is required.

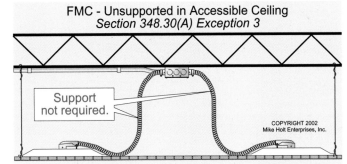

FMC - Unsupported in Accessible Ceiling
*Section 348.30(A) Exception 3*

Support not required.

COPYRIGHT 2002
Mike Holt Enterprises, Inc.

FMC for a luminaire tap connection not exceeding 6 ft is not required to be secured or supported.

**Figure 348-3**

*Exception No. 3:* FMC used for a luminaire whip in lengths not exceeding 6 ft in accordance with 410.67(C) does not need to be secured or supported. Figure 348-3

**(B) Horizontal Runs.** FMC installed horizontally in bored or punched holes in wood or metal framing members, or notches in wooden members, is considered supported, but the raceway shall be secured within 1 ft of termination. Figure 348-4

### 348.42 Fittings

Angle connector fittings cannot be installed in concealed locations.

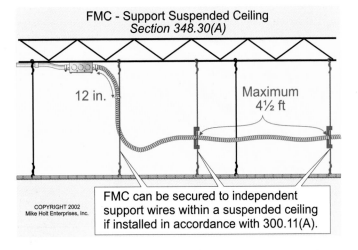

FMC - Support Suspended Ceiling
*Section 348.30(A)*

12 in.

Maximum
4½ ft

COPYRIGHT 2002
Mike Holt Enterprises, Inc.

FMC can be secured to independent support wires within a suspended ceiling if installed in accordance with 300.11(A).

**Figure 348-2**

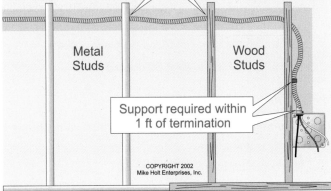

FMC - Horizontal Support
*Section 348.30(B)*

FMC run horizontally though a framing member is considered supported if secured within 1 ft of termination.

Metal Studs

Wood Studs

Support required within 1 ft of termination

COPYRIGHT 2002
Mike Holt Enterprises, Inc.

**Figure 348-4**

## 348.60 Grounding

Where used to connect equipment where flexibility is required, an equipment grounding conductor shall be installed within the raceway with the circuit conductors, except as permitted by 250.102(E). See 250.134(B) and 300.3(B).

> **AUTHOR'S COMMENT:** See 250.118(6) for the conditions when FMC is suitable to be used as an effective ground-fault current path. Figure 348-5

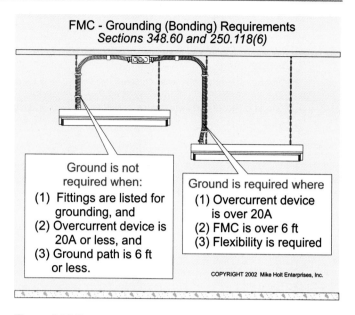

**FMC - Grounding (Bonding) Requirements**
*Sections 348.60 and 250.118(6)*

Ground is not required when:
(1) Fittings are listed for grounding, and
(2) Overcurrent device is 20A or less, and
(3) Ground path is 6 ft or less.

Ground is required where
(1) Overcurrent device is over 20A
(2) FMC is over 6 ft
(3) Flexibility is required

COPYRIGHT 2002 Mike Holt Enterprises, Inc.

**Figure 348-5**

# Article 348

1.  FMC cannot be installed _____.
    (a) underground                      (b) embedded in poured concrete
    (c) where subject to physical damage (d) all of these

2.  FMC can be installed exposed or concealed where not subject to physical damage.
    (a) True            (b) False

3.  The largest size THHN conductor permitted in $^3/_8$ in. FMC is _____.
    (a) 12 AWG          (b) 16 AWG          (c) 14 AWG          (d) 10 AWG

4.  How many 12 AWG XHHW conductors, not counting a bare ground wire, are allowed in 3/8 in. FMC (maximum of 6 ft) with outside fittings?
    (a) 4               (b) 3               (c) 2               (d) 5

5.  FMC shall be secured _____.
    (a) at intervals not exceeding $4^1/_2$ ft
    (b) within 12 in. on each side of a box
    (c) a and b
    (d) none of these

# Article 350
# Liquidtight Flexible Metal Conduit (LFMC)

## Part I. General

### 350.1 Scope

This article covers the use, installation, and construction specifications of liquidtight flexible metal conduit (LFMC) and associated fittings.

### 350.2 Definition

LFMC is a listed raceway of circular cross section having an outer liquidtight, nonmetallic, sunlight-resistant jacket over an inner flexible metal core with associated couplings, connectors and fittings and approved for the installation of electric conductors. LFMC is commonly called Sealtight® or simply "liquidtight." LFMC is of similar construction to FMC, but has an outer liquidtight thermoplastic covering.

### 350.6 Listing Requirement

LFMC and its associated fittings shall be listed for the purpose. Figure 350-1

## Part II. Installation

### 350.10 Uses Permitted

**(A) Permitted Use.** Listed LFMC can be installed either exposed or concealed at any of the following locations:

(1) Where flexibility or protection from liquids, vapors, or solids is required.

LFMC - Listing Required
Section 350.6

LFMC and associated fittings must be listed for the purpose.

COPYRIGHT 2002
Mike Holt Enterprises, Inc.

**Figure 350-1**

(2) In hazardous (classified) locations, as permitted in 501.4(B), 502.4(A)(2), 502.4(B)(2), or 503.3(A)(2).

(3) For direct burial, if listed and marked for this purpose.

### 350.12 Uses Not Permitted

(1) Where subject to physical damage.

(2) Where the combination of the ambient and conductor operating temperatures exceeds the rating of the material.

### 350.20 Size

**(A) Minimum.** LFMC in sizes less than $1/2$ in. shall not be used.

*Exception:* LFMC can be smaller than $1/2$ in. where installed in accordance with 348.20(A).

(1) For enclosing the leads of motors.

(2) In lengths not in excess of 6 ft:

    a. For utilizing equipment,

    b. As part of a listed assembly, or

    c. For tap connections to luminaires as permitted by 410.67(C).

(3) In manufactured wiring systems, 604.6(A).

(4) In hoistways, 620.21(A)(1).

(5) As part of a listed assembly to connect wired luminaire sections, 410.77(C).

**(B) Maximum.** LFMC larger than 4 in. shall not be used.

### 350.22 Number of Conductors

**(A) Raceway $1/2$ in. and Larger.** Raceways shall be large enough to permit the installation and removal of conductors without damaging the conductors' insulation. When all conductors in a raceway are the same size and insulation, the number of conductors permitted can be found in Annex C for the raceway type.

**Question:** How many 6 AWG THHN conductors can be installed in 1 in. LFMC? Figure 350-2

**Answer:** Seven conductors, Annex C, Table C7.

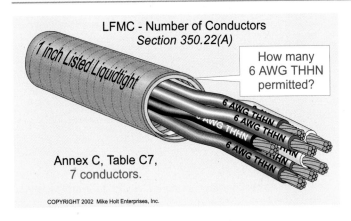

**Figure 350-2**

---

**AUTHOR'S COMMENT:** See 300.17 for additional examples on how to size raceways when conductors are not all the same size.

---

**(B) Raceway ³/₈ in.** The number and size of conductors in a ³/₈ in. LFMC shall be in accordance with Table 348.22.

## 350.24 Bends

Bends shall be made so the conduit will not be damaged and the internal diameter of the conduit will not be effectively reduced. The radius of the curve of the inner edge of any field bend shall not be less than shown in Table 344.22.

## 350.26 Number of Bends (360°)

Bends between pull points, such as boxes and conduit bodies, cannot exceed 360°, and the raceway shall be installed so the conductors can be pulled or removed without damaging the conductor insulation.

## 350.30 Secured and Supports

LFMC shall be installed as a complete system [300.10, 300.12, and 300.18(A)], and it shall be securely fastened in place and supported in accordance with the following:

**(A) Securely Fastened.** LFMC shall be securely fastened by an approved means within 1 ft of termination and at intervals not exceeding 4¹/₂ ft.

*Exception No. 1:* Where fished, that portion of the raceway that cannot be reached is not required to be secured.

*Exception No. 2:* Securing the raceway is not required for lengths up to 3 ft from the last support to the termination, where flexibility is required.

*Exception No. 3:* LFMC used for a luminaire whip in lengths not exceeding 6 ft in accordance with 410.67(C) does not need to be secured or supported.

**(B) Horizontal Runs.** LFMC installed horizontally in bored or punched holes in wood or metal framing members, or notches in wooden members, is considered supported, but the raceway shall be secured within 1 ft of termination.

## 350.42 Fittings

Angle connector fittings shall not be installed in concealed locations.

## 350.60 Grounding

Where used to connect equipment where flexibility is required, an equipment grounding conductor shall be installed within the raceway with the circuit conductors, except as permitted by 250.102(E). See 250.134(B) and 300.3(B).

---

**AUTHOR'S COMMENT:** See 250.118(7) for the conditions when LFMC is considered suitable as an effective ground-fault current path, in accordance with 250.4(A)(5). Figure 350-3

---

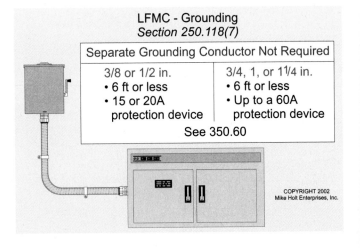

**Figure 350-3**

## Article 350

1. The use of listed and marked LFMC shall be permitted for _____.
   (a) direct burial where listed and marked for the purpose
   (b) exposed work
   (c) concealed work
   (d) all of these

2. The maximum number of 14 THHN permitted in 3/8 in. LFMC with outside fittings is _____.
   (a) 4                (b) 7                (c) 5                (d) 6

3. Where flexibility is necessary, securing LFMC is not required for lengths not exceeding _____ at terminals.
   (a) 2 ft             (b) 3 ft             (c) 4 ft             (d) 6 ft

4. When LFMC is used as a fixed raceway, it must be secured within _____ on each side of the box and shall be at intervals not exceeding _____.
   (a) 12 in., $4^1/_2$ ft     (b) 18 in., 3 ft        (c) 12 in., 3 ft        (d) 18 in., 4 ft

5. When LFMC is used to connect equipment requiring flexibility, a separate _____ conductor must be installed.
   (a) bond jumper      (b) bonding      (c) equipment grounding   (d) none of these

# Article 352
# Rigid Nonmetallic Conduit (RNC)

## Part I. General

### 352.1 Scope

This article covers the use, installation, and construction specifications of rigid nonmetallic conduit (RNC) and associated fittings.

### 352.2 Definition

RNC is a listed nonmetallic raceway of circular cross section with integral or associated couplings, approved for the installation of electrical conductors.

## Part II. Installation

### 352.10 Uses Permitted

> FPN: In extreme cold, RNC can become brittle and is susceptible to physical damage.

(A) **Concealed.** RNC can be concealed within walls, floors, or ceilings, directly buried or embedded in concrete in buildings of any height.

(B) **Corrosive Influences.** RNC can be installed in areas subject to severe corrosive influences as covered in 300.6 for which the material is specifically approved.

(C) **Wet Locations.** RNC can be installed in wet locations such as dairies, laundries, canneries, car washes, and other areas that are frequently washed or outdoors. Supporting fittings such as straps, screws, and bolts shall be made of corrosion-resistant materials, or shall be protected with a corrosion-resistant coating [300.6].

(E) **Dry and Damp Locations.** RNC can be installed in dry and damp locations, except where limited in 352.12.

(F) **Exposed.** Schedule 40 RNC is approved for exposed locations where not subject to physical damage [352.12(C)]. In areas exposed to physical damage, Schedule 80 RNC, IMC, or RMC shall be used.

(G) **Underground.** RNC installed underground shall comply with the burial requirements of 300.5.

(H) **Support of Conduit Bodies.** Rigid nonmetallic conduit shall be permitted to support nonmetallic conduit bodies not larger than the largest trade size of an entering raceway. The conduit bodies shall not contain devices or support luminaires (fixtures) or other equipment.

### 352.12 Uses Not Permitted

(A) **Hazardous (Classified) Locations.** RNC shall not be installed in hazardous (classified) locations, except as permitted by 501.4(A)(1)(a) Ex. 1, 503.3(A), 514.8 Ex. 2, or 515.5. Figure 352-1

(B) **Support of Luminaires.** RNC shall not be used for the support of luminaires or other equipment not described in 352.10(H).

---

**AUTHOR'S COMMENT:** RNC can be used to support conduit bodies. See 314.23(E) Ex.

---

(C) **Physical Damage.** Schedule 40 RNC shall not be installed where subject to physical damage, but heavier and thicker Schedule 80 RNC can be subject to physical damage. See 300.5(D).

(D) **Ambient Temperature.** RNC shall not be installed where the ambient temperature is in excess of 50°C/122°F.

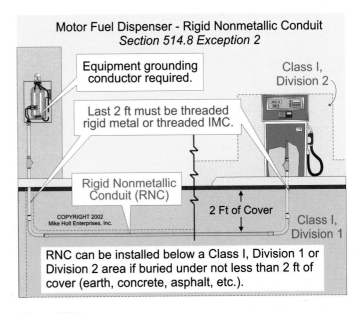

**Figure 352-1**

**(E)  Conductor Insulation.** RNC shall not contain conductors whose operating temperature would exceed the conduit temperature rating.

**(F)  Places of Assembly.** Because of toxicity, RNC shall not be installed in places of assembly and theaters, unless encased in not less than 2 in. of concrete as permitted in 518.4(B) and 520.5(C).

> **AUTHOR'S COMMENT:** RNC is prohibited as a wiring method for patient care areas in health care facilities [517.13(A)] and in ducts, plenums, and other spaces used for environmental air handling [300.22].

## 352.20 Size

**(A)  Minimum Size.** $1/2$ in.

**(B)  Maximum Size.** 6 in.

## 352.22 Number of Conductors

Raceways shall be large enough to permit the installation and removal of conductors without damaging the conductors' insulation. When all conductors in a raceway are the same size and insulation, the number of conductors permitted can be found in Annex C for the raceway type.

**Question:** How many 4/0 AWG THHN conductors can be installed in 2 in. RNC Schedule 40?

**Answer:** Four conductors, Annex C, Table C10.

> **AUTHOR'S COMMENT:** Schedule 80 RNC has the same outside diameter as Schedule 40 RNC, but the wall thickness of Schedule 80 is greater, which results in a reduced interior area for conductor fill.

**Question:** How many 4/0 AWG THHN conductors can be installed in 2 in. RNC Schedule 80?

**Answer:** Three conductors, Annex C, Table C9.

> **AUTHOR'S COMMENT:** See 300.17 for additional examples on how to size raceways when conductors are not all the same size.

## 352.24 Bends

Raceway bends shall not be made in any manner that would damage the raceway or significantly change its internal diameter (no kinks). This is accomplished by complying with the bending radius requirements in Table 344.24.

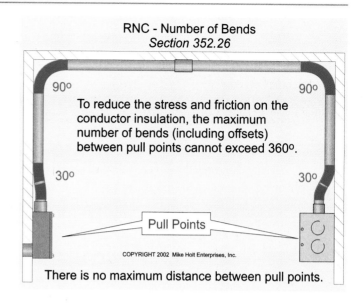

RNC - Number of Bends
*Section 352.26*

90°        90°

To reduce the stress and friction on the conductor insulation, the maximum number of bends (including offsets) between pull points cannot exceed 360°.

30°        30°

Pull Points

COPYRIGHT 2002 Mike Holt Enterprises, Inc.

There is no maximum distance between pull points.

**Figure 352-2**

## 352.26 Number of Bends (360°)

To reduce the stress and friction on the conductor insulation, the number of bends (including offsets) between pull points cannot exceed 360°. Figure 352-2

> **AUTHOR'S COMMENT:** Be sure to use equipment that is designed to heat the raceway so it is pliable for bending. An open-flame torch should not be used.

## 352.28 Trimming

The cut ends of RNC shall be trimmed (inside and out) to remove the burrs and rough edges. Trimming nonmetallic raceway is very easy; most of the burrs will rub off with fingers, and a knife can be used to smooth the rough edges.

## 352.30 Secured and Supported

RNC shall be installed as a complete system [300.10, 300.12, and 300.18(A)], and it shall be securely fastened in place and supported in accordance with the following:

**(A)  Secured.** RNC shall be secured within 3 ft of every box, cabinet, or termination fitting, such as a conduit body.

**(B)  Supports.** RNC shall be supported at intervals not exceeding the values in the following table, and the raceway shall be fastened in a manner that permits movement from thermal expansion or contraction.

| Raceway Size In. | Support Spacing Ft. |
|---|---|
| $\frac{1}{2}$ – 1 | 3 |
| $1\frac{1}{4}$ – 2 | 5 |
| $2\frac{1}{2}$ – 3 | 6 |
| $3\frac{1}{2}$ – 5 | 7 |
| 6 | 8 |

RNC installed horizontally in bored or punched holes in wood or metal framing members, or notches in wooden members, is considered supported, but the raceway shall be secured within 3 ft of termination.

## 352.44 Expansion Fittings

Expansion fittings shall be provided to compensate for thermal expansion and contraction of the raceway or building structure in accordance with Tables 342.44(A) and (B), where the length change is expected to be $\frac{1}{4}$ in. or greater in a straight run between securely mounted items such as boxes, cabinets, elbows, or other conduit terminations. Figure 352-3

**AUTHOR'S COMMENT:** Listing instructions with expansion fittings indicate that you should add 30°F to the ambient temperature when the raceway is in direct sunlight. Assuming a high ambient temperature of 90°F (plus 30°F due to solar heating) and a low temperature of 0°F (no solar exposure), the temperature change will be 120°F. Table 352.44 indicates that the total expansion and contraction length change would be approximately 4.9 in. for a 100 ft run. Be sure you follow

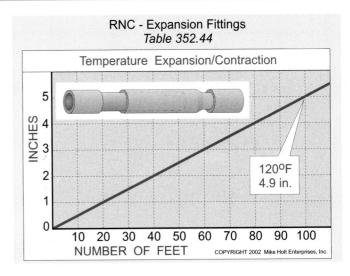

**Figure 352-4**

manufacturer's instructions when you install these fittings. Figure 352-4

*CAUTION: If the ambient temperature during installation is high, you should realize that the conduit is at its expanded range and will contract when the temperature drops. Of course, the opposite applies if the ambient temperature is low.*

## 352.46 Bushings

Where conductors 4 AWG or larger enter a box, fitting, or other enclosure, a fitting that provides a smooth, rounded insulating surface, such as a bushing or adapter, shall be provided to protect the wire from abrasion during and after installation. See 300.4(F).

Some RNC adapters (connectors) and all bell-ends provide the conductor protection required in this section. Figure 352-5

**AUTHOR'S COMMENT:** When RNC is stubbed into an open-bottom switchboard or other apparatus, the raceway, including the end fitting (bell-ends), shall not rise more than 3 in. above the bottom of the switchboard enclosure. See 300.16(B) and 408.10.

## 352.48 Joints

Joints, such as couplings and connectors, shall be made in an approved manner. This means you shall follow the manufacturer's instructions for the raceway, fitting, and glue.

**AUTHOR'S COMMENT:** Some PVC glues require the raceway surface to be cleaned with a solvent before the application of the glue. After the application of the glue to both surfaces, a quarter turn of the fitting is required.

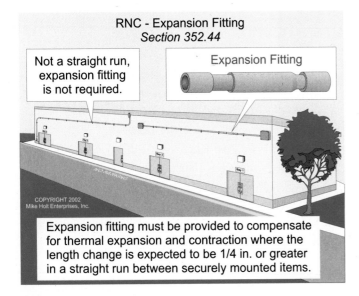

Expansion fitting must be provided to compensate for thermal expansion and contraction where the length change is expected to be 1/4 in. or greater in a straight run between securely mounted items.

**Figure 352-3**

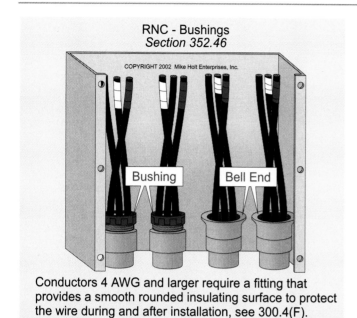

**RNC - Bushings**
*Section 352.46*

COPYRIGHT 2002  Mike Holt Enterprises, Inc.

Bushing     Bell End

Conductors 4 AWG and larger require a fitting that provides a smooth rounded insulating surface to protect the wire during and after installation, see 300.4(F).

*Figure 352-5*

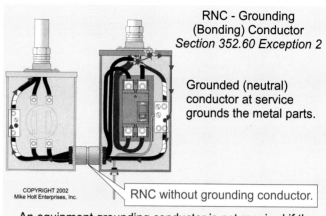

**RNC - Grounding
(Bonding) Conductor**
*Section 352.60 Exception 2*

Grounded (neutral) conductor at service grounds the metal parts.

COPYRIGHT 2002
Mike Holt Enterprises, Inc.

RNC without grounding conductor.

An equipment grounding conductor is not required if the grounded conductor is used for grounding, see 250.142(A).

*Figure 352-6*

## 352.60 Grounding

Where an equipment grounding conductor is required, it shall be installed within the raceway. See 300.2(B).

*Exception No. 2:* An equipment grounding conductor is not required in RNC, if the grounded (neutral) conductor is used to ground service equipment [250.24(B)], as permitted in 250.142(A). Figure 352-6

## Article 352

1. Extreme _____ may cause RNCs to become brittle and therefore more susceptible to damage from physical contact.
   (a) sunlight          (b) corrosive conditions     (c) heat          (d) cold

2. RNC and fittings can be used in areas of dairies, laundries, canneries, or other wet locations and in locations where walls are frequently washed. However, the entire conduit system including boxes and _____ shall be installed & equipped to prevent water from entering the conduit.
   (a) luminaries         (b) fittings          (c) supports          (d) all of these

3. RNC can be installed exposed in buildings of _____, where not subject to physical damage.
   (a) three floors        (b) twelve floors     (c) six floors        (d) any height

4. RNC can be used to support nonmetallic conduit bodies, but the conduit bodies shall not contain devices, luminaires, or other equipment.
   (a) True                (b) False

5. RNC shall not be used _____.
   (a) in hazardous (classified) locations
   (b) for the support of luminaires or other equipment
   (c) where subject to physical damage unless identified for such use
   (d) all of these

6. When installing RNC _____.
   (a) all cut ends shall be trimmed inside and outside to remove rough edges
   (b) there shall be a support within 2 ft of each box and cabinet
   (c) all joints shall be made by an approved method
   (d) a and c

7. RNC shall be securely fastened within _____ of each box.
   (a) 6 in.              (b) 24 in.            (c) 12 in.           (d) 36 in.

8. One-inch RNC must be supported every _____.
   (a) 2 ft               (b) 3 ft             (c) 4 ft             (d) 6 ft

9. Expansion fittings for RNC shall be provided to compensate for thermal expansion and contraction when the length change in a straight run between securely mounted boxes, cabinets, elbows, or other conduit terminations is expected to be _____ or greater.
   (a) $1/4$ in.          (b) $1/2$ in.         (c) 1 in.            (d) none of these

10. An equipment grounding conductor is not required in the conduit if the grounded conductor is used to ground equipment as permitted in 250.142.
    (a) True               (b) False

# Article 354
# Nonmetallic Underground Conduit with Conductors (NUCC)

## Part I. General

### 354.1 Scope

This article covers the use, installation, and construction specifications of nonmetallic underground conduit with conductors (NUCC).

### 354.2. Definition

NUCC is a factory assembly of conductors or cables inside a nonmetallic, smooth-wall conduit with a circular cross section.

> **AUTHOR'S COMMENT:** The nonmetallic conduit for this wiring method is composed of a material that is resistant to moisture and corrosive agents. It is capable of being supplied on reels without damage or distortion and shall be of sufficient strength to withstand abuse in handling and during installation without damage to conduit or conductors.

### 354.6 Listing Requirement

NUCC and its associated fittings shall be listed for the purpose. Figure 354-1

## Part II. Installation

### 354.10. Uses Permitted

(1) For direct burial underground installation. See Table 300.5 for minimum cover requirements.

(2) Encased or embedded in concrete.

(3) In cinder fill.

(4) In underground locations subject to severe corrosive influences.

### 354.12. Uses Not Permitted

(1) In exposed locations.

(2) Inside buildings.

*Exception:* The conductor or the cable portion of the assembly can extend within the building for termination purposes in accordance with 300.3. Figure 354-2

(3) In hazardous (classified) locations, except as permitted by 503.3(A), 504.20, 514.8, and 501.4(B).

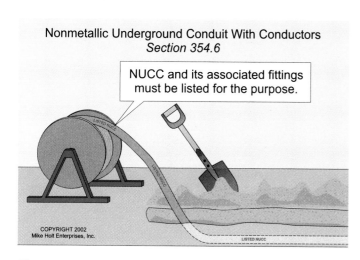

Nonmetallic Underground Conduit With Conductors
Section 354.6

NUCC and its associated fittings must be listed for the purpose.

COPYRIGHT 2002
Mike Holt Enterprises, Inc.

**Figure 354-1**

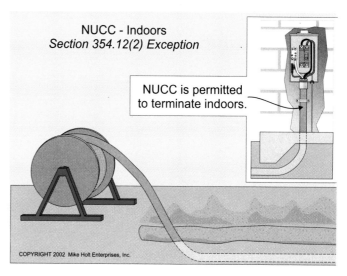

NUCC - Indoors
Section 354.12(2) Exception

NUCC is permitted to terminate indoors.

COPYRIGHT 2002 Mike Holt Enterprises, Inc.

**Figure 354-2**

## 354.20. Size

**(A) Minimum.** $1/2$ in.

**(B) Maximum.** 4 in.

## 354.24. Bends

Raceway bends shall be done manually and shall not be made in any manner that would damage the raceway or significantly change its internal diameter (no kinks). This is accomplished by complying with the bending radius requirements contained in the following:

| Trade Size In. | Minimum Bending Radius In. |
|---|---|
| $1/2$ | 10 |
| $3/4$ | 12 |
| 1 | 14 |
| $1 1/4$ | 18 |
| $1 1/2$ | 20 |
| 2 | 26 |
| $2 1/2$ | 36 |
| 3 | 48 |
| 4 | 60 |

## 354.26. Bends — Number in One Run

To reduce the stress and friction on the conductor insulation, the number of bends (including offsets) between pull points shall not exceed 360°.

## 354.28. Trimming

At terminations, the conduit shall be trimmed inside and out to remove rough edges using an approved method that will not damage the conductor or cable insulation or jacket.

## 354.46. Bushings

Where the NUCC terminates, a bushing or adapter shall be provided to protect the conductor from abrasion, unless the design of the box, fitting, or enclosure provides equivalent protection. See 300.4(F) for bushing requirements for raceway terminations.

## 354.48. Joints

All joints between the conduit, fittings, and boxes shall be made by an approved method.

## 354.50. Conductor Terminations

All terminations between the conductors or cables and equipment shall be made by an approved method.

# Article 354

1. NUCC and its associated fittings must be _____ for the purpose.
   (a) listed          (b) approved          (c) identified          (d) none of these

2. The use of NUCC shall be permitted _____.
   (a) for direct burial underground installation          (b) to be encased or embedded in concrete
   (c) in cinder fill          (d) all of these

3. NUCC shall not be used _____.
   (a) in exposed locations          (b) inside buildings
   (c) in hazardous (classified) locations          (d) all of these

4. NUCC larger than _____ shall not be used.
   (a) 1 in.          (b) 2 in.          (c) 3 in.          (d) 4 in.

5. NUCC shall be capable of being supplied on reels without damage or _____ and shall be of sufficient strength to withstand abuse, such as impact or crushing in handling and during installation, without damage to conduit or conductors.
   (a) distortion          (b) breakage          (c) shattering          (d) all of these

# Article 356
# Liquidtight Flexible Nonmetallic Conduit (LFNC)

## Part I. General

### 356.1 Scope

This article covers the use, installation, and construction specifications of liquidtight flexible nonmetallic conduit (LFNC) and associated fittings.

### 356.2 Definition

LFNC is a listed raceway of circular cross section having an outer liquidtight, nonmetallic, sunlight-resistant jacket over a flexible inner core with associated couplings, connectors, and fittings and approved for the installation of electrical conductors.

**(1) Type LFNC-A (orange color).** A smooth seamless inner core and cover bonded together and having one or more reinforcement layers between the core and cover.

**(2) Type LFNC-B (gray color).** A smooth inner surface with integral reinforcement within the conduit wall.

**(3) Type LFNC-C (black color).** A corrugated internal and external surface without integral reinforcement within the conduit wall.

### 356.6 Listing Requirement

LFNC and its associated fittings shall be listed for the purpose.

## Part II. Installation

### 356.10 Uses Permitted

Listed LFNC can be installed either exposed or concealed at any of the following locations:

(1) Where flexibility is required.

(2) Where protection from liquids, vapors, or solids is required.

(3) Outdoors, if listed and marked for this purpose.

(4) Directly buried in the earth, if listed and marked for this purpose.

(5) LFNC-B (gray color) can be installed in lengths over 6 ft secured according to 356.30.

(6) LFNC-B (black color) as a listed manufactured pre-wired assembly.

### 356.12 Uses Not Permitted

(1) Where subject to physical damage.

(2) Where the combination of ambient and conductor temperature will produce an operating temperature in excess of the rating of the material.

(3) In lengths longer than 6 ft, except if approved as essential for a required degree of flexibility.

### 356.20 Size

**(A) Minimum.** LFNC smaller than $1/2$ in. shall not be used except for:

(1) Enclosing the leads of motors, 430.145(B).

(2) Tap connections to lighting fixtures as permitted by 410.67(C).

(3) Electric sign connections, 600.32(A).

**(B) Maximum.** 4 in.

### 356.22 Number of Conductors

Raceways shall be large enough to permit the installation and removal of conductors without damaging the insulation. When all conductors in a raceway are the same size and insulation, the number of conductors permitted can be found in Annex C for the raceway type.

**Question:** How many 8 AWG THHN conductors can be installed in $3/4$ in. LFNC-B?

**Answer:** Six conductors, Annex C, Table C5.

**(B) Raceway $3/8$ in.** The number and size of conductors in a $3/8$ in. LFNC shall be in accordance with *NEC* Table 348.22.

### 356.24 Bends

Raceway bends shall not be made in any manner that would damage the raceway or significantly change its internal diameter (no kinks). This is accomplished by complying with the bending radius requirements in Table 344.24.

## 356.26 Number of Bends (360°)

Bends between pull points such as boxes and conduit bodies cannot exceed 360°, and the raceway shall be installed so the conductors can be pulled or removed without damaging the conductor insulation.

## 356.30 Secured and Supports

Type LFNC-B (gray color) shall be securely fastened and supported in accordance with one of the following:

(1) The conduit shall be securely fastened at intervals not exceeding 3 ft and within 1 ft of termination.

(2) Securing or supporting is not required where it is fished, installed in lengths not exceeding 3 ft at terminals where flexibility is required, or installed in lengths not exceeding 6 ft for tap conductors to luminaires as permitted in 410.67(C).

(3) Horizontal runs of LFNC may be supported by openings through framing members at intervals not exceeding 3 ft and securely fastened within 1 ft of termination.

## 356.42 Fittings

Angle connector fittings shall not be installed in concealed locations.

## 356.60 Grounding

Where an equipment grounding conductor is required, it shall be installed within the raceway. See 300.2(B).

## Article 356

1.  Type LFNC-B shall be permitted to be installed in lengths longer than _____ where secured in accordance with 356.30.
    (a) 2 ft      (b) 3 ft      (c) 6 ft      (d) 10 ft

2.  The use of listed and marked LFMC shall be permitted for _____.
    (a) direct burial where listed and marked for the purpose
    (b) exposed work
    (c) concealed work
    (d) all of these

3.  When LFNC is used as a fixed raceway, it must be secured within _____ on each side of the box and shall be at intervals not exceeding _____.
    (a) 12 in., $4^1/_2$ ft      (b) 18 in., 3 ft      (c) 12 in., 3 ft      (d) 18 in., 4 ft

4.  Where flexibility is necessary, securing LFNC is not required for lengths not exceeding _____ at terminals.
    (a) 2 ft      (b) 3 ft      (c) 4 ft      (d) 6 ft

5.  When LFNC is used to connect equipment requiring flexibility, a separate _____ conductor must be installed.
    (a) bond jumper      (b) bonding      (c) equipment grounding      (d) none of these

# Article 358
# Electrical Metallic Tubing (EMT)

## Part I. General

### 358.1 Scope

This article covers the use, installation, and construction specifications of electrical metallic tubing (EMT).

### 358.2 Definition

EMT is a listed metallic tubing of circular cross section approved for the installation of electrical conductors when joined together with listed fittings. Compared to RMC and IMC, EMT is relatively easy to bend, cut, and ream, and because it is not threaded, all connectors and couplings are of the threadless type.

### 358.6 Listing Requirement

EMT, factory elbows, and associated fittings shall be listed.

## Part II. Installation

### 358.10 Use

(A) **Exposed and Concealed.** EMT can be installed exposed or concealed.

(B) **Corrosion Protection.** EMT, elbows, couplings, and fittings can be installed in concrete, in direct contact with the earth, or in areas subject to severe corrosive influences where protected by corrosion protection and judged suitable for the condition. See 300.6.

> CAUTION: Supplementary coatings for corrosion protection have not been investigated by a product testing and listing agency, and these coatings are known to cause cancer in laboratory animals.

(C) **Wet Locations.** All support fittings, such as screws, straps, etc. installed in a wet location shall be made of corrosion-resistant material, or they shall be protected by a corrosion-resistant coating. See 300.6.

### 358.12 Uses Not Permitted

(1) Where, during installation or afterward, it will be subject to severe physical damage.

(2) Where protected from corrosion solely by enamel.

(3) In cinder concrete or cinder fill where subject to permanent moisture, unless protected on all sides by a layer of noncinder concrete at least 2 in. thick, or unless the tubing is at least 18 in. under the fill.

(4) In any hazardous location, except as permitted by 502.4, 503.3, and 504.20.

(5) For the support of luminaires or other equipment, except conduit bodies no larger than the largest trade size of the tubing can be supported to the raceway. Figure 358-1

(6) Where practicable, dissimilar metals in contact anywhere in the system shall be avoided to eliminate the possibility of galvanic action.

*Exception:* Aluminum fittings on steel EMT and steel fittings on aluminum EMT are permitted.

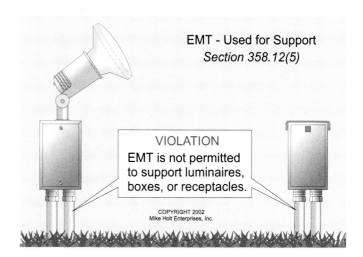

EMT - Used for Support
*Section 358.12(5)*

VIOLATION
EMT is not permitted to support luminaires, boxes, or receptacles.

COPYRIGHT 2002
Mike Holt Enterprises, Inc.

**Figure 358-1**

## 358.20 Size

**(A)  Minimum Size.** ½ in.

**(B)  Maximum Size.** 4 in.

## 358.22 Conductor Fill

Raceways shall be large enough to permit the installation and removal of conductors without damaging the conductors' insulation. When all conductors in a raceway are the same size and insulation, the number of conductors permitted can be found in Annex C for the raceway type.

**Question:** How many 12 AWG THHN conductors can be installed in 1 in. EMT? Figure 358-2

**Answer:** 26 conductors, Annex C, Table C1.

**AUTHOR'S COMMENT:** See 300.17 for additional examples on how to size raceways when conductors are not all the same size.

## 358.24 Bends

Raceway bends shall not be made in any manner that would damage the raceway or significantly change its internal diameter (no kinks). This is accomplished by complying with the bending radius requirements in Table 344.24.

**AUTHOR'S COMMENT:** This is not a problem, because most benders are made to comply with this table.

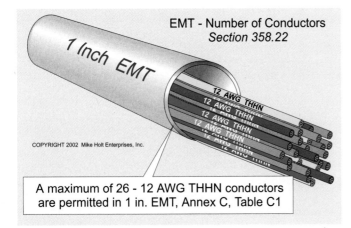

**Figure 358-2**

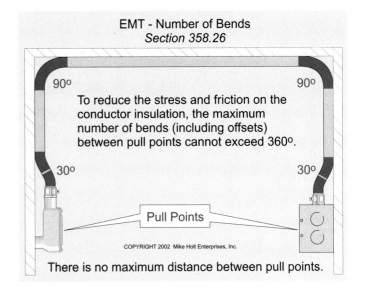

**Figure 358-3**

## 358.26 Number of Bends (360°)

To reduce the stress and friction on the conductor insulation, the number of bends (including offsets) between pull points shall not exceed 360°. Figure 358-3

## 358.28 Reaming and Threading

**(A)  Reaming.** Reaming to remove the burrs and rough edges is required when the raceway is cut.

**AUTHOR'S COMMENT:** It's considered an accepted practice to ream small raceways with a screwdriver or the backside of Channelocks®.

**(B)  Threading.** EMT shall not be threaded.

## 358.30 Secured and Supported

EMT shall be installed as a complete system [300.10, 300.12, and 300.18(A)], and it shall be securely fastened in place and supported in accordance with the following:

**(A)  Securely Fastened.** EMT shall be securely fastened within 3 ft of every box, cabinet, or termination fitting, and at intervals not exceeding 10 ft. Figure 358-4

**AUTHOR'S COMMENT:** Support is not required within 3 ft of a coupling.

*Exception No. 1:* When structural members do not permit the raceway to be secured within 3 ft of a box or termination fitting, an unbroken raceway is permitted to be secured within 5 ft of a box or termination fitting. Figure 358-5

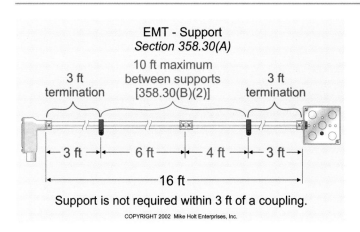

**Figure 358-4**

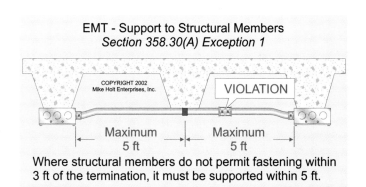

**Figure 358-5**

**(B) Horizontal Runs.** EMT installed horizontally in bored or punched holes in wood or metal framing members, or notches in wooden members, is considered supported, but the raceway shall be secured within 3 ft of termination.

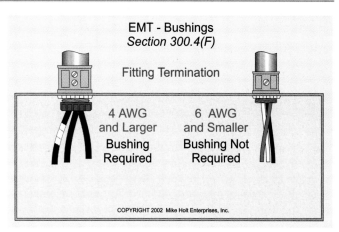

**Figure 358-6**

### 358.42 Coupling and Connectors

Threadless couplings and connectors shall be made up wrenchtight to maintain an effective low-impedance path to conduct safely fault current. See 250.4(A)(5), 250.96(A), and 300.10. When EMT is installed in concrete, the fittings shall be concrete tight, and when installed in a wet location, the raceway shall be the raintight type.

**AUTHOR'S COMMENT:** To protect conductors from abrasion, a metal or plastic bushing shall be installed on raceway termination, unless the design of the box, fitting, or enclosure is such as to afford equivalent protection. EMT connectors without bushings are permitted for conductors 6 AWG and smaller. EMT connectors for 4 AWG and larger require bushings. See 300.4(F). Figure 358-6

# Article 358

1.  The minimum and maximum size of EMT is _____, except for special installations.
    (a) $5/16$ to 3 in.          (b) $3/8$ to 4 in.          (c) $1/2$ to 3 in.          (d) $1/2$ to 4 in.

2.  A run of EMT between outlet boxes shall not exceed _____ offsets close to the box.
    (a) 360° plus                               (b) 360° total including
    (c) four quarter bends plus                 (d) 180° total including

3.  EMT shall not be threaded.
    (a) True                   (b) False

4.  EMT must be supported within 3 ft of each coupling.
    (a) True                   (b) False

5.  Couplings and connectors used with EMT shall be made up _____.
    (a) of metal                                (b) in accordance with industry standards
    (c) tight                                   (d) none of these

# Article 362
# Electrical Nonmetallic Tubing (ENT)

## Part I. General

### 362.1 Scope

This article covers the use, installation, and construction specifications of electrical nonmetallic tubing (ENT) and associated fittings.

### 362.2 Definition

ENT is a pliable corrugated raceway of circular cross section with integral or associated couplings, connectors, and fittings listed for the installation of electrical conductors. It is composed of a material that is resistant to moisture and chemical atmospheres and is flame retardant. The field name for ENT is "smurf pipe," because it was available only in blue when originally it came out at the height of popularity of the children's cartoon characters, the "Smurfs." Today, the raceway is available in a rainbow of colors such as white, yellow, red, green, and orange, and is sold in both fixed lengths and on reels.

ENT is a pliable raceway that can be bent by hand with a reasonable force, but without other assistance.

## Part II. Installation

### 362.10 Uses Permitted

(1)  In buildings not exceeding three floors. Figure 362-1

   a. Exposed, where not prohibited by 362.12.

   b. Concealed within walls, floors, and ceilings.

(2)  In buildings exceeding three floors, ENT shall be installed concealed in walls, floors, or ceilings that provide a thermal barrier having a 15-minute finish rating as identified in listings of fire-rated assemblies. Figure 362-2

**Definition of First Floor:** The first floor of a building is the floor that has 50 percent or more of the exterior wall surface area level with or above finished grade. One additional level not designed for human habitation and used only for vehicle parking, storage, or similar use is permitted.

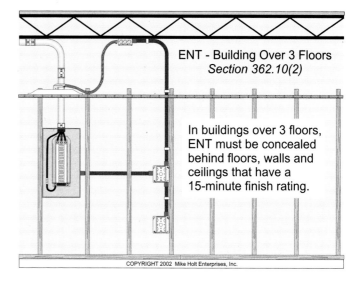

ENT - Building not Over 3 Floors
*Section 362.10(1)*

Ring for Service

In a building not over 3 floors, ENT can be run exposed, concealed or above a suspended ceiling.

COPYRIGHT 2002 Mike Holt Enterprises, Inc.

**Figure 362-1**

ENT - Building Over 3 Floors
*Section 362.10(2)*

In buildings over 3 floors, ENT must be concealed behind floors, walls and ceilings that have a 15-minute finish rating.

COPYRIGHT 2002 Mike Holt Enterprises, Inc.

**Figure 362-2**

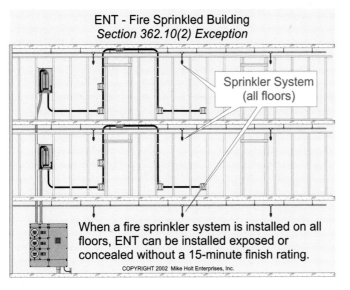

**Figure 362-3**

*Exception:* Where a fire sprinkler system is installed on all floors, in accordance with NFPA 13-1999, *Standard for the Installation of Sprinkler Systems*, ENT can be installed exposed or concealed in buildings of any height. Figure 362-3

(3) ENT can be installed in severe corrosive and chemical locations when identified for this use.

(4) ENT can be installed in dry and damp concealed locations, where not prohibited by 362.12.

(5) Above Suspended Ceiling. ENT can be installed above a suspended ceiling if the suspended ceiling provides a thermal barrier having a 15-minute finish rating as identified in listings of fire-rated assemblies. Figure 362-4

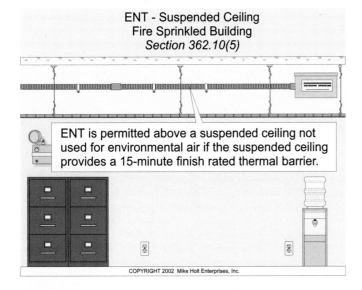

**Figure 362-4**

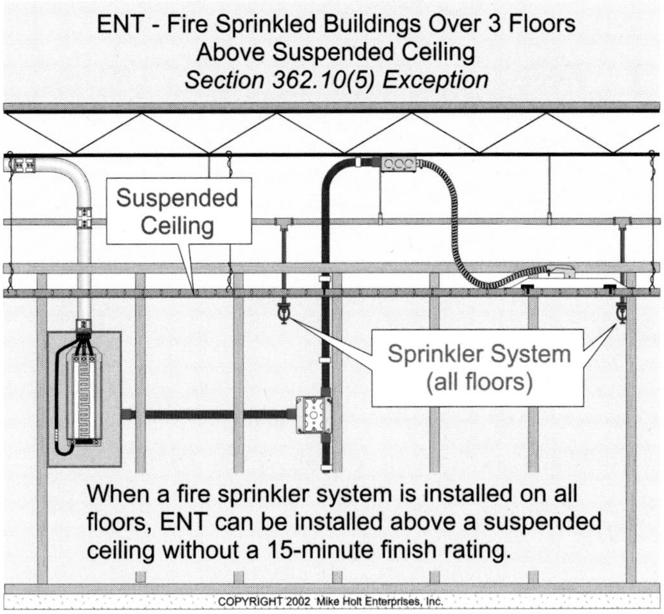

**Figure 362-5**

*Exception:* Where a fire sprinkler system is installed on all floors, in accordance with NFPA 13-1999, *Standard for the Installation of Sprinkler Systems*, ENT can be installed above a suspended ceiling that does not have a 15-minute finish rated thermal barrier material. Figure 362-5

(6) ENT can be encased or embedded in a concrete slab on grade where the tubing is placed on sand or approved screenings, provided fittings identified for the purpose are used.

**Note.** ENT shall not be buried in the earth. See 362.12(5).

(7) ENT can be installed in wet locations indoors or in a concrete slab on or below grade, with fittings listed for the purpose.

(8) Listed prewired ENT is permitted in sizes from $1/2$ in. to 1 in.

### 362.12 Uses Not Permitted

(1) In hazardous (classified) locations, except as permitted by 504.20 and 505.15(A)(1).

(2) For the support of luminaires or equipment. See 314.2.

(3) Where the ambient temperature is in excess of 50°C/122°F.

(4) To contain conductors that operate at a temperature above the temperature rating of the raceway.

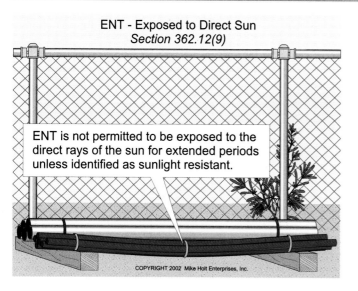

**ENT - Exposed to Direct Sun**
*Section 362.12(9)*

ENT is not permitted to be exposed to the direct rays of the sun for extended periods unless identified as sunlight resistant.

COPYRIGHT 2002 Mike Holt Enterprises, Inc.

**Figure 362-6**

(5) For direct earth burial, though it can be encased in concrete. See 362.10(6).

(6) As a wiring method for systems over 600V.

(7) Exposed in buildings over three floors, except as permitted 362.10(2) and (5) Ex.

(8) In places of assembly or theaters, except as permitted by 518.4 and 520.5.

(9) Exposed to the direct rays of the sun for an extended period, unless listed as sunlight-resistant.

**AUTHOR'S COMMENT:** Exposing ENT to direct rays of the sun for an extended time may result in the product becoming brittle, unless it has specific compounds to resist the effects of ultraviolet (UV) radiation. Figure 362-6

(10) Where subject to physical damage.

**AUTHOR'S COMMENT:** ENT is prohibited as a wiring method in ducts, plenums, and other spaces used for environmental air handling [300.22] and for patient care area circuits in health care facilities [517.13(A)].

## 362.20 Sizes

**(A) Minimum Size.** 1/2 in.

**(B) Maximum Size.** 2 in.

## 362.22 Number of Conductors

Raceways shall be large enough to permit the installation and removal of conductors without damaging the conductors' insulation. When all conductors in a raceway are the same size and insulation, the number of conductors permitted can be found in Annex C for the raceway type.

**Question:** How many 12 AWG THHN conductors can be installed in 1/2 in. ENT?

**Answer:** Seven conductors, Annex C, Table C2.

**AUTHOR'S COMMENT:** See 300.17 for additional examples on how to size raceways when conductors are not all the same size.

## 362.24 Bends

Raceway bends shall not be made in any manner that would damage the raceway or significantly change its internal diameter (no kinks). This is accomplished by complying with the bending radius requirements in Table 344.24.

## 362.26 Number of Bends (360°)

To reduce the stress and friction on the conductor insulation, the number of bends (including offsets) between pull points shall not exceed 360°.

## 362.28 Trimming

The cut ends of ENT shall be trimmed (inside and out) to remove the burrs and rough edges. Trimming nonmetallic raceway is very easy; most of the burrs will rub off with fingers, and a knife can be used to smooth the rough edges.

## 362.30 Secured and Supported

ENT shall be installed as a complete system [300.10, 300.12, and 300.18(A)], and it shall be securely fastened in place and supported in accordance with the following:

**(A) Secured.** ENT shall be secured within 3 ft of every box, cabinet, or termination fitting, such as a conduit body and at intervals not exceeding 3 ft. Figure 362-7

*Exception:* ENT used for a luminaire whip in lengths not exceeding 6 ft in accordance with 410.67(C) does not need to be secured or supported.

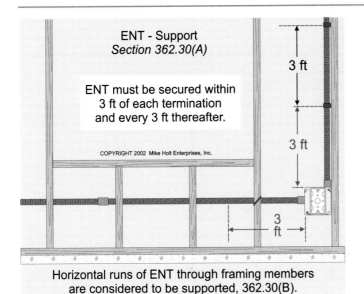

Horizontal runs of ENT through framing members are considered to be supported, 362.30(B).

**Figure 362-7**

(B) **Horizontal Runs.** ENT installed horizontally in bored or punched holes in wood or metal framing members, or notches in wooden members, is considered supported, but the raceway shall be secured within 3 ft of terminations.

## 362.46 Bushings

Where conductors 4 AWG or larger enter a box, fitting, or other enclosure, a fitting that provides a smooth, rounded insulating surface, such as a bushing or adapter, shall be provided to protect the wire from abrasion during and after installation. See 300.4(F).

## 362.48 Joints

Joints, such as couplings and connectors, shall be made in an approved manner. This means you shall follow the manufacturer's instructions for the raceway, fittings, and glue.

> **AUTHOR'S COMMENT:** According to testing laboratory publications, RNC fittings can be used with ENT.

> *CAUTION: When using glue on ENT, it shall be approved for the tubing and the fittings. Glue for RNC (PVC) generally cannot be used with ENT because it damages the plastic from which ENT is manufactured.*

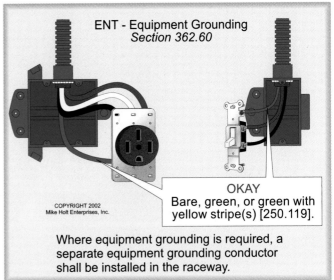

Where equipment grounding is required, a separate equipment grounding conductor shall be installed in the raceway.

**Figure 362-8**

## 362.60 Grounding

Where equipment grounding is required, a separate equipment grounding conductor shall be installed in the raceway. Figure 362-8

## Article 362

1.  When ENT is installed concealed in walls, floors, and ceilings of buildings exceeding three floors above grade, a thermal barrier must be provided having a minimum _____ minute finish rating as listed for fire-rated assemblies.
    (a) 5                (b) 10                (c) 15                (d) 30

2.  When a building is supplied with a(n) _____ fire sprinkler system, ENT can be installed exposed or concealed in buildings of any height.
    (a) listed           (b) identified        (c) approved          (d) none of these

3.  ENT can be installed above a suspended ceiling if the suspended ceiling provides a thermal barrier having a _____-minute finish rating as identified in listings of fire-rated assemblies.
    (a) 5                (b) 10                (c) 15                (d) none of these

4.  When a building is supplied with an approved fire sprinkler system(s), ENT can be installed above any suspended ceiling.
    (a) True             (b) False

5.  ENT and fittings shall be permitted to be _____ in a concrete slab on grade where the tubing is placed upon sand or approved screenings.
    (a) encased          (b) embedded          (c) either a or b     (d) none of these

6.  ENT is permitted for direct earth burial when used with fittings listed for this purpose.
    (a) True             (b) False

7.  ENT is not permitted in places of assembly, unless it is encased in at least _____ of concrete.
    (a) 1 in.            (b) 2 in.             (c) 3 in.            (d) 4 in.

8.  In electrical nonmetallic tubing, the maximum number of bends between pull points cannot exceed _____, including any offsets.
    (a) 320°             (b) 270°              (c) 360°             (d) unlimited

9.  ENT must be secured in place every _____
    (a) 12 in.           (b) 18 in.            (c) 24 in.           (d) 36 in.

10. Bushings or adapters are required at ENT terminations to protect the conductors from abrasion.
    (a) True             (b) False

# Article 376
# Metal Wireways

## Part I. General

### 376.1 Scope

This article covers the use, installation, and construction specifications of metallic wireways and associated fittings.

### 376.2 Definition

A metal wireway is a sheet metal raceway with hinged or removable covers for housing and protecting electric wires and cable, and in which conductors are placed after the wireway has been installed. Figure 376-1

> **AUTHOR'S COMMENT:** Some call this wiring method an auxiliary gutter [Article 366], which it is not.

## Part II. Installation

### 376.10 Uses Permitted

(1) Exposed.

(2) Concealed, as permitted by 376.10(4).

(3) In hazardous (classified) locations as permitted by 501.4(B), 502.4(B), or 504.20.

(4) Unbroken through walls, partitions, and floors.

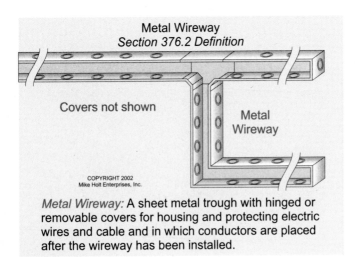

Metal Wireway
Section 376.2 Definition

Covers not shown

Metal Wireway

COPYRIGHT 2002
Mike Holt Enterprises, Inc.

*Metal Wireway:* A sheet metal trough with hinged or removable covers for housing and protecting electric wires and cable and in which conductors are placed after the wireway has been installed.

**Figure 376-1**

### 376.12 Not Permitted

(1) Subject to severe physical damage.

(2) Subject to corrosive environments.

### 376.21 Conductor – Maximum Size

The maximum size conductor permitted in a wireway shall not be larger than that for which the wireway is designed.

### 376.22 Conductors – Maximum Number

The maximum number of conductors permitted in a wireway is limited to 20 percent of the cross-sectional area of the wireway. Figure 376-2

> **AUTHOR'S COMMENT:** Splices and taps shall not fill the wiring space at any cross section to more than 75 percent. See 376.56.

**Ampacity Adjustment.** When more than 30 current-carrying conductors are installed in any cross-sectional area of the wireway, the conductor ampacity, as listed in Table 310.16, shall be reduced according to the adjustment factor listed in Table 310.15(B)(2).

> **AUTHOR'S COMMENT:** Signaling and motor control conductors between a motor and its starter and used only for starting duty are not considered current-carrying.

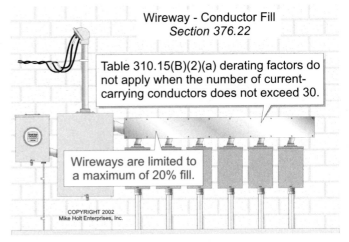

Wireway - Conductor Fill
Section 376.22

Table 310.15(B)(2)(a) derating factors do not apply when the number of current-carrying conductors does not exceed 30.

Wireways are limited to a maximum of 20% fill.

COPYRIGHT 2002
Mike Holt Enterprises, Inc.

**Figure 376-2**

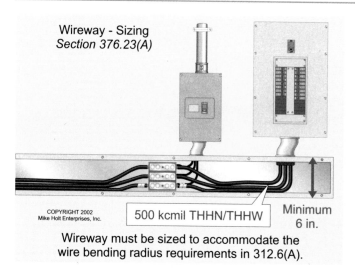

Wireway - Sizing
*Section 376.23(A)*

COPYRIGHT 2002
Mike Holt Enterprises, Inc.

500 kcmil THHN/THHW

Minimum
6 in.

Wireway must be sized to accommodate the
wire bending radius requirements in 312.6(A).

**Figure 376-3**

## 376.23 Wireway Sizing

**(A) Bending Radius.** Where conductors are bent within a wireway, the wireway shall be sized to meet the bending radius requirements contained in 312.6(A), based on one conductor per terminal. For example, a wireway shall be sized no less than 6 in. to permit sufficient bending radius for 500 kcmil conductors. Figure 376-3

**(B) Wireway Used as Pull Box.** Where insulated conductors 4 AWG or larger are pulled through a wireway, the distance between raceway and cable entries enclosing the same conductor shall not be less than required by 314.28. Figure 376-4

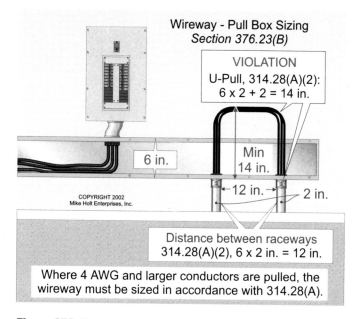

Wireway - Pull Box Sizing
*Section 376.23(B)*

VIOLATION
U-Pull, 314.28(A)(2):
6 x 2 + 2 = 14 in.

6 in.

Min
14 in.

COPYRIGHT 2002
Mike Holt Enterprises, Inc.

12 in.

2 in.

Distance between raceways
314.28(A)(2), 6 x 2 in. = 12 in.

Where 4 AWG and larger conductors are pulled, the
wireway must be sized in accordance with 314.28(A).

**Figure 376-4**

**Straight Pulls.** The minimum distance from where the conductors enter to the opposite wall shall not be less than eight times the trade size of the largest raceway [314.28(A)(1)].

**Angle Pulls.** The distance from the raceway entry to the opposide wall shall not be less than six times the trade diameter of the largest raceway, plus the sum of the diameters of the remaining raceways on the same wall [314.28(A)(2)].

**U Pulls.** When a conductor enters and leaves from the same wall, the distance from where the raceways enter to the opposite wall shall not be less than six times the trade diameter of the largest raceway, plus the sum of the diameters of the remaining raceways on the same wall [314.28(A)(2)].

The distance between raceways enclosing the same conductor shall not be less than six times the trade diameter of the largest raceway, measured from raceway edge-to-edge [314.28(A)(2)].

## 376.30 Supports

Wireways shall be supported in accordance with the following:

**(A) Horizontal Support.** Where run horizontally, wireways shall be supported at each end and at intervals not to exceed 5 ft.

**(B) Vertical Support.** Where run vertically, wireways shall be securely supported at intervals not exceeding 15 ft, with no more than one joint between supports.

## 376.56 Splices and Taps

Splices and taps in wireways shall be accessible, and they shall not fill the wireway to more than 75 percent of its cross-sectional area. Figure 376-5

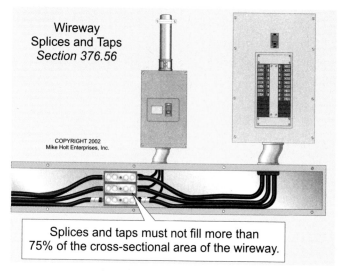

Wireway
Splices and Taps
*Section 376.56*

COPYRIGHT 2002
Mike Holt Enterprises, Inc.

Splices and taps must not fill more than
75% of the cross-sectional area of the wireway.

**Figure 376-5**

# Article 376

1. Wireways shall be permitted for _____.
   (a) exposed work
   (b) concealed work
   (c) wet locations if listed for the purpose
   (d) a and c

2. Wireways can pass transversely through a wall _____.
   (a) if the length passing through the wall is unbroken
   (b) if the wall is not fire rated
   (c) in hazardous locations
   (d) if the wall is fire rated

3. Where a wireway is used as a pull box for insulated conductors 4 AWG or larger, the distance between raceway and cable entries enclosing the same conductor shall not be less than that required in 314.28(A)(1) for straight pulls and 314.28(A)(2) for angle pulls.
   (a) True
   (b) False

4. Wireways shall be supported where run horizontally at each end and at intervals not to exceed _____ or for individual lengths longer than _____ at each end or joint, unless listed for other support intervals.
   (a) 5 ft
   (b) 10 ft
   (c) 3 ft
   (d) 6 ft

5. Splices and taps shall be permitted within a wireway provided they are accessible. The conductors, including splices and taps, shall not fill the wireway to more than _____ percent of its area at that point.
   (a) 25
   (b) 80
   (c) 125
   (d) 75

# Article 378
# Nonmetallic Wireways

## Part I. General

### 378.1 Scope

This article covers the use, installation, and construction specifications of nonmetallic wireways and associated fittings.

*CAUTION: Nonmetallic wireways do not have the same heat-transfer characteristics as metal wireways. The ampacity adjustment factors of Table 310.15(B)(2)(a) apply any time there are four or more current-carrying conductors in any cross-sectional area of a nonmetallic wireway [378.22]. Effectively, this severely limits the use of nonmetallic wireways.*

### 378.2 Definition

A nonmetallic wireway is a flame-retardant raceway with hinged or removable covers for housing and protecting electric wires and cable and in which conductors are placed after the wireway has been installed as a complete system.

## Part II. Installation

### 378.10 Uses Permitted

(1) Exposed, except as permitted in 378.10(4).

(2) Subject to corrosive environments, where identified for the use.

(3) In wet locations, where listed for the purpose.

(4) In unbroken extensions passing transversely through walls.

### 378.12 Uses Not Permitted

(1) Where subject to severe physical damage.

(2) In hazardous (classified) locations as permitted by 504.20.

(3) Where exposed to sunlight, unless listed and marked as suitable for this purpose.

(4) Where subject to ambient temperatures in excess of the wireway listing.

(5) Where the conductors operate at a temperature higher than the raceway temperature rating.

### 378.21 Conductor – Maximum Size

The maximum size conductor permitted in a wireway shall not be larger than that for which the wireway is designed.

### 378.22 Conductors – Maximum Number

The maximum number of conductors permitted in a wireway is limited to 20 percent of the cross-sectional area of the wireway, except that splices and taps shall not fill the nonmetallic wireway to more than 75 percent of its area at that point [378.56].

The ampacity adjustment factors of Table 310.15(B)(2)(a) apply any time there are four or more current-carrying conductors in any cross-sectional area of the wireway, but signaling and motor control conductors are not considered current-carrying.

### 378.23 Wireway Sizing

(A) **Bending Radius.** Where conductors are bent within a wireway, the wireway shall be sized to meet the bending radius requirements in 312.6(A), based on one conductor per terminal. For example, a wireway shall be sized no less than 6 in. to permit sufficient bending radius for 500 kcmil conductors.

(B) **Wireway Used as Pull Box.** Where insulated conductors 4 AWG or larger are pulled through a wireway, the distance between raceway and cable entries enclosing the same conductor shall not be less than required by 314.28.

**Straight Pulls.** The minimum distance from where the conductors enter to the opposite wall shall not be less than eight times the trade size of the largest raceway [314.28(A)(1)].

**Angle Pulls.** The distance from the raceway entry to the opposite wall shall not be less than six times the trade diameter of the largest raceway, plus the sum of the diameters of the remaining raceways on the same wall [314.28(A)(2)].

**U Pulls.** When a conductor enters and leaves from the same wall, the distance from where the raceways enter to the opposite wall shall not be less than six times the trade diameter of the largest raceway, plus the sum of the diameters of the remaining raceways on the same wall [314.28(A)(2)].

The distance between raceways enclosing the same conductor shall not be less than six times the trade diameter of the largest raceway, measured from raceway edge-to-edge [314.28(A)(2)].

## 378.30 Supports

Wireways shall be supported in accordance with the following:

**(A) Horizontal Support.** Where run horizontally, wireways shall be supported at each end and at intervals not to exceed 3 ft.

**(B) Vertical Support.** Where run vertically, wireways shall be securely supported at intervals not exceeding 4 ft, with no more than one joint between supports.

## 378.44 Expansion Fittings

Expansion fittings for nonmetallic wireways shall be provided to compensate for thermal expansion and contraction where the length change is expected to be $1/4$ in. or greater in a straight run. The expansion characteristics of PVC nonmetallic wireway are identical to those for PVC rigid nonmetallic conduit in Table 352.44.

## 378.56 Splices and Taps

Splices and taps in wireways shall be accessible and shall not fill the wireway to more than 75 percent of its cross-sectional area.

**AUTHOR'S COMMENT:** The maximum number of conductors permitted in a wireway is limited to 20 percent of its cross-sectional area. See 378.22.

## 378.60 Grounding

Where an equipment grounding conductor is required, it shall be installed within the raceway. See 250.102(E) and 300.3(B).

*Exception No. 2:* An equipment grounding conductor is not required if the grounded (neutral) conductor is used to ground service equipment [250.24(B)], as permitted in 250.142(A).

# Article 378

1. Nonmetallic wireways shall be permitted for _____.
   (a) exposed work
   (b) concealed work
   (c) wet locations if listed for the purpose
   (d) a and c

2. Nonmetallic wireways can pass transversely through a wall _____.
   (a) if the length passing through the wall is unbroken
   (b) if the wall is not fire rated
   (c) in hazardous locations
   (d) if the wall is fire rated

3. Where a nonmetallic wireway is used as a pull box for insulated conductors 4 AWG or larger, the distance between raceway and cable entries enclosing the same conductor shall not be less than that required in 314.28(A)(1) for straight pulls and 314.28(A)(2) for angle pulls.
   (a) True
   (b) False

4. Wireways shall be supported where run horizontally at each end and at intervals not to exceed _____ or for individual lengths longer than _____ at each end or joint, unless listed for other support intervals.
   (a) 5 ft
   (b) 10 ft
   (c) 3 ft
   (d) 6 ft

5. Splices and taps shall be permitted within a nonmetallic wireway provided they are accessible. The conductors, including splices and taps, shall not fill the wireway to more than _____ percent of its area at that point.
   (a) 25
   (b) 80
   (c) 125
   (d) 75

**Article 380**
**Multioutlet Assembly**

## 380.1 Scope

This article covers the use, installation, and construction specifications of multioutlet assemblies.

> **AUTHOR'S COMMENT:** According to Article 100, a multioutlet assembly is a surface, flush, or freestanding raceway designed to hold conductors and receptacles, assembled in the field or at the factory. Figure 380-1

## 380.2 Uses

**(A) Permitted.** Dry locations only.

**(B) Not Permitted.**

(1) Concealed.

(2) Where subject to severe physical damage.

(3) Where the voltage is 300V or more between conductors, unless the assembly is metal having a thickness of not less than 0.040 in.

(4) Where subject to corrosive vapors.

(5) In hoistways.

(6) In hazardous (classified) locations, except as permitted by 501.4(B).

## 380.3 Through Partitions

Metal multioutlet assemblies can pass through a dry partition, provided no receptacle is concealed in the wall and the cover of the exposed portion of the system can be removed.

> **AUTHOR'S COMMENT:** The feeder and service load for multioutlet assemblies shall be according to the requirements in 220.3(B)(8).

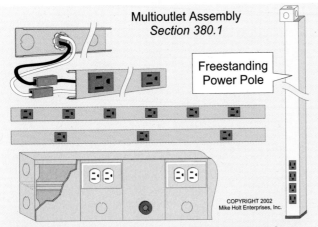

*Multioutlet Assembly:* (Article 100 Definition) A surface, flush or freestanding raceway designed to hold conductors and receptacles assembled in the field or at the factory.

**Figure 380-1**

## Article 380

1.  A multioutlet assembly is a surface, flush, or freestanding raceway designed to hold conductors and receptacles, assembled in the field or at the factory.
    (a) True          (b) False

2.  A multioutlet assembly can be installed in _____.
    (a) dry locations          (b) wet locations          (c) a and b          (d) none of these

3.  A multioutlet assembly cannot be installed _____.
    (a) in concealed locations          (b) where subject to severe physical damage
    (c) where subject to corrosive vapors          (d) all of these

4.  Metal multioutlet assemblies can pass through a dry partition, provided no receptacle is concealed in the wall and the cover of the exposed portion of the system can be removed.
    (a) True          (b) False

# Article 384
# Strut-Type Channel Raceway

## Part I. General

### 384.1 Scope

This article covers the use, installation, and construction specifications of strut-type channel raceways and associated fittings.

### 384.2 Definition

Strut-type channel raceway is a metallic raceway intended to be mounted to the surface or suspended, with associated accessories in which conductors are placed after the raceway has been installed as a complete system. Figure 384-1

## PART II. Installation

### 384.10 Uses Permitted

(1) Exposed.

(2) In dry locations.

(3) Where subject to corrosive environments, where protected by finishes suitable for the condition.

(4) Where the voltage is 600V or less.

(5) As power poles.

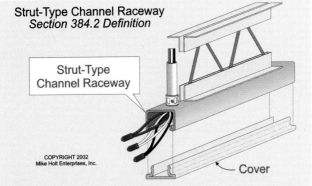

Strut-Type Channel Raceway
*Section 384.2 Definition*

Strut-Type
Channel Raceway

COPYRIGHT 2002
Mike Holt Enterprises, Inc.

Cover

*Strut-Type Channel Raceway:* A metallic raceway mounted to the surface or suspended, in which conductors are laid in place after the raceway has been installed.

**Figure 384-1**

(6) In hazardous (classified) locations as permitted by 501.4(B)(3).

(7) Unbroken through walls, partitions, and floors.

(8) Indoors only, if protected by enamel.

### 384.12 Uses Not Permitted

(1) Concealed.

(2) Where subject to severe corrosive influences.

### 384.21 Conductor – Maximum Size

The maximum size conductor permitted shall not be larger than that for which the wireway is designed.

### 384.22 Number of Conductors Permitted

The number of conductors permitted in strut-type channel raceways shall not exceed the percentage fill values in Table 384.22.

The ampacity adjustment factors of 310.15(B)(2)(a) do not apply to conductors installed in strut-type channel raceways where all of the following conditions are met:

(1) The cross-sectional area of the raceway is at least 4 square inches.

(2) The number of current-carrying conductors does not exceed 30.

(3) The sum of the cross-sectional areas of all contained conductors does not exceed 20 percent of the interior cross-sectional area of the strut-type channel raceways.

### 384.30 Securing and Supporting

**(A) Surface Mounted.** A surface mount strut-type raceway shall be secured to the mounting surface with retention straps external to the channel at intervals not exceeding 10 ft and within 3 ft of each outlet box, cabinet, junction box, or other channel raceway termination.

**(B) Suspension Mount.** Strut-type channel raceways can be suspension mounted with approved appropriate methods designed for the purpose, at intervals not to exceed 10 ft.

## 384.56 Splices and Taps

Splices and taps shall be accessible and shall not fill the wireway to more than 75 percent of its cross-sectional area.

## 384.60 Grounding

Strut-type channel raceway enclosures shall have a means for connecting an equipment grounding conductor. The raceway is considered suitable as an equipment grounding conductor in accordance with 250.118(14).

# Article 384

1. A strut-type channel raceway is a metallic raceway intended to be mounted to the surface or suspended, with associated accessories in which conductors are placed after the raceway has been installed as a complete system.
   (a) True      (b) False

2. A strut-type channel raceway can be installed _____.
   (a) where exposed
   (b) as power poles
   (c) unbroken through walls, partitions, and floors
   (d) all of these

3. A strut-type channel raceway cannot be installed _____.
   (a) in concealed locations
   (b) where subject to corrosive vapors
   (c) a or b
   (d) none of these

4. A surface mount strut-type channel raceway shall be secured to the mounting surface with retention straps external to the channel at intervals not exceeding _____ and within 3 ft of each outlet box, cabinet, junction box, or other channel raceway termination.
   (a) 3 ft      (b) 5 ft      (c) 6 ft      (d) 10 ft

5. Splices and taps shall be permitted within a strut-type channel raceway provided they are accessible. The conductors, including splices and taps, shall not fill the wireway to more than _____ percent of its area at that point.
   (a) 25      (b) 80      (c) 125      (d) 75

# Article 386
# Surface Metal Raceways

## Part I. Surface Metal Raceways

### 386.1 Scope

This article covers the use, installation, and construction specifications of surface metal raceways and associated fittings.

### 386.2 Definition

**Surface Metal Raceway.** A surface metal raceway is a metallic raceway intended to be mounted to the surface, with associated accessories, in which conductors are placed after the raceway has been installed as a complete system. Figure 386-1

> **AUTHOR'S COMMENT:** Surface raceways are available in different shapes and sizes and can be mounted on walls, ceilings, or floors. They have removable covers, which eliminates the need for wire pulling. Some surface raceways have two or more separate compartments, which permit the separation of power, lighting, control, signal, and communications cables and conductors. See 386.70.

### 386.6 Listing Requirements

Surface metal raceways and associated fittings shall be listed.

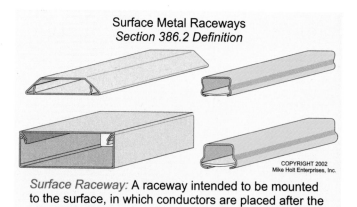

Surface Metal Raceways
Section 386.2 Definition

COPYRIGHT 2002
Mike Holt Enterprises, Inc.

*Surface Raceway:* A raceway intended to be mounted to the surface, in which conductors are placed after the raceway has been installed as a complete system.

*Figure 386-1*

> **AUTHOR'S COMMENT:** Enclosures for switches, receptacles, luminaires, and other devices are identified by the markings on their packaging, which identify the type of surface raceway with which the enclosure can be used.

## Part II. Installation

### 386.10 Uses Permitted

(1)  In dry locations.

(2)  In Class I, Division 2 hazardous (classified) locations as permitted in 501.4(B)(3).

(3)  Under raised floors, as permitted in 645.5(D)(2).

(4)  Extension through walls and floors, if the length passing through is unbroken and access to the conductors shall be maintained on both sides of the wall, partition, or floor.

### 386.12 Uses Not Permitted

(1)  Where subject to severe physical damage, unless otherwise approved.

(2)  Where the voltage is 300V or more between conductors, unless the metal has a thickness of not less than 0.040 in., nominal.

(3)  Where subject to corrosive vapors.

(4)  In hoistways.

(5)  Where concealed, except as permitted in 386.10.

### 386.21 Size of Conductors

The maximum size conductor permitted in a wireway shall not be larger than that for which the wireway is designed.

> **AUTHOR'S COMMENT:** Because partial packages are often purchased, you may not always get this information.

**Surface Raceway - Conductor Fill**
*Section 386.22*

Table 310.15(B)(2)(a) ampacity adjustment factors do not apply if:
(1) The cross-sectional area of the raceway exceeds 4 sq in.,
(2) the number of current-carrying conductors does not exceed 30, and
(3) the conductor fill does not exceed 20% of the total cross-sectional area of the raceway.

COPYRIGHT 2002  Mike Holt Enterprises, Inc.

**Figure 386-2**

### 386.22 Number of Conductors Permitted

The raceway is considered suitable as an equipment grounding conductor in accordance with 250.118(14).

The ampacity adjustment factors of 310.15(B)(2)(a) do not apply to conductors installed in raceways where all of the following conditions are met: Figure 386-2

(1) The cross-sectional area of the raceway is at least 4 sq in., and

(2) The number of current-carrying conductors does not exceed 30, and

(3) The sum of the cross-sectional areas of all contained conductors does not exceed 20 percent of the interior cross-sectional area of the raceways.

### 386.56 Splices and Taps

Splices and taps shall be accessible and shall not fill the raceway to more than 75 percent of its cross-sectional area.

### 386.60 Grounding

Surface metal raceway fittings shall be mechanically and electrically joined together in a manner that does not subject the conductors to abrasion. Surface raceways that allow a transition to another wiring method, such as knockouts for connecting conduits, shall have a means for the termination of an equipment grounding conductor. The raceway is considered suitable as an equipment grounding conductor in accordance with 250.118(14).

### 386.70 Separate Compartments

Where surface raceways have separate compartments within a single raceway, one compartment can be used for power and lighting, and the other can be used for control, signal, or communications as required by:

- CATV                          820.10(F)(1)
- Communications                800.52(A)(1)
- Control and Signaling         725.55(A)
- Fire Alarm                    760.55(A)
- Intrinsically Safe Systems    504.30(A)(2)
- Instrument Tray Cable         727.5
- Network Broadband             830.58(C)
- Radio and Television          810.18(C)
- Sound Systems                 640.9(C)

# Article 386

1.  It is permissible to run unbroken lengths of surface metal raceways through dry _____.
    (a) walls       (b) partitions       (c) floors       (d) all of these

2.  The derating factors of 310.15(B)(2)(a), (Notes to Ampacity Tables of 0 through 2,000V), shall not apply to conductors installed in surface metal raceways where _____.
    (a) the cross-sectional area exceeds 4 sq in.
    (b) the current-carrying conductors do not exceed 30 in number
    (c) the total cross-sectional area of all conductors does not exceed 20 percent of the interior cross-sectional area of the raceway
    (d) all of these

3.  The maximum number of conductors permitted in any surface raceway shall be _____.
    (a) no more than 30 percent of the inside diameter
    (b) no greater than the number for which it was designed
    (c) no more than 75 percent of the cross-sectional area
    (d) that which is permitted in Table 312.6(A)

4.  The conductors, including splices and taps, in a surface metal raceway, shall not fill the raceway to more than _____ percent of its cross-sectional area at that point.
    (a) 75       (b) 40       (c) 38       (d) 53

5.  Where combination surface metal raceways are used for both signaling and for lighting and power circuits, the different systems shall be run in separate compartments identified by _____ of the interior finish, and the same relative position of compartments shall be maintained throughout the premises.
    (a) brilliant colors       (b) etching       (c) identification       (d) contrasting colors

# Article 388
# Surface Nonmetallic Raceways

## Part I. General

### 388.1 Scope

This article covers the use, installation, and construction specifications of surface metal raceways and associated fittings.

### 388.2 Definition

A surface nonmetallic raceway is a nonmetallic raceway intended to be mounted to the surface, with associated accessories, in which conductors are placed after the raceway has been installed as a complete system.

### 388.6 Listing Requirements

Surface nonmetallic raceways and associated fittings shall be listed.

## Part II. Installation

### 388.10 Uses Permitted

   (1)  In dry locations.

   (2)  Unbroken through walls, partitions, and floors.

### 388.12 Uses Not Permitted

   (1)  Concealed, except as permitted by 388.10(2).

   (2)  Where subject to severe physical damage.

   (3)  Where the voltage is 300V or more between conductors, unless listed for higher voltage.

   (4)  In hoistways.

   (5)  In hazardous (classified) locations as permitted by 501.4(B)(3).

   (6)  Where subject to ambient temperatures in excess of the wireway listing.

   (7)  Where the conductors operate at a temperature higher than the raceway temperature rating.

### 388.21 Size of Conductors

The maximum size conductor permitted in a wireway shall not be larger than that for which the wireway is designed.

### 388.22 Number of Conductors

The number of conductors permitted in a surface raceway shall not be more than the number for which the raceway has been listed.

> **AUTHOR'S COMMENT:** The ampacity adjustment factors contained in Table 310.15(B)(2)(a) applies whenever four or more current-carrying conductors are in any cross-sectional area of a nonmetallic surface raceway.

### 388.56 Splices and Taps

Splices and taps shall be accessible and shall not fill the wireway to more than 75 percent of its cross-sectional area.

### 388.60 Grounding

Where an equipment grounding conductor is required, it shall be installed within the raceway. See 250.102E and 300.3(B).

### 388.70 Separate Compartments

Where surface raceways have separate compartments within a single raceway, one compartment can be used for power and lighting, and the other can be used for control, signal, or communications as required by:

| | |
|---|---|
| • CATV, | 820.10(F)(1) |
| • Communications, | 800.52(A)(1) |
| • Control and Signaling, | 725.55(A) |
| • Fire Alarm, | 760.55(A) |
| • Intrinsically Safe Systems, | 504.30(A)(2) |
| • Instrument Tray Cable, | 727.5 |
| • Network Broadband, | 830.58(C) |
| • Radio and Television, | 810.18(C) |
| • Sound Systems, | 640.9(C) |

## Article 388

1. It is permissible to run unbroken lengths of surface nonmetallic raceways through dry _____.
   (a) walls      (b) partitions      (c) floors      (d) all of these

2. The maximum number of conductors permitted in any surface raceway shall be _____.
   (a) no more than 30 percent of the inside diameter
   (b) no greater than the number for which it was designed
   (c) no more than 75 percent of the cross-sectional area
   (d) that which is permitted in Table 312.6(A)

3. The conductors, including splices and taps, in a nonmetallic surface raceway shall not fill the raceway to more than _____ percent of its cross-sectional area at that point.
   (a) 75      (b) 40      (c) 38      (d) 53

4. Where combination surface nonmetallic raceways are used for both signaling and for lighting and power circuits, the different systems shall be run in separate compartments identified by _____ of the interior finish, and the same relative position of compartments shall be maintained throughout the premises.
   (a) brilliant colors      (b) etching      (c) identification      (d) contrasting colors

# Article 392
# Cable Trays

## 392.1 Scope

This article covers cable tray systems, including ladder, ventilated trough, ventilated channel, solid bottom, and other similar structures.

## 392.2 Definition

**Cable Tray System.** A unit or assembly of units or sections and associated fittings forming a rigid structural system used to securely fasten or support cables and raceways.

**AUTHOR'S COMMENT:** A cable tray is not a raceway

## 392.3 Uses Permitted

Cable trays can be used as a support system for:

Services, feeders, and branch circuits.

Communications circuits, control circuits, and signaling circuits.

Cable tray installations are not limited to industrial establishments.

Where exposed to direct rays of the sun, insulated conductors and jacketed cables shall be identified as being sunlight-resistant.

Cable trays and their associated fittings shall be identified by the manufacturer for the intended use. Figure 392-1

Cable Tray - Uses Permitted
Section 392.3

Cable trays can be used as a support system for:
- services
- feeders
- branch circuits
- communication circuits
- control circuits
- signaling circuits

COPYRIGHT 2002 Mike Holt Enterprises, Inc.

**Figure 392-1**

**AUTHOR'S COMMENT:** Cable trays are manufactured in many forms, from a simple hanger or wire mesh to a substantial, rigid, steel support system. Cable trays are designed and manufactured to support specific wiring methods. To ensure a safe support system, it is necessary that cable trays be identified for their intended use.

**(A) Wiring Methods.** Any wiring method in Table 392.3(A) can be installed in a cable tray.

**AUTHOR'S COMMENT:** Control, signal, and communications cables shall maintain 2 in. separation from single power conductors.
- Coaxial Cable                     820.52(A)(2) Ex. 1
- Class 2 and 3 Cables        725.55(J)
- Communications Cable     800.52(A)(2) Ex. 1
- Fire Alarm Cables             760.55(G)
- Intrinsically Safe
- Systems Cables                  504.30(A)(2) Ex. 1
- Network-Powered
- Broadband Cables            830.58(A)(2) Ex. 1
- Radio and Television Cable   810.18(B) Ex. 1

**(D) Hazardous (classified) Locations.** Cable trays in hazardous (classified) locations shall contain only the cable types permitted in 501.4, 502.4, 503.3, 504.20, and 505.15.

**(E) Nonmetallic Cable Tray.** In addition to the uses permitted elsewhere in Article 392, nonmetallic cable trays can be used in corrosive areas and in areas requiring voltage isolation.

## 392.4 Uses Not Permitted

- Hoistways.
- Where subject to severe physical damage.
- Environmental airspaces, except as permitted by 300.22, to support wiring methods recognized for use in such spaces.

## 392.6 Installation

**(A) Complete System.** Cable trays shall be installed as a complete system, except that mechanically discontinuous segments between cable tray runs or between cable tray runs and equipment are permitted. A bonding jumper sized in accordance with 250.102 and installed in accordance with 250.96 shall bond the sections of cable tray, or the cable tray and the raceway or equipment.

**(B) Completed Before Installation.** Each run of cable tray shall be completed before the installation of cables.

**(C) Support.** Supports for cable trays shall be provided to prevent stress on cables where they enter raceways or other enclosures from cable tray systems. Cable trays shall be supported in accordance with the manufacturer's installation instructions.

**(G) Through Partitions and Walls.** Cable trays can extend through partitions and walls or vertically through platforms and floors where the installation is made in accordance with the fire seal requirements of 300.21.

**(H) Exposed and Accessible.** Cable trays shall be exposed and accessible, except as permitted by 392.6(G).

**(I) Adequate Access.** Sufficient space shall be provided and maintained about cable trays to permit adequate access for installing and maintaining the cables.

**(J) Raceways, Cables, and Boxes Supported from Cable Trays.** In industrial facilities where conditions of maintenance and supervision ensure that only qualified persons will service the installation, and where the cable tray system is designed and installed to support the load, such cable tray systems can be used to support raceways, cables, boxes and conduit bodies. Figure 392-2

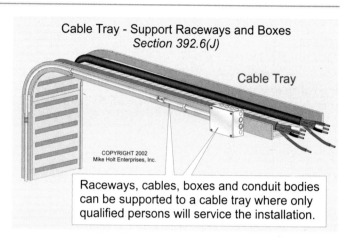

Cable Tray - Support Raceways and Boxes
Section 392.6(J)

Cable Tray

COPYRIGHT 2002
Mike Holt Enterprises, Inc.

Raceways, cables, boxes and conduit bodies can be supported to a cable tray where only qualified persons will service the installation.

**Figure 392-2**

For raceways terminating at the tray, a listed cable tray clamp or adapter shall be used to securely fasten the raceway to the cable tray system. The raceway shall be supported in accordance with the requirements of the appropriate raceway article.

Raceways or cables running parallel to the cable tray system can be attached to the bottom or side of a cable tray system. The raceway or cable shall be fastened and supported in accordance with the requirements of the appropriate raceway or cable article.

Boxes and conduit bodies attached to the bottom or side of a cable tray system shall be in accordance with 314.23.

# Article 392

1. A cable tray is a unit or assembly of units or sections and associated fittings forming a _____ system used to securely fasten or support cables and raceways.
   (a) structural      (b) flexible      (c) movable      (d) secure

2. Cable trays can be used as a support system for _____.
   (a) services, feeders, branch circuits      (b) communications circuits
   (c) control and signaling circuits      (d) all of these

3. The intent of Article 392 is to limit the use of cable trays to industrial establishments only.
   (a) True      (b) False

4. Where exposed to direct rays of the sun, insulated conductors and jacketed cables must be _____ as being sunlight resistant.
   (a) listed      (b) approved      (c) identified      (d) none of these

5. Cable trays and their associated fittings must be _____ for the intended use.
   (a) listed      (b) approved      (c) identified      (d) none of these

6. Cable trays can be used in corrosive areas and in areas requiring voltage isolation.
   (a) True      (b) False

7. Cable tray systems shall not be used _____.
   (a) in hoistways      (b) where subject to severe physical damage
   (c) in hazardous locations      (d) a and b

8. Each run of cable tray shall be _____ before the installation of cables.
   (a) tested for 25-ohms resistance      (b) insulated
   (c) completed      (d) all of these

9. Supports for cable trays must be provided in accordance with the _____.
   (a) installation instructions      (b) NEC
   (c) a or b      (d) none of these

10. In industrial facilities where conditions of maintenance and supervision ensure that only qualified persons will service the installation, cable tray systems can be used to support _____.
    (a) raceways      (b) cables
    (c) boxes and conduit bodies      (d) all of these

11. For raceways terminating at the tray, a(n) _____ cable tray clamp or adapter must be used to securely fasten the raceway to the cable tray system.
    (a) listed      (b) approved      (c) identified      (d) none of these

12. Steel or aluminum cable tray systems can be used as an equipment grounding conductor provided the cable tray sections and fittings are identified for _____ purposes.
    (a) grounding      (b) special      (c) industrial      (d) all

13. Steel cable trays shall not be used as equipment grounding conductors for circuits with ground-fault protection above _____ amperes.
    (a) 200      (b) 300      (c) 600      (d) 800

# 2002 Code Change Library

This library is a must have for everyone in the electrical industry. Mike's all new Illustrated Changes to the NEC, 2002 Edition is a detailed review of the most important changes to the 2002 NEC. Printed in full-color, Illustrated Changes to the NEC contains over 200 pages with more than 175 detailed graphics. This library also includes 9 hours of video taped from a live class in both high-quality DVD format and VHS video tape, the interactive Code Change CD-Rom and Mike's 2002 Code Tabs for your Code book.

**Call Today 1.888.NEC.CODE or visit us online at www.NECcode.com for the latest information and pricing.**

# Chapter 4 – Equipment for General Use

## Article 400 – Flexible Cords

This article covers the general requirements, applications, and construction specifications for flexible cords and flexible cables.

## Article 404 – Switches

The requirements of Article 404 apply to switches of all types. These include snap (toggle) switches, dimmers, fan switches, knife switches, circuit breakers used as switches, and automatic switches such as time clocks and timers, including switches and circuit breakers used for disconnecting means.

## Article 406 – Receptacles, Cord Connectors, and Attachment Plugs (Caps)

This article covers the rating, type, and installation of receptacles, cord connectors, and attachment plugs (cord caps).

## Article 408 – Switchboards and Panelboards

Article 408 covers the specific requirements for switchboards, panelboards, and distribution boards that control light and power circuits.

## Article 422 – Appliances

Article 422 covers electric appliances used in any occupancy.

## Article 424 – Fixed Electric Space-Heating Equipment

This article covers fixed electric equipment used for space heating. For the purpose of this article, heating equipment shall include heating cable, unit heaters, boilers, central systems, or other approved fixed electric space-heating equipment. This article shall not apply to process heating and room air conditioning.

## Article 430 – Motors And Motor Control Centers

Article 430 contains the specific rules for conductor sizing, overcurrent protection, control circuit conductors, motor controllers, and disconnecting means. The installation requirements for motor control centers are covered in 110.26(f) and air-conditioning and refrigerating equipment are covered in Article 440.

## Article 440 – Air-Conditioning and Refrigerating Equipment

This article applies to electrically driven air-conditioning and refrigeration equipment that has a hermetic refrigerant motor compressor. The rules in this article are in addition to, or amend, the rules in Article 430 and other articles.

## Article 445 – Generators

This article contains the electrical installation requirements for generators such as where generators can be installed, nameplate markings, conductor ampacity, and disconnecting means.

## Article 450 – Transformers

This article covers the installation of all transformers.

## Article 460 – Capacitors

This article covers the installation of capacitors, including those in hazardous (classified) locations as modified by Articles 501 through 503.

# Article 400
# Flexible Cords and Cables

## 400.1 Scope

This article covers the general requirements, applications, and construction specifications for flexible cords.

> **AUTHOR'S COMMENT:** Article 400 applies to the cords and cables shown in Table 400.4. Because of how the flexible cords and cables covered by Article 400 are used, they are not considered a wiring method. It does not apply to the cables in Chapter 3 such as NM, AC, or MC cable.

## 400.3 Suitability

Flexible cords, cables, and their fittings shall be approved for the use and shall be suitable for the location. For example, when using cords in wet locations, the cord and the fittings shall be approved for the wet location. Figure 400-1

## 400.4 Types

Flexible cords and flexible cables shall conform to the description in Table 400.4.

## 400.5 Ampacity of Flexible Cords and Cables

Tables 400.5(A) and 400.5(B) list the allowable ampacity for copper conductors in cords and cables. See 400.13 for overcurrent protection requirements.

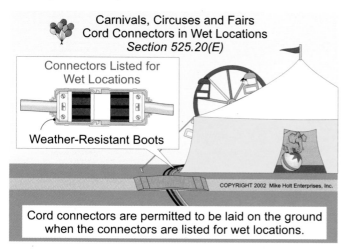

Carnivals, Circuses and Fairs
Cord Connectors in Wet Locations
*Section 525.20(E)*

Connectors Listed for Wet Locations

Weather-Resistant Boots

COPYRIGHT 2002 Mike Holt Enterprises, Inc.

Cord connectors are permitted to be laid on the ground when the connectors are listed for wet locations.

**Figure 400-1**

## 400.7 Uses Permitted

**(A) Uses Permitted.**

(1) Pendants. See 210.50(A) and 314.23(H).

(2) Wiring of luminaires. See 410.14 and 410.30(B).

(3) Connection of portable lamps, portable and mobile signs, or appliances. See 422.16.

(4) Elevator cables.

(5) Wiring of cranes and hoists.

(6) Connection of utilization equipment to facilitate frequent interchange. See 422.16. Figure 400-2

(7) Prevention of the transmission of noise or vibration. See 422.16.

(8) Appliances where the fastening means and mechanical connections are specifically designed to permit ready removal for maintenance and repair, and the appliance is intended or identified for flexible cord connections. See 422.16.

(9) Data-processing cables. See 645.5.

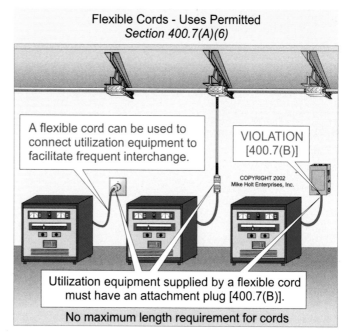

Flexible Cords - Uses Permitted
*Section 400.7(A)(6)*

A flexible cord can be used to connect utilization equipment to facilitate frequent interchange.

VIOLATION
[400.7(B)]

COPYRIGHT 2002
Mike Holt Enterprises, Inc.

Utilization equipment supplied by a flexible cord must have an attachment plug [400.7(B)].

No maximum length requirement for cords

**Figure 400-2**

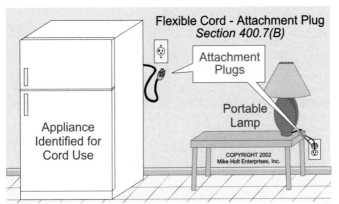

Flexible cords must have an attachment plug for:
• Portable lamps or appliances [400.7(3)]
• Equipment to facilitate frequent interchange [400.7(6)]
• Appliances identified for flexible cord usage [400.7(8)]

**Figure 400-3**

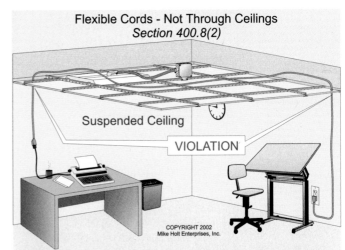

Cords are not permitted to be run through walls, structural ceilings, suspended ceilings, dropped ceilings or floors.

**Figure 400-4**

(10) Connection of moving parts.

(11) Temporary wiring. See 527.4(B) and 527.4(C).

**(B) Attachment Plugs.** Attachment plugs are required for cords used for any of the following applications: Figure 400-3

Portable lamps, portable and mobile signs, or appliances [400.7(3)].

Stationary equipment to facilitate their frequent interchange. See 422.16 [400.7(6)].

Appliances specifically designed to permit ready removal for maintenance and repair and identified for flexible cord connection [400.7(8)].

**AUTHOR'S COMMENT:** An attachment plug can serve as the disconnecting means for stationary appliances [422.33] and room air conditioners [440.63].

### 400.8 Uses Not Permitted

Unless specifically permitted in 400.7, flexible cords and cables shall not be:

(1) Used as a substitute for the fixed wiring of a structure.

(2) Run through holes in walls, structural ceilings, suspended/dropped ceilings, or floors. Figure 400-4

**AUTHOR'S COMMENT:** According to an article published in the Electrical Inspectors' magazine IAEI News, a cord run through a cabinet for an appliance is not considered as being run through a wall. Figure 400-5

(3) Run through doorways, windows, or similar openings.

(4) Attached to building surfaces.

(5) Where concealed by walls, floors or ceilings, or located above suspended or dropped ceilings. Figure 400-6

**AUTHOR'S COMMENT:** Cords within a raised floor that are not used for environmental air are permitted, because this space is not considered a concealed space. See Article 100 for the definition for "exposed."

(6) Installed in raceways.

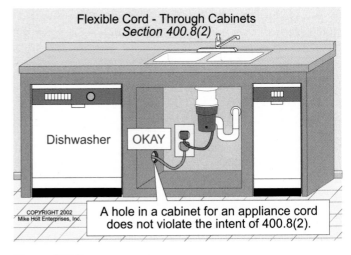

A hole in a cabinet for an appliance cord does not violate the intent of 400.8(2).

**Figure 400-5**

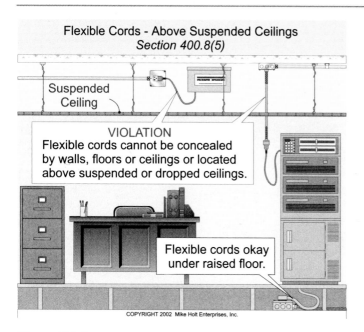

Flexible Cords - Above Suspended Ceilings
*Section 400.8(5)*

Suspended
Ceiling

VIOLATION
Flexible cords cannot be concealed
by walls, floors or ceilings or located
above suspended or dropped ceilings.

Flexible cords okay
under raised floor.

COPYRIGHT 2002 Mike Holt Enterprises, Inc.

**Figure 400-6**

## 400.10 Pull at Joints and Terminals

Flexible cords shall be installed so tension will not be transmitted to the conductor terminals, by knotting the cord, winding with tape, or using fittings that are designed for the purpose, such as a strain-relief fitting. Figure 400-7

> **AUTHOR'S COMMENT:** When critical health and economic activities are engaged in, the best method is a factory-made stress-relieving listed device; not an old-timer's knot. See Figure 400-7

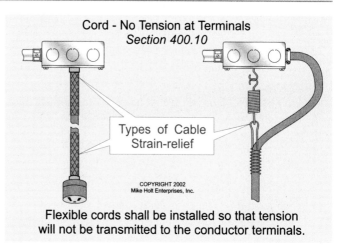

Cord - No Tension at Terminals
*Section 400.10*

Types of Cable
Strain-relief

COPYRIGHT 2002
Mike Holt Enterprises, Inc.

Flexible cords shall be installed so that tension
will not be transmitted to the conductor terminals.

**Figure 400-7**

## 400.13 Overcurrent Protection

Cords shall be protected against overcurrent in accordance with 240.5.

## Article 400

1. Flexible cords and cables can be used for _____.
   (a) wiring of luminaires
   (b) connection of portable lamps or appliances
   (c) connection of utilization equipment to facilitate frequent interchange
   (d) all of these

2. Flexible cords shall not be used as a substitute for _____ wiring.
   (a) temporary          (b) fixed          (c) overhead          (d) none of these

3. Unless specifically permitted in 400.7, flexible cords and cables shall not be used where _____.
   (a) run through holes in walls, ceilings, or floors
   (b) run through doorways, windows, or similar openings
   (c) attached to building surfaces
   (d) all of these

4. Flexible cords and cables shall not be concealed behind building _____, or run through doorways, windows, or similar openings.
   (a) structural ceilings                    (b) suspended or dropped ceilings
   (c) floors or walls                        (d) all of these

5. Flexible cords and cables shall be connected to devices and to fittings so that tension will not be transmitted to joints or terminal screws. This shall be accomplished by _____.
   (a) knotting the cord                      (b) winding with tape
   (c) the use of fittings designed for that purpose    (d) all of these

# Article 402
# Fixture Wires

## 402.1 Scope

This article covers the general requirements and construction specifications for fixture wires.

## 402.3 Types

Fixture wires shall be of a type contained in Table 402-3.

## 402.5 Allowable Ampacity of Fixture Wires

| AWG | Ampacity |
| --- | --- |
| 18 | 6A |
| 16 | 8A |
| 14 | 17A |
| 12 | 23A |
| 10 | 28A |

## 402.6 Minimum Size

Fixture wires shall not be smaller than 18 AWG.

## 402.7 Raceway Size

Raceways shall be large enough to permit the installation and removal of conductors without damaging the conductors' insulation. The number of fixture wires permitted in a single conduit or tubing shall not exceed the percentage fill specified in Table 1, Chapter 9. See 300.17 for additional details.

**Same Size Conductors.** When all conductors in a raceway are the same size and insulation, the number of conductors permitted can be found in Annex C for the raceway type.

**Question:** How many 18 AWG TFFN conductors can be installed in $1/2$ in. EMT? Figure 402-1

(a) 12          (b) 14

(c) 19          (d) 22

**Answer:** (d) 22 conductors, Annex C, Table C1

## 402.8 Grounded (neutral) Conductor

Grounded (neutral) conductors for luminaires shall be connected to the screw shell of the lampholder [200.10(C) and 410.23].

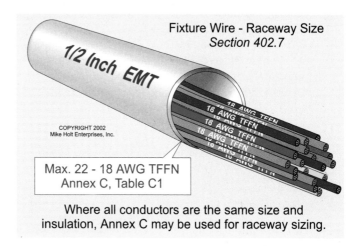

Fixture Wire - Raceway Size
Section 402.7

COPYRIGHT 2002
Mike Holt Enterprises, Inc.

Max. 22 - 18 AWG TFFN
Annex C, Table C1

Where all conductors are the same size and insulation, Annex C may be used for raceway sizing.

**Figure 402-1**

## 402.10 Uses Permitted

Fixture wires can be used for the connection of luminaires. Figure 402-2

**AUTHOR'S COMMENT:** Fixture wires can also be used for elevators and escalators [620.11(C)], Class 1 control and power-limited circuits [725.27(B)], and non-power limited fire alarm circuits [760.27(B)].

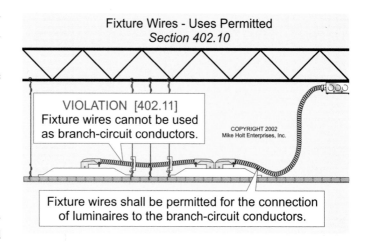

Fixture Wires - Uses Permitted
Section 402.10

VIOLATION [402.11]
Fixture wires cannot be used as branch-circuit conductors.

COPYRIGHT 2002
Mike Holt Enterprises, Inc.

Fixture wires shall be permitted for the connection of luminaires to the branch-circuit conductors.

**Figure 402-2**

## 402.11 Uses Not Permitted

Fixture wires cannot be used for branch-circuit wiring.

Figure 402-3

## 402.12 Overcurrent Protection

Fixture wires shall be protected against overcurrent according to the requirements of 240.5.

---

**AUTHOR'S COMMENT:** Fixture wires used for motor control circuit taps shall have overcurrent protection in accordance with 430.72(A), and Class 1 control circuits shall have overcurrent protection in accordance with 725.23.

---

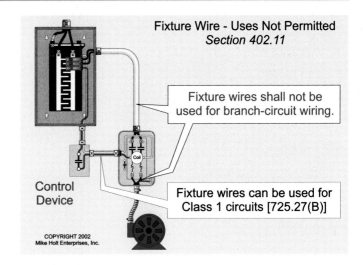

Fixture Wire - Uses Not Permitted
*Section 402.11*

Fixture wires shall not be used for branch-circuit wiring.

Fixture wires can be used for Class 1 circuits [725.27(B)]

Control Device

COPYRIGHT 2002
Mike Holt Enterprises, Inc.

**Figure 402-3**

## Article 402

1.  Fixture wires 18 AWG TFFN are rated for _____.
    (a) 14A            (b) 10A            (c) 8A            (d) 6A

2.  The smallest size fixture wire permitted in the *NEC* is _____ .
    (a) 22 AWG            (b) 20 AWG            (c) 18 AWG            (d) 16 AWG

3.  Fixture wires shall be permitted for connecting luminaires to the _____ conductors supplying the luminaires.
    (a) service            (b) branch-circuit            (c) supply            (d) none of these

4.  Fixture wires shall be permitted for installation in luminaires and in similar equipment where enclosed or protected and not subject to _____ in use, or for connecting luminaires to the branch-circuit conductors supplying the luminaires.
    (a) bending or twisting            (b) knotting            (c) stretching or straining    (d) none of these

# Article 404
# Switches

## 404.1 Scope

The requirements of Article 404 apply to all types of switches, such as snap (toggle) switches, knife switches, circuit breakers used as switches, and automatic switches such as time clocks, including switches and circuit breakers used for disconnecting means.

## 404.2 Switch Connections

**(A)** **Three-Way and Four-Way Switches.** All three-way and four-way switching shall be done with the ungrounded (hot) conductor. Figure 404-1

> **AUTHOR'S COMMENT:** The white or gray conductor within a cable assembly can be used for single-pole, three-way or four-way switch loops if it is permanently reidentified to indicate its use as an ungrounded (hot) conductor at each location where the conductor is visible and accessible [200.7(C)(2)].

Where metal raceway or metal-clad cable contains the conductors for switches, the wiring shall be arranged to avoid heating the surrounding metal by induction. This is accomplished by installing all circuit conductors in the same raceway. See 300.3(B) and 300.20(A).

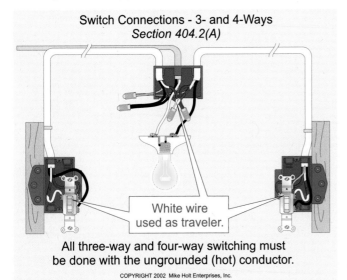

Switch Connections - 3- and 4-Ways
Section 404.2(A)

White wire used as traveler.

All three-way and four-way switching must be done with the ungrounded (hot) conductor.

COPYRIGHT 2002  Mike Holt Enterprises, Inc.

*Figure 404-1*

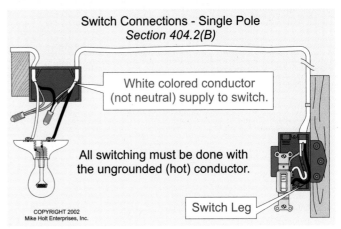

Switch Connections - Single Pole
Section 404.2(B)

White colored conductor (not neutral) supply to switch.

All switching must be done with the ungrounded (hot) conductor.

Switch Leg

COPYRIGHT 2002
Mike Holt Enterprises, Inc.

*Figure 404-2*

*Exception:* A grounded (neutral) conductor is not required in the same raceway or cable with travelers and switch leg (switch loops) conductors.

**(B)** **Switching Grounded (neutral) Conductors.** Only the ungrounded conductor can be used for switching. Figure 404-2

## 404.3 Switch Enclosures

**(A)** **General.** Switches and circuit breakers shall be of the externally operable type mounted in an enclosure listed for the intended use.

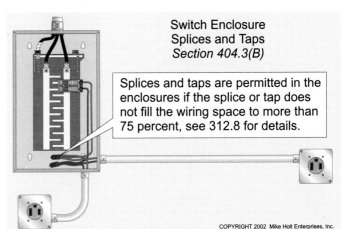

Switch Enclosure
Splices and Taps
Section 404.3(B)

Splices and taps are permitted in the enclosures if the splice or tap does not fill the wiring space to more than 75 percent, see 312.8 for details.

COPYRIGHT 2002  Mike Holt Enterprises, Inc.

*Figure 404-3*

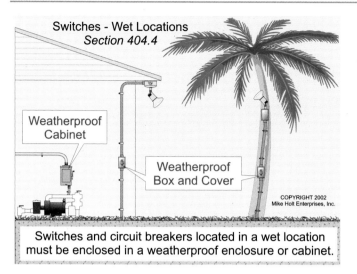

Switches - Wet Locations
Section 404.4

Weatherproof Cabinet

Weatherproof Box and Cover

COPYRIGHT 2002 Mike Holt Enterprises, Inc.

Switches and circuit breakers located in a wet location must be enclosed in a weatherproof enclosure or cabinet.

**Figure 404-4**

**(B) Used for Raceway or Splices.** Switch or circuit-breaker enclosures can contain splices and taps and can have conductors feed through them, if the splices or taps do not fill the wiring space at any cross section to more than 75 percent, and the wiring at any cross section does not exceed 40 percent in accordance with 312.8. Figure 404-3

### 404.4 Wet Locations

Switches and circuit breakers installed in wet locations shall be installed in a weatherproof enclosure. The enclosure shall be installed so that at least $^1/_4$ in. airspace is provided between the enclosure and the wall or other supporting surface. See 312.2(A). Figure 404-4

Switches can be located next to, but not within, a bathtub or shower space, unless the switch and its assembly have been listed for this purpose. Figure 404-5

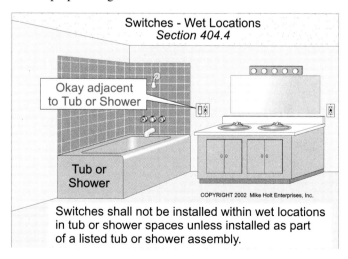

Switches - Wet Locations
Section 404.4

Okay adjacent to Tub or Shower

Tub or Shower

COPYRIGHT 2002 Mike Holt Enterprises, Inc.

Switches shall not be installed within wet locations in tub or shower spaces unless installed as part of a listed tub or shower assembly.

**Figure 404-5**

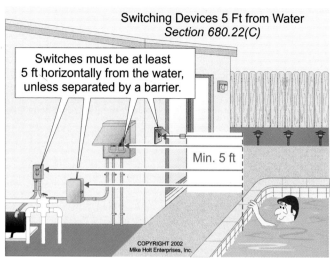

Switching Devices 5 Ft from Water
Section 680.22(C)

Switches must be at least 5 ft horizontally from the water, unless separated by a barrier.

Min. 5 ft

COPYRIGHT 2002 Mike Holt Enterprises, Inc.

**Figure 404-6**

**AUTHOR'S COMMENT:** Switches shall be located at least 5 ft from pools [680.22(C)], outdoor spas and hot tubs [680.41], and indoor spas or hot tubs [680.43(C)]. Figure 404-6

This 5 ft rule does not apply to switches located adjacent to bathtubs, shower stalls, or hydromassage bathtubs [680.70 and 680.72]. Figure 404-7

### 404.6 Position of Knife Switches

**(A) Single-Throw Knife Switch.** Single-throw knife switches shall be installed so that gravity will not tend to close them.

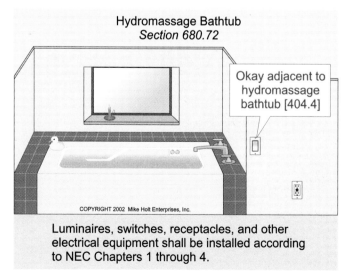

Hydromassage Bathtub
Section 680.72

Okay adjacent to hydromassage bathtub [404.4]

COPYRIGHT 2002 Mike Holt Enterprises, Inc.

Luminaires, switches, receptacles, and other electrical equipment shall be installed according to NEC Chapters 1 through 4.

**Figure 404-7**

## 404.7 Indicating

Switches, motor circuit switches, and circuit breakers shall be marked to indicate whether they are in the "on" or "off" position. When the switch is operated vertically, it shall be installed so the "up" position is the "on" position [240.81].

*Exception:* Double-throw switches such as three-way and four-way switches are not required to be marked "on" or "off."

## 404.8 Accessibility and Grouping

**(A)  Location.** All switches and circuit breakers used as switches shall be capable of being operated from a readily accessible location. They shall be installed so the center of the grip of the operating handle of the switch or circuit breaker, when in its highest position, is not more than 6 ft 7 in. above the floor or working platform. Figure 404-8

---

**AUTHOR'S COMMENT:** There is no minimum height requirement for switches and circuit breakers. However, the service disconnecting means for mobile and manufactured homes shall be mounted a minimum of 2 ft from the finish grade [550.23(F)]. Figure 404-9

---

*Exception No. 2:* Switches and circuit breakers used as switches can be mounted higher than 6 ft 7 in. if they are next to the equipment they supply and are accessible by portable means. Figure 404-10

**(B)  Voltage Between Devices.** Snap switches shall not be grouped or ganged in enclosures with other snap switches, receptacles, or similar devices if the voltage between devices exceeds 300V. Figure 404-11 and Figure 404-12

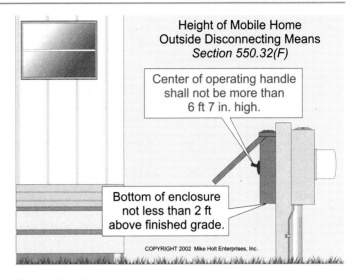

Height of Mobile Home
Outside Disconnecting Means
*Section 550.32(F)*

Center of operating handle shall not be more than 6 ft 7 in. high.

Bottom of enclosure not less than 2 ft above finished grade.

COPYRIGHT 2002  Mike Holt Enterprises, Inc.

**Figure 404-9**

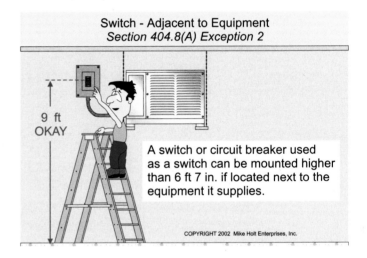

Switch - Adjacent to Equipment
*Section 404.8(A) Exception 2*

9 ft
OKAY

A switch or circuit breaker used as a switch can be mounted higher than 6 ft 7 in. if located next to the equipment it supplies.

COPYRIGHT 2002  Mike Holt Enterprises, Inc.

**Figure 404-10**

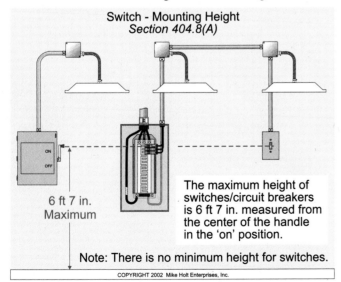

Switch - Mounting Height
*Section 404.8(A)*

ON
OFF

6 ft 7 in.
Maximum

The maximum height of switches/circuit breakers is 6 ft 7 in. measured from the center of the handle in the 'on' position.

Note: There is no minimum height for switches.

COPYRIGHT 2002  Mike Holt Enterprises, Inc.

**Figure 404-8**

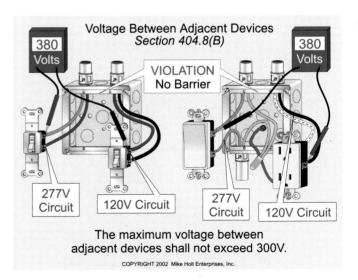

Voltage Between Adjacent Devices
*Section 404.8(B)*

380 Volts

VIOLATION
No Barrier

380 Volts

277V Circuit      120V Circuit      277V Circuit      120V Circuit

The maximum voltage between adjacent devices shall not exceed 300V.

COPYRIGHT 2002  Mike Holt Enterprises, Inc.

**Figure 404-11**

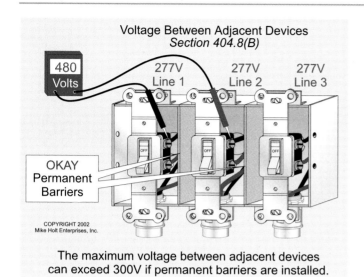

**Voltage Between Adjacent Devices**
*Section 404.8(B)*

480 Volts

277V Line 1    277V Line 2    277V Line 3

OKAY
Permanent
Barriers

COPYRIGHT 2002
Mike Holt Enterprises, Inc.

The maximum voltage between adjacent devices can exceed 300V if permanent barriers are installed.

*Figure 404-12*

## 404.9 Switch Faceplates

**(A) Mounting.** Faceplates for switches shall be installed so they are completely covered, and where flush mounted, the faceplate shall seat against the wall surface.

**(B) Grounding.** Switches, including dimmer and similar control switches, shall be effectively grounded (bonded) and shall provide a means to ground (bond) metal faceplates, whether or not a metal faceplate is installed. A switch is considered effectively grounded (bonded) if: Figure 404-13

(1) The switch is mounted with metal screws to a metal box or to a nonmetallic box with integral means for grounding devices.

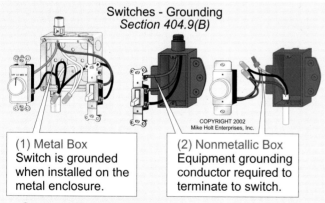

**Switches - Grounding**
*Section 404.9(B)*

COPYRIGHT 2002
Mike Holt Enterprises, Inc.

**(1) Metal Box**
Switch is grounded when installed on the metal enclosure.

**(2) Nonmetallic Box**
Equipment grounding conductor required to terminate to switch.

Snap switches, dimmers and similar control switches must be effectively grounded to provide a means to ground metal faceplates, whether or not a metal faceplate is installed.

*Figure 404-13*

(2) An equipment grounding conductor or equipment bonding jumper is connected to an equipment grounding termination of the snap switch. See 404.12.

***Exception:*** An existing snap switch installed where no grounding (bonding) means exist in the outlet box can be replaced without grounding (bonding) the switch yoke, if covered with a nonmetallic faceplate.

## 404.10 Mounting Snap Switches

**(A) Mounting of Snap Switches.** Snap switches installed in recessed boxes shall have the ears of the yoke seated against the finished wall surface. Because drywall installers are often very aggressive when cutting out the opening for electrical outlet boxes, this rule is difficult to comply with.

> **AUTHOR'S COMMENT:** In walls or ceilings of noncombustible material, boxes shall not be set back more than $1/4$ in. from the finished surface. In combustible walls or ceilings, boxes shall be flush with the finished surface [314.20]. There shall not be any gaps greater than $1/8$ in. at the edge of the box [314.21].

## 404.11 Circuit Breakers Used as Switches

A manually operable circuit breaker used as a switch shall show when it is in the "on" (closed) or "off" (open) position. See 404.6(C).

> **AUTHOR'S COMMENT:** Circuit breakers used to switch 120V or 277V fluorescent lighting circuits shall be listed and marked "SWD" or "HID." Circuit breakers used to switch high-intensity discharge lighting circuits shall be listed and shall be marked as "HID." See 240.83(D). Figure 404-14

## 404.12 Grounding

Metal enclosures for switches and circuit breakers shall be grounded (bonded) in accordance with 250.110 and 250.148. Nonmetallic enclosures containing switches and circuit breakers are permitted if the wiring method includes an equipment grounding conductor as required by 404.9(B).

## 404.14 Rating and Use of Snap Switches

> **AUTHOR'S COMMENT:** See Article 100 for the definition of a snap switch.

**(A) AC General-Use Snap Switch.** Alternating-current general-use snap switches can be used to control:

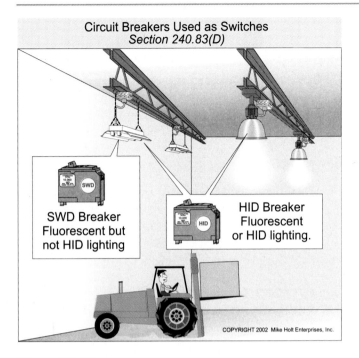

**Circuit Breakers Used as Switches**
*Section 240.83(D)*

SWD Breaker Fluorescent but not HID lighting

HID Breaker Fluorescent or HID lighting.

COPYRIGHT 2002 Mike Holt Enterprises, Inc.

**Figure 404-14**

(1) Resistive and inductive loads, including electric-discharge lamps not exceeding the ampere rating of the switch, at the voltage involved.

(2) Tungsten-filament lamp loads not exceeding the ampere rating of the switch at 120V.

(3) Motor loads, 2-hp or less, that do not exceed 80 percent of the ampere rating of the switch. See 430.109(C).

**(C) CO/ALR Snap Switches.** Snap switches rated 15A or 20A shall be marked CO/ALR when connected to aluminum wire. See 406.2(C).

**AUTHOR'S COMMENT:** According to UL requirements, aluminum conductors cannot terminate onto screwless (push-in) terminals of a snap switch.

**(E) Dimmer.** General-use dimmer switches shall only be used to control permanently installed incandescent luminaires, unless listed for the control of other loads.

**AUTHOR'S COMMENT:** Dimmers are not listed to control a receptacle. Figure 404-15

## 404.15 Switch Marking

**(A) Markings.** Switches shall be marked with the current, voltage, and if horsepower rated, the maximum rating for which they are designed.

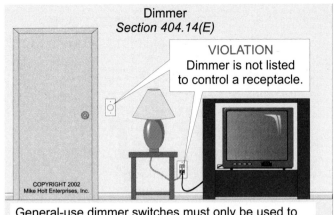

**Dimmer**
*Section 404.14(E)*

VIOLATION
Dimmer is not listed to control a receptacle.

General-use dimmer switches must only be used to control permanently installed incandescent luminaires, unless otherwise listed for control of other loads.

**Figure 404-15**

**(B) Off Indication.** Where in the off position, a switching device with a marked "Off" position shall completely disconnect all ungrounded conductors of the load it controls.

**AUTHOR'S COMMENT:** This rule requires a switch with an "off" marking to disconnect all power. Where an electronic occupancy sensor is used for switching, voltage will be present and a small current of 0.05 mA can flow when the switch is in the expected "off" position. This small amount of current can startle a person, perhaps causing a fall. To solve this problem, manufacturers have simply removed the word "off" from the switch. Figure 404-16

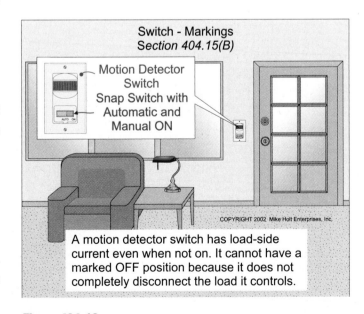

**Switch - Markings**
*Section 404.15(B)*

Motion Detector Switch Snap Switch with Automatic and Manual ON

COPYRIGHT 2002 Mike Holt Enterprises, Inc.

A motion detector switch has load-side current even when not on. It cannot have a marked OFF position because it does not completely disconnect the load it controls.

**Figure 404-16**

## Article 404

1.  Switches or circuit breakers shall not disconnect the grounded conductor of a circuit unless the switch or circuit breaker _____.

    (a) can be opened and closed by hand levers only
    (b) simultaneously disconnects all conductors of the circuit
    (c) opens the grounded conductor before it disconnects the ungrounded conductors
    (d) none of these

2.  Switch or circuit-breaker enclosures can be used as a junction box or raceway for conductors feeding through splices, or taps, when installed in accordance with 312.8.
    (a) True              (b) False

3.  Which of the following switches must indicate whether they are in the (open) OFF or (closed) ON position?
    (a) General-use switches.              (b) Motor-circuit switches.
    (c) Circuit breakers.                  (d) all of these

4.  All switches and circuit breakers used as switches must be installed so that they may be operated from a readily accessible place. They shall be installed so that the center of the grip of the operating handle of the switch or circuit breaker, when in its highest position, is not more than 6 ft 7 in. above the floor or working platform.
    (a) True              (b) False

5.  Snap switches shall not be grouped or ganged in enclosures with other _____, if the voltage between adjacent devices exceeds 300V, unless they are installed in enclosures equipped with permanently installed barriers between adjacent devices.
    (a) snap switches        (b) receptacles        (c) similar devices        (d) all of these

6.  Snap switches shall not be grouped or ganged in enclosures unless they can be arranged so that the voltage between adjacent switches does not exceed _____ volts, or unless they are installed in enclosures equipped with permanently installed barriers between adjacent switches.
    (a) 100             (b) 200             (c) 300             (d) 400

7.  All snap switches, including dimmer and similar control switches, must be effectively grounded so that they can provide a means to ground metal faceplates, whether or not a metal faceplate is installed.
    (a) True              (b) False

8.  Snap switches installed in recessed boxes must have the _____ seated against the finished wall surface.
    (a) mounting yoke       (b) body        (c) toggle        (d) all of these

9.  A form of general-use snap switches, suitable only for use on ac circuits, can control _____.
    (a) resistive and inductive loads that do not exceed the ampere and voltage rating of the switch
    (b) tungsten-filament lamp loads that do not exceed the ampere rating of the switch at 120V
    (c) motor loads that do not exceed 80 percent of the ampere and voltage rating of the switch
    (d) all of these

10. A form of general-use snap switch, suitable for use on either ac or dc circuits, may be used for control of inductive loads not exceeding _____ percent of the ampere rating of the switch at the applied voltage.
    (a) 75             (b) 90             (c) 100             (d) 50

11. Snap switches rated _____ amperes or less directly connected to aluminum conductors must be listed and marked CO/ALR.
    (a) 15             (b) 20             (c) 25             (d) 30

13. Where in the off position, a switching device with a marked "OFF" position must completely disconnect all _____ conductors of the load it controls.
    (a) grounded        (b) ungrounded        (c) grounding        (d) all of these

# Article 406
## Receptacles, Cord Connectors, and Attachment Plugs (Caps)

## 406.1 Scope

This article covers the rating, type, and installation of receptacles, cord connectors, and attachment plugs (cord caps).

## 406.2 Receptacle Rating and Type

**(B)  Rating.** Receptacles and cord connectors shall be rated not less than 15A, 125V or 15A, 250V.

> FPN: Single receptacles shall have an ampere rating not less than the rating of the branch circuit. [210.21(B)(1)]. Multioutlet receptacles shall have a rating in accordance with 210.21(B)(2) and (B)(3).

**(C)  Receptacles for Aluminum Conductors.** Receptacles rated 20A or less and designed for direct connection to aluminum wire shall be marked CO/ALR.

> **AUTHOR'S COMMENT:** According to UL requirements, aluminum conductors cannot terminate onto screwless (push-in) terminals of a receptacle.

**(D)  Isolated Ground Receptacles.** Isolated ground receptacles used for the reduction of electric noise [250.146(D)] shall be identified by having an orange triangle marking on the face of the receptacle. Figure 406-1

(1)  Isolated ground receptacles can only be used with an insulated grounding conductor installed with the circuit conductors in accordance with 250.146(D).

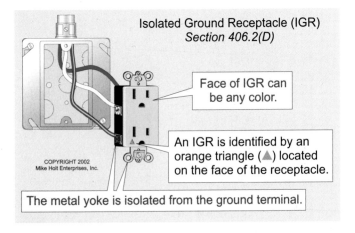

Isolated Ground Receptacle (IGR)
*Section 406.2(D)*

Face of IGR can be any color.

An IGR is identified by an orange triangle (▲) located on the face of the receptacle.

COPYRIGHT 2002 Mike Holt Enterprises, Inc.

The metal yoke is isolated from the ground terminal.

**Figure 406-1**

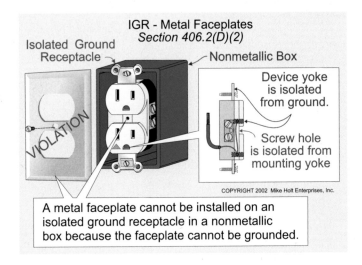

IGR - Metal Faceplates
*Section 406.2(D)(2)*

Isolated Ground Receptacle

Nonmetallic Box

Device yoke is isolated from ground.

Screw hole is isolated from mounting yoke

COPYRIGHT 2002  Mike Holt Enterprises, Inc.

A metal faceplate cannot be installed on an isolated ground receptacle in a nonmetallic box because the faceplate cannot be grounded.

**Figure 406-2**

(2)  Isolated ground receptacles installed in nonmetallic boxes shall be covered with a nonmetallic faceplate, because a metal faceplate cannot be grounded (bonded). Figure 406-2

## 406.3 General Installation Requirements

> **AUTHOR'S COMMENT:** The orientation (position) of the ground terminal of a receptacle is not specified in the *NEC*. The ground terminal can be up, down, or to the side. Proposals to specify the mounting position of the ground were all rejected. For more information on this subject, visit http://www.mikeholt.com/Newsletters/9-23-99.htm. Figure 406-3

**(B)  Receptacle Grounding.** Receptacles shall have their grounding terminals grounded (bonded) to a low-impedance fault current path in accordance with 250.146, 250.148, and 406.3(C).

*Exception No. 2:* Replacement receptacles are not required to have their grounding contacts grounded (bonded) if they are GFCI protected and installed in accordance with 406.3(D).

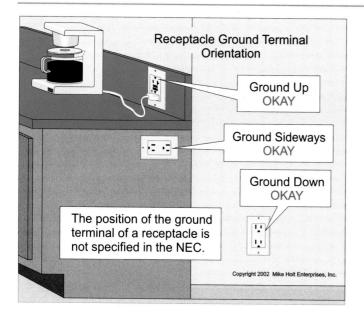

Receptacle Ground Terminal Orientation

**Ground Up** OKAY

**Ground Sideways** OKAY

**Ground Down** OKAY

The position of the ground terminal of a receptacle is not specified in the NEC.

Copyright 2002 Mike Holt Enterprises, Inc.

*Figure 406-3*

**(C) Methods of Grounding Receptacles.** The grounding terminals for receptacles shall be connected to the branch-circuit equipment grounding conductor [250.146]. Figure 406-4

> FPN: See 250.146(D) for the installation requirements for isolated ground receptacles. See 406.2(D). Figure 406-5

**(D) Receptacle Replacement.**

**(1) Where Grounding Means Exist.** Where a grounding means exists in the receptacle enclosure, grounding-type receptacles shall replace nongrounding type receptacles, and the receptacle's grounding terminal shall be grounded (bonded) in accordance with 406.3(C).

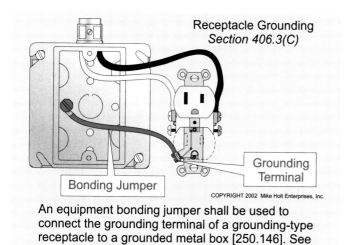

Receptacle Grounding
*Section 406.3(C)*

Bonding Jumper

Grounding Terminal

COPYRIGHT 2002 Mike Holt Enterprises, Inc.

An equipment bonding jumper shall be used to connect the grounding terminal of a grounding-type receptacle to a grounded metal box [250.146]. See 250.146(A) through (D) for other other connections.

*Figure 406-4*

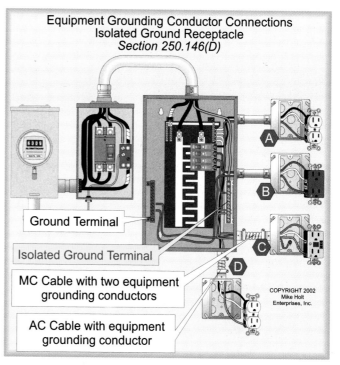

Equipment Grounding Conductor Connections
Isolated Ground Receptacle
*Section 250.146(D)*

Ground Terminal

Isolated Ground Terminal

MC Cable with two equipment grounding conductors

AC Cable with equipment grounding conductor

COPYRIGHT 2002 Mike Holt Enterprises, Inc.

*Figure 406-5*

**(2) GFCI Protection Required.** When receptacles are replaced in locations where GFCI protection is required, the replacement receptacles shall be GFCI-protected. This includes the replacement of receptacles in dwelling unit bathrooms, garages, outdoors, crawl spaces, unfinished basements, kitchens, wet bar sinks, rooftops, etc. See 210.8 for GFCI protection requirements.

**(3) Where No Ground Exists.** Where no grounding (bonding) means exists in the outlet box, such as old 2-wire NM cable without a ground, nongrounding type receptacles can be replaced with: Figure 406-6

(a) Another nongrounding type receptacle

(b) A GFCI-receptacle if marked "No Equipment Ground."

(c) A grounding-type receptacle, if GFCI protected and marked "GFCI Protected" and "No Equipment Ground."

**AUTHOR'S COMMENT:** GFCI protection functions properly on a 2-wire circuit without an equipment grounding conductor, because the equipment grounding conductor serves no purpose in the operation of the GFCI protection device. See Article 100 definition of Ground-Fault Circuit-Interrupter. Figure 406-7

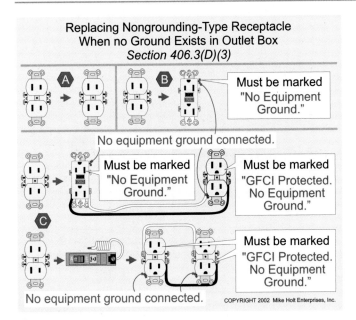

Replacing Nongrounding-Type Receptacle
When no Ground Exists in Outlet Box
*Section 406.3(D)(3)*

Must be marked "No Equipment Ground."

No equipment ground connected.

Must be marked "No Equipment Ground."

Must be marked "GFCI Protected. No Equipment Ground."

Must be marked "GFCI Protected. No Equipment Ground."

No equipment ground connected.

COPYRIGHT 2002 Mike Holt Enterprises, Inc.

**Figure 406-6**

*CAUTION: Permission to replace nongrounding type receptacles with GFCI-protected grounding type receptacles does not apply to new receptacle outlets that extend from an existing ungrounded outlet box. Once you add a receptacle outlet (branch-circuit extension), the receptacle shall be of the grounding type and it shall have its grounding terminal grounded in accordance with 250.130(C). Figure 406-8*

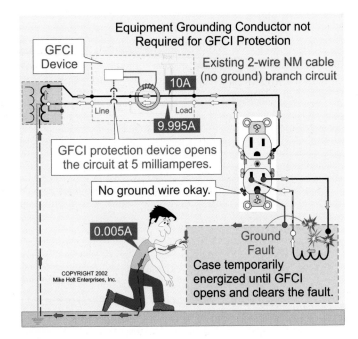

Equipment Grounding Conductor not Required for GFCI Protection

GFCI Device

Existing 2-wire NM cable (no ground) branch circuit

10A

Line        Load

9.995A

GFCI protection device opens the circuit at 5 milliamperes.

No ground wire okay.

0.005A

COPYRIGHT 2002 Mike Holt Enterprises, Inc.

Ground Fault

Case temporarily energized until GFCI opens and clears the fault.

**Figure 406-7**

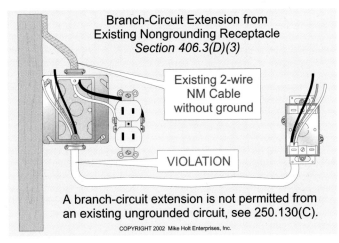

Branch-Circuit Extension from
Existing Nongrounding Receptacle
*Section 406.3(D)(3)*

Existing 2-wire NM Cable without ground

VIOLATION

A branch-circuit extension is not permitted from an existing ungrounded circuit, see 250.130(C).

COPYRIGHT 2002 Mike Holt Enterprises, Inc.

**Figure 406-8**

## 406.4 Receptacle Mounting

Receptacles shall be installed in boxes designed for the purpose, and the box shall be securely fastened in place in accordance with 314.23.

**(A) Boxes Set Back.** Receptacles in boxes that are set back from the wall surface shall be installed so the mounting yoke of the receptacle is held rigidly to the surface of the wall. Because drywall installers are often very aggressive when cutting out the opening for electrical outlet boxes, this rule is difficult to comply with.

**AUTHOR'S COMMENT:** In walls or ceilings of noncombustible material, boxes shall not be set back more than $1/4$ in. from the finished surface. In walls or ceilings of combustible material, boxes shall be flush with the finished surface [314.20]. There shall not be any gaps greater than $1/8$ in. at the edge of the box [314.21].

**(B) Boxes that are Flush with the Surface.** Receptacles mounted in boxes that are flush shall be installed so the mounting yoke of the receptacle is held rigidly against the box or raised box cover.

**(C) Receptacles Mounted on Covers.** Receptacles supported by a cover shall be held rigidly to the cover with two screws. Figure 406-9

**(E) Receptacles in Countertops and Similar Work Surfaces in Dwelling Units.** Receptacles cannot be installed in a face-up position in countertops or similar work surfaces. Figure 406-10

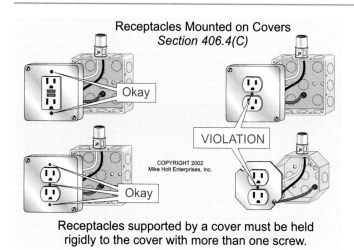

**Receptacles Mounted on Covers**
*Section 406.4(C)*

Okay

VIOLATION

COPYRIGHT 2002
Mike Holt Enterprises, Inc.

Okay

Receptacles supported by a cover must be held
rigidly to the cover with more than one screw.

*Figure 406-9*

## 406.5 Receptacle Faceplates (Cover Plates)

Faceplates for receptacles shall completely cover the outlet openings, and they shall seat firmly against the mounting surface.

**(B) Grounding.** Metal faceplates for receptacles shall be grounded (bonded). This can be accomplished by securing the metal faceplate to a receptacle. See 517.13 Ex. 1. Figure 406-11

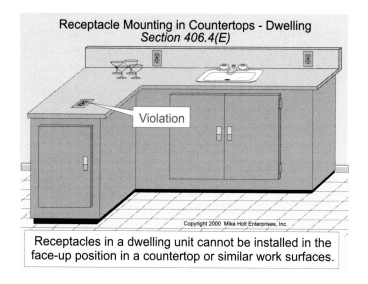

**Receptacle Mounting in Countertops - Dwelling**
*Section 406.4(E)*

Violation

Copyright 2000  Mike Holt Enterprises, Inc.

Receptacles in a dwelling unit cannot be installed in the
face-up position in a countertop or similar work surfaces.

*Figure 406-10*

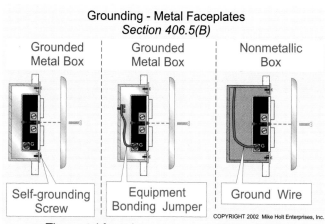

**Grounding - Metal Faceplates**
*Section 406.5(B)*

| Grounded Metal Box | Grounded Metal Box | Nonmetallic Box |

Self-grounding Screw

Equipment Bonding Jumper

Ground Wire

COPYRIGHT 2002 Mike Holt Enterprises, Inc.

The metal faceplate screw grounds the metal
faceplate to the receptacle's grounding terminal.

*Figure 406-11*

## 406.6 Attachment Plugs

Attachment plugs and cord connectors shall be listed for the purpose.

**(A) Exposed Live Parts.** Attachment plugs and cord connectors shall have no exposed current-carrying parts, except the prongs, blades, or pins.

**(B) No Energized Parts.** Attachment plugs shall be installed so their prongs, blades, or pins are not energized unless inserted into an energized receptacle. Figure 406-12

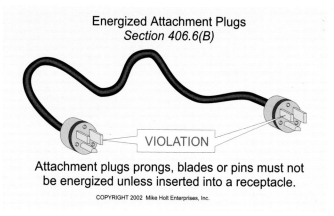

**Energized Attachment Plugs**
*Section 406.6(B)*

VIOLATION

Attachment plugs prongs, blades or pins must not
be energized unless inserted into a receptacle.

COPYRIGHT 2002  Mike Holt Enterprises, Inc.

*Figure 406-12*

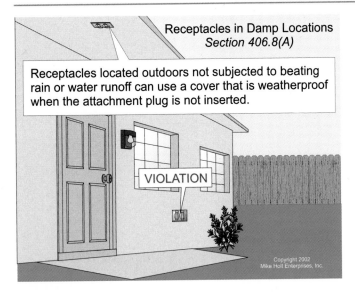

Receptacles in Damp Locations
Section 406.8(A)

Receptacles located outdoors not subjected to beating rain or water runoff can use a cover that is weatherproof when the attachment plug is not inserted.

VIOLATION

Copyright 2002 Mike Holt Enterprises, Inc.

**Figure 406-13**

### 406.8 Receptacles in Damp or Wet Locations

**(A) Damp Locations.** A receptacle installed outdoors under roofed open porches, canopies, marquees, and the like, and not subjected to beating rain or water runoff or in other damp locations, shall have an enclosure for the receptacle that is weatherproof when the attachment plug cap not inserted and receptacle covers closed. Figure 406-13

**(B) Wet Locations.**

**(1) 15A and 20A Outdoor Receptacles.** All 15A and 20A, 125V and 250V receptacles installed outdoors in a wet location shall be within an enclosure and cover that is weatherproof at all times, even when an attachment plug is inserted. Figure 406-14

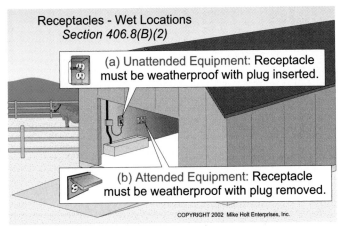

Receptacles - Wet Locations
Section 406.8(B)(2)

(a) Unattended Equipment: Receptacle must be weatherproof with plug inserted.

(b) Attended Equipment: Receptacle must be weatherproof with plug removed.

COPYRIGHT 2002 Mike Holt Enterprises, Inc.

**Figure 406-15**

**(2) Other Receptacles.** Other receptacles installed in a wet location shall comply with (a) or (b): Figure 406-15

**(a) Wet Location Cover.** A receptacle installed in a wet location shall have an enclosure that is weatherproof with the attachment plug cap inserted, if the equipment plugged into it is not attended while in use.

**(b) Damp Location Cover.** A receptacle installed in a wet location can have an enclosure that is weatherproof only when the attachment plug is removed, if equipment like portable tools will be attended while in use.

**(C) Bathtub and Shower Space.** Receptacles can be installed next to, but not within a bathtub or shower space. Figure 406-16

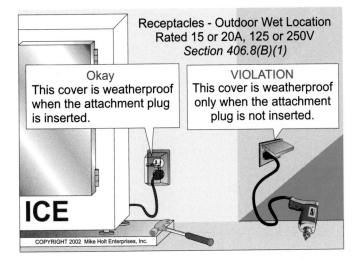

Receptacles - Outdoor Wet Location
Rated 15 or 20A, 125 or 250V
Section 406.8(B)(1)

Okay
This cover is weatherproof when the attachment plug is inserted.

VIOLATION
This cover is weatherproof only when the attachment plug is not inserted.

ICE

COPYRIGHT 2002 Mike Holt Enterprises, Inc.

**Figure 406-14**

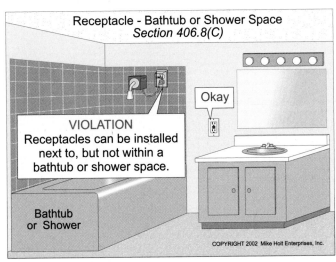

Receptacle - Bathtub or Shower Space
Section 406.8(C)

Okay

VIOLATION
Receptacles can be installed next to, but not within a bathtub or shower space.

Bathtub or Shower

COPYRIGHT 2002 Mike Holt Enterprises, Inc.

**Figure 406-16**

**(E) Flush Mounting with Faceplate.** The enclosure for a receptacle installed in an outlet box flush-mounted on a wall surface shall be made weatherproof by a weatherproof faceplate assembly that provides a watertight connection between the plate and the wall surface.

## 406.10 Connecting Receptacle Grounding Terminal to Box

The grounding terminal of receptacles shall be grounded (bonded) in accordance with 250.146.

## Article 406

1.  Isolated ground receptacles installed in nonmetallic boxes must be covered with a nonmetallic faceplate because a metal faceplate cannot be grounded to the circuit equipment grounding conductor.
    (a) True                    (b) False

2.  When replacing an ungrounded type receptacle in a bedroom of a dwelling unit and no grounding means exists in the receptacle enclosure, you must use a _____.
    (a) nongrounding receptacle            (b) grounding receptacle
    (c) GFCI-type receptacle               (d) a or c

3.  Receptacles connected to circuits having different voltages, frequencies, or types of current (ac or dc) on the _____ shall be of such design that the attachment plugs used on these circuits are not interchangeable.
    (a) building         (b) interior         (c) same premises         (d) exterior

4.  Receptacles mounted in boxes that are set back of the wall surface shall be installed so that the mounting _____ of the receptacle is held rigidly at the surface of the wall.
    (a) screws or nails      (b) yoke or strap      (c) face plate      (d) none of these

5.  Receptacles mounted to and supported by a cover shall be secured by more than one screw.
    (a) True                    (b) False

6.  Receptacles, cord connectors, and attachment plugs shall be constructed so that the receptacle or cord connectors will not accept an attachment plug with a different _____ or current rating than that for which the device is intended.
    (a) voltage         (b) amperage         (c) heat         (d) all of these

7.  A receptacle shall be considered to be in a location protected from the weather when located under roofed open porches, canopies, marquees, and the like, and will not be subjected to _____.
    (a) spray from a hose                   (b) a direct lightning hit
    (c) beating rain or water runoff        (d) falling or wind-blown debris

8.  Receptacles installed outdoors in a location protected from the weather or other damp locations, shall be in an enclosure that is _____ when the receptacle is covered.
    (a) raintight         (b) weatherproof         (c) rainproof         (d) weathertight

19.  A receptacle installed outdoors in a location protected from the weather or in other damp locations must have an enclosure for the receptacle that is weatherproof when the receptacle is _____.
    (a) covered         (b) enclosed         (c) protected         (d) none of these

10.  A receptacle shall be considered to be in a location protected from the weather (damp location) where _____.
    (a) located under a roofed open porch
    (b) not subjected to beating rain or water runoff
    (c) a or b
    (d) a and b

11.  _____, 125 and 250V receptacles installed outdoors in a wet location must have an enclosure that is weatherproof.
    (a) 15A         (b) 20A         (c) a and b         (d) none of these

12.  Which of the following statements is/are true for receptacle covers in an outdoor wet location?
    1. The receptacle cover shall be listed as weatherproof while the attachment plug is inserted for stationary or fixed loads that are intended to have an attachment plug inserted into the receptacle.
    2. The receptacle cover shall be listed as weatherproof while the attachment plug is not inserted for portable loads such as appliances and power tools.
    (a) 1 only         (b) 2 only         (c) 1 and 2         (d) none of these

13. An enclosure that is weatherproof only when the receptacle cover is closed can be used for receptacles in an indoor wet location when the receptacle is used for _____ while attended.

(a) portable equipment      (b) portable tools      (c) fixed equipment      (d) a and b

14. The enclosure for a receptacle installed in an outlet box flush-mounted on a wall surface in a damp or wet location shall be made weatherproof by means of a weatherproof faceplate that provides a _____ connection between the plate and the wall surface.

(a) sealed      (b) weathertight      (c) sealed and protected      (d) watertight

15. Receptacles and cord connectors having grounding terminals must have those terminals _____.

(a) grounded      (b) connected      (c) labeled      (d) listed

# Article 408
# Switchboards and Panelboards

## Part I. General

### 408.1 Scope

Article 408 covers the specific requirements for switchboards, panelboards, and distribution boards that control light and power circuits. For the purposes of this book, we will only cover the requirements for panelboards. In the trade, the nickname for a panelboard is "guts." Figure 408-1

### 408.3 Arrangement of Busbars and Conductors

**(E) Phase Arrangement Panelboards.** Panelboards supplied by a 4-wire, 3Ø, delta system shall have the phase having the higher voltage-to-ground connected to the "B" (center) phase of the equipment. Figure 408-2

> FPN: Orange identification is required for the high-leg conductor. See 110.15, 215.8, and 230.56 for details.

> *WARNING: The ANSI standard for meter equipment requires the high-leg conductor to terminate on the "C" (right) phase of the meter enclosure.*

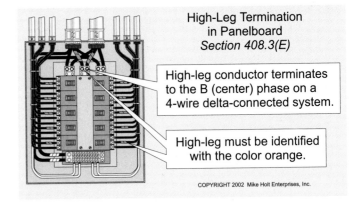

**Figure 408-2**

### 408.4 Circuit Identification

All circuits and circuit modifications shall be legibly identified as to purpose or use on a circuit directory located on the face or inside of the panel door of a panelboard. See 110.22. Figure 408-3

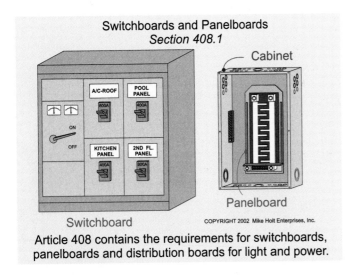

**Figure 408-1**

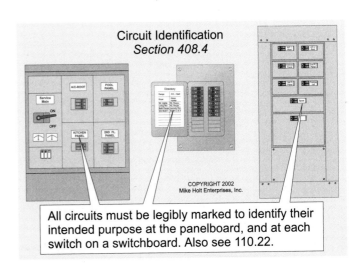

**Figure 408-3**

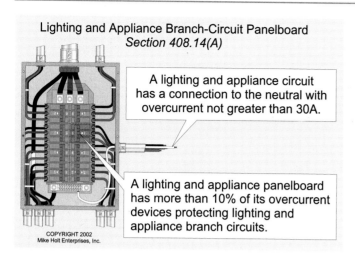

Lighting and Appliance Branch-Circuit Panelboard
*Section 408.14(A)*

A lighting and appliance circuit has a connection to the neutral with overcurrent not greater than 30A.

A lighting and appliance panelboard has more than 10% of its overcurrent devices protecting lighting and appliance branch circuits.

COPYRIGHT 2002
Mike Holt Enterprises, Inc.

**Figure 408-4**

## Part III. Panelboards

### 408.14 Classification of Panelboards

**(A)** **Lighting and Appliance Panelboard.** A lighting and appliance branch-circuit panelboard is one having more than 10 percent of its overcurrent protection devices protecting "lighting and appliance branch circuits." Figure 408-4

**Definition:** A lighting and appliance branch circuit is a branch circuit that has a connection to the neutral of the panelboard and overcurrent protection of 30A or less.

**(B)** **Power Panelboard.** A power panelboard is one having 10 percent or fewer of its overcurrent protection devices protecting "lighting and appliance branch circuits." Figure 408-5

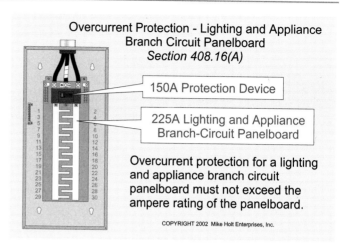

Overcurrent Protection - Lighting and Appliance
Branch Circuit Panelboard
*Section 408.16(A)*

150A Protection Device

225A Lighting and Appliance Branch-Circuit Panelboard

Overcurrent protection for a lighting and appliance branch circuit panelboard must not exceed the ampere rating of the panelboard.

COPYRIGHT 2002 Mike Holt Enterprises, Inc.

**Figure 408-6**

### 408.15 Number of Overcurrent Protection Devices

Not counting the main breaker, the maximum number of overcurrent protection devices that can be installed in a lighting and appliance branch-circuit panelboard is 42. A 2-pole circuit breaker counts as two overcurrent protection devices, and a 3-pole circuit breaker counts as three overcurrent protection devices.

### 408.16 Overcurrent Protection of Panelboard

**(A)** **Lighting and Appliance Branch-Circuit Panelboard.** Each lighting and appliance branch-circuit panelboard shall be provided with overcurrent protection having a rating not greater than that of the panelboard. Figure 408-6

**Exception No. 1:** Individual protection for a lighting and appliance branch-circuit panelboard is not required if the panelboard feeder has overcurrent protection not greater than the rating of the panelboard. Figure 408-7

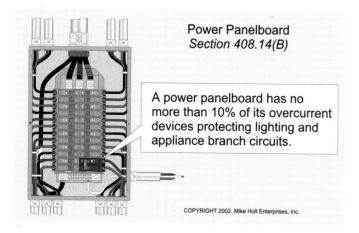

Power Panelboard
*Section 408.14(B)*

A power panelboard has no more than 10% of its overcurrent devices protecting lighting and appliance branch circuits.

COPYRIGHT 2002 Mike Holt Enterprises, Inc.

**Figure 408-5**

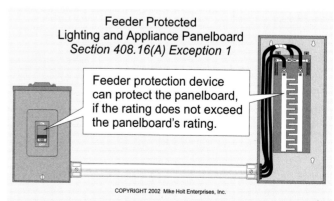

Feeder Protected
Lighting and Appliance Panelboard
*Section 408.16(A) Exception 1*

Feeder protection device can protect the panelboard, if the rating does not exceed the panelboard's rating.

COPYRIGHT 2002 Mike Holt Enterprises, Inc.

**Figure 408-7**

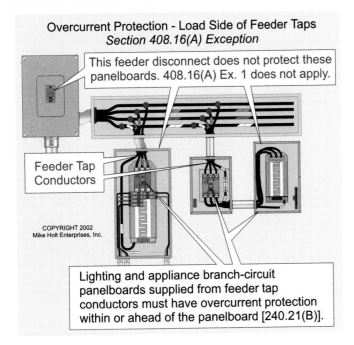

Overcurrent Protection - Load Side of Feeder Taps
*Section 408.16(A) Exception*

This feeder disconnect does not protect these panelboards. 408.16(A) Ex. 1 does not apply.

Feeder Tap Conductors

COPYRIGHT 2002
Mike Holt Enterprises, Inc.

Lighting and appliance branch-circuit panelboards supplied from feeder tap conductors must have overcurrent protection within or ahead of the panelboard [240.21(B)].

**Figure 408-8**

*CAUTION: When tap conductors supply a lighting and appliance branch-circuit panelboard, as permitted in 240.21(B), overcurrent protection shall be installed within or ahead of the panelboard. The required overcurrent protection can be in a separate enclosure ahead of the panelboard, or it can be a main breaker within the panelboard cabinet. Figure 408-8*

**(D) Panelboards Supplied through a Transformer.** When a lighting and appliance branch-circuit panelboard is supplied from a transformer as permitted in 240.21(C), the overcurrent protection for the panelboard shall be on the secondary

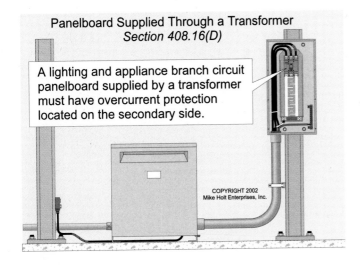

Panelboard Supplied Through a Transformer
*Section 408.16(D)*

A lighting and appliance branch circuit panelboard supplied by a transformer must have overcurrent protection located on the secondary side.

COPYRIGHT 2002
Mike Holt Enterprises, Inc.

**Figure 408-9**

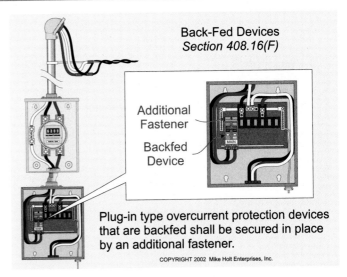

Back-Fed Devices
*Section 408.16(F)*

Additional Fastener

Backfed Device

Plug-in type overcurrent protection devices that are backfed shall be secured in place by an additional fastener.

COPYRIGHT 2002  Mike Holt Enterprises, Inc.

**Figure 408-10**

side of the transformer. The required overcurrent protection can be in a separate enclosure ahead of the panelboard, or it can be a main breaker within the panelboard cabinet. Figure 408-9

**Exception:** Panelboards supplied by a 2-wire transformer can be protected by the primary protection device. See 240.4(F) and 240.21(C)(1) for details.

**(F) Back-Fed Devices.** Plug-in circuit breakers that are back-fed shall be secured in place by an additional fastener that requires other than a pull to release the breaker from panel. The purpose of the fastener is to prevent the circuit breaker from being accidentally released from the panelboard, thereby exposing energized parts. Figure 408-10

**AUTHOR'S COMMENT:** Circuit breakers are often back-fed to provide overcurrent protection for lighting and appliance panelboards as required in 408.16(A).

*CAUTION: Circuit breakers marked "Line" and "Load" shall be installed in accordance with listed or labeled instructions [110.3(B)], therefore they cannot be back-fed. Figure 408-11*

### 408.20 Grounding of Panelboards

Metal panelboard cabinets and frames shall be grounded (bonded) in accordance with 250.134. Where the panelboard is used with nonmetallic raceways or cables or where separate grounding conductors are provided, a terminal bar for the equipment grounding conductors shall be bonded to the metal cabinet enclosure. Figure 408-12

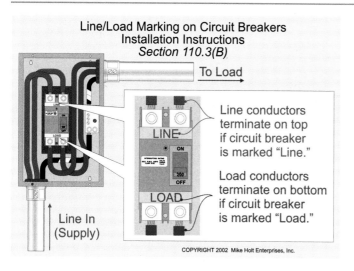

Line/Load Marking on Circuit Breakers
Installation Instructions
Section 110.3(B)

Line conductors terminate on top if circuit breaker is marked "Line."

Load conductors terminate on bottom if circuit breaker is marked "Load."

COPYRIGHT 2002 Mike Holt Enterprises, Inc.

**Figure 408-11**

**Exception:** Isolated equipment grounding conductors installed for the reduction of electrical noise as permitted in 250.146(D) can pass through the panelboard without terminating onto the equipment grounding terminal of the panelboard. Figure 408-13

Equipment grounding conductors shall not terminate on the grounded (neutral) terminal bar, and grounded (neutral) conductors shall not terminate on the equipment grounding terminal, except as permitted by 250.142 for services, separately derived systems, or separate building disconnects. Figure 408-14

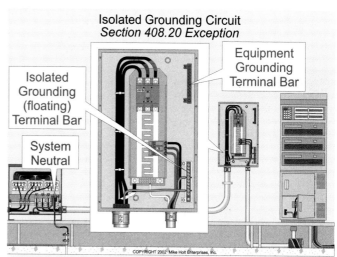

Isolated Grounding Circuit
Section 408.20 Exception

Equipment Grounding Terminal Bar

Isolated Grounding (floating) Terminal Bar

System Neutral

COPYRIGHT 2002 Mike Holt Enterprises, Inc.

An isolated grounding conductor can pass through a metal enclosure, but it must terminate to system neutral.

**Figure 408-13**

CAUTION: Most panelboards are rated as "suitable for use as service equipment," which means that they are supplied with a neutral-to-case bonding screw or bonding strap. This screw or strap shall not be installed except when the panelboard is used for service equipment, separately derived systems and remote building disconnects as permitted by 250.142. In addition, a panelboard marked "suitable only for use as service equipment" means that the neutral has been bonded to the case at the factory, and this panelboard cannot be used in nonservice-type applications.

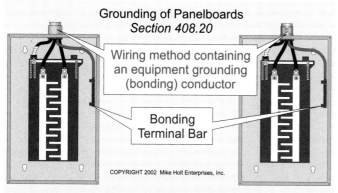

Grounding of Panelboards
Section 408.20

Wiring method containing an equipment grounding (bonding) conductor

Bonding Terminal Bar

COPYRIGHT 2002 Mike Holt Enterprises, Inc.

Where a panelboard is used with a nonmetallic raceway or cable, or where separate equipment grounding conductors are provided, a terminal bar for the equipment grounding conductors shall be bonded to the cabinet.

**Figure 408-12**

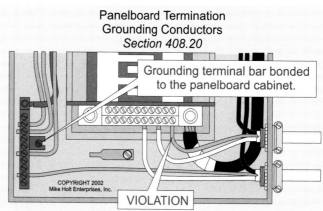

Panelboard Termination
Grounding Conductors
Section 408.20

Grounding terminal bar bonded to the panelboard cabinet.

COPYRIGHT 2002 Mike Holt Enterprises, Inc.

VIOLATION

Grounding conductors must not terminate on the same terminal bar with the grounded (neutral) conductor, except as permitted by 250.142 for services, separately derived systems or separate buildings.

**Figure 408-14**

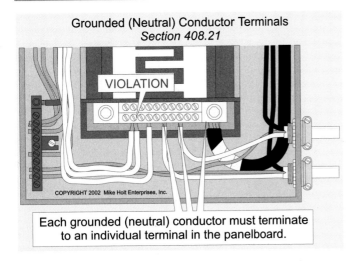

**Grounded (Neutral) Conductor Terminals**
*Section 408.21*

VIOLATION

COPYRIGHT 2002 Mike Holt Enterprises, Inc.

Each grounded (neutral) conductor must terminate
to an individual terminal in the panelboard.

**Figure 408-15**

## 408.21 Grounded Conductor Terminations

Each grounded (neutral) conductor shall terminate within the panelboard in an individual terminal. Figure 408-15

> **AUTHOR'S COMMENT:** If two grounded (neutral) conductors are in the same terminal, and someone removes one of the neutrals, the other neutral may unintentionally be removed as well. If that happens to the grounded (neutral) conductor of a multiwire circuit, it could result in excessive line-to-neutral voltage for one of the circuits. For example, if it was a 120/240V multiwire circuit, with one circuit having a 24Ω load and the other circuit having an 12Ω load, a loose neutral could result in as much as 160V across a 120V rated load. See 300.13(B) for additional details. Figure 408-16
>
> This requirement does not apply to equipment grounding (bonding) conductors because it doesn't affect the voltage of the circuit if an equipment grounding conductor is temporarily removed. Figure 408-17

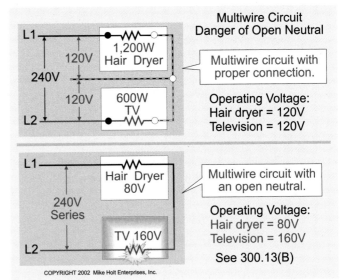

Multiwire Circuit
Danger of Open Neutral

L1

120V
240V
120V

1,200W
Hair Dryer

600W
TV

L2

Multiwire circuit with
proper connection.

Operating Voltage:
Hair dryer = 120V
Television = 120V

L1

240V
Series

Hair Dryer
80V

TV 160V

L2

Multiwire circuit with
an open neutral.

Operating Voltage:
Hair dryer = 80V
Television = 160V

See 300.13(B)

COPYRIGHT 2002 Mike Holt Enterprises, Inc.

**Figure 408-16**

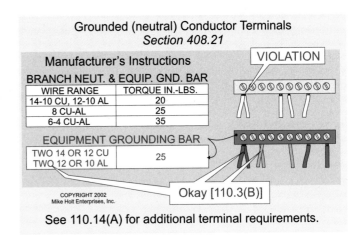

**Grounded (neutral) Conductor Terminals**
*Section 408.21*

Manufacturer's Instructions

VIOLATION

**BRANCH NEUT. & EQUIP. GND. BAR**

| WIRE RANGE | TORQUE IN.-LBS. |
|---|---|
| 14-10 CU, 12-10 AL | 20 |
| 8 CU-AL | 25 |
| 6-4 CU-AL | 35 |

**EQUIPMENT GROUNDING BAR**

| TWO 14 OR 12 CU TWO 12 OR 10 AL | 25 |
|---|---|

Okay [110.3(B)]

COPYRIGHT 2002
Mike Holt Enterprises, Inc.

See 110.14(A) for additional terminal requirements.

**Figure 408-17**

## Article 408

1. Panelboards supplied by a 3-phase, 4-wire, delta-connected system, shall have that phase having the higher voltage-to-ground (high-leg) connected to the _____ phase.
   (a) A
   (b) B
   (c) C
   (d) any of these

2. All panelboard circuits and circuit _____ shall be legibly identified as to purpose or use on a circuit directory located on the face or inside of the panel doors.
   (a) manufacturers
   (b) conductors
   (c) feeders
   (d) modifications

3. To qualify as a lighting and appliance branch-circuit panelboard, the number of circuits rated at 30A or less and having a neutral conductor must be _____ of the total.
   (a) more than 10 percent
   (b) 10 percent
   (c) 20 percent
   (d) 40 percent

4. A lighting and appliance branch-circuit panelboard shall be provided with physical means to prevent the installation of more _____ devices than that number for which the panelboard was designed, rated, and approved.
   (a) overcurrent
   (b) equipment
   (c) breaker
   (d) all of these

5. A lighting and appliance branch-circuit panelboard contains six 3-pole breakers and eight 2-pole breakers. The maximum allowable number of single-pole breakers that can be added to this panelboard is _____.
   (a) 8
   (b) 16
   (c) 28
   (d) 42

6. A lighting and appliance branch-circuit panelboard is considered protected by the _____ conductor protection device if the protection device rating is not greater than the panelboard rating.
   (a) grounded
   (b) feeder
   (c) branch circuit
   (d) none of these

7. When equipment grounding conductors are installed in panelboards, a _____ is required for the proper termination of the equipment grounding conductors.
   (a) neutral
   (b) grounded terminal bar
   (c) grounding terminal bar
   (d) none of these

8. Each _____ conductor must terminate within the panelboard in an individual terminal that is not also used for another conductor.
   (a) grounded
   (b) ungrounded
   (c) grounding
   (d) all of these

# Article 410
# Luminaires, Lampholders, and Lamps

## Part I. General

### 410.1 Scope

Article 410 contains the requirements for luminaires, lampholders, and lamps. Because of the many types and applications of luminaires, manufacturer's instructions are very important and helpful for proper installation. UL produces a pamphlet called the "Luminaire Marking Guide," which provides information for properly installing common types of incandescent, fluorescent, and high-intensity discharge (HID) luminaires. Figure 410-1

> **AUTHOR'S COMMENT:** Incandescent lighting includes quartz, tungsten halogen. Electric-discharge lighting includes fluorescent, neon, argon, and low-pressure sodium. High-intensity discharge lighting applies to a group of light sources, which include high-pressure sodium, mercury vapor, and metal halide.

## Part II. Luminaire Locations

### 410.4 Specific Locations

**(A) Wet or Damp Locations.** Luminaires in wet or damp locations shall be installed in a manner that will prevent water from accumulating in any part of the luminaire. Luminaires marked "Suitable for Dry Locations Only" shall be installed only in a dry location, luminaires marked "Suitable for

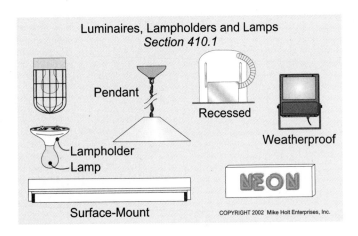

**Figure 410-1**

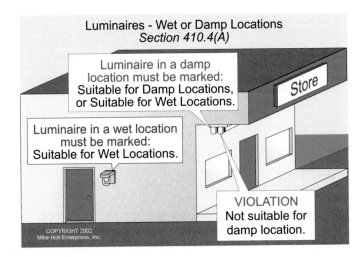

**Figure 410-2**

Damp Locations" can be installed in either a damp or dry location, and luminaires marked "Suitable for Wet Locations" can be installed in a dry, damp or wet location. Figure 410-2

> **AUTHOR'S COMMENT:** A dry location can be subjected to occasional dampness or wetness; see Article 100 definition of "location, dry."

**(B) Corrosive Locations.** Luminaires installed in corrosive locations shall be approved for the location.

**(C) In Ducts or Hoods.** Luminaires in commercial cooking hoods are permitted where all of the following conditions are met: Figure 410-3

(1) The luminaire is identified for use within commercial cooking hoods.

(2) The luminaire is constructed so all exhaust vapors, grease, oil, or cooking vapors are excluded from the lamp and wiring compartment.

(3) The luminaire shall be corrosion resistant or protected against corrosion, and the surface shall be smooth so as not to collect deposits and to facilitate cleaning.

(4) Wiring methods and materials supplying the luminaire(s) shall not be exposed within the cooking hood.

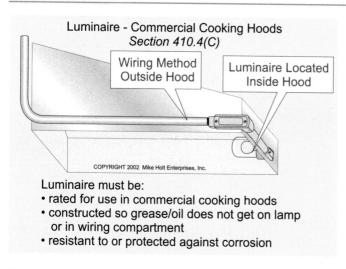

Luminaire must be:
• rated for use in commercial cooking hoods
• constructed so grease/oil does not get on lamp or in wiring compartment
• resistant to or protected against corrosion

**Figure 410-3**

---

**AUTHOR'S COMMENT:** Standard gasketed luminaires are not permitted, because they permit the accumulation of grease and oil deposits that can result in a fire because of high temperatures on the glass globe.

---

**(D)  Above Bathtubs.** No part of a cord-connected luminaire, hanging luminaire, lighting track, pendants or suspended ceiling paddle fans can be located within 3 ft horizontally and 8 ft vertically from the top of the bathtub rim or shower stall threshold. Figure 410-4

---

**AUTHOR'S COMMENT:** This rule does not apply to recessed or surface-mounted luminaires.

---

## 410.8 Clothes Closets

The *Code* does not require a luminaire to be installed in a clothes closet.

**(A)  Definition of Storage Space.** Storage space is defined as a volume bounded by the sides and back closet walls extending from the closet floor vertically to a height of 6 ft or the highest clothes-hanging rod at a horizontal distance of 2 ft from the sides and back of the closet walls. Storage space continues vertically to the closet ceiling for a distance of 1 ft or the width of the shelf, whichever is greater. Figure 410-5

**(B)  Luminaire Types Permitted in Clothes Closets.** The following types of luminaires can be installed in a clothes closet:

(1)  A surface or recessed incandescent luminaire with enclosed lamp.

(2)  A surface or recessed fluorescent luminaire.

**(C)  Luminaire Types not Permitted in Clothes Closets.** Incandescent luminaires that have open or partially open lamps and pendant-type luminaires cannot be installed in a clothes closet. Figure 410-6

**(D)  Installation of Luminaires in Clothes Closets.**

(1)  Surface-mounted totally enclosed incandescent luminaires shall maintain a minimum clearance of 1 ft from the storage space. Figure 410-7

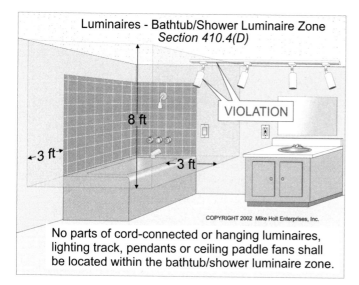

No parts of cord-connected or hanging luminaires, lighting track, pendants or ceiling paddle fans shall be located within the bathtub/shower luminaire zone.

**Figure 410-4**

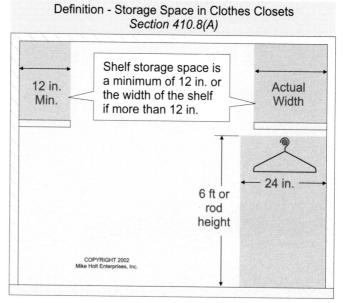

**Figure 410-5**

---

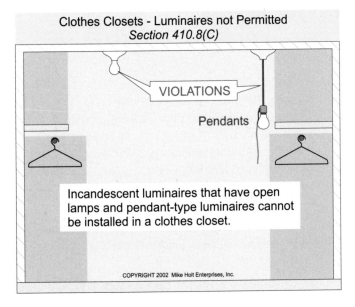

Figure 410-6

(2) Surface-mounted fluorescent luminaires shall maintain a minimum clearance of 6 in. from the storage space. See Figure 410-7

(3) Recessed incandescent luminaires with a completely enclosed lamp shall maintain a minimum clearance of 6 in. from storage space. Figure 410-8

(4) Recessed fluorescent luminaires shall maintain a minimum clearance of 6 in. from the storage space. See Figure 410-8

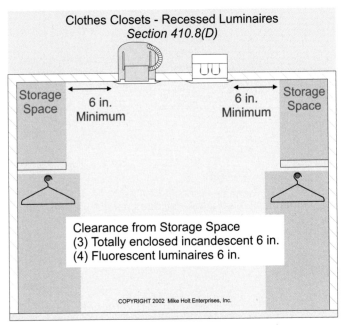

Figure 410-8

## Part III. Luminaire Outlet Boxes and Covers

### 410.12 Outlet Box to Be Covered

All outlet boxes used for luminaires shall be covered with a luminaire, lampholder, or blank cover. See 314.25. Figure 410-9

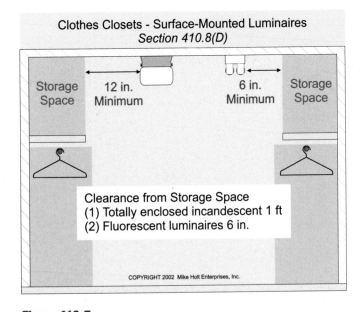

Figure 410-7

Figure 410-9

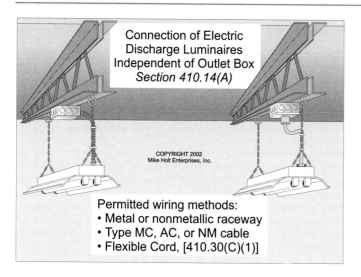

Connection of Electric
Discharge Luminaires
Independent of Outlet Box
Section 410.14(A)

COPYRIGHT 2002
Mike Holt Enterprises, Inc.

Permitted wiring methods:
• Metal or nonmetallic raceway
• Type MC, AC, or NM cable
• Flexible Cord, [410.30(C)(1)]

**Figure 410-10**

## 410.14 Connection of Electric-Discharge Luminaires

**(A) Luminaire Supported Independent of the Outlet Box.** Electric-discharge luminaires supported independently of the outlet box shall be connected to the branch circuit with a raceway, MC, AC or NM cable. Figure 410-10

Electric-discharge luminaires can be cord-connected if the luminaires are provided with internal adjustments to position the lamp to facilitate aiming [410.30(B)] .

> **AUTHOR'S COMMENT:** Electric-discharge luminaries can be cord connected if the cord is visible for its entire length and is plugged into a receptacle, and the installation is in accordance with 410.30(C).

**(B) Access to Outlet Box.** When an electric-discharge luminaire is mounted over an outlet box, the luminaire shall permit access to the branch-circuit wiring within the outlet box. Figure 410-11

## Part IV. Luminaire Supports

### 410.15 Metal Poles

**(A) General Support Requirement.** Luminaires and lampholders shall be securely supported.

**(B) Metal Poles.** Metal poles can be used to support luminaires.

(1) The metal pole shall have an accessible 2 in. × 4 in. handhole with a raintight cover to provide access to the supply conductors within the pole.

*Exception No. 1:* The handhole can be omitted on poles 8 ft or less in height, if the supply conductors for the luminaire are accessible by removing the luminaire.

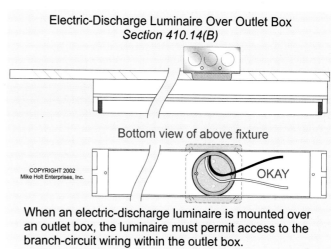

Electric-Discharge Luminaire Over Outlet Box
Section 410.14(B)

Bottom view of above fixture

COPYRIGHT 2002
Mike Holt Enterprises, Inc.

OKAY

When an electric-discharge luminaire is mounted over an outlet box, the luminaire must permit access to the branch-circuit wiring within the outlet box.

**Figure 410-11**

*Exception No. 2:* The handhole can be omitted on poles 20 ft or less in height, if the pole is provided with a hinged base.

(2) When the supply raceway or cable does not enter the pole, a threaded fitting or nipple shall be welded, brazed, or tapped opposite the handhole opening for the supply conductors.

(3) An accessible grounding terminal shall be installed inside the pole with a handhole or hinged base.

*Exception:* The grounding terminal can be omitted on metal poles 8 ft or less in height, if the supply conductors for the luminaire are accessible by removing the luminaire.

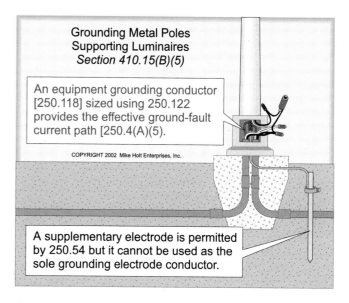

Grounding Metal Poles
Supporting Luminaires
Section 410.15(B)(5)

An equipment grounding conductor
[250.118] sized using 250.122
provides the effective ground-fault
current path [250.4(A)(5)].

COPYRIGHT 2002 Mike Holt Enterprises, Inc.

A supplementary electrode is permitted
by 250.54 but it cannot be used as the
sole grounding electrode conductor.

**Figure 410-12**

**(5) Grounding the Metal Pole.** The metal pole shall be grounded to an effective ground-fault current path as recognized by 250.118. Figure 410-12

**AUTHOR'S COMMENT:** We cannot use the earth as a return path back to the power source for fault current because the earth is a poor conductor of electricity. For more information, see 250.4(A)(5). Figure 410-13

**(6) Conductor Vertical Supports.** Conductors in vertical metal poles shall be supported if the vertical rise exceeds the values in Table 300.19(A).

**AUTHOR'S COMMENT:** When provided by the manufacturer, so-called J-hooks for support of conductors typical of roadway light poles shall be used, as they are part of the listing instructions.

## 410.16 Support

**(A) Outlet Boxes.** Outlet boxes or fittings specifically designed and listed for the support of luminaires shall be supported in accordance with 314.23. Outlet boxes can be used to support luminaires that weigh up to 50 lbs., unless the box is listed for the luminaire's weight [314.27(B)].

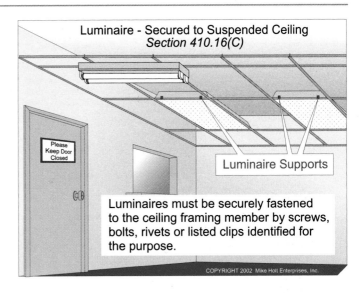

Figure 410-14

**(C) Suspended-Ceiling Framing Members.**

To protect firefighters, luminaires shall be attached to the suspended-ceiling framing with screws, bolts, rivets, or clips listed and identified for such use. Figure 410-14

Independent support wires for suspended-ceiling luminaries are not required by the *NEC*. Figure 410-15

**AUTHOR'S COMMENT:** Raceways and cables within a suspended ceiling shall be supported in accordance with 300.11(A). Outlet boxes can be secured to the ceiling framing members by bolts, screws, rivets, clips, or independent support wires that are taut and secured at both ends [314.23(D)].

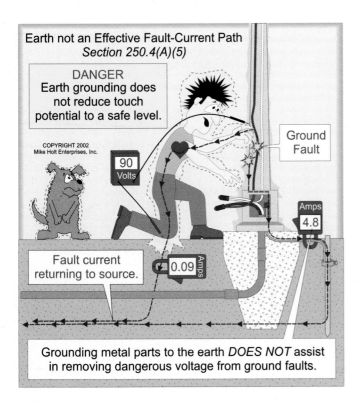

Figure 410-13

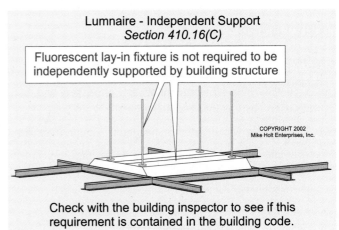

Figure 410-15

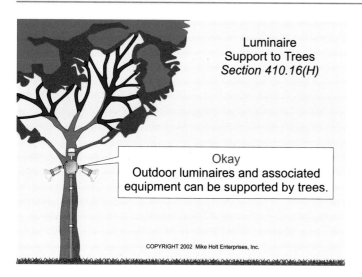

**Figure 410-16**

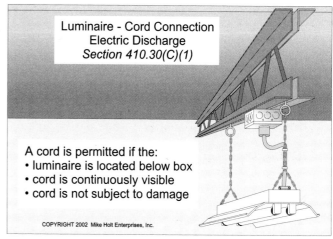

**Figure 410-18**

**(H)** **Luminaires Supported to Trees.** Trees can be used to support luminaires, but they cannot be used for the support of overhead conductor spans [225.26]. Figure 410-16

## Part VI. Wiring of Luminaires

### 410.23 Polarization of Luminaires

Luminaires shall have the grounded (neutral) conductor connected to the screw shell of the lampholder [200.10(C)], and the grounded (neutral) conductor shall be properly identified as required in 200.6.

### 410.30 Cord-Connected Luminaires

**(B)** **Adjustable Luminaires.** Luminaires that require adjusting or aiming after installation can be cord-connected with or

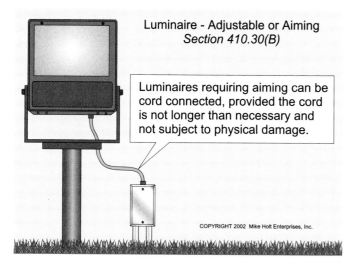

**Figure 410-17**

without an attachment plug, provided the exposed cord is of the hard usage or extra-hard usage type. The cord shall not be longer than necessary for luminaire adjustment, and it shall not be subject to strain or physical damage [400.10]. Figure 410-17

**(C)** **Electric-Discharge Luminaires.** A listed luminaire or listed luminaire assembly can be cord-connected if: Figure 410-18

(1) The luminaire is mounted directly below the outlet box or busway, and

(2) The flexible cord:

    a. Is visible its entire length,

    b. Is not subject to strain or physical damage [400.10], and

    c. Terminates in a grounding-type attachment plug, busway plug, or luminaire assembly with a strain relief and canopy.

**AUTHOR'S COMMENT:** The *Code* does not require twist-lock plugs and receptacles for this application.

### 410.31 Luminaires Used As Raceway

Luminaires shall not be used as a raceway for circuit conductors unless the luminaire is listed and marked for use as a raceway. Figure 410-19

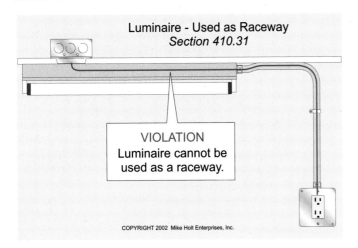

*Figure 410-19*

## 410.32 Wiring Luminaires Connected Together

Luminaires designed for end-to-end connection, or luminaires connected together by recognized wiring methods, can contain the conductors of a 2-wire branch circuit, or one multiwire branch circuit, supplying the connected luminaires. One additional 2-wire branch circuit separately supplying one or more of the connected luminaires can be installed. Figure 410-20

## 410.33 Branch-Circuit Conductors and Ballast

Conductors within 3 in. of a ballast shall have at least a 90°C insulation rating, unless the fixture is marked suitable for conductors with a different insulation temperature rating.

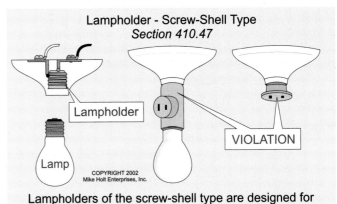

*Figure 410-21*

## Part VII. Lampholders

### 410.47 Screw-Shell Lampholder

Lampholders of the screw-shell type shall be installed for the use with lampholders only. Figure 410-21

**AUTHOR'S COMMENT:** A receptacle adapter cannot be installed in the lampholder, but UL lists this product! These are often installed for Christmas and other decorative lighting in violation of this Section.

## Part XI. Recessed Luminaires

### 410.65 Thermally Protected

**(C)  Recessed Incandescent Luminaires.** Recessed incandescent luminaires shall be thermally protected and shall be identified as such.

**AUTHOR'S COMMENT:** When higher-wattage lamps or improper trims are installed, the lampholder contained in a recessed luminaire can overheat, activating the thermal protection device and causing the luminaire to cycle on and off.

**Exception No. 2:** Thermal protection is not required for recessed luminaires whose design, construction, and thermal performance characteristics are equivalent to a thermally protected fixture, and are identified as inherently protected (Type IC).

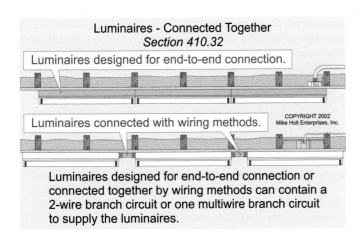

*Figure 410-20*

## 410.66 Recessed Luminaire Clearances

**(A) Clearances From Combustible Materials.**

**(1) Non-Type IC Luminaire.** A recessed luminaire that is not identified for contact with insulation shall have all recessed parts, except the points of supports, spaced at least $1/2$ in. from combustible materials.

**(2) Type IC Luminaire.** A Type IC luminaire (identified for contact with insulation) can be in contact with combustible materials.

**(B) Clearances From Thermal Insulation.** Thermal insulation cannot be located within 3 in. of a recessed luminaire, wiring compartment, or ballast, unless the luminaire is listed for direct contact with insulation (Type IC).

**AUTHOR'S COMMENT:** Since it is beyond the control of the electrical installer where insulation is installed, it is advisable that you always use Type IC rated recessed luminaires.

## 410.67 Wiring

**(C) Tap Conductors.** Luminaire tap conductors are permitted when not over 6 ft in length if they are installed in a raceway or AC or MC cable and the outlet box is at least 1 ft away.

## Part XIII. Electric-Discharge Lighting (1,000V or Less)

## 410.76 Luminaire Mounting

**(B) Surface-Mounted Luminaires with Ballast.** Surface-mounted luminaires containing ballast shall have a minimum of $1^1/_2$ in. clearance from combustible low-density fiberboard. Figure 410-22

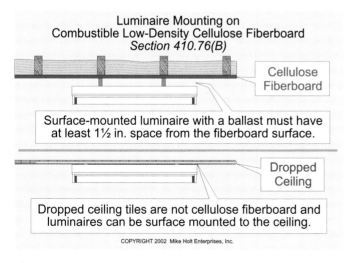

**Figure 410-22**

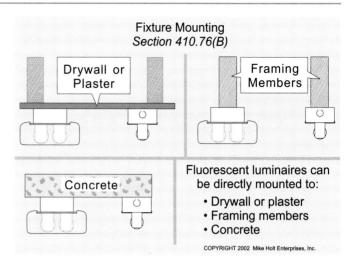

**Figure 410-23**

**AUTHOR'S COMMENT:** This rule does not apply to the mounting of fluorescent luminaires directly onto wood, plaster, concrete, or drywall. Figure 410-23

## Part XV. Track Lighting

## 410.100 Definition

Track Lighting is a manufactured assembly designed to support and energize luminaires that are capable of being readily repositioned on the track. Its length may be altered by the addition or subtraction of sections of track.

## 410.101 Installation

**(A) Lighting Track.** Lighting tracks shall be permanently installed and permanently connected to the branch-circuit wiring. Lampholders for track lighting are designed for lamps only, and a receptacle adapter is not permitted [410.47].

**(B) Circuit Rating.** The connected load on lighting track shall not exceed the rating of the track, and the track shall not be supplied by a circuit whose rating exceeds that of the track.

**AUTHOR'S COMMENT:** This means 15A lighting track cannot be connected to a 20A circuit. Figure 410-24

**(C) Uses.** Track lighting cannot be installed at any of the following locations:

(1) Where likely to be subjected to physical damage.

(2) In wet or damp locations.

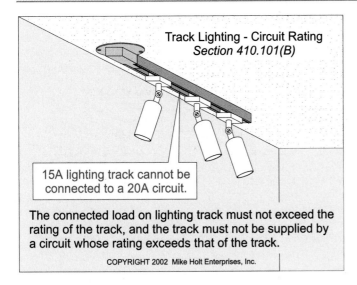

**Track Lighting - Circuit Rating**
*Section 410.101(B)*

15A lighting track cannot be connected to a 20A circuit.

The connected load on lighting track must not exceed the rating of the track, and the track must not be supplied by a circuit whose rating exceeds that of the track.

COPYRIGHT 2002 Mike Holt Enterprises, Inc.

*Figure 410-24*

(3)  Where subject to corrosive vapors.

(4)  In storage battery rooms.

(5)  In hazardous (classified) locations.

(6)  Where concealed.

(7)  Unbroken through walls, partitions, and floors.

(8)  Less than 5 ft above the finished floor, except where protected from physical damage or the track operates at less than 30V RMS open-circuit voltage.

(9)  Within the zone measured 3 ft horizontally and 8 ft vertically from the top of a bathtub rim or shower space [410.(D)].

## 410.104 Fastening

Track lighting shall be securely mounted to support the weight of the luminaïres. A single track section 4 ft or shorter in length shall have two supports, and, where installed in a continuous row, each individual track section of not more than 4 ft in length shall have one additional support.

# Article 410

1. A luminaire marked "Suitable for Damp Locations" _____ be used in a wet location.
   (a) can                 (b) cannot

2. Luminaires can be installed in a cooking hood if the luminaire is identified for use within a _____ cooking hood.
   (a) nonresidential      (b) commercial        (c) multifamily        (d) all of these

3. No part of cord-connected luminaires, hanging luminaires, lighting track, pendants or suspended-ceiling fans shall be located within a zone measured 3 ft horizontally and _____ vertically from the top of the bathtub rim or shower stall threshold.
   (a) 4 ft                (b) 6 ft              (c) 8 ft               (d) none of these

4. The *NEC* requires a lighting outlet in clothes closets.
   (a) True                (b) False

5. Incandescent luminaires that have open lamps, and pendant-type luminaires, can be installed in clothes closets where proper clearance is maintained from combustible products.
   (a) True                (b) False

6. Surface-mounted fluorescent luminaires in clothes closets can be installed on the wall above the door or on the ceiling, provided there is a minimum clearance of _____ between the luminaire and the nearest point of a storage space.
   (a) 3 in.               (b) 6 in.             (c) 9 in.              (d) 12 in.

7. In clothes closets, recessed incandescent luminaires with a completely enclosed lamp shall be permitted to be installed in the wall or on the ceiling, provided there is a minimum clearance of _____ between the luminaire and the nearest point of a storage space.
   (a) 3 in.               (b) 6 in.             (c) 9 in.              (d) 12 in.

8. When an electric-discharge luminaire is mounted directly over an outlet box, the luminaire must provide access to the conductor wiring within the outlet box.
   (a) True                (b) False

9. The maximum weight of a luminaire that may be mounted by the screw shell of a brass socket is _____
   (a) 2 lbs.              (b) 6 lbs.            (c) 3 lbs.             (d) 50 lbs.

10. The handhole of metal luminaire poles can be omitted for metal poles _____ or less in height above finish grade. This is only permitted if the pole is provided with a hinged base and the grounding terminal is accessible within the hinged base.
    (a) 8 ft               (b) 18 ft            (c) 20 ft             (d) none of these

11. Metal poles that support luminaires must meet which of the following requirements?
    (a) They must have an accessible handhole (sized 2 X 4 in.) with a raintight cover.
    (b) An accessible grounding terminal must be installed accessible from the handhole.
    (c) a and b
    (d) none of these

12. Metal poles used for the support of luminaires must be bonded to a(n) _____.
    (a) grounding electrode    (b) grounded conductor    (c) equipment grounding conductor    (d) any of these

13. Luminaires attached to the suspended-ceiling framing shall be secured to the framing member with screws, bolts, rivets, or clips _____ for use with the type of ceiling framing member(s) and luminaires involved.
    (a) marked             (b) labeled          (c) approved           (d) listed

14. Trees can be used to support luminaires.
    (a) True               (b) False

15. Luminaires that require adjustment or aiming after installation can be cord-connected without an attachment plug.
    (a) True               (b) False

16. A listed luminaire or a listed fixture assembly shall be permitted to be cord-connected if located _____ the outlet box and the cord is continuously visible for its entire length outside the fixture and is not subject to strain or physical damage.
    (a) within     (b) directly below     (c) directly above     (d) adjacent to

17. Luminaires designed for end-to-end connection to form a continuous assembly, or luminaires connected together by recognized wiring methods, shall be permitted to contain the conductors of a 2-wire branch circuit, or one _____ branch circuit, supplying the connected luminaires and need not be listed as a raceway.
    (a) small appliance     (b) appliance     (c) multiwire     (d) industrial

18. Branch-circuit conductors within _____ in. of a ballast shall have an insulation temperature rating not lower than 90°C (194°F) unless supplying a luminaire is listed and marked as suitable for a different insulation temperature.
    (a) 1     (b) 3     (c) 6     (d) none of these

19. Recessed incandescent luminaires shall have _____ protection and shall be identified as thermally protected.
    (a) physical     (b) corrosion     (c) thermal     (d) all of these

20. The minimum distance that an outlet box containing tap supply conductors can be placed from a recessed luminaire (fixture) is _____.
    (a) 1 ft     (b) 2 ft     (c) 3 ft     (d) 4 ft

21. The raceway or cable for tap conductors to recessed luminaires shall have a minimum length of _____
    (a) 6 in.     (b) 12 in.     (c) 18 in.     (d) 24 in.

22. Surface-mounted luminaires with a ballast must have a minimum clearance of _____ from combustible low-density cellulose fiberboard, unless the fixture is marked "Suitable for Surface Mounting on Combustible Low-Density Cellulose Fiberboard."
    (a) $1/2$ in.     (b) 1 in.     (c) $1^1/_2$ in.     (d) 2 in.

23. Lighting track is a manufactured assembly and its length may not be altered by the addition or subtraction of sections of track.
    (a) True     (b) False

24. Lighting tracks shall not be installed less than _____ above the finished floor except where protected from physical damage or track operating at less than 30V rms, open-circuit voltage.
    (a) 4 ft     (b) 5 ft     (c) $5^1/_2$ ft     (d) 6 ft

25. Track lighting shall not be installed within the zone measured 3 ft horizontally and _____ vertically from the top of the bathtub rim.
    (a) 2 ft     (b) 3 ft     (c) 4 ft     (d) 8 ft

# Article 411
## Lighting Systems Operating at 30V or Less

## 411.1 Scope

This article covers lighting systems operating at 30V or less and their associated components.

## 411.2 Definition

A lighting system consisting of an isolating power supply operating at 30V or less, with each secondary circuit limited to 25A, supplying luminaires and associated equipment identified for the use. Figure 411-1

## 411.3 Listing Required

Lighting systems shall be listed for the purpose.

## 411.4 Locations Not Permitted

Lighting systems shall not be installed: (1) where concealed or extended through a building wall, unless using a wiring method specified in Chapter 3, or (2) within 10 ft of pools, spas, fountains, or similar locations, except as permitted by Article 680.

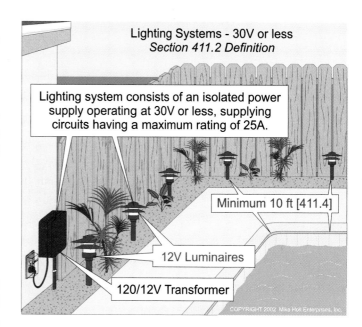

Lighting Systems - 30V or less
Section 411.2 Definition

Lighting system consists of an isolated power supply operating at 30V or less, supplying circuits having a maximum rating of 25A.

Minimum 10 ft [411.4]

12V Luminaires

120/12V Transformer

COPYRIGHT 2002 Mike Holt Enterprises, Inc.

**Figure 411-1**

# Article 422
# Appliances

## Part I. General

### 422.1 Scope

The scope of Article 422 includes appliances that are fastened in place, permanently connected, or cord-and-plug-connected in any occupancy. Figure 422-1

**Appliance [Article 100 Definition].** Utilization equipment, generally other than industrial, that is normally built in standardized sizes or types and is installed or connected as a unit to perform one or more functions, such as clothes washing, air conditioning, food mixing, deep frying, and so forth.

### 422.3 Other Articles

Appliances used in hazardous (classified) locations shall comply with the requirements of Articles 500 through 517. Motor-operated appliances shall comply with Article 430, and appliances containing hermetic refrigerant motor compressor(s) shall comply with Article 440.

> **AUTHOR'S COMMENT:** Room air-conditioning equipment shall be installed in accordance with Part VII of Article 440.

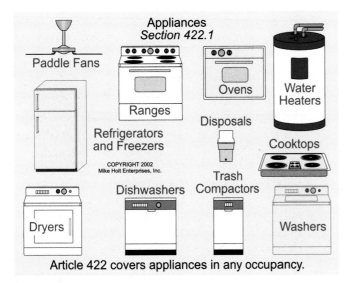

Article 422 covers appliances in any occupancy.

*Figure 422-1*

## Part II. Branch-Circuit Requirements

### 422.10 Branch-Circuit Rating

**(A) Individual Circuits.** The branch-circuit rating for an individual appliance shall not be less than the marked rating of the appliance [110.3(B)].

> **AUTHOR'S COMMENT:** Appliances rarely have the branch-circuit conduct rating marked on them.

**(B) Circuits Supplying Two or More Loads.** Branch circuits supplying appliance and other loads, shall be sized in accordance with 210.23(A).

**Portable Equipment.** Cord-and-plug-connected equipment shall not be rated more than 80 percent of the branch-circuit rating [210.23(A)(1)].

> **AUTHOR'S COMMENT:** UL and other approved testing laboratories list portable equipment up to 100 percent of the circuit rating. The *NEC* is an installation standard, not a product standard.

**Fixed Equipment.** Equipment fastened in place shall not be rated more than 50 percent of the branch-circuit ampere rating, if this circuit supplies both luminaires and receptacles [210.23(A)(2)].

### 422.11 Overcurrent Protection

**(A) Branch-Circuit.** Branch-circuit conductors shall have overcurrent protection in accordance with 240.4, and the protective device rating shall not exceed the rating marked on the appliance.

**(E) Nonmotor Appliances.** The rating of the appliance overcurrent protection device shall:

(1) Not exceed the rating marked on the appliance.

(2) For appliances rated 13.3A or less, the protection device shall not exceed 20A.

(3) For appliances rated over 13.3A, the protection device shall not exceed 150 percent of the appliance current rating. When 150 percent rating does not correspond with the standard ratings of overcurrent protection

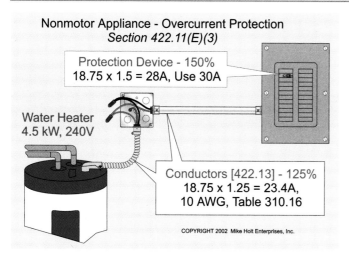

Nonmotor Appliance - Overcurrent Protection
Section 422.11(E)(3)

Protection Device - 150%
18.75 x 1.5 = 28A, Use 30A

Water Heater
4.5 kW, 240V

Conductors [422.13] - 125%
18.75 x 1.25 = 23.4A,
10 AWG, Table 310.16

COPYRIGHT 2002 Mike Holt Enterprises, Inc.

**Figure 422-2**

devices as listed in 240.6(A), the next higher standard rating is permitted.

**Question:** What is the maximum size overcurrent protection for a 4,500W, 240V water heater? Figure 422-2

**Answer:** 30A

4,500W/240V = 18.75A $\times$ 1.5 = 28A,
next size up, 30A [240.6(A)]

## 422.12 Fossil Fuel Heating Equipment (Furnaces)

Central heating equipment, such as gas, oil, or coal furnaces, shall be supplied by an individual branch circuit.

> **AUTHOR'S COMMENT:** This rule is not intended to apply to a listed wood-burning fireplace with a fan.

*Exception:* Auxiliary equipment to the fossil fuel heating equipment, such as pumps, valves, humidifiers, and electrostatic air cleaners, can be connected to the central heater circuit.

> **AUTHOR'S COMMENT:** Electric space-heating equipment shall be installed in accordance with Article 424 – Electric Space-Heating Equipment.

## 422.13 Water Heaters

Branch-circuit conductors for electric water heaters having a capacity of no more than 120 gallons shall be sized at no less than 125 percent of the appliance rating. See 422.11(E)(3) for overcurrent protection requirements.

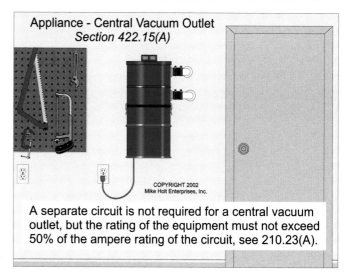

Appliance - Central Vacuum Outlet
Section 422.15(A)

COPYRIGHT 2002
Mike Holt Enterprises, Inc.

A separate circuit is not required for a central vacuum outlet, but the rating of the equipment must not exceed 50% of the ampere rating of the circuit, see 210.23(A).

**Figure 422-3**

## 422.15 Central Vacuum

(A) **Circuit Loading.** A separate 15A circuit is required for a central vacuum receptacle outlet if the rating of the central vacuum exceeds 7.5A, and a separate 20A circuit is required for a central vacuum receptacle outlet if the rating of the central vacuum exceeds 10A. See 210.23(A)(2). Figure 422-3

> **AUTHOR'S COMMENT:** Section 210.23(A)(2) specifies that equipment fastened in place, other than luminaires, shall not be rated more than 50 percent of the branch-circuit ampere rating, if this circuit supplies both luminaires and receptacles.

## 422.16 Flexible Cords

(A) **General.** Flexible cord can be used for connection of appliances, if the appliance is identified for flexible cord use [400.7(A)]:

(1) Facilitate frequent interchange or to prevent the transmission of noise and vibration.

(2) Facilitate the removal of appliances fastened in place, where the fastening means and mechanical connections are specifically designed to permit ready removal.

> **AUTHOR'S COMMENT:** Flexible cords cannot be used for the connection of water heaters, kitchen exhaust hoods, furnaces, and other appliances fastened in place, unless the appliances are specifically identified to be used with a flexible cord. Figure 422-4

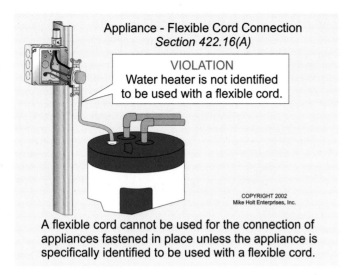

**Appliance - Flexible Cord Connection**
*Section 422.16(A)*

**VIOLATION**
Water heater is not identified to be used with a flexible cord.

COPYRIGHT 2002
Mike Holt Enterprises, Inc.

A flexible cord cannot be used for the connection of appliances fastened in place unless the appliance is specifically identified to be used with a flexible cord.

*Figure 422-4*

**(B)  Specific Appliances.**

**(1)  Waste (Garbage) Disposal.** A flexible cord is permitted for a waste disposal if:

(1) The cord has a grounding-type attachment plug.

(2) The cord length is not less than 18 in. and not longer than 3 ft.

(3) The waste disposal receptacle is located to avoid damage to the cord.

(4) The waste disposal receptacle is accessible.

**(2)  Dishwasher and Trash Compactor.** A cord is permitted for a dishwasher or trash compactor if:

(1) The cord has a grounding-type attachment plug.

(2) The cord length is not less than 3 ft and no longer than 4 ft, measured from the rear plane of the appliance.

**AUTHOR'S COMMENT:** This means that the minimum cord length for a dishwasher would be between 5 and 6 ft

(3) The appliance receptacle is located to avoid damage to the cord.

(4) The receptacle is located in the space occupied by the appliance or in the space adjacent to the appliance.

(5) The receptacle is accessible.

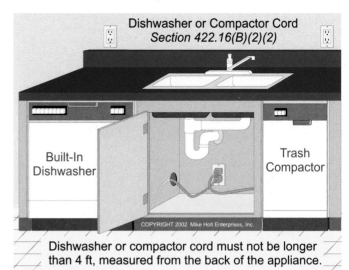

**Dishwasher or Compactor Cord**
*Section 422.16(B)(2)(2)*

Built-In Dishwasher

Trash Compactor

COPYRIGHT 2002 Mike Holt Enterprises, Inc.

Dishwasher or compactor cord must not be longer than 4 ft, measured from the back of the appliance.

*Figure 422-5*

**AUTHOR'S COMMENT:** According to publications in the Electrical Inspectors' magazine, *IAEI News*, a cord run through a cabinet for an appliance is not considered as being run through a wall. Figure 422-5

**(3)  Wall-Mounted Ovens and Counter-Mounted Cooking Units.** Wall-mounted ovens and counter-mounted cooking units can be cord-and-plug-connected.

## 422.18 Paddle Fans

**(A)  Paddle Fan Not Over 35 Lbs.** A paddle fan that weighs 35 lbs. or less, with or without accessories, can be supported directly to a listed paddle fan outlet box that is securely fastened in accordance with 314.23 and 314.27(D).

**(B)  Paddle Fan Over 35 Lbs.** A paddle fan that weighs more than 35 lbs., with or without accessories, shall be supported independently of the paddle fan outlet box. See 314.27(D).

*Exception:* Paddle fan outlet boxes that are identified for the purpose can support a paddle fan, with or without accessories, that weighs up to 70 lbs. Figure 422-6

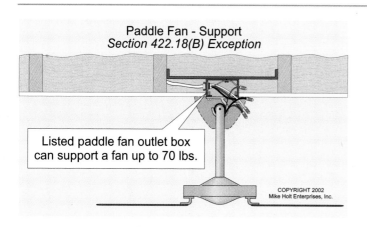

**Figure 422-6**

## Part III. Disconnect

### 422-30 Disconnecting Means

A disconnecting means that opens all ungrounded conductors shall be provided for each appliance in accordance with the requirements of 422.31 through 422.35.

### 422.31 Permanently Connected Appliance Disconnect

**(A)** **Appliance Rated at Not Over 300 VA or $\frac{1}{8}$-hp.** The branch-circuit overcurrent device, such as a plug fuse or circuit breaker, can serve as the appliance disconnect.

**(B)** **Appliance Rated Over 300 VA or $\frac{1}{8}$-hp.** A switch or circuit breaker located within sight from the appliance can serve as the appliance disconnect, or the switch or circuit breaker shall be capable of being locked in the open position. Figure 422-7

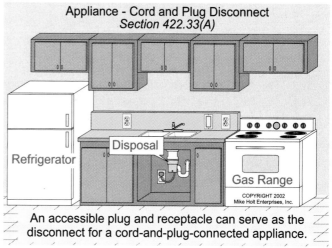

An accessible plug and receptacle can serve as the disconnect for a cord-and-plug-connected appliance.

**Figure 422-8**

**AUTHOR'S COMMENT:** Within sight is visible and not more than 50 ft from the other [Article 100].

### 422.33 Cord-and-Plug–Connected Appliance Disconnect

**(A)** **Attachment Plug and Receptacle.** An accessible plug and receptacle can serve as the disconnecting means for a cord-and-plug-connected appliance. Figure 422-8

**(B)** **Cord-and-Plug-Connected Range.** The cord and receptacle of a cord-and-plug-connected household electric range serve as the range disconnect if it is accessible from the front by the removal of a drawer. Figure 422-9

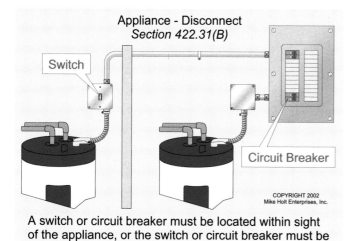

A switch or circuit breaker must be located within sight of the appliance, or the switch or circuit breaker must be capable of being locked in the open position.

**Figure 422-7**

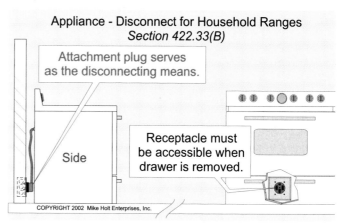

**Figure 422-9**

## 422.34 Unit Switch as Disconnect

A unit switch with a marked "Off" position that is a part of the appliance can serve as the appliance disconnect, if it disconnects all ungrounded conductors of the circuit. Figure 422-10

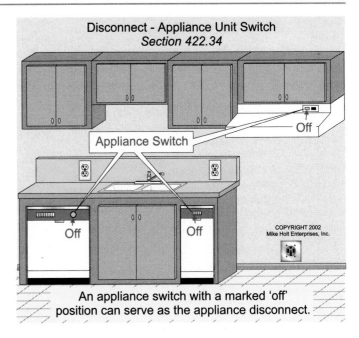

**Disconnect - Appliance Unit Switch**
*Section 422.34*

Appliance Switch

Off

Off

Off

COPYRIGHT 2002
Mike Holt Enterprises, Inc.

An appliance switch with a marked 'off'
position can serve as the appliance disconnect.

**Figure 422-10**

## Article 422

1   If a protective device rating is marked on an appliance, the branch-circuit overcurrent protection device rating shall not be greater than _____ percent of the protective device rating marked on the appliance.
    (a) 100             (b) 50             (c) 80             (d) 115

2.  The rating or setting of an overcurrent protection device for a 16.3A single nonmotor-operated appliance should not exceed _____ amperes.
    (a) 15              (b) 35             (c) 25             (d) 45

3.  Central heating equipment, other than fixed electric space-heating equipment, must be supplied by a(n) _____ branch circuit.
    (a) multiwire       (b) individual     (c) multipurpose   (d) small-appliance

4.  Water heaters having a capacity of _____ gallons or less shall have a branch-circuit rating not less than 125 percent of the nameplate rating of the water heater.
    (a) 60              (b) 75             (c) 90             (d) 120

5.  A waste disposal can be cord-and-plug-connected, but the cord must not be less than 18 in. or more than _____ and it must be protected from physical damage.
    (a) 30 in.          (b) 36 in.         (c) 42 in.         (d) 48 in.

6.  The cord for a dishwasher and trash compactor shall not be longer than _____ measured from the back of the appliance.
    (a) 2 ft            (b) 4 ft           (c) 6 ft           (d) 8 ft

7.  Ceiling-suspended (paddle) fans that do not exceed _____ in weight, with or without accessories, must be supported by outlet boxes identified for such use and supported in accordance with 314.23 and 314.27.
    (a) 20 lbs.         (b) 25 lbs.        (c) 30 lbs.        (d) 35 lbs.

8.  Listed outlet boxes or outlet box systems that are identified for the purpose shall be permitted to support ceiling-suspended fans that weigh no more than _____
    (a) 50 lbs.         (b) 60 lbs.        (c) 70 lbs.        (d) all of these

9.  For permanently connected appliances rated over _____ or $1/_8$-hp, the branch-circuit switch or circuit breaker shall be permitted to serve as the disconnecting means where the switch or circuit breaker is within sight from the appliance or is capable of being locked in the open position.
    (a) 200 VA          (b) 300 VA         (c) 400 VA         (d) 500 VA

10. Appliances that have a unit switch with an _____ setting that disconnects all the ungrounded conductors can serve as the disconnecting means for the appliance.
    (a) on              (b) off            (c) on/off         (d) all of these

# Article 424
# Fixed Electric Space Heating Equipment

## Part I. General

### 424.1 Scope

This article contains the installation requirements for fixed electric equipment used for space heating, such as heating cable, unit heaters, boilers, or central systems.

> **AUTHOR'S COMMENT:** Wiring for central heating equipment, such as gas, oil, or coal furnaces, shall be installed in accordance with Article 422 – Appliances, specifically 422.12. Room air-conditioning equipment shall be installed in accordance with Part VII of Article 440.

### 424.3 Branch Circuits

**(B) Conductors and Overcurrent Protection.** The branch-circuit conductor and overcurrent protection for fixed electric space heating equipment shall be no sized smaller than 125 percent of the total heating load.

**Question:** What size THHN conductor (75°C terminals) and protection device is required for a 9 kW 240V fixed electric space heater with a 3A motor? Figure 424-1

**Answer:** 6 AWG, 60A protection

Step 1. Determine the total load

$$I = 10,000 \text{ VA}/240\text{V} = 41.67\text{A} + 3\text{A} = 44.67\text{A} \times 1.25 = 56\text{A}$$

Step 2. Size conductor at 125% [110.14(C) and Table 310.16]

41.67A + 3A = 44.67A × 1.25 = 56A, 6 AWG, rated 65A at 75°C

Step 3. Size protection [240.4(B) and 240.6(A)]

41.67A + 3A = 44.67A × 1.25 = 56A, 60A overcurrent device

### 424.9 Permanently Installed Baseboard with Receptacles

Permanently installed electric baseboard heaters that have factory-installed receptacle outlets, or outlets provided as a separate listed assembly, can be used at the receptacle outlet(s) required by 210.50(B). Such receptacle outlets cannot be connected to the heater circuits.

> FPN: Listed baseboard heaters include instructions that may not permit their installation below receptacle outlets.

## Part III. Electric Space Heating Equipment

### 424.19 Disconnecting Means

Means shall be provided to disconnect the heater, motor controller(s), and supplementary overcurrent protective device(s) of all fixed electric space-heating equipment from all ungrounded conductors. The disconnecting means shall be within sight from the equipment, or it shall be capable of being locked in the open position. Figure 424-2

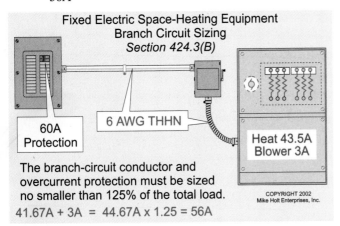

Fixed Electric Space-Heating Equipment
Branch Circuit Sizing
*Section 424.3(B)*

60A Protection

6 AWG THHN

Heat 43.5A
Blower 3A

The branch-circuit conductor and overcurrent protection must be sized no smaller than 125% of the total load.
41.67A + 3A = 44.67A x 1.25 = 56A

COPYRIGHT 2002
Mike Holt Enterprises, Inc.

**Figure 424-1**

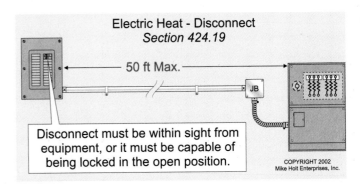

Electric Heat - Disconnect
*Section 424.19*

50 ft Max.

JB

Disconnect must be within sight from equipment, or it must be capable of being locked in the open position.

COPYRIGHT 2002
Mike Holt Enterprises, Inc.

**Figure 424-2**

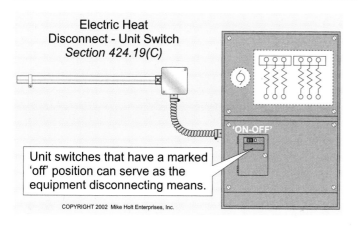

**Electric Heat**
**Disconnect - Unit Switch**
*Section 424.19(C)*

Unit switches that have a marked 'off' position can serve as the equipment disconnecting means.

COPYRIGHT 2002 Mike Holt Enterprises, Inc.

**Figure 424-3**

**AUTHOR'S COMMENT:** Within sight is visible and not more than 50 ft from each other [Article 100].

**(C)  Unit Switch as Disconnect.** A unit switch with a marked "Off" position that is an integral part of the equipment can serve as the appliance disconnect if it disconnects all ungrounded conductors of the circuit. Figure 424-3

## Part V. Electric Space-Heating Cables

### 424.44 Installation of Cables in Concrete or Poured Masonry Floors

**(G)  Ground-Fault Circuit Interrupter Protection.** GFCI protection for personnel shall be provided for electrically heated floors in bathroom, hydromassage bathtub, spa, and hot tub locations. Figure 424-4

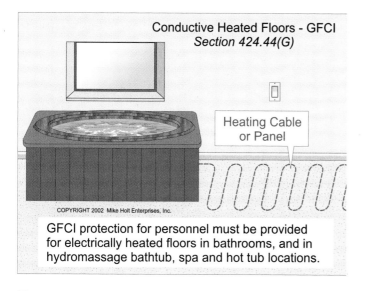

**Conductive Heated Floors - GFCI**
*Section 424.44(G)*

Heating Cable or Panel

COPYRIGHT 2002 Mike Holt Enterprises, Inc.

GFCI protection for personnel must be provided for electrically heated floors in bathrooms, and in hydromassage bathtub, spa and hot tub locations.

**Figure 424-4**

## Part VI. Duct Heaters

### 424.65 Disconnect for Electric Duct Heater Controller

Means shall be provided to disconnect the heater, motor controller(s), and supplementary overcurrent protective device from all ungrounded conductors of the circuit. The disconnecting means shall be within sight from the equipment, or it shall be capable of being locked in the open position.

**AUTHOR'S COMMENT:** Some installations use 120V control circuits that are fed from a foreign source that must be turned off (lockout/tagout) at another location.

The disconnect for the duct heater can be above or within the suspended ceiling adjacent to utilization equipment, and it can be accessed by portable means. See 240.24(A)(4), 404.8(A) Ex 2. Figure 424-5

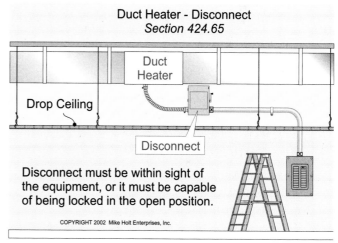

**Duct Heater - Disconnect**
*Section 424.65*

Duct Heater

Drop Ceiling

Disconnect

Disconnect must be within sight of the equipment, or it must be capable of being locked in the open position.

COPYRIGHT 2002 Mike Holt Enterprises, Inc.

**Figure 424-5**

## *Article 424*

1.  The branch-circuit conductor and overcurrent protection device for fixed electric space-heating equipment loads shall not be smaller than _____ percent of the total load.
    (a) 80              (b) 100              (c) 125              (d) 150

2.  Permanently installed electric baseboard heaters equipped with factory-installed receptacle outlets can be used as the outlets required by 210.50(B).
    (a) True            (b) False

3   If the disconnect is not within sight of the fixed electric space heater (without supplementary overcurrent protection devices), it must be capable of being _____.
    (a) locked                          (b) locked in the closed position
    (c) locked in the open position     (d) within sight

4.  GFCI protection for personnel shall be provided for electrically heated floors in _____ locations.
    (a) bathroom        (b) spa          (c) hot tub          (d) all of these

# Article 430
# Motors, Motor Circuits, and Controllers

## Part I. General

### 430.1 Scope  Figure 430-1

Part I. General

Part II. Motor Circuit Conductors

Part III. Overload Protection

Part IV. Motor Branch-Circuit Short-Circuit and Ground-Fault Protection

Part V. Motor Feeder Short-Circuit and Ground-Fault Protection

Part VI. Motor Control Circuits

Part VII. Motor Controllers

Part IX. Disconnecting Means

> FPN No. 1: Article 440 contains the installation requirements for air-conditioning and refrigerating equipment.

### 430.6 Table FLC versus Motor Nameplate Current Rating

**(A) General Requirements**

**(1) Table Full Load Current (FLC).** The motor full load current ratings listed in Tables 430.147, 430.148, or 430.150 are used when determining conductor

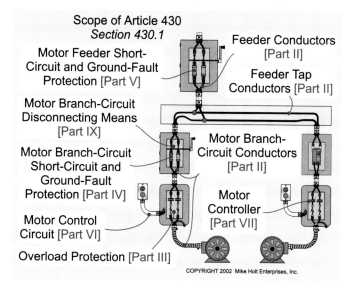

Figure 430-1

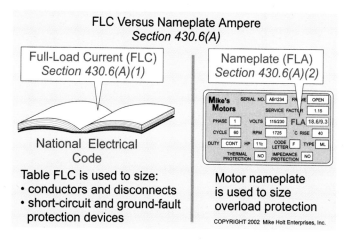

**Figure 430-2**

ampacity [430.22], the branch-circuit short-circuit and ground-fault protection device [430.52 and 62], and the ampere rating of switches [430.110].

The actual current rating on the motor nameplate is not to be used for this purpose. Figure 430-2

**(2) Motor Nameplate Current Rating.** Overload devices intended to protect motors, motor control apparatus, and motor branch-circuit conductors against excessive heating due to motor overloads and failure to start shall be based on the motor nameplate current rating. See 430.31.

### 430.9 Motor Controllers Terminal Requirements

**(B) Copper Conductors.** Motor controllers and terminals of control circuit devices shall be connected with copper conductors unless identified otherwise.

**(C) Torque Requirements.** Motor control conductors 14 AWG and smaller shall be torqued at a minimum of 7 lb-in. for screw-type pressure terminals, unless identified otherwise. See 110.3(B) and 110.14 FPN.

### 430.14 Location of Motors

**(A) Ventilation and Maintenance.** Motors shall be located so that adequate ventilation is provided and maintenance can be readily accomplished.

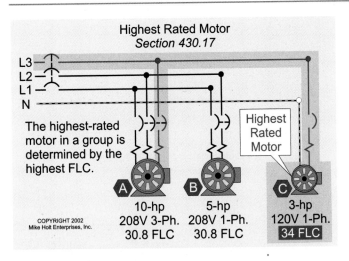

Figure 430-3

## 430.17 The Highest-Rated Motors

When sizing motor circuit conductors in accordance with 430.24 and 430.25, the highest-rated motor will be the motor with the highest-rated FLC.

**Question:** Which of the following motors has the highest FLC rating? Figure 430-3

(a) 10-hp, 3Ø, 208V           (b) 5-hp, 1Ø, 208V

(c) 3-hp, 1Ø, 120V            (d) none of these

**Answer:** (C) 3-hp, 1Ø, 120V

10-hp = 30.8A [Table 430.150]

5-hp = 30.8A [Table 430.148]

3-hp = 34.0A [Table 430.148]

## Part II. Conductor Size

### 430.22 Single Motor Conductor Size

**(A) Branch-Circuit Conductor Size.** Motor branch-circuit conductors to a single motor shall be sized no smaller than 125 percent of the motor FLC rating as listed in Table 430.147 or 430.148. Figure 430-4

**Question:** What size branch-circuit conductor is required for a $7\frac{1}{2}$-hp, 230V, 3Ø motor? Figure 430-5

(a) 14 AWG                    (b) 12 AWG

(c) 10 AWG                    (d) none of these

**Answer:** (c) 10 AWG

Branch-Circuit Conductor [430.22 and Table 430.150]

22A × 1.25 = 27.5A, 10 AWG, rated 30A at 60°C [110.14(C) and Table 310.16]

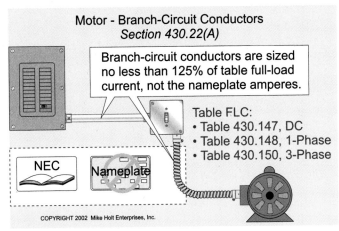

Figure 430-4

*Note: The branch-circuit short-circuit and ground-fault protection device is sized in accordance with 240.6(A), and 430.52(C)(1) Ex. 1. For an inverse-time breaker: 22A × 2.5 = 55A, next size up = 60A*

**AUTHOR'S COMMENT:** The equipment grounding conductor for a motor is sized based on the motor's short-circuit and ground-fault protection device rating, but it is not required to be larger than the branch-circuit conductors [250.122].

**Question:** What size equipment grounding conductor is required for a motor with a 45A protection device and 12 AWG branch-circuit conductors. Figure 430-6

(a) 14 AWG                    (b) 12 AWG

(c) 10 AWG                    (d) none of these

**Answer.** (b) 12 AWG.

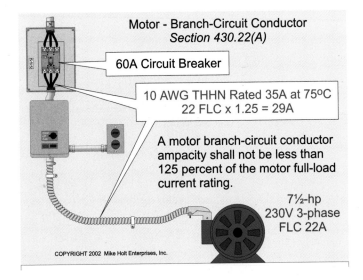

Figure 430-5

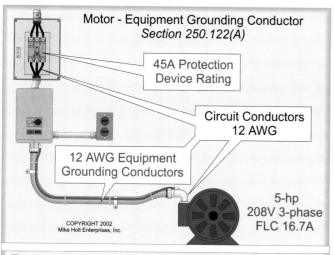

Motor - Equipment Grounding Conductor
Section 250.122(A)

45A Protection Device Rating

Circuit Conductors 12 AWG

12 AWG Equipment Grounding Conductors

COPYRIGHT 2002 Mike Holt Enterprises, Inc.

5-hp 208V 3-phase FLC 16.7A

Equipment grounding conductor is sized to Table 250.122 based on protection device rating, but it is not required to be larger than the branch-circuit conductors [250.122].

**Figure 430-6**

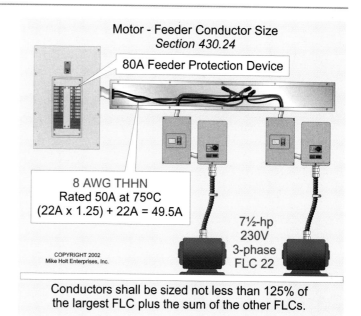

Motor - Feeder Conductor Size
Section 430.24

80A Feeder Protection Device

8 AWG THHN
Rated 50A at 75°C
(22A x 1.25) + 22A = 49.5A

COPYRIGHT 2002 Mike Holt Enterprises, Inc.

7½-hp 230V 3-phase FLC 22

Conductors shall be sized not less than 125% of the largest FLC plus the sum of the other FLCs.

**Figure 430-7**

Table 250.122 specifies that a 10 AWG equipment grounding conductor is required based on 45A short-circuit and ground-fault protection device, but the equipment grounding conductor is not required to be larger then the 12 AWG circuit conductors.

### 430.24 Motor Feeder Conductor Size

Conductors that supply several motors shall not be sized smaller than 125 percent of the largest motor FLC, plus the sum of the FLCs of the other motors [250.122].

**Question:** What size 75°C feeder conductor is required for two 7½-hp, 230V, 3Ø motors? Figure 430-7

(a) 14 AWG    (b) 12 AWG

(c) 10 AWG    (d) 8 AWG

**Answer:** (d) 8 AWG

Feeder Conductor [430.24 and Table 430.150]

(22A × 1.25) + 22A = 49.5A, 8 AWG rated 50A at 75°C [110.14(C) and Table 310.16 at 75°C]

**AUTHOR'S COMMENT:** The feeder protection device is sized in accordance with 430.62 as follows:

Step 1. Determine the branch-circuit protection rating in accordance with 240.6(A), 430.52(C)(1) Ex. 1, Table 430.150. Inverse-Time Breaker: 22A × 2.5 = 55A, next size up 60A

Step 2. Size the feeder protection device in accordance with 240.6(A) and 430.62.

Inverse-Time Breaker: 60A + 22A = 82A, next size down, 80A

### 430.28 Motor Tap Conductors

Motor conductors tapped from a feeder shall have an ampacity according to 430.22(A), and they shall terminate in a branch-circuit protective device sized in accordance with 430.52.

**(1) 10 ft Tap.** Tap conductors not over 10 ft in length shall have an ampacity of at least $^1/_{10}$ the rating of the feeder protection device.

**(2) 25 ft Tap.** Tap conductors over 10 ft but not over 25 ft shall have an ampacity of at least one-third the ampacity of the feeder conductor.

## Part III. Overload Protection

This specifies overload devices intended to protect motors, motor control apparatus, and motor branch-circuit conductors against excessive heating due to motor overloads and failure to start. Overload is the operation of equipment in excess of normal, full-load current rating, which if it persists for a sufficient amount of time, would cause damage or dangerous overheating of the apparatus.

**AUTHOR'S COMMENT:** Because of the difference between starting and running current, the overcurrent protection for motors is generally accomplished by having the overload device separate from the motor's short-circuit and ground-fault protection device. Figure 430-8

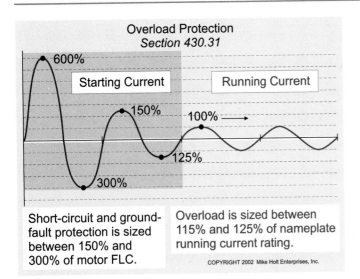

Figure 430-8

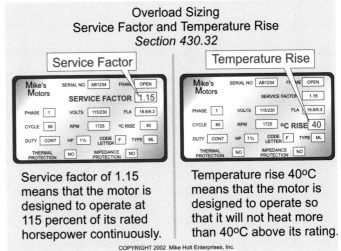

Figure 430-10

## 430.31 Overload

Overload protection devices, often called "heaters," are intended to provide overload protection for the motor, the motor control equipment, and branch-circuit conductors. They come in a variety of configurations, and they can be conventional (motor starter with heaters) or electronic. In addition, a fuse sized in accordance with 430.22 can be used. Figure 430-9

> **AUTHOR'S COMMENT:** In reality, motor overload protection sizing is accomplished by simply installing an overload protection device in accordance with the controller instruction based on motor nameplate current rating.

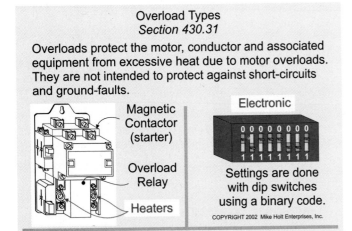

Figure 430.9

## 430.32 Overload Sizing

**(A)  Motors Rated More Than 1-Hp.** Motors rated more than 1-hp without integral thermal protection shall have an overload device sized no more than 115 percent of the motor "nameplate current rating," except for the following: Figure 430-10

**Service Factor.** Motors with a marked service factor (SF) of 1.15 or 1.25 on the nameplate shall have the overload device sized no more than 125 percent of the motor "nameplate current rating."

> **AUTHOR'S COMMENT:** A service factor of 1.15 means that the motor is designed to operate at 115 percent of its rated horsepower continuously. Studies have shown that when the operating temperature of a motor is increased 10°C, the motor winding insulating material's anticipated life is reduced by 50 percent.

**Temperature Rise.** Motors with a nameplate temperature rise 40°C or less shall have the overload device sized no more than 125 percent of motor "nameplate current rating."

> **AUTHOR'S COMMENT:** A motor with a nameplate temperature rise of 40°C means that the motor is designed to operate so it will not heat up 40°C above its rated ambient when operated at rated load and voltage.

## 430.36 Use of Fuses for Overload Protection

If remote control is not necessary for the motor, considerable savings in installation cost can be achieved by using fuses (preferably dual-element) sized in accordance with 430.32 to protect the motor and the circuit conductors against overcurrent (overload, short circuit and ground fault).

## 430.37 Number of Overload Devices

An overload protection device shall be installed in each ungrounded (hot) conductor.

## Part IV. Branch-Circuit Short-Circuit and Ground-Fault Protection

### 430.51 General

A branch-circuit short-circuit and ground-fault protection device protects the motor, the motor control apparatus, and the conductors against short circuits or ground faults, but not overload. Figure 430-11

### 430.52 Branch-Circuit Short-Circuit and Ground-Fault Protection

**(B) All Motors.** A motor branch-circuit short-circuit and ground-fault protective device shall be capable of carrying the motor's starting current.

**(C) Rating or Setting.**

(1) Each motor branch circuit shall be protected against short circuit and ground fault by a protection device sized no greater than the following percentages listed in Table 430.52.

| Motor Type | Nontime Delay Fuse | Dual Element Fuse | Inverse-Time Circuit Breaker |
|---|---|---|---|
| Wound Rotor | 150% | 150% | 150% |
| Direct Current | 150% | 150% | 150% |
| All Other Motors | 300% | 175% | 250% |

**Question:** What size conductor and inverse-time circuit breaker are required for a 2-hp, 230V, 1Ø motor? Figure 430-12

(a) 14 AWG, 30A breaker  (b) 14 AWG, 35A breaker

(c) 14 AWG, 40A breaker  (d) none of these

**Answer:** (a) 14 AWG, 30A breaker

Branch-Circuit Conductor [Table 310.16, 430.22(A), and Table 430.148]

12A × 1.25 = 15A, 14 AWG, rated 20A at 60°C [110.14(C) and Table 310.16]

Branch-Circuit Protection [240.6(A), 430.52(C)(1), Table 430.148]

12A × 2.5 = 30A

**AUTHOR'S COMMENT:** I know it bothers many in the electrical industry to see a 14 AWG conductor protected by a 30A circuit breaker, but branch-circuit conductors are protected against overcurrent by overloads sized at 115–125 percent of motor nameplate current rating [430.32].

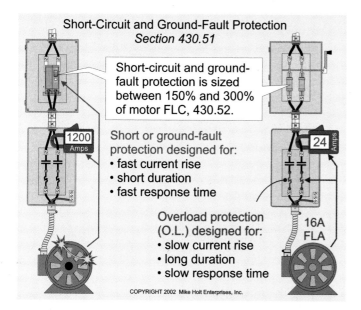

Figure 430-11

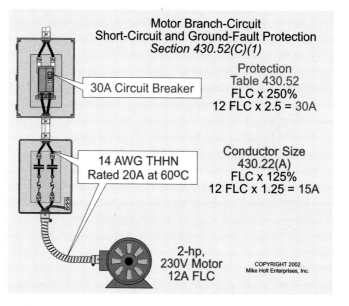

Figure 430-12

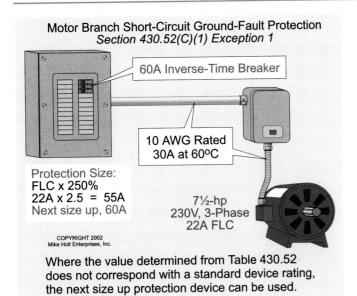

Motor Branch Short-Circuit Ground-Fault Protection
*Section 430.52(C)(1) Exception 1*

60A Inverse-Time Breaker

10 AWG Rated
30A at 60°C

Protection Size:
FLC x 250%
22A x 2.5 = 55A
Next size up, 60A

7½-hp
230V, 3-Phase
22A FLC

COPYRIGHT 2002
Mike Holt Enterprises, Inc.

Where the value determined from Table 430.52
does not correspond with a standard device rating,
the next size up protection device can be used.

*Figure 430-13*

**Exception No. 1:** Where the motor short-circuit and ground-fault protection device values derived from Table 430.52 do not correspond with the standard overcurrent device ratings listed in 240.6(A), the next higher protection device rating can be used.

**Question:** What size conductor and inverse-time circuit breaker are required for a $7^1/_2$-hp, 230V, 3Ø motor? Figure 430-13

(a) 10 AWG, 50A breaker    (b) 10 AWG, 60A breaker

(c) A or B    (d) none of these

**Answer.** (c) 10 AWG, 50 or 60A breaker

Step 1.  Branch-Circuit Conductor [Table 310.16, 430.22(A), and Table 430.150]

$22A \times 1.25 = 27.5A$, 10 AWG, rated 30A at 60°C [110.14(C) and Table 310.16]

Step 2.  Branch-Circuit Protection [240.6(A), 430.52(C)(1) Ex. 1, Table 430.150]

$22A \times 2.5 = 55A$, next size up = 60A

## 430.55 Single Overcurrent Protective Device

A motor can be protected against overload, short circuit and ground fault by a single protection device sized in accordance with the overload rules (115–125%) in 430.32.

**Question:** What size dual-element fuse can be used to protect a 5-hp, 230V, 1Ø motor with a service factor of 1.16 with a nameplate current rating of 28A? Figure 430-14

(a) 20A    (b) 25A

(c) 30A    (d) 35A

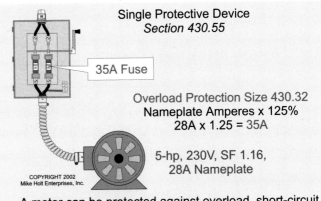

Single Protective Device
*Section 430.55*

35A Fuse

Overload Protection Size 430.32
Nameplate Amperes x 125%
28A x 1.25 = 35A

5-hp, 230V, SF 1.16,
28A Nameplate

COPYRIGHT 2002
Mike Holt Enterprises, Inc.

A motor can be protected against overload, short-circuit and ground-fault by a single protection device sized in accordance with the overload rules of 430.32.

*Figure 430-14*

**Answer:** (d) 35A

Overload Protection [430-32(A)(1)] 125 percent of motor nameplate current [SF 1.16]

$28A \times 1.25 = 35A$

## Part V. Feeder Short-Circuit and Ground-Fault Protection

### 430.62 Feeder Protection

(A)  **Motors Only.** Feeder conductors shall be protected against short circuits and ground faults by a protection device sized not greater than the largest rating of the branch-circuit short-circuit and ground-fault protective device for any motor, plus the sum of the full-load currents of the other motors in the group.

**Question:** What size feeder protection (inverse-time breakers with 75°C terminals) and conductor are required for the following two motors? Figure 430-15

Motor 1 – 20-hp, 460V, 3Ø = 27A

Motor 2 – 10-hp, 460V, 3Ø = 14A

(a) 8 AWG/70A    (b) 8 AWG/80A

(c) 8 AWG/90A    (d) none of these

**Answer:** (b) 8 AWG/80A

Feeder Conductor Size [430.24]

$(27A \times 1.25) + 14A = 48A$

8 AWG rated 50A at 75°C [110.14(C) and Table 310.16]

Feeder Protection [430.63(A)] not greater than largest branch-circuit protection device + other motor FLC

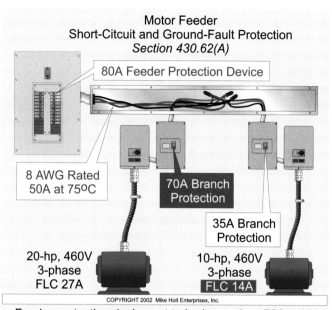

Motor Feeder
Short-Circuit and Ground-Fault Protection
*Section 430.62(A)*

80A Feeder Protection Device

8 AWG Rated 50A at 75°C

70A Branch Protection

35A Branch Protection

20-hp, 460V 3-phase FLC 27A

10-hp, 460V 3-phase FLC 14A

COPYRIGHT 2002 Mike Holt Enterprises, Inc.

Feeder protection device not to be larger than 70A + 14A

**Figure 430-15**

Step 1. Determine Largest Branch-Circuit Protection Device [430-52(C)(1)]

20-hp Motor = 27A × 2.5 = 68, next size up = 70A [430.52(C)(1) Ex.]

10-hp Motor = 14A × 2.5 = 35A

Step 2. Size Feeder Protection = 70A + 14A, = 84A, next size down = 80A

## Part VI. Motor Control Circuits

### 430.71 Definition of Motor Control Circuit

A control circuit that carries the electrical signals directing the performance of the motor controller. Figure 430-16

### 430.72 Overcurrent Protection for Control Circuits

**(A) Class 1 Control Conductors.** Motor control conductors not tapped from the branch-circuit protection device are Class 1 remote-control conductors, and they shall have overcurrent protection in accordance with 725.23.

#### 725.23 Class 1 Circuit Overcurrent Protection

Overcurrent protection for conductors 14 AWG and larger shall be in accordance with the conductor ampacity as listed in Table 310.16. Overcurrent protection for 18 AWG shall not exceed 7A, and 16 AWG conductors shall be protected by a 10A device.

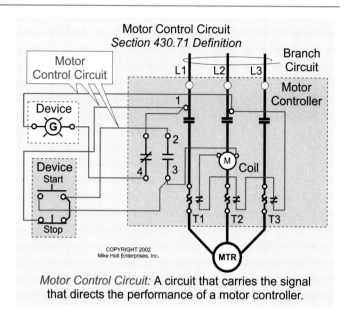

Motor Control Circuit
*Section 430.71 Definition*

COPYRIGHT 2002
Mike Holt Enterprises, Inc.

*Motor Control Circuit:* A circuit that carries the signal that directs the performance of a motor controller.

**Figure 430-16**

**(B) Motor Control Conductors.**

(2) Motor control circuit conductors tapped from the motor branch-circuit protection device that extend beyond the tap enclosure shall have overcurrent protection in accordance with the following:

| Conductor | Protection |
|---|---|
| 18 AWG | 7A |
| 16 AWG | 10A |
| 14 AWG | 45A |
| 12 AWG | 60A |
| 10 AWG | 90A |

**AUTHOR'S COMMENT:** The above limitations do not apply to internal wiring of industrial control panels listed to UL 508.

*Exception No. 2:* Two-wire Transformers. Conductors on the secondary side of a 2-wire transformer are considered protected by the primary protection device, if the primary protection rating protects the secondary conductors according to the secondary-to-primary voltage ratio of 430.72(C).

**(C) Control Circuit Transformers Protection.** Transformers for motor control circuit conductors shall have overcurrent protection on the primary side.

**AUTHOR'S COMMENT:** Many control transformers have very little iron, which results in a very high inrush (excitation) current when the coil is energized. Because of this high inrush current, standard fuses might blow, so you should use the fuses recommended by the control transformer manufacturer.

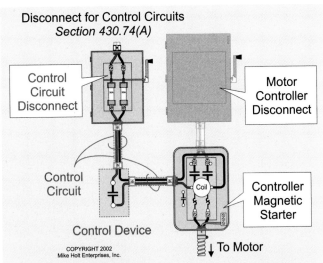

Disconnect for Control Circuits
*Section 430.74(A)*

Control Circuit Disconnect

Motor Controller Disconnect

Control Circuit

Controller Magnetic Starter

Control Device

↓ To Motor

COPYRIGHT 2002
Mike Holt Enterprises, Inc.

A disconnect for the motor control circuit is required when the control circuit is not tapped from the controller disconnect. Both the control circuit disconnect and the controller disconnect must be adjacent to each other.

**Figure 430-17**

### 430.74 Disconnect for Control Circuit

Motor control circuit conductors shall have a disconnecting means that opens all conductors of the motor control circuit.

**(A) Control Circuit Disconnect.** The controller disconnect can serve as the disconnecting means for control circuit conductors, if the control circuit conductors are tapped from the controller disconnect [430.102(A)].

If the control circuit conductors are not tapped from the controller disconnect, a separate disconnect is required for the control circuit conductors, and it shall be located adjacent to the controller disconnect. Figure 430-17

*CAUTION: The control circuit disconnect shall not be higher than 6 ft 7 in. above the floor or working platform unless located adjacent to the equipment it supplies. See 404.8(A).*

## Part VII. Motor Controllers

### 430.81 General

**(A) Controller Definition.**

**Controller:** A switch or device that is used for the purpose of starting and stopping the motor.

**AUTHOR'S COMMENT:** A controller could be a horsepower-rated switch, snap switch, or circuit breaker. A pushbutton that operates an electromechanical relay is not a controller, because it does not meet the controller rating requirements of 430.83. Figure 430-18

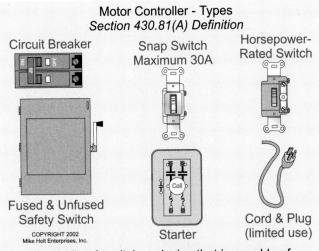

Motor Controller - Types
*Section 430.81(A) Definition*

Circuit Breaker

Snap Switch
Maximum 30A

Horsepower-Rated Switch

Fused & Unfused Safety Switch

COPYRIGHT 2002
Mike Holt Enterprises, Inc.

Coil

Starter

Cord & Plug (limited use)

*Controller:* A switch or device that is capable of interrupting the motor locked-rotor current used for the purpose of starting and stopping the motor.

**Figure 430-18**

### 430.83 Controller Rating

**(A) General.**

**(1) Horsepower Rating.** Controllers other than circuit breakers and molded case switches shall have a horsepower rating no less than that of the motor.

**(2) Circuit Breakers.** A circuit breaker can serve as a motor controller. See 430.111.

**AUTHOR'S COMMENT:** Circuit breakers are not required to be horsepower rated.

**(3) Molded Case Switch.** A molded case switch, rated in amperes, can serve as a motor controller.

**(C) Stationary Motors of 2 Horsepower or Less.** For stationary motors rated at 2-hp or less and 300V or less, the controller shall be permitted to be either of the following:

(1) A general-use switch having an ampere rating not less than twice the full-load current rating of the motor.

(2) A general-use snap switch where the motor full-load current rating is not more than 80 percent of the ampere rating of the switch.

**AUTHOR'S COMMENT:** See Article 100 for the definition.

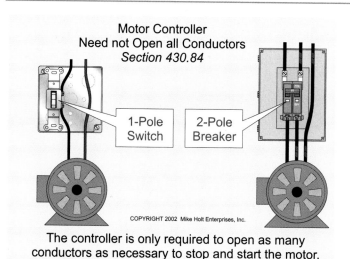

Motor Controller
Need not Open all Conductors
*Section 430.84*

1-Pole Switch

2-Pole Breaker

COPYRIGHT 2002 Mike Holt Enterprises, Inc.

The controller is only required to open as many conductors as necessary to stop and start the motor.

*Figure 430-19*

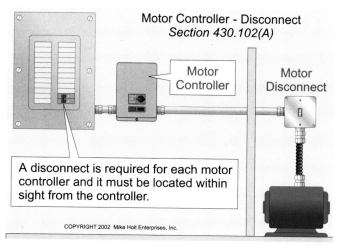

Motor Controller - Disconnect
*Section 430.102(A)*

Motor Controller

Motor Disconnect

A disconnect is required for each motor controller and it must be located within sight from the controller.

COPYRIGHT 2002 Mike Holt Enterprises, Inc.

*Figure 430-20*

## 430.84 Need not Open all Conductors

The motor controller is required to open only as many conductors of the circuit as necessary to start and stop the motor. For example, one conductor shall be opened to control a 2-wire, 1Ø motor, and only two conductors are required to control a 3-wire, 3Ø motor. Figure 430-19

**AUTHOR'S COMMENT:** The controller is only required to start and stop the motor; it is not a disconnecting means. See 430.103.

## 430.87 Controller for Each Motor

Each motor requires its own individual controller.

## 430.91 Motor Controller Enclosure Types

Motor controllers are installed in a variety of environments: rain, snow, sleet, corrosive agents, submersion, dirt, liquids, dust, oil, and so on. The selection of the enclosure for the environment shall be in accordance with Table 430.91 of the *NEC*.

## Part IX. Disconnecting Means

## 430.102 Disconnect Requirement

(A) **Controller Disconnect.** A disconnect is required for each motor controller, and it shall be located within sight from the controller. Figure 430-20

**AUTHOR'S COMMENT:** Within sight is visible and not more than 50 ft from each other [Article 100].

The controller disconnect shall open all circuit conductors simultaneously [430-103]. Figure 430-21

**AUTHOR'S COMMENT:** The controller disconnect can serve as the disconnect for motor control circuit conductors [430.74] and the motor [430.102(B) Ex.].

(B) **Motor Disconnect.** A disconnecting means is required for each motor, and it shall be located in sight from the motor location and the driven machinery location.

The controller disconnecting means [430-102(A)] can be used for the motor disconnecting means, if the controller disconnect is located in sight from the motor location and the driven machinery location. Figure 430-22

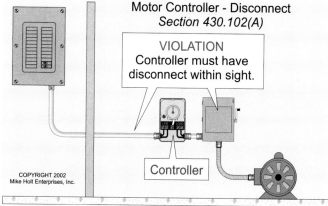

Motor Controller - Disconnect
*Section 430.102(A)*

VIOLATION
Controller must have disconnect within sight.

Controller

COPYRIGHT 2002
Mike Holt Enterprises, Inc.

The controller disconnect must disconnect all circuit conductors of the controller simultaneously [430.103].

*Figure 430-21*

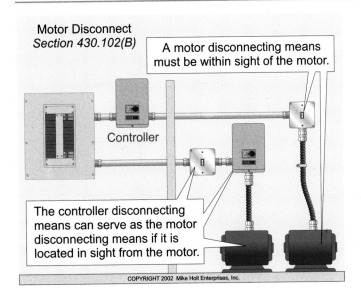

**Figure 430-22**

*Exception:* The motor disconnecting means is not required to be in sight from the motor and the driven machinery location under either condition (a) or (b) below, provided the controller disconnecting means [430.102(A)] is capable of being individually locked in the open position by a permanently installed fitting. Figure 430-23

(a)   Where locating the disconnecting means within sight of the motor is impracticable or introduces additional or increased hazards to persons or property.

(b)   In industrial installations, with written safety procedures, where conditions of maintenance and supervision ensure that only qualified persons will service the equipment.

FPN No. 1: Increased or additional hazards include, but are not limited to: motors rated in excess of 100-hp, multimotor equipment, submersible motors, motors associated with variable-frequency drives and motors in hazardous (classified) locations.

FPN No. 2: For information on lockout/tagout procedures, see NFPA 70E-2000 Standard for Electrical Safety Requirements for Employee Workplaces.

**AUTHOR'S COMMENT:** For more information, visit www.mikeholt.com/Newsletters/430.102.pdf or www.mikeholt.com/Newsletters/430.102.doc .

## 430.103 Disconnect Opens All Conductors

The disconnecting means for the motor controller and the motor shall open all ungrounded supply conductors simultaneously. Figure 430-24

## 430.104 Marking and Mounting

The controller and motor disconnect shall indicate whether they are in the "on" or "off" position.

**AUTHOR'S COMMENT:** The disconnecting means shall be legibly marked to identify its intended purpose [110.22 and 408.4], and when operated vertically, the "up" position shall be the "on" position [240.81 and 404.6(C)].

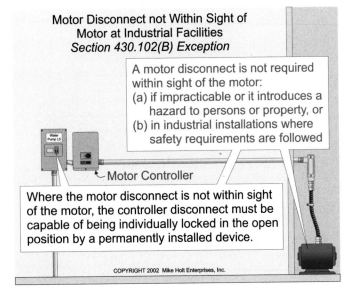

**Figure 430-23**

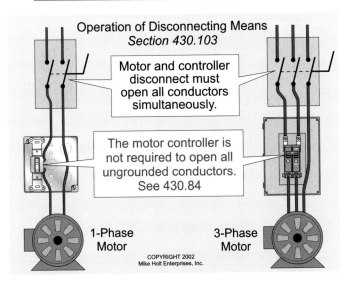

**Figure 430-24**

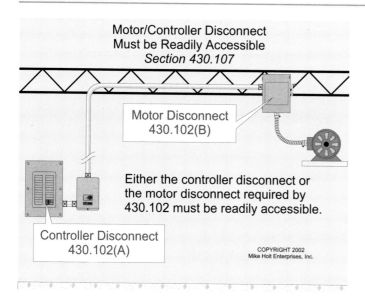

**Figure 430-25**

## 430.107 Readily Accessible

Either the controller disconnect or motor disconnect required by 430.102 shall be readily accessible. Figure 430-25

## 430.109 Disconnect Rating

**(A) General.** The disconnecting means for the motor controller and/or the motor shall be a:

   **(1) Motor-Circuit Switch.** A listed horsepower-rated motor-circuit switch.

   **(2) Molded Case Circuit Breaker.** A listed molded case circuit breaker.

   **(3) Molded Case Switch.** A listed molded case switch.

   **(6) Manual Motor Controller.** Listed manual motor controllers marked "Suitable as Motor Disconnect.."

**(C) Stationary Motors of 2-hp or Less.**

   **(2) General-Use Snap Switch.** A general-use ac snap switch can be used as the motor disconnect for ac motors rated 2-hp or less [430.83(C)]. Figure 430-26

**(F) Cord-and-Plug–Connected Motors.** Listed cord-and-plug connected motors rated $1/3$-hp or less can use a horsepower-rated attachment plug and a receptacle for the disconnecting means.

## 430.111 Combination Controller-Disconnect

A horsepower-rated switch or circuit breaker can serve as a motor controller and a disconnecting means if it opens all ungrounded conductors to the motor as required by 430.103.

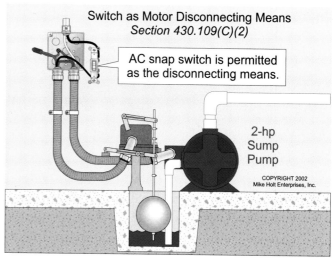

**Figure 430-26**

**AUTHOR'S COMMENT:** Circuit breakers are not required to be horsepower rated.

## Table 430.148 Full-Load Current, Single-Phase Motors

Table 430.148 lists the full-load current for single-phase alternating-current motors. These values are used to determine motor conductor sizing, ampere ratings of disconnects, controller rating, and branch-circuit and feeder protection, but not overload protection [430.6(A)(1)].

## Table 430.150 Full-Load Current, Three-Phase Motors

Table 430.150 lists the full-load current for 3-phase alternating-current motors. The values are used to determine motor conductor sizing, ampere ratings of disconnects, controller rating, and branch-circuit and feeder protection, but not overload protection [430.6(A)(1)].

## Table 430.151 Locked-Rotor Currents

Table 151(A) contains locked-rotor current for single-phase motors, and Table 151(B) contains the locked-rotor current for 3-phase motors. These values are used in the selection of controllers and disconnecting means when the horsepower rating is not marked on the motor nameplate.

## Article 430

1.  For general motor applications, the motor branch-circuit, short-circuit and ground-fault protection device must be sized based on the _____ amperes.
    (a) motor nameplate        (b) NEMA standards        (c) *NEC* Table        (d) Factory Mutual

2.  The motor _____ current as listed in Tables 430.147 through 430.150 must be used for sizing motor branch-circuit conductors and short-circuit, ground-fault protection devices.
    (a) nameplate        (b) full-load        (c) load        (d) none of these

3.  Motor controllers and terminals of control circuit devices are required to be connected to copper conductors unless identified as approved for use with a different type of conductor.
    (a) True        (b) False

4   Torque requirements for motor control circuit device terminals shall be a minimum of _____ pound-inches (unless otherwise indicated) for screw-type pressure terminals used for 14 AWG and smaller copper conductors.
    (a) 7        (b) 9        (c) 10        (d) 15

5   Branch-circuit conductors supplying a single continuous duty motor shall have an ampacity not less than _____.
    (a) 125 percent of the motor's nameplate current rating
    (b) 125 percent of the motor's full-load current as listed in the *NEC*
    (c) 125 percent of the motor's full locked-rotor rating
    (d) 80 percent of the motor's full-load current rating

6.  Feeder tap conductors supplying motor circuits, with an ampacity of at least one-third that of the feeder, shall not exceed _____ in length.
    (a) 10 ft        (b) 15 ft        (c) 20 ft        (d) 25 ft

7.  Overload devices are intended to protect motors, motor control apparatus, and motor branch-circuit conductors against _____.
    (a) excessive heating due to motor overloads        (b) excessive heating due to failure to start
    (c) short circuits and ground faults        (d) a and b

8.  Motor overload protection is not required where _____.
    (a) conductors are oversized by 125 percent        (b) conductors are part of a limited-energy circuit
    (c) it might introduce additional hazards        (d) short-circuit protection is provided

9.  An overload device used to protect continuous-duty motors (rated more than 1-hp) shall be selected to trip or be rated at no more than _____ percent of the motor nameplate full-load current rating for motors with a marked service factor not less than 1.15.
    (a) 110        (b) 115        (c) 120        (d) 125

10. The motor branch-circuit short-circuit and ground-fault protective device shall be capable of carrying the _____ current of the motor.
    (a) varying        (b) starting        (c) running        (d) continuous

11. The maximum rating or setting of motor branch-circuit short-circuit and ground-fault protective devices for a single-phase motor is _____ percent for an inverse-time breaker.
    (a) 125        (b) 175        (c) 250        (d) 300

12. A feeder shall have a protective device with a rating or setting _____ branch-circuit short-circuit and ground-fault protective device for any motor in the group, plus the sum of the full-load currents of the other motors of the group.
    (a) not greater than the largest        (b) 125 percent of the largest rating
    (c) equal to the largest rating        (d) none of these

13. Motor control circuits shall be arranged so that they will be disconnected from all sources of supply when the disconnecting means is in the open position. Where separate devices are used for the motor and control circuit, they shall be located immediately adjacent to each other.
    (a) True                 (b) False

14. If the control circuit transformer is located in the controller enclosure, the transformer must be connected to the _____ side of the control circuit disconnect.
    (a) line            (b) load            (c) adjacent            (d) none of these

15. The branch-circuit protective device shall be permitted to serve as the controller for a stationary motor rated at _____-hp or less that is normally left running and cannot be damaged by overload or failure to start.
    (a) $1/8$            (b) $1/4$            (c) $3/8$            (d) $1/2$

16. The motor controller shall have horsepower ratings at the application voltage not _____ the horsepower rating of the motor.
    (a) lower than       (b) higher than       (c) equal to       (d) none of these

17. Each motor shall be provided with an individual controller.
    (a) True                 (b) False

18. A disconnecting means is required to disconnect the _____ from all ungrounded supply conductors.
    (a) motor       (b) motor or controller       (c) controller       (d) motor and controller

19. A _____ shall be located in sight from the motor location and the driven machinery location.
    (a) controller       (b) protection device       (c) disconnecting means       (d) all of these

20. The motor disconnecting means is not required to be in sight from the motor and the driven machinery location provided _____.
    (a) the controller disconnecting means is capable of being individually locked in the open position.
    (b) the provisions for locking are permanently installed on, or at the switch or circuit breaker used as the controller disconnecting means.
    (c) locating the motor disconnecting within sight of the motor is impractical or introduces additional or increased hazards to people or property.
    (d) all of these

21. The disconnecting means for the controller and motor must open all ungrounded supply conductors and shall be designed so that no pole can be operated independently.
    (a) True                 (b) False

22. Where more than one motor disconnecting means is provided in the same motor branch circuit, only one of the disconnecting means is required to be readily accessible.
    (a) True                 (b) False

23. The motor disconnecting means can be a _____.
    (a) circuit breaker                 (b) motor-circuit switch rated in horsepower
    (c) molded case switch              (d) any of these

24. If the motor disconnecting means is a motor-circuit switch, it shall be rated in _____.
    (a) horsepower       (b) watts       (c) amperes       (d) locked-rotor current

25. A branch-circuit overcurrent protection device, such as a plug fuse, may serve as the disconnecting means for a stationary motor of $1/8$-hp or less.
    (a) True                 (b) False

# Article 440
## Air-Conditioning and Refrigeration Equipment

## Part I. General

### 440.1 Scope

This article applies to electrically driven air-conditioning and refrigeration equipment that has a hermetic refrigerant motor-compressor.

### 440.2 Definitions

A combination consisting of a compressor and motor, both of which are enclosed in the same housing, with no external shaft or shaft seals and the motor operating in the refrigerant.

### 440.3 Other Articles

(B) **Equipment with No Hermetic Motor-Compressors.** Air-conditioning and refrigeration equipment that does not have hermetic refrigerant motor-compressors, such as furnaces with evaporator coils, shall comply with Article 422 for appliances, Article 424 for electric space heating, and in some cases, Article 430 for motors.

(C) **Household Refrigerant Motor-Compressor Appliances.** Room air conditioners, household refrigerators and freezers, drinking water coolers, and beverage dispensing machines are appliances, and their installation shall comply with the requirements in Article 422. Figure 440-1

## Part II. Disconnecting Means

### 440.13 Cord-and-Plug-Connected Equipment

Listed cord-and-plug-connected equipment, such as room air conditioners [440.63], household refrigerators and freezers, drinking water coolers, and beverage dispensers, can use the attachment plug and receptacle as the disconnecting means.

### 440.14 Location

The disconnecting means shall be located within sight from and readily accessible from the air-conditioning or refrigerating equipment. Figure 440-2

The disconnecting means can be installed on or within the air-conditioning or refrigerating equipment, but it cannot be located on panels that are designed to allow access to the air-conditioning or refrigeration equipment. Figure 440-3

> **AUTHOR'S COMMENT:** Within sight is visible and not more than 50 ft from each other [Article 100]. Figure 440-4

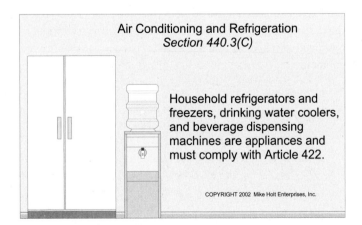

Air Conditioning and Refrigeration
*Section 440.3(C)*

Household refrigerators and freezers, drinking water coolers, and beverage dispensing machines are appliances and must comply with Article 422.

COPYRIGHT 2002 Mike Holt Enterprises, Inc.

**Figure 440-1**

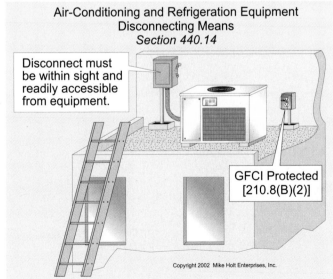

Air-Conditioning and Refrigeration Equipment
Disconnecting Means
*Section 440.14*

Disconnect must be within sight and readily accessible from equipment.

GFCI Protected
[210.8(B)(2)]

Copyright 2002 Mike Holt Enterprises, Inc.

**Figures 440-2**

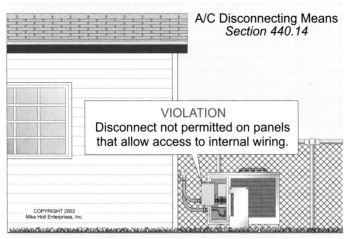

A/C Disconnecting Means
*Section 440.14*

**VIOLATION**
Disconnect not permitted on panels
that allow access to internal wiring.

COPYRIGHT 2002
Mike Holt Enterprises, Inc.

It can be installed on or within the equipment, but not over panels designed to allow access to internal wiring.

**Figure 440-3**

**Exception No. 2:** Where an attachment plug and receptacle serve as the disconnecting means, they shall be accessible, but they are not required to be readily accessible.

## Part III. Circuit Protection

**AUTHOR'S COMMENT:** The size and type of short-circuit and ground-fault protection device for air-conditioning and refrigeration equipment is often marked on the equipment nameplate. This is calculated by the manufacturer in accordance with the requirements in 440.22 and 440.32. Figure 440-5

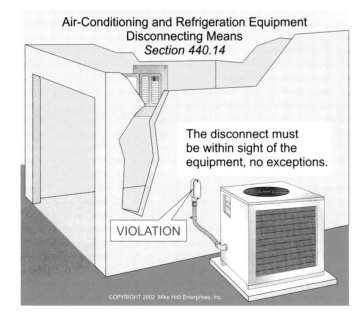

Air-Conditioning and Refrigeration Equipment
Disconnecting Means
*Section 440.14*

The disconnect must be within sight of the equipment, no exceptions.

**VIOLATION**

COPYRIGHT 2002  Mike Holt Enterprises, Inc.

**Figure 440-4**

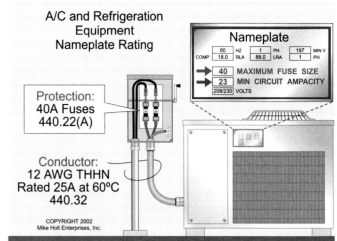

A/C and Refrigeration
Equipment
Nameplate Rating

**Nameplate**

| | | HZ | | PH | | MIN V |
|---|---|---|---|---|---|---|
| COMP | 60 | | 1 | | 197 | |
| | 18.0 | RLA | 88.0 | LRA | 1 | PH |

→ **40** MAXIMUM FUSE SIZE
→ **23** MIN CIRCUIT AMPACITY
208/230 VOLTS

Protection:
40A Fuses
440.22(A)

Conductor:
12 AWG THHN
Rated 25A at 60°C
440.32

COPYRIGHT 2002
Mike Holt Enterprises, Inc.

Protection and conductor size marked on equipment nameplate have been determined by the manufacturer in accordance with 440.22 and 440.32.

**Figure 440-5**

### 440.21 General

The branch-circuit conductors, control apparatus, and circuits supplying hermetic refrigerant motor-compressors shall be protected against short circuits and ground faults in accordance with 440.22.

**AUTHOR'S COMMENT:** If the equipment nameplate specifies "Maximum Fuse Size," then it shall be protected by a one-time or dual-element fuse. If the nameplate specifies "HACR Circuit Breaker," then the equipment shall be protected by an HACR-rated circuit breaker [110.3(B)].

### 440.22 Short-Circuit and Ground-Fault Protection Device Size

Short-circuit and ground-fault protection for air-conditioning and refrigeration equipment that has a hermetic refrigerant motor-compressor shall be sized no larger than identified on the equipment's nameplate. If the equipment does not have a nameplate specifying the size and type of protection device, then the protection device shall be sized as follows:

**(A) One Motor-Compressor.** The short-circuit and ground-fault protection device for motor-compressor conductors shall be capable of carrying the starting current of the motor. In addition, the protection device shall not be greater than 175 percent of the equipment load current rating.

If the protection device sized at 175 percent, is not capable of carrying the starting current of the motor-compressor, the

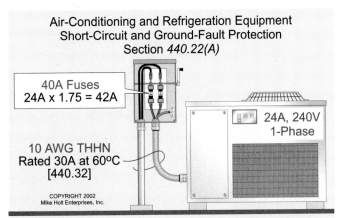

**Air-Conditioning and Refrigeration Equipment**
**Short-Circuit and Ground-Fault Protection**
**Section 440.22(A)**

40A Fuses
24A x 1.75 = 42A

10 AWG THHN
Rated 30A at 60°C
[440.32]

24A, 240V
1-Phase

COPYRIGHT 2002
Mike Holt Enterprises, Inc.

Short-circuit and ground-fault protection device must not
be greater than 175% of the equipment load current rating.

**Figure 440-6**

next larger protection device can be used, but it shall not
exceed 225 percent of the motor-compressor current rating.

**Question:** What size THHN conductor and fuse protection
are required for a 24A, 240V motor-compressor? Figure
440-6

(a) 10 AWG, 40A fuse        (b) 10 AWG, 50A fuse

(c) A or B                   (d) none of these

**Answer:** (a) 10 AWG, 40A

Branch-Circuit Conductor [Table 310.16 and 440.32]

24A × 1.25 = 30A, 10 AWG, rated 30A at 60°C
[110.14(C) and Table 310.16]

Branch-Circuit Protection [240.6(A) and 440.22(A)]

24A × 1.75 = 42A, next size down = 40A

If a 40A protection device is not capable of carrying the
starting current, then the protection device can be sized up
to 225 percent of the equipment load current rating (24A ×
2.25 = 54A, next size down 50A).

**(B)  Rating for Equipment.** Where the equipment incorporates
more than one hermetic refrigerant motor-compressor, or a
hermetic refrigerant motor-compressor and other motors or
other loads, the equipment short-circuit and ground-fault
protection shall be sized as follows:

**(1)  Motor-Compressor Largest Load.** The rating of the
branch-circuit short-circuit and ground-fault protective
device shall not be greater than the largest motor-compressor short-circuit ground-fault protection device,
plus the sum of the rated-load currents of the other
compressors.

**AUTHOR'S COMMENT:** The branch-circuit conductors are sized at 125 percent of the larger motor-compressor current, plus the sum of the rated-load currents
of the other compressors [440.33].

## Part IV. Conductor Sizing

The branch-circuit conductors for air-conditioning and
refrigeration equipment that has a hermetic refrigerant motor-
compressor shall be sized no smaller than identified on the equipment's nameplate. If the equipment does not have a nameplate
specifying the branch-circuit conductors, the conductors shall be
sized in accordance with 440.32. See Figure 440-5

### 440.32 Conductor Size – One Motor-Compressor

If equipment is not marked with minimum circuit ampacity,
the branch-circuit conductors to a single motor-compressor shall
have an ampacity of not less than 125 percent of the motor-compressor current.

**AUTHOR'S COMMENT:** Branch-circuit conductors
are protected against short circuits and ground faults
between 175 percent and 225 percent of the rated-load
current. See 440.22(A).

### 440.33 Conductor Size - Several Motor-Compressors

Conductors that supply several motor-compressors shall
have an ampacity of not less than 125 percent of the highest-rated
motor-compressor current of the group, plus the sum of the rated-
load currents of the other compressors.

**AUTHOR'S COMMENT:** These conductors shall be
protected against short circuits and ground faults in
accordance with 440.22(B)(1).

## Part VII. Room Air Conditioners

### 440.60 General

The rules in this part apply to window or in-wall type room
air conditioner units that incorporate a hermetic refrigerant motor-
compressor rated not over 40A, 250V, 1Ø.

### 440.62 Branch-Circuit Requirements

**(A)  Conductor and Protection Size.** Branch-circuit conductors
for a cord-and-plug–connected room air conditioner rated
not over 40A at 250V shall have an ampacity of not less than
125 percent of the rated-load currents [440.32].

**(B) Maximum Load on Circuit.** Where the room air conditioner is the only load on a circuit, the marked rating of the air conditioner shall not exceed 80 percent of the rating of the circuit [210.23(A)].

**(C) Other Loads on Circuit.** The total rating of a cord-and-attachment-plug-connected room air conditioner shall not exceed 50 percent of the rating of a branch circuit where lighting outlets, other appliances, or general-use receptacles are also supplied [210.23(B)].

### 440.63 Disconnecting Means

An attachment plug and receptacle can serve as the disconnecting means for a room air conditioner, provided: Figure 440-7

(1) The manual controls on the room air conditioner are readily accessible and within 6 ft of the floor, or

(2) An approved manually operable switch is installed in a readily accessible location within sight from the room air conditioner.

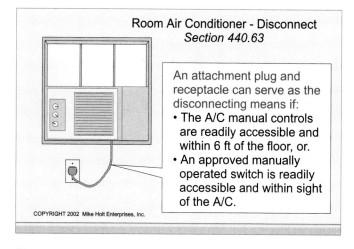

**Room Air Conditioner - Disconnect**
*Section 440.63*

An attachment plug and receptacle can serve as the disconnecting means if:
• The A/C manual controls are readily accessible and within 6 ft of the floor, or.
• An approved manually operated switch is readily accessible and within sight of the A/C.

COPYRIGHT 2002 Mike Holt Enterprises, Inc.

**Figure 440-7**

### 440.64 Supply Cord

The supply cord for room air-conditioning equipment shall not exceed 10 ft in length for 120V units or 6 ft for equipment rated 208 through 240V.

## Article 440

1.  Article 440 applies to electric motor-driven air-conditioning and refrigerating equipment that has a hermetic refrigerant motor compressor.
    (a) True                    (b) False

2.  The disconnecting means for air-conditioning and refrigerating equipment must be _____ from the air-conditioning or refrigerating equipment.
    (a) readily accessible       (b) within sight        (c) a or b            (d) a and b

3   Disconnecting means must be located within sight from and readily accessible from the air-conditioning or refrigerating equipment. The disconnecting means can be installed _____ the air-conditioning or refrigerating equipment, but not on panels that are designed to allow access to the air-conditioning or refrigeration equipment.
    (a) on                      (b) within              (c) a or b           (d) none of these

4.  The rating of the branch-circuit, short-circuit and ground-fault protection device for an individual motor-compressor shall not exceed _____ percent of the rated-load current or branch-circuit selection current, whichever is greater, if the protection device will carry the starting current of the motor.
    (a) 100                     (b) 125                 (c) 175              (d) none of these

5.  The attachment plug and receptacle shall not exceed _____ amperes at 250V for a cord-and-plug-connected air-conditioning motor-compressor.
    (a) 15                      (b) 20                  (c) 30               (d) 40

6.  The total rating of a plug-connected room air conditioner, connected to the same branch circuit where lighting units or other appliances are also supplied, shall not exceed _____ percent of the branch-circuit rating.
    (a) 80                      (b) 70                  (c) 50               (d) 40

7.  When supplying a 120V rated (nominal) room air conditioner, the length of the flexible supply cord shall not exceed _____.
    (a) 4 ft                    (b) 6 ft                (c) 8 ft             (d) 10 ft

# Article 445
# Generators

## 445.1 Scope

This article covers the installation of generators.

## 445.3 Other Articles

Generators, associated wiring, and equipment shall also be installed in accordance with the requirements in Articles 695 – Fire Pumps, 700 – Emergency Systems, 701 – Legally Required Standby Systems, 702 – Optional Standby Systems, and 705 – Interconnected Electric Power Production Sources.

## 445.11 Marking

Each generator shall be provided with a nameplate giving the manufacturer's name, rated frequency, power factor, number of phases if alternating current, rating in kilowatts or kilovolt amperes, normal volts and amperes corresponding to the rating, rated revolutions per minute, insulation system class and rated ambient temperature or rated temperature rise, and time rating.

## 445.12 Overcurrent Protection

(A) **Constant-Voltage Generators.** Constant-voltage generators shall be protected from overloads by inherent design, circuit breakers, fuses, or other suitable acceptable overcurrent protective means.

## 445.13 Ampacity of Conductors

The ampacity of the conductors from the generator terminals to the first distribution device(s) containing overcurrent protection cannot be less than 115 percent of the nameplate current rating of the generator. Figure 445-1

The grounded (neutral) conductor shall be sized no smaller than required by 220.22, but if the grounded (neutral) conductor is required to carry fault current, it shall be sized no smaller than required by 250.24(B).

**AUTHOR'S COMMENT:** If the conductors from the generator terminate in a transfer switch that does not switch the grounded (neutral) conductor, then the grounded (neutral) conductor will be required to carry fault current back to the generator.

Where the transfer switch does switch the grounded

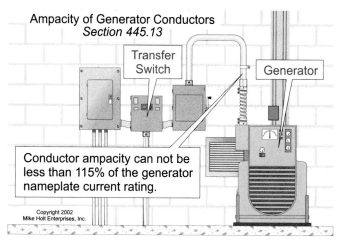

**Figure 445-1**

(neutral) conductor, the generator is considered a separately derived system (Article100 definition). A neutral-to-case connection is made at the generator, which will permit the effective ground-fault current path (equipment grounding conductor) to carry fault current instead of the grounded (neutral) conductor. The grounded (neutral) conductor in this case is sized to the maximum unbalanced load using 220.22. Figure 445-2

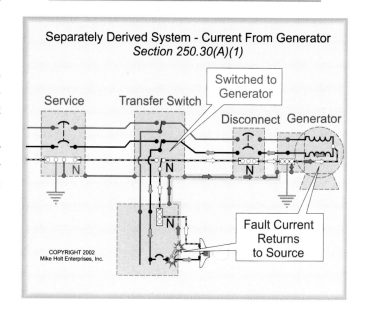

**Figure 445-2**

## 445.18 Disconnecting Means Required for Generators

The generator shall have a disconnect that disconnects power to all protective devices and control apparatus, except where:

(1) The driving means for the generator can be readily shut down, and

(2) The generator is not arranged to operate in parallel with another generator or other source of voltage.

# Article 450
# Transformers

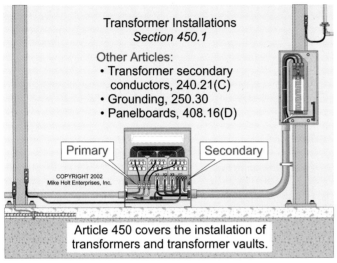

Transformer Installations
*Section 450.1*

Other Articles:
• Transformer secondary
  conductors, 240.21(C)
• Grounding, 250.30
• Panelboards, 408.16(D)

COPYRIGHT 2002
Mike Holt Enterprises, Inc.

Primary    Secondary

Article 450 covers the installation of
transformers and transformer vaults.

*Figure 450-1*

## 450.1 Scope

Article 450 covers the installation requirements of power
and lighting transformers and transformer vaults. Figure 450-1

## 450.3 Overcurrent Protection

FPN No. 2: Three-phase, 4-wire wye systems, such as
208Y/120V or 480Y/277V that supply nonlinear line-to-neutral loads, can overheat because of circulating odd triplen harmonic currents (3rd, 9th, 15th, 21st, etc.) within the primary
winding. See FPN No. 2 to 450.9. Figure 450-2

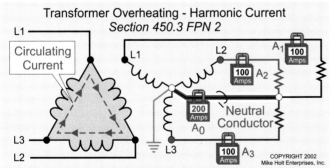

Transformer Overheating - Harmonic Current
*Section 450.3 FPN 2*

L1

Circulating
Current

L1    L2    A₁ 100 Amps
           100 A₂
           Amps
      200 A₀
      Amps  Neutral
            Conductor
L3
L2    L3   100 A₃
           Amps
COPYRIGHT 2002
Mike Holt Enterprises, Inc.

In 3-phase, 4-wire delta/wye transformers, odd
triplen harmonic currents from nonlinear loads can
cause excessive heating of the primary winding.

*Figure 450-2*

**AUTHOR'S COMMENT:** For more information on this
subject, visit www.NECCode.com and click on the
"Power Quality" link.

**(B) Overcurrent Protection for Transformers not Over 600V**

The primary winding of transformers shall be protected
against overcurrent in accordance with the percentages listed in
Table 450.3 and all applicable notes.

**Table 450.3**

| Primary Current Rating | Maximum Protection |
|---|---|
| 9A or More | 125 % (Note 1) |
| Less Than 9A | 167% |
| Less Than 2A | 300% |

Note 1. Where 125 percent of the primary current does not correspond to a standard rating of a fuse or nonadjustable circuit breaker, the next higher rating can be used [240.6(A)].

**Question:** What is the maximum primary protection device
rating permitted for a 45 kVA, 3Ø, 480V transformer?
Figure 450-3

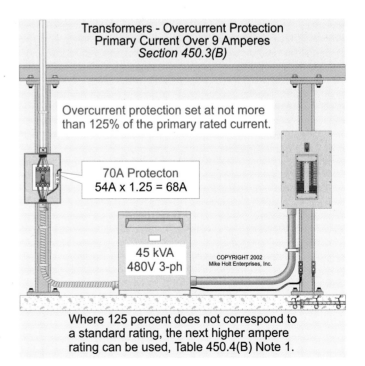

Transformers - Overcurrent Protection
Primary Current Over 9 Amperes
*Section 450.3(B)*

Overcurrent protection set at not more
than 125% of the primary rated current.

70A Protecton
54A x 1.25 = 68A

45 kVA
480V 3-ph

COPYRIGHT 2002
Mike Holt Enterprises, Inc.

Where 125 percent does not correspond to
a standard rating, the next higher ampere
rating can be used, Table 450.4(B) Note 1.

*Figure 450-3*

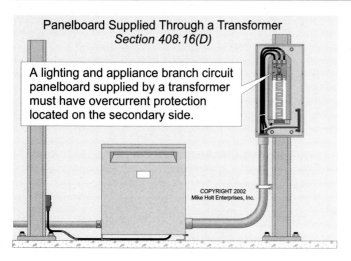

Panelboard Supplied Through a Transformer
*Section 408.16(D)*

A lighting and appliance branch circuit panelboard supplied by a transformer must have overcurrent protection located on the secondary side.

COPYRIGHT 2002
Mike Holt Enterprises, Inc.

**Figure 450-4**

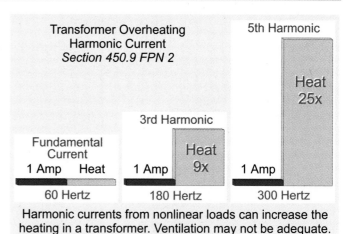

Transformer Overheating
Harmonic Current
*Section 450.9 FPN 2*

5th Harmonic

Heat 25x

3rd Harmonic

Fundamental Current
1 Amp   Heat

Heat 9x
1 Amp

1 Amp

60 Hertz        180 Hertz       300 Hertz

Harmonic currents from nonlinear loads can increase the heating in a transformer. Ventilation may not be adequate.

COPYRIGHT 2002 Mike Holt Enterprises, Inc.

**Figure 450-5**

(a) 40A            (b) 50A

(c) 60A            (d) 70A

**Answer:** (d) 70A

Step 1.  Determine primary current

$I = VA/(E \times \sqrt{3})$,

$I = 45,000 \text{ VA}/(480V \times 1.732), = 54A$

Step 2.  Determine primary protection device rating [240.6(A)]

$54A \times 1.25 = 68A$, next size up 70A

**AUTHOR'S COMMENT:** Secondary conductors shall be protected in accordance with 240.21(C), and over-current protection for lighting and appliance branch-circuit panelboards shall be located on the secondary side of the transformer [408.16(D)]. Figure 450-4

## 450.9 Ventilation

Transformer ventilating openings shall not be blocked, and they shall be installed in accordance with the manufacturer's instructions [110.3(B)].

FPN No. 2: Additional losses may occur in some transformers where harmonic currents are present, resulting in increased heat in the transformer above its rating. The heating from harmonic currents is proportionate to the square of the harmonic current. Figure 450-5

**AUTHOR'S COMMENT:** For more information on this subject, visit www.NECCode.com and go to the Power Quality link.

## 450.13 Transformer Accessibility

Transformers rated 600V or less shall be installed so they are readily accessible to qualified personnel for inspection and maintenance, except as permitted by (A) or (B) below:

**(A)  Open Installations.** Dry-type transformers can be located in the open on walls, columns, or structures. Figure 450-6

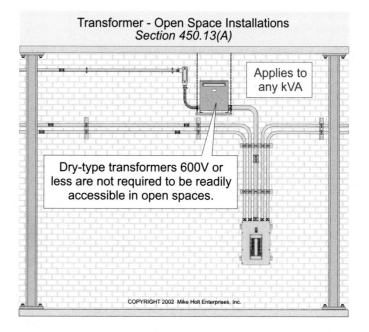

Transformer - Open Space Installations
*Section 450.13(A)*

Applies to any kVA

Dry-type transformers 600V or less are not required to be readily accessible in open spaces.

COPYRIGHT 2002 Mike Holt Enterprises, Inc.

**Figure 450-6**

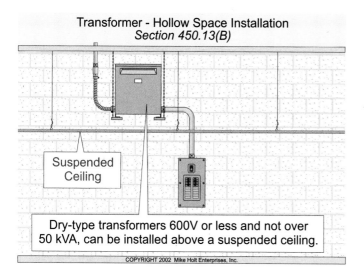

**Transformer - Hollow Space Installation**
*Section 450.13(B)*

Suspended
Ceiling

Dry-type transformers 600V or less and not over
50 kVA, can be installed above a suspended ceiling.

COPYRIGHT 2002 Mike Holt Enterprises, Inc.

*Figure 450-7*

**(B) Suspended Ceilings.** Dry-type transformers rated not more
than 50 kVA can be installed above suspended ceilings or
other hollow spaces of buildings not permanently closed in
by the structure. Figure 450-7

## Article 450

1. What size "primary only" overcurrent protection is required for a 45 kVA transformer, rated 600V, nominal, or less, that has a primary current rating of 54A?

   (a) 70          (b) 80          (c) 90          (d) 100

# Article 460
# Capacitors

## 460.1 Scope

This article covers the installation of capacitors, including those in hazardous (classified) locations as modified by Articles 501 through 503.

## 460.2 Enclosing and Guarding

**(A) Containing More than 3 gallons of Flammable Liquid.** Capacitors (single unit) containing more than 3 gallons of flammable liquid shall be enclosed in vaults or outdoor fenced enclosures complying with Article 110, Part III.

**(B) Accidental Contact.** Where accessible to unauthorized and unqualified persons, capacitors shall be enclosed, located, or guarded so accidental contact with exposed energized parts, terminals, or buses associated with them will not occur. This requirement does not apply where enclosures are accessible only to authorized and qualified persons.

## Part I. 600V, Nominal, and Under

## 460.8 Conductors

**(A) Ampacity.** Capacitor circuit conductors shall have an ampacity of not less than 135 percent of the rated current of the capacitor. Conductors that connect a capacitor to a motor circuit shall have an ampacity of not less than one-third the ampacity of the motor circuit conductors, but not less than 135 percent of the rated current of the capacitor.

**(B) Overcurrent Protection.** The rating or setting of the overcurrent device for a capacitor bank shall be as low as practicable.

*Exception:* A separate overcurrent device shall not be required for a capacitor connected on the load side of a motor overload protective device.

**(C) Disconnecting Means.** A disconnect shall be provided for each capacitor bank in accordance with the following:

(1) The disconnect shall open all ungrounded conductors simultaneously.

(2) The disconnect is permitted to disconnect the capacitor from the line as a regular operating procedure.

(3) The disconnect shall not be rated less than 135 percent of the rated current of the capacitor.

*Exception:* A separate disconnect shall not be required for a capacitor connected on the load side of a motor controller.

## 460.9 Rating or Setting of Motor Overload Device

Where a capacitor is connected on the load side of the motor overload device, the rating or setting of the motor overload device shall be based on the improved power factor of the motor circuit.

The effect of reduced current because of the capacitor shall be disregarded when determining the motor circuit conductor size in accordance with 430.22.

## Article 460

1.  Capacitors shall be _____ so that persons cannot come into accidental contact or bring conducting materials into accidental contact with exposed energized parts, terminals, or buses associated with them.
    (a) enclosed          (b) located          (c) guarded          (d) any of these

2.  The ampacity of capacitor circuit conductors shall not be less than _____ percent of the rated current of the capacitor.
    (a) 100          (b) 115          (c) 125          (d) 135

# Index

Mike Holt Enterprises, Inc. • www.NECcode.com • 1.888.NEC.CODE

Mike Holt Enterprises, Inc.  •  www.NECcode.com  •  1.888.NEC.CODE

# Commercial Wire and Raceway Chart

| Overcurrent Protection Size | Copper (1) Wire 60°C Terminal | Copper (2) Wire 75°C Terminal | Maximum (3) Continuous Ampere Load | Raceway (4) | Copper (5) Ground Wire | Max. Continuous 1-Phase VA Load (3) | | | | | Max. Continuous 3-Phase VA Load (3) | | |
|---|---|---|---|---|---|---|---|---|---|---|---|---|---|
| | | | | | | 120 V | 208 V | 240 V | 277 V | 480 V | 208 V | 240 V | 480 V |
| 15 | 14 | 14 | 12 | 1/2" | 14 | 1,440 | 2,496 | 2,880 | 3,324 | 5,760 | 4,323 | 4,988 | 9,976 |
| 20 | 12 | 12 | 16 | 1/2" | 12 | 1,920 | 3,328 | 3,840 | 4,432 | 7,680 | 5,764 | 6,651 | 13,302 |
| 25 | 10 | 10 | 20 | 3/4" | 10 | 2,400 | 4,160 | 4,800 | 5,540 | 9,600 | 7,205 | 8,314 | 16,627 |
| 30 | 10 | 10 | 24 | 3/4" | 10 | 2,880 | 4,992 | 5,760 | 6,648 | 11,520 | 8,646 | 9,976 | 19,953 |
| 35 | 8 | 8 | 28 | 1" | 10 | 3,360 | 5,824 | 6,720 | 7,756 | 13,440 | 10,087 | 11,639 | 23,278 |
| 40 | 8 | 8 | 32 | 1" | 10 | 3,840 | 6,656 | 7,680 | 8,864 | 15,360 | 11,528 | 13,302 | 26,604 |
| 45 | 6 | 8 | 36 | 1" | 10 | 4,320 | 7,488 | 8,640 | 9,972 | 17,280 | 12,969 | 14,964 | 29,929 |
| 50 | 6 | 8 | 40 | 1" | 10 | 4,800 | 8,320 | 9,600 | 11,080 | 19,200 | 14,410 | 16,627 | 33,254 |
| 60 | 4 | 6 | 48 | 1" | 10 | 5,760 | 9,984 | 11,520 | 13,296 | 23,040 | 17,292 | 19,953 | 39,905 |
| 70 | 4 | 4 | 56 | 1 1/4" | 8 | 6,720 | 11,648 | 13,440 | 15,512 | 26,880 | 20,174 | 23,278 | 46,556 |
| 80 | 3 | 4 | 64 | 1 1/4" | 8 | 7,680 | 13,312 | 15,360 | 17,728 | 30,720 | 23,056 | 26,604 | 53,207 |
| 90 | 2 (1 1/2") | 3 | 72 | 1 1/4"(7) | 8 | 8,640 | 14,976 | 17,280 | 19,944 | 34,560 | 25,938 | 29,929 | 59,858 |
| 100 | 1 (2") | 3 | 80 | 1 1/4"(7) | 8 | 9,600 | 16,640 | 19,200 | 22,160 | 38,400 | 28,820 | 33,254 | 66,509 |
| 110 | | 2 | 88 | 1 1/2" | 6 | 10,560 | 18,304 | 21,120 | 24,376 | 42,240 | 31,703 | 36,580 | 73,160 |
| 125 | | 1 | 100 | 2" | 6 | 12,000 | 20,800 | 24,000 | 27,700 | 48,000 | 36,026 | 41,568 | 83,136 |
| 150 | | 1/0 | 120 | 2" | 6 | 14,400 | 24,960 | 28,800 | 33,240 | 57,600 | 43,231 | 49,882 | 99,763 |
| 175 | | 2/0 | 140 | 2" | 6 | 16,800 | 29,120 | 33,600 | 38,780 | 67,200 | 50,436 | 58,195 | 116,390 |
| 200 | | 3/0 | 160 | 2 1/2" | 6 | 19,200 | 33,280 | 38,400 | 44,320 | 76,800 | 57,641 | 66,509 | 133,018 |
| 225 | | 4/0 | 180 | 2 1/2" | 4 | 21,600 | 37,440 | 43,200 | 49,860 | 86,400 | 64,846 | 74,822 | 149,645 |
| 250 | | 250 kcmil | 200 | 3" | 4 | 24,000 | 41,600 | 48,000 | 55,400 | 96,000 | 72,051 | 83,136 | 166,272 |
| 300 | | 350 kcmil | 240 | 3 1/2" | 4 | 28,800 | 49,920 | 57,600 | 66,480 | 115,200 | 86,461 | 99,763 | 199,526 |
| 350 | | 400 kcmil | 268(6) | 3 1/2" | 3 | 32,160 | 55,744 | 64,320 | 74,236 | 128,640 | 96,549 | 111,402 | 222,804 |
| 400 | | 500 kcmil | 304(6) | 4" | 3 | 36,480 | 63,232 | 72,960 | 84,208 | 145,920 | 109,518 | 126,367 | 252,733 |
| 400 | | 600 kcmil | 320 | 4" | 3 | 38,400 | 66,560 | 76,800 | 88,640 | 153,600 | 115,282 | 133,108 | 266,035 |

(1) Conductor size based on 60°C terminal rating. Ampacity based on four 90°C THHN current-carrying conductors [110.14(C), 310.15, Table 310.16].
(2) Conductor size based on 75°C terminal rating. Ampacity based on four 90°C THHN current-carrying conductors [110.14(C), 310.15, Table 310.16].
(3) Maximum continuous nonlinear load in an ambient temperature of 30°C limited to 80 percent of the overcurrent device rating [210.19(A), 240.6(A)].
(4) To ensure ease of installation, raceways are sized to six THHN conductors (based on 75°C column, Note 3) in rigid nonmetallic conduit [Annex C10].
(5) Copper equipment grounding conductor is sized in accordance with Table 250.122.
(6) Maximum continuous load limited to 80 percent of 75°C conductor ampacity, because conductor ampacity is lower than the overcurrent protection device rating.
(7) Raceway size is based on 75°C conductor size, not the 60°C conductor size.

---

# Formulas

## Conversion Formulas
Area of Circle = $\pi r^2$
Breakeven Dollars = Overhead Cost $/Gross Profit %
Busbar Ampacity AL = 700A Sq. in. and CU = 1000A Sq. in.
Centimeters = Inches x 2.54
Inch = 0.0254 Meters
Inch = 2.54 Centimeters
Inch = 25.4 Millimeters
Kilometer = 0.6213 Miles
Length of Coiled Wire = Diameter of Coil (average) x Number of Coils x $\pi$
Lightning Distance in Miles = Seconds between flash and thunder/4.68
Meter = 39.37 Inches
Mile = 5280 ft, 1760 yards, 1609 meters, 1.609 km
Millimeter = 0.03937 Inch
Selling Price = Estimated Cost $/(1 - Gross Profit %)
Speed of Sound (Sea Level) = 1128 fps or 769 mph
Temp C = (Temp F - 32)/1.8
Temp F = (Temp C x 1.8) + 32
Yard = 0.9144 Meters

## Electrical Formulas Based on 60 Hz
Capacitive Reactance ($X_C$) in Ohms = $1/(2\pi f\, C)$
Effective (RMS) AC Amperes = Peak Amperes x 0.707

Effective (RMS) AC Volts = Peak Volts x 0.707
Efficiency (percent) = Output/Input x 100
Efficiency = Output/Input
Horsepower = Output Watts/746
Inductive Reactance ($X_L$) in Ohms = $2\pi f\, L$
Input = Output/Efficiency
Neutral Current (Wye)
= $\sqrt{A^2 + B^2 + C^2 - (AB + BC + AC)}$
Output = Input x Efficiency
Peak AC Volts = Effective (RMS) AC Volts x $\sqrt{2}$
Peak Amperes = Effective (RMS) Amperes x $\sqrt{2}$
Power Factor (PF) = Watts/VA
VA (apparent power) = Volts x Ampere or Watts/Power Factor
VA 1-Phase = Volts x Amperes
VA 3-Phase = Volts x Amperes x $\sqrt{3}$
Watts (real power) Single-Phase = Volts x Amperes x Power Factor
Watts (real power) Three-Phase = Volts x Amperes x Power Factor x $\sqrt{3}$

## Parallel Circuits
Note 1: Total resistance is always less than the smallest resistor
RT = 1/(1/R1 + 1/R2 + 1/R3 +...)
Note 2: Total current is equal to the sum of the currents of all parallel resistors
Note 3: Total power is equal to the sum of power of all parallel resistors

Note 4: Voltage is the same across each of the parallel resistors

## Series Circuits
Note 1: Total resistance is equal to the sum of all the resistors
Note 2: Current in the circuit remains the same through all the resistors
Note 3: Voltage source is equal to the sum of voltage drops of all resistors
Note 4: Power of the circuit is equal to the sum of the power of all resistors

## Transformer Amperes
Secondary Amperes 1-Phase = VA/Volts
Secondary Amperes 3-Phase = VA/Volts x $\sqrt{3}$
Secondary Available Fault 1-Phase = VA/(Volts x %impedance)
Secondary Available Fault 3-Phase = VA/(Volts x $\sqrt{3}$ x %Impedance)
Delta 4-Wire: Line Amperes = Phase (one winding) Amperes x $\sqrt{3}$
Delta 4-Wire: Line Volts = Phase (one Winding) Volts
Delta 4-Wire: High-Leg Voltage (L-to-G) = Phase (one winding) Volts x 0.5 x $\sqrt{3}$
Wye: Line Volts = Phase (one winding) Volts x $\sqrt{3}$
Wye: Line Amperes = Phase (one winding) Amperes

## Voltage Drop
VD (1-Phase) = 2KID/CM
VD (3-Phase) = $\sqrt{3}$ KID/CM

CM (1-Phase) = 2KID/VD
CM (3-Phase) = $\sqrt{3}$ KID/VD

## Code Rules
Breaker/Fuse Ratings – 240.6(A)
Conductor Ampacity – 310.15 and Table 310.16
Equipment Grounding Conductor – 250.122
Grounding Electrode Conductor – 250.66
Motor Conductor Size – 430.22 (Single) 430.24 (Multiple)
Motor Short-Circuit Protection – 430.52
Transformer Overcurrent Protection – 450.3

To order any of
Mike's other fine
products visit
NECcode.com

ISBN 097103071-5

9 780971 030718

MIKE HOLT ENTERPRISES, INC.

# 1.888.NEC.CODE
## www.NECcode.com